Functional An

Functional Analysis

by

Harro G. Heuser
University of Karlsruhe,
Federal Republic of Germany

translated by

John Horváth
University of Maryland

A Wiley-Interscience Publication

175 YEARS OF PUBLISHING
1807 1982

JOHN WILEY & SONS
Chichester · New York · Brisbane · Toronto · Singapore

This edition is published by permission of Verlag B G Teubner, Stuttgart, and is the sole authorized English translation of the original German edition.

British Library Cataloguing in Publication Data:

Heuser, Harro
Functional analysis.
1. Functional analysis
I. Title II. Funktionanalysis. *English*
515.7 QA320

ISBN 0 471 28052 6 (cloth)
0 471 10069 2 (paper)

Filmset by Composition House Limited, Salisbury, Wiltshire, England.

Printed in the United States of America.

To my mother

Preface

A science as vitally important as functional analysis can fortunately not be defined. But its great 'Leitmotiv' can nevertheless be indicated: It is the fusion of algebraic and topological structures. This seemingly abstract and anaemic subject has developed in such a rich and lively manner that nowadays functional analysis has a strong influence on a great number of completely different fields inside and outside of mathematics: systems engineers and atomic physicists cannot do without it, just like mathematicians working in numerical analysis, differential or integral equations, the theory of approximation or representation theory—just to name a few.

The purpose of this book is to give a lively and thorough exposition of the basic concepts, the essential statements, the main methods and finally also of the way of thinking of functional analysis and to satisfy the needs of a large circle of users, in content as well as didactically.

To achieve this purpose I chose an orientation toward problems. The present book starts out, whenever possible, from questions and facts of classical analysis and algebra, and tries to get to their core by leaving aside what is accidental in order to obtain general concepts and assertions. Conversely, it aims at making the newly acquired tools fruitful for the classical fields. Naturally in the course of this process so many questions of a 'purely functional analytic nature' accumulate that finally functional analysis is propelled by its own problems. But the main text and the exercises return constantly to the familiar world of analysis and algebra. I hope to enhance in this way the motivation and the intuitive understanding of the reader and to save him from that particular feeling of being lost, which occurs so easily and annoyingly when studying an abstract theory. I think that in this way I also acquit myself best of the duty imposed on an author by the word 'Leitfaden'[1], i.e., leading thread. The first leading thread was that of Ariadne; Plutarch reports about it in his biography of Theseus the following[2]: '. . . after he (Theseus) arrived in Creta, he slew the

[1] The original German edition was published in the 'Mathematische Leitfäden' series of B. G. Teubner.

[2] Plutarch: The Lives of the Noble Grecians and Romans, translated by Thomas North, the Nonesuch Press, London, 1929.

Minotaur . . . by the means and help of Ariadne, who being fallen in fancy with him, did give him a clue of thread, by the help whereof she taught him how he might easily wind out of the turnings and cranks of the Labyrinth'.

An organization directed towards problems does not try to represent a science as it evolved—this is the purpose of the genetic method—but how it could also have evolved. It is the coming together of fortunate circumstances—maybe even more—that one of the foundations of this book, the concept of a bilinear system, already is tacitly the basis of the pioneering works of Fredholm concerning integral equations, which started functional analysis. The problem of the solvability of Fredholm's integral equations will therefore play a central role in this book. It leads us through the Neumann series to the theory of Banach algebras and through the concept of a bilinear system and of normal solvability to the extension principle of Hahn–Banach and to the duality theory which follows from it, a crown jewel of functional analysis. It should be observed emphatically that the investigation of Fredholm's integral equation originates from a very concrete situation: many boundary value problems of physics and of technology can be transformed into an equation of this kind.

The table of contents gives detailed information about the subjects treated and their interdependence. I want to indicate here only a few major blocks. The first seven chapters are dominated by the problem of equations: under what conditions are equations in general spaces solvable, how can they be effectively solved, and how do the solutions depend on certain initial data? Mainly linear problems will be considered; the non-linear domain will be represented by the fixed-point theorems of Banach and Schauder; the latter occur, however, first only in §106. In Chapter VIII approximation problems will arise. The great subject of orthogonality will emerge here, and will be developed in the following two chapters (orthogonal decomposition and spectral theory in Hilbert spaces). Chapter XI to XIII reach the summit of the progress towards abstraction: they show how linear and topological structures are fusioned in the concept of a topological vector space. This fusion leads, if the structures are rich enough, back to a quite concrete situation: commutative, complex B^*-algebras are nothing else but algebras of continuous functions. With this theorem of Gelfand and Neumark the book concludes.

The monumental work of Dunford and Schwartz 'Linear Operators' has 2592 pages; the theory of topological vector spaces is only touched in it. It is needless to say that my book had to sacrifice many important or only attractive subjects. I did not want to sacrifice, however, a copious motivation, an illustration from several angles of the central facts, and the details of the proofs.

For long stretches only the elements of analysis and linear algebra are required as prerequisites. The concept of a metric space and its continuous maps will be developed in the book. I listed in §81 and §101 without proofs the few topological facts which will be needed from §82 on. The spectral theory in Chapter VII cannot be studied profitably without some knowledge of the theory of analytic functions of a complex variable: one needs Cauchy's integral theorem, Liouville's theorem, the Taylor and the Laurent expansions. At a few places theorems from real and complex analysis will be used which might not be familiar to every reader (e.g., the Stone–Weierstrass theorem); in these cases I have indicated references to the literature, where the proofs can easily be read. The Lebesgue integral and the theory of partial differential equations will not be used.

I have to thank cordially the Universidad de los Andes in Bogotá (Colombia) and the University of Toronto in Toronto (Canada): they gave me through visiting positions the possibility to work intensively on this book. Mrs. Mia Münzel put her home, which lies quietly above Lake Garda, at my disposition for the last chapters; I am deeply indebted to her. My special thanks go to Dr. H. Kroh, Dr. U. Mertins and Mr. G. Schneider, further to Mrs. Y. Paasche and Mrs. K. Zeder. The three gentlemen were at my side during the preparation of this book with advice and help, they improved it and cleaned it up through valuable indications, and read all the proofs; the two ladies transformed with unbelievable patience a miserable manuscript into a clean typescript. I thank the Teubner–Verlag cordially for the pleasant collaboration and for its watchful help and assistance.

Nastätten in Taunus, August 1975 HARRO HEUSER

Table of contents

Preface vii

List of symbols xv

Introduction 1

I. Banach's fixed-point theorem

§1 Metric spaces 6

§2 Banach's fixed-point theorem 15

§3 Some applications of Banach's fixed-point theorem . . 17

II. Normed spaces

§4 Vector spaces 24

§5 Linear maps 29

§6 Normed spaces 33

§7 Continuous linear maps 41

§8 The Neumann series 46

§9 Normed algebras 51

§10 Finite-dimensional normed spaces 55

§11 The Neumann series in non-complete normed spaces . . 60

§12 The completion of metric and normed spaces 65

§13 Compact operators 69

III. Bilinear systems and conjugate operators

§14 Bilinear systems 75

§15 Dual systems 78

§16 Conjugate operators 83

§17 The equation $(I - K)x = y$ with finite-dimensional K . 90

§18 The equation $(R - S)x = y$ with a bijective R and finite-dimensional S 95

§19 The Fredholm integral equation with continuous kernel . 98

§20 Quotient spaces 100

§21 The quotient norm 103

§22 Quotient algebras 105

IV. Fredholm operators

§23 Operators with finite deficiency 107
§24 Fredholm operators on normed spaces 110
§25 Fredholm operators in saturated operator algebras . . 114
§26 Representation theorems for Fredholm operators . . . 122
§27 The equation $Ax = y$ with a Fredholm operator A . . 124

V. Four principles of functional analysis and some applications

§28 The extension principle of Hahn–Banach 128
§29 Normal solvability 133
§30 The normal solvability of the operators $I - K$ with compact K 136
§31 The Baire category principle 137
§32 The open mapping principle and the closed graph theorem 138
§33 The principle of uniform boundedness 142
§34 Some applications of the principles of functional analysis to analysis 144
§35 Analytic representation of continuous linear forms . . 151
§36 Operators with closed image space 155
§37 Fredholm operators on Banach spaces 158

VI. The Riesz–Schauder theory of compact operators

§38 Operators with finite chains 160
§39 Chain-finite Fredholm operators 164
§40 The Riesz theory of compact operators 166
§41 The bidual of a normed space. Reflexivity 169
§42 The dual transformation of a compact operator . . . 173
§43 Singular values and eigenvalues of a compact operator . 176

VII. Spectral theory in Banach spaces and Banach algebras

§44 The resolvent 181
§45 The spectrum 183
§46 Vector-valued holomorphic functions. Weak convergence 186
§47 Power series in Banach algebras 194
§48 The functional calculus 200
§49 Spectral projectors 204
§50 Isolated points of the spectrum 207
§51 The Fredholm region 209
§52 Riesz operators 217
§53 Essential spectra 221
§54 Normaloid operators 223

VIII. Approximation problems in normed spaces

§55 An approximation problem 230
§56 Strictly convex spaces 233
§57 Inner product spaces 235
§58 Orthogonality 239
§59 The Gauss approximation 244
§60 The general approximation problem 247
§61 Approximation in uniformly convex spaces 249
§62 Approximation in reflexive spaces 252

IX. Orthogonal decomposition in Hilbert spaces

§63 Orthogonal complements 254
§64 Orthogonal series 255
§65 Orthonormal bases 258
§66 The dual of a Hilbert space 260
§67 The adjoint transformation 263

X. Spectral theory in Hilbert spaces

§68 Symmetric operators 265
§69 Orthogonal projectors 270
§70 Normal operators and their spectra 272
§71 Normal meromorphic operators 277
§72 Symmetric compact operators 280
§73 The Sturm–Liouville eigenvalue problem 282
§74 Wielandt operators 286
§75 Determination and estimation of eigenvalues 291
§76 General eigenvalue problems for differential operators . 296
§77 Preliminary remarks concerning the spectral theorem for symmetric operators 301
§78 Functional calculus for symmetric operators 303
§79 The spectral theorem for symmetric operators on Hilbert spaces 305

XI. Topological vector spaces

§80 Metric vector spaces 309
§81 Basic notions from topology 312
§82 The weak topology 318
§83 The concept of a topological vector space. Examples . . 321
§84 The neighborhoods of zero in topological vector spaces . 327
§85 The generation of vector space topologies 330
§86 Subspaces, product spaces and quotient spaces . . . 332
§87 Continuous linear maps of topological vector spaces . . 334
§88 Finite-dimensional topological vector spaces 336
§89 Fredholm operators on topological vector spaces . . . 338

XII. Locally convex vector spaces

§90 Bases of neighborhoods of zero in locally convex vector spaces 340
§91 The generation of locally convex topologies by semi-norms 342
§92 Subspaces, products and quotients of locally convex spaces 345
§93 Normable locally convex spaces. Bounded sets . . . 346

XIII. Duality and compactness

§94 The Hahn–Banach theorem 349
§95 The topological characterization of normal solvability . 350
§96 Separation theorems 351
§97 Three applications to normed spaces 353
§98 Admissible topologies 355
§99 The bipolar theorem 356
§100 Locally convex topologies are $\mathfrak{S}$-topologies 358
§101 Compact sets 360
§102 The Alaoglu–Bourbaki theorem 363
§103 The characterization of the admissible topologies . . 364
§104 Bounded sets in admissible topologies 365
§105 Barrelled spaces. Reflexivity 367
§106 Convex, compact sets: The theorems of Krein–Milman and Schauder 372

XIV. The representation of commutative Banach algebras

§107 Preliminary remarks on the representation problem . . 381
§108 Multiplicative linear forms and maximal ideals . . . 384
§109 The Gelfand representation theorem 387
§110 The representation of commutative B^*-algebras . . . 389

Bibliography 393

Index 401

List of symbols

A 86, 336
A^+ 85
A^* 84, 263
$A(E)$ 31

$B(T)$ 39
$BV[a, b]$ 39

$\mathbf{C}$ 1
$C(a, b)$ 310
$C[a, b]$ 39
$C^{(n)}[a, b]$ 39, 310
$C_0(\mathbf{R})$ 39
$C(T)$ 383
(c) 39
(c_0) 39
codim G 28
co(M) 341

$d(x, y)$ 7
dim E 26

E' 73, 334
E'' 169, 369
E^* 73
E^{**} 81
$E^\times$ 384
E'_β 369
$\mathscr{E}(E, F)$, $\mathscr{E}(E)$ 72

$\mathscr{F}(\mathscr{A}(E))$ 114
$\mathscr{F}(E, F)$, $\mathscr{F}(E)$ 72

$\mathscr{H}(x)$ 200

ind(A) 108

$\mathbf{K}$ 1
$K_r(x)$ 12
$K_r[x]$ 12
$\mathbf{K}^n$ 25
$\mathscr{K}(E, F)$, $\mathscr{K}(E)$ 70

$L^p(a, b)$ 40
$\mathscr{L}(E, F)$, $\mathscr{L}(E)$ 42, 334
$l^p(n)$ 39
l 39

$\overline{M}$ 14, 314
$M^\perp$ 133, 243
$M^{\perp\perp}$ 134, 244
M^0 356
M^{00} 357
$[M]$, $[x_1, x_2, \ldots]$ 25

$\mathbf{N}$ 1
$N(A)$ 31

$p(A)$ 160

$q(A)$ 160

$\mathbf{R}$ 1
Re α, Im α 1
rad(E) 386
$r_\Phi(A)$ 215
$r(x)$ 184

$\mathscr{S}(E, F)$, $\mathscr{S}(E)$ 29
(s) 25, 309
sgn α 153

$\alpha(A)$ 95
$|\alpha_{\nu\mu}|$ 1

$\beta(A)$ 95
$\beta(E^+, E)$ 325

$\Delta(E)$ 107
$\Delta_\alpha(E)$, $\Delta_\beta(E)$ 110
δ_{ik} 1

Φ_A 211
$\Phi(\mathscr{A}(E))$ 114
$\Phi(E)$ 112

$\rho(A)$ 181

$\sigma(A)$ 181
$\sigma_e(A)$ 222
$\sigma_\pi(A)$ 225
$\sigma_p(A)$ 181
$\sigma(E, E^+)$ 318

$\tau(E, E^+)$ 364

$\|\cdot\|$ 34
$\rightharpoonup$ 187, 193
$\to$ 9, 42
$\Rightarrow$ 1, 42
$\Leftrightarrow$ 1
$\perp$ 242
$\prec$ 4, 313
$\circ$ 2
$:=$, $=:$ 1
$\oplus$ 27

Introduction

In this section some notation and facts will be listed, which are of fundamental importance for everything that follows.

General notation

N, **R**, **C** denote the set of natural, real and complex numbers, respectively. **K** stands for either the field **R** or the field **C**; elements of **K** are also called *scalars* and they will usually be denoted by lower case Greek characters. Re α and Im α denote the real and the imaginary part of α, respectively. $\bar{\alpha}$ is the conjugate of the complex number α. We denote by $|\alpha_{\nu\mu}|$ the determinant of the n by n matrix $(\alpha_{\nu\mu})$; there is no risk of confusion with the absolute value of the number $\alpha_{\nu\mu}$. In equations which are definitions we use the symbols '$:=$' or '$=:$', where the colon stands on the side of the symbol which is to be defined.

Examples: 1. $f(x) := x^2$; 2. $\{1, 2, 3\} =: M$; 3. *Kronecker's symbol*

$$\delta_{ik} := \begin{cases} 1 & \text{for} \quad i = k \\ 0 & \text{for} \quad i \neq k. \end{cases}$$

$A \Rightarrow B$ means that from statement A the statement B follows; $A \Leftrightarrow B$ says that the statements A and B are equivalent (each follows from the other). The end of a proof is usually marked by ■.

Sets

$\varnothing$ is the empty set. $A \subset B$ means that A is a subset of B (where $A = B$ is permitted). For the formation of unions and intersections the signs $\cup$ and $\cap$ will be used, respectively. The *difference set* $E \setminus M$ is the set of all elements of E which do not belong to M; if M is a subset of E then $E \setminus M$ is also called the *complement* of M in E.

If the set M consists of all elements of a set E which possess a certain property P, we write $M = \{x \in E: x \text{ possesses property } P\}$. Examples (where at the same time certain symbols are defined): $[\alpha, \beta] := \{\xi \in \mathbf{R}: \alpha \leqq \xi \leqq \beta\}$ is the closed, $(\alpha, \beta) := \{\xi \in \mathbf{R}: \alpha < \xi < \beta\}$ is the open interval of the real line with endpoints α, β.

Maps

Let E, F be nonempty sets. A *map* f from E into F associates with each $x \in E$ one and only one element $y \in F$, which is also denoted $f(x)$ and is called the *image* of x (with respect to f). E is the *set of definition*, F the *target set* of f. In order to exhibit clearly the three components of a map (rule of correspondence f, set of definition E, target set F), we write

$$f: E \to F \quad \text{or} \quad f: \begin{cases} E \to F \\ x \mapsto f(x) \end{cases}$$

($x \mapsto f(x)$ means that with the element x one associates the image $f(x)$). The notation $f: x \mapsto f(x)$ or even simpler $x \mapsto f(x)$ is also used; then the set of definition and the target set must be given separately, if they are not evident. At times it will be handy—and harmless—to infringe on these notational conventions. Thus, for instance, we shall speak of the function $\sin x$ (instead of $x \mapsto \sin x$), of the polynomial x^2 (instead of $x \mapsto x^2$), and of the kernel $k(s, t)$ (instead of $(s, t) \mapsto k(s, t)$) of an integral equation. A *self-map* of E is a map from E into E. The *identity map* i_E of E is the map $x \mapsto x$ $(x \in E)$.

Two maps $f_1: E_1 \to F_1$, $f_2: E_2 \to F_2$ are said to be *equal* if $E_1 = E_2$, $F_1 = F_2$ and $f_1(x) = f_2(x)$ for all $x \in E_1$.

Let $f: E \to F$ be given. For $A \subset E$, $B \subset F$ the set $f(A) := \{f(x) \in F: x \in A\}$ is the *image* of A, $f^{-1}(B) := \{x \in E: f(x) \in B\}$ is the *preimage* of B. f is said to be *surjective* if $f(E) = F$, *injective* if $f(x) = f(y)$ implies $x = y$, and *bijective* if it is both surjective and injective. The locution 'f maps E onto F' means that f is surjective. A *family* $(a_\iota: \iota \in J)$ is only another name and writing for the map $\iota \mapsto a_\iota$ from a *set* J *of indices* into a set A. When $J = \mathbf{N}$ one rather speaks of a *sequence* than of a family.

A sequence $a_1, a_2, \ldots$ of elements of A will be denoted briefly by (a_n) or $(a_n) \subset A$ and occasionally also by $(a_1, a_2, \ldots)$.

The word *function* (and also *functional*) will in general be used only for maps from a set E into the field of scalars $\mathbf{K}$ (scalar-valued or $\mathbf{K}$-valued maps).

If the maps $f: E \to F$, $g: F \to G$ are given (observe that the target set of f is the set of definition of g), then their *composition* (*product*) is the map $g \circ f$ from E into G which associates with each $x \in E$ the image $(g \circ f)(x) := g(f(x))$ in G. For a self-map f of E, the *iterates* (*powers*) f^n are defined by $f^0 := i_E$, $f^n := f \circ f^{n-1}$ $(n = 1, 2, \ldots)$.

Every injective map $f: E \to F$ has an *inverse* map

$$f^{-1}: \begin{cases} f(E) \to E \\ f(x) \mapsto x. \end{cases}$$

One has $f^{-1} \circ f = i_E$, $f \circ f^{-1} = i_{f(E)}$.

$f: E \to F$ is then and only then bijective, if there exist maps $g: F \to E, h: F \to E$ with $g \circ f = i_E, f \circ h = i_F$. In this case $g = h = f^{-1}$.

The following rules will be used frequently for the study of maps $f: E \to F$ (let A, A_ι be subsets of E, while B, B_ι are subsets of F; $f^{-1}(B)$ is the above defined preimage, where f is not assumed to be injective):

$$A_1 \subset A_2 \Rightarrow f(A_1) \subset f(A_2);$$

$$f\left(\bigcap_{\iota \in J} A_\iota\right) \subset \bigcap_{\iota \in J} f(A_\iota), \qquad f\left(\bigcup_{\iota \in J} A_\iota\right) = \bigcup_{\iota \in J} f(A_\iota);$$

$$B_1 \subset B_2 \Rightarrow f^{-1}(B_1) \subset f^{-1}(B_2);$$

$$f^{-1}\left(\bigcup_{\iota \in J} B_\iota\right) = \bigcup_{\iota \in J} f^{-1}(B_\iota), \qquad f^{-1}\left(\bigcap_{\iota \in J} B_\iota\right) = \bigcap_{\iota \in J} f^{-1}(B_\iota);$$

$$f^{-1}(F \setminus B) = E \setminus f^{-1}(B);$$

$$f^{-1}(f(A)) \supset A; f \text{ injective} \Leftrightarrow f^{-1}(f(A)) = A \text{ for every } A \subset E;$$

$$f(f^{-1}(B)) \subset B; f \text{ surjective} \Leftrightarrow f(f^{-1}(B)) = B \text{ for every } B \subset F.$$

If the map $g: F \to G$ is also given, then

$$(g \circ f)^{-1}(C) = f^{-1}(g^{-1}(C))$$

for every $C \subset G$.

The *cartesian product* of a family $(E_\iota;\ \iota \in J)$ of non-empty sets E_ι is the set of all families $(x_\iota \in E_\iota: \iota \in J)$, i.e., the set of all maps $\iota \mapsto x_\iota \in E_\iota$ defined on J. It will be denoted $\prod_{\iota \in J} E_\iota$, and in the case of a finite or countable index set also by $E_1 \times \cdots \times E_n$ or $E_1 \times E_2 \times \ldots$, respectively. $E_1 \times \cdots \times E_n$ is the set of all n-tuples $(x_1, \ldots, x_n)$, $x_k \in E_k$ for $k = 1, \ldots, n$, while $E_1 \times E_2 \times \ldots$ is the collection of all sequences $(x_1, x_2, \ldots)$ with $x_k \in E_k$ for $k \in \mathbf{N}$. There is no risk of confusing a couple $(x_1, x_2) \in E_1 \times E_2$ with an open interval.

For a given $x := (x_\iota: \iota \in J) \in \prod_{\iota \in J} E_\iota$ the element x_ι is called the *component* of x in E_ι; the maps $\pi_\iota: (x_\iota: \iota \in J) \mapsto x_\iota$ are called the *projections* or *projectors* onto the components.

Rules of complementation

Let $(A_\iota: \iota \in J)$ be a family of subsets of E and $M' := E \setminus M$ the complement of $M \subset E$ in E. Then

$$\left(\bigcup_{\iota \in J} A_\iota\right)' = \bigcap_{\iota \in J} A'_\iota, \qquad \left(\bigcap_{\iota \in J} A_\iota\right)' = \bigcup_{\iota \in J} A'_\iota.$$

Zorn's lemma

For certain pairs x, y of elements of a set $\mathfrak{M} \neq \emptyset$ let a relation '$x \prec y$' be defined, which satisfies the following axioms:

1. $x \prec x$ for every $x \in \mathfrak{M}$.
2. If $x \prec y$ and $y \prec x$, then $x = y$.
3. If $x \prec y$ and $y \prec z$, then $x \prec z$.

Such a relation is called an *order* on $\mathfrak{M}$ and $\mathfrak{M}$ itself is said to be an *ordered set*. An order is called *total*, and $\mathfrak{M}$ a *totally ordered set*, if two elements x, y of $\mathfrak{M}$ are always *comparable*, i.e., if either $x \prec y$ or $y \prec x$ holds. Every subset of $\mathfrak{M}$ becomes through the order on $\mathfrak{M}$ an ordered, possibly even a totally ordered set. $y \in \mathfrak{M}$ is called an *upper bound* for $\mathfrak{N} \subset \mathfrak{M}$ if $x \prec y$ for every $x \in \mathfrak{N}$. $z \in \mathfrak{M}$ is a *maximal element* if $z \prec x$ holds only for $x = z$. Zorn's lemma is as follows:

If every totally ordered *subset of an ordered set $\mathfrak{M}$ has an upper bound in $\mathfrak{M}$, then there exists in $\mathfrak{M}$ at least one* maximal *element.*

Inequalities

The quantities α_k, β_k, $f(x)$, $g(x)$ which occur in what follows are complex numbers. The sums are finite or infinite; in the latter case it will be supposed that every series, which is on the right hand side of an inequality, converges. In the integral inequalities it is enough to assume for our purposes that the integrands are continuous functions on a finite interval of the real line. Proofs can be found in [180].

Hölder's inequalities: If $p > 1$ and $1/p + 1/q = 1$, then

$$\sum |\alpha_k \beta_k| \leqq (\sum |\alpha_k|^p)^{1/p} (\sum |\beta_k|^q)^{1/q},$$

$$\int_a^b |f(x)g(x)|\,dx \leqq \left(\int_a^b |f(x)|^p\,dx\right)^{1/p} \left(\int_a^b |g(x)|^q\,dx\right)^{1/q};$$

for $p = q = 2$ one obtains the

Cauchy–Schwarz inequalities:

$$\sum |\alpha_k \beta_k| \leqq (\sum |\alpha_k|^2)^{1/2} (\sum |\beta_k|^2)^{1/2},$$

$$\int_a^b |f(x)g(x)|\,dx \leqq \left(\int_a^b |f(x)|^2\,dx\right)^{1/2} \left(\int_a^b |g(x)|^2\,dx\right)^{1/2}.$$

Minkowski's inequalities: For $p \geqq 1$ one has

$$(\sum |\alpha_k + \beta_k|^p)^{1/p} \leqq (\sum |\alpha_k|^p)^{1/p} + (\sum |\beta_k|^p)^{1/p},$$

$$\left(\int_a^b |f(x) + g(x)|^p\,dx\right)^{1/p} \leqq \left(\int_a^b |f(x)|^p\,dx\right)^{1/p} + \left(\int_a^b |g(x)|^p\,dx\right)^{1/p}.$$

For $p > 1$ equality holds in the Minkowski inequalities if and only if one of the sequences (α_k), (β_k) or one of the functions f, g is a non-negative multiple of the other.

Labeling of results, references

To indicate the significance of the results, we use the hierarchy 'Lemma, Proposition, Theorem, Principle'. The statements within each class will be numbered consecutively in each section, thus Lemma 27.1 is the first lemma in §27, Proposition 27.1 the first proposition in §27, and Theorem 27.1 the first theorem in §27. The examples are numbered correspondingly. The exercises are at the end of each section; cross-references like '§16 Exercise 6' or 'Exercise 6 in §16' need no explanation. The sections (paragraphs) are numbered consecutively (§1 to §110). Square brackets refer to the bibliography. For the easier orientation of the reader it is divided into several sections, which are of course not completely independent of each other. We call the reader's attention especially to the section 'expository articles'; in them he will be made familiar easily and thoroughly with historical development, the essential problems, the fundamental notions and methods of certain fields of functional analysis; most of these articles also contain a very detailed bibliography.

Exercises

The exercises form an essential part of this book. They serve to get practice in the knowledge and methods of the main text (and give thereby also an opportunity for the reader to check whether he has understood it), they prepare for subsequent developments and communicate further interesting facts from functional analysis. Some exercises will be needed in the course of the main text, they are marked with a star in front of their number (e.g., *5). The reader is earnestly urged to work the exercises, the starred ones as well as the unstarred ones; the very hasty user of this book should at least glance through them. Those unstarred exercises which contain particularly interesting facts from functional analysis, which are not treated in the main text, are indicated by a plus sign in front of the number (e.g., $^{+}2$).

Guide

Basic are Chapters I–III and sections 28, 31–33. The following division according to subjects does not contain them any more. The division is to be understood only as an orientation; it does not show the existing logical interdependence:

(A) Geometry of spaces:
   1. Normed spaces: Chapter VIII (§41 must be read before §61)
   2. Inner product spaces: §§57–59, Chapter IX
   3. Topological vector spaces: Chapters XI–XIII

(B) Banach algebras: §§44–48, Chapter XIV

(C) Operators on normed spaces: §§29, 36, 41, 44–50, 53, 54, 97

(D) Operators on inner product spaces: §67, Chapter IX

(E) Fredholm operators: Chapter IV, §§37, 39, 51, 89

(F) Compact and Riesz operators: §30, Chapter VI, §§52, 72, 75.

I

Banach's fixed point theorem

§1. Metric spaces

One of the basic concepts of classical analysis is the concept of *convergence* of a sequence of numbers. This in turn is based on the concept of *distance*: in fact the convergence of the sequence (x_k) to x means that *the distance* $|x_k - x|$ *of the k-th term* x_k *from the limit x will be arbitrarily small when k increases beyond all bounds.* A corresponding definition is given for sequences of elements of $\mathbf{K}^n$, where the distance $d(x, y)$ between $x = (\xi_1, \ldots, \xi_n)$ and $y = (\eta_1, \ldots, \eta_n)$ is for instance defined by

$$(1.1) \qquad d(x, y) := \left(\sum_{\nu=1}^{n} |\xi_\nu - \eta_\nu|^2 \right)^{1/2}.$$

If one wants to build a theory of convergence, which is valid for sequences of numbers as well as for sequences of vectors, and which eventually disregards completely the nature of the terms of the sequences, then one cannot use definitions of a distance like (1.1) (which make sense only for concretely given objects), but rather one will have to work with some properties, which intuitively every reasonable concept of distance must have. Such properties are for instance the following, where we call 'points' the objects between which the distances are defined: The distance of a point from itself and only from itself is 0, the distance of a point x from a point y is exactly as large as conversely the distance of the point y from the point x (symmetry property of distance), and finally a 'deviation property': If one does not go from the point x directly to the point y but first to the point z and then from there to x, then one has not made a shortcut, possibly one covered a larger distance. We need now only to express these properties precisely in the language of mathematical formulas to obtain the fundamental concepts of a metric and of a metric space:

A function d, which associates with any two elements x, y of a non-empty set E a real number $d(x, y)$, is called a *metric* on E, if it possesses the following properties:

(M1) $d(x, y) \geqq 0$, where $d(x, y) = 0$ if and only if $x = y$,
(M2) $d(x, y) = d(y, x)$,
(M3) $d(x, y) \leqq d(x, z) + d(z, y)$.

A *metric space* (E, d) is a (non-empty) set E on which a metric is defined.

The elements of a metric space are usually called *points*, the number $d(x, y)$ is the *distance* between the points x and y. (M3) is called the *triangle inequality*.

As we shall soon see, many different metrics can be defined on a set E. The notation (E, d) for a metric space takes this fact into account: it indicates not only the underlying set E but also the metric d given on it. This careful notation is not always necessary, we shall take therefore frequently the liberty to speak simply of the metric space E instead of (E, d). The metric of the metric space E will then always be denoted by d.

The best known metric space is the field $\mathbf{K}$ with the distance $d(x, y) := |x - y|$. The examples which now follow are only the first samples, many others will be presented later. Basically we have to show for the distances defined below that axioms (M1)–(M3) are satisfied. However, if an axiom can be trivially verified, then we will not mention it at all.

Example 1.1. Let $p \geqq 1$ be a fixed real number. We define a metric d on $\mathbf{K}^n$ by associating with two points $x = (\xi_1, \ldots, \xi_n)$, $y = (\eta_1, \ldots, \eta_n)$ of $\mathbf{K}^n$ the distance

$$d(x, y) := \left(\sum_{\nu=1}^{n} |\xi_\nu - \eta_\nu|^p\right)^{1/p}. \tag{1.2}$$

The triangle inequality (M3) follows directly from Minkowski's inequality, since with $z = (\zeta_1, \ldots, \zeta_n)$ one has

$$\begin{aligned} d(x, y) &= \left(\sum_{\nu=1}^{n} |(\xi_\nu - \zeta_\nu) + (\zeta_\nu - \eta_\nu)|^p\right)^{1/p} \\ &\leqq \left(\sum_{\nu=1}^{n} |\xi_\nu - \zeta_\nu|^p\right)^{1/p} + \left(\sum_{\nu=1}^{n} |\zeta_\nu - \eta_\nu|^p\right)^{1/p} = d(x, z) + d(z, y). \end{aligned}$$

Our example shows that infinitely many metrics can be defined on $\mathbf{K}^n$. For $p = 2$ we obtain the euclidean metric already mentioned in (1.1).

If on $\mathbf{K}^n$ we introduce the distance (1.2), then we denote the metric space so obtained by $l^p(n)$.

Example 1.2. We obtain the metric space $l^\infty(n)$ by defining on $\mathbf{K}^n$, with the notations of Example 1.1, the so-called *maximum-metric* through

$$d(x, y) := \max_{\nu=1}^{n} |\xi_\nu - \eta_\nu|. \tag{1.3}$$

Thus the spaces $l^p(n)$ have been introduced for $1 \leqq p \leqq \infty$.

Example 1.3. Let T be a non-empty set and $B(T)$ the set of all bounded functions $x: T \to \mathbf{K}$. $B(T)$ becomes a metric space by means of the *supremum-metric*, which we define by

$$d(x, y) := \sup_{t \in T} |x(t) - y(t)|. \tag{1.4}$$

If $T = \mathbf{N}$, then $B(T)$ is the set l^∞ of all bounded sequences $x = (\xi_n)$, $y = (\eta_n), \ldots$ with the definition of distance

$$d(x, y) := \sup_{n=1}^{\infty} |\xi_n - \eta_n|. \tag{1.5}$$

For $T = \{1, \ldots, n\}$ the space $B(T)$ is obviously the $l^\infty(n)$ of Example 1.2.

Example 1.4. Let $C[a, b]$ be the set of all functions x with values in $\mathbf{K}$ which are continuous on the (finite closed) interval $[a, b]$. On $C[a, b]$ we introduce the *maximum-metric* by

$$d(x, y) := \max_{a \leqq t \leqq b} |x(t) - y(t)|. \tag{1.6}$$

Example 1.5. On the set (s) of all sequences with terms in $\mathbf{K}$ we define a metric by

$$d(x, y) := \sum_{n=1}^{\infty} \frac{1}{2^n} \frac{|\xi_n - \eta_n|}{1 + |\xi_n - \eta_n|}, \tag{1.7}$$

where $x = (\xi_1, \xi_2, \ldots)$, $y = (\eta_1, \eta_2, \ldots)$. The convergence of the series $\sum 1/2^n$ ensures that $d(x, y)$ exists for all points x, y in (s). To prove (M3) we observe that the function $t \mapsto t/(1 + t)$ has a positive derivative for $t > -1$, and is therefore increasing. From here follows the estimate

$$\frac{|\alpha + \beta|}{1 + |\alpha + \beta|} \leqq \frac{|\alpha| + |\beta|}{1 + |\alpha| + |\beta|} \leqq \frac{|\alpha|}{1 + |\alpha|} + \frac{|\beta|}{1 + |\beta|} \tag{1.8}$$

for arbitrary scalars α, β, and so for any $z = (\zeta_1, \zeta_2, \ldots)$ the triangle inequality

$$d(x, y) = \sum \frac{1}{2^n} \frac{|(\xi_n - \zeta_n) + (\zeta_n - \eta_n)|}{1 + |(\xi_n - \zeta_n) + (\zeta_n - \eta_n)|}$$

$$\leqq \sum \frac{1}{2^n} \frac{|\xi_n - \zeta_n|}{1 + |\xi_n - \zeta_n|} + \sum \frac{1}{2^n} \frac{|\zeta_n - \eta_n|}{1 + |\zeta_n - \eta_n|} = d(x, z) + d(z, y).$$

It is noteworthy that $d(x, y) \leqq 1$ for all x, y in (s).

The metrics introduced in the above examples are canonical in the sense that e.g., $B(T)$ will *always* be equipped with the supremum-metric—unless the contrary is explicitly stated.

The reader will have noticed that it does not appear from the symbols $l^p(n)$, $B(T)$, $C[a, b]$ and (s) whether the field $\mathbf{R}$ or the field $\mathbf{C}$ was used to construct these spaces. Indeed, our results do not depend on whether e.g., the functions in $B(T)$ have real or complex values, hence a distinction is unnecessary.

The next example shows that every non-empty set can be made into a metric space.

Example 1.6. On any set $E \neq \varnothing$ the so-called *discrete metric* can be defined by

$$d(x, y) := \begin{cases} 1 & \text{if} \quad x \neq y, \\ 0 & \text{if} \quad x = y. \end{cases} \tag{1.9}$$

Metric spaces should serve to obtain the concept of convergence in its pure form. This can be achieved through the following definition.

A sequence (x_k) in a metric space E *converges* to a point $x \in E$—in symbols: $x_k \to x$ or $\lim x_k = x$ –, if $d(x_k, x)$ converges to 0, i.e., if given $\varepsilon > 0$ there exists an index $k_0(\varepsilon)$ such that $d(x_k, x) \leqq \varepsilon$ for all $k \geqq k_0(\varepsilon)$. The point x is called the *limit* of the sequence (x_k).

The limit x is uniquely determined. Indeed, if (x_k) converges also to y, then it follows from $0 \leqq d(x, y) \leqq d(x, x_k) + d(x_k, y) = d(x_k, x) + d(x_k, y) \to 0$ that $d(x, y) = 0$, hence $x = y$.

The spaces $l^p(n)$, $B(T)$, $C[a, b]$ and (s) are *function spaces*, i.e., their elements are (scalar-valued) functions with a common set of definition D. If $D = \mathbf{N}$, then the function space is most often called a *sequence space*; thus l^∞ and (s) are sequence spaces. In a function space there exists a natural concept of convergence, that of *pointwise convergence* (in a sequence space or in $\mathbf{K}^n$ it is preferable to speak of *componentwise convergence*): The sequence of functions (x_k) converges *pointwise* to the function x if $x_k(t) \to x(t)$ for every t in the common set of definition D. It is a useful exercise, which we will perform at once, to compare the pointwise convergence in the above listed function spaces with the convergence in the sense of their canonical metric—the *metric convergence*.

One sees without difficulty that in $l^p(n)$ *metric convergence is equivalent to componentwise convergence*, i.e., $x_k = (\xi_1^{(k)}, \ldots, \xi_n^{(k)})$ converges to $x = (\xi_1, \ldots, \xi_n)$ in the sense of the metric of $l^p(n)$ if and only if $\xi_\nu^{(k)} \to \xi_\nu$ for $\nu = 1, \ldots, n$. This remark shows furthermore, that convergence in the sense of the metric of $l^p(n)$ is equivalent to convergence in the sense of the metric in $l^q(n)$, more precisely: $x_k \to x$ in $l^p(n)$ if and only if $x_k \to x$ in $l^q(n)$. From the point of view of a pure theory of convergence, all the metrics of $\mathbf{K}^n$ defined by (1.2) or (1.3) perform the same service; we shall see soon that it can, however, be useful to consider them separately.

If $x_k \to x$ in $B(T)$, then to every $\varepsilon > 0$ there exists a $k_0 = k_0(\varepsilon)$ so that for $k \geqq k_0$ one has $d(x_k, x) = \sup_{t \in T} |x_k(t) - x(t)| \leqq \varepsilon$, and so a fortiori

$$|x_k(t) - x(t)| \leqq \varepsilon \qquad \text{for all} \quad k \geqq k_0 \qquad \text{and all} \quad t \in T. \tag{1.10}$$

Thus the sequence of functions x_k converges not only pointwise but even uniformly on T to the function x, since the index k_0 depends only on ε but not on t. Conversely, if the sequence (x_k) does converge uniformly to x, i.e., to every $\varepsilon > 0$ there exists k_0 such that (1.10) holds, then obviously

$$d(x_k, x) = \sup_{t \in T} |x_k(t) - x(t)| \leqq \varepsilon$$

for $k \geqq k_0$, i.e., $x_k - x$ in the sense of the metric of $B(T)$. Thus *in $B(T)$ metric convergence is equivalent to uniform convergence on T. The same is true in* $C[a, b]$; we only have to observe that $C[a, b]$ is a subset of $B([a, b])$ and that the maximum-metric on $C[a, b]$ is nothing but the supremum-metric of $B([a, b])$ restricted to $C[a, b]$.—In the future we shall abbreviate $B([a, b])$ by $B[a, b]$.

Next we show that *metric convergence in* (s) *is equivalent to componentwise convergence*, i.e. that

$$x_k = (\xi_1^{(k)}, \xi_2^{(k)}, \ldots) \to x = (\xi_1, \xi_2, \ldots)$$

in the sense of the metric of (s) if and only if

$$\xi_n^{(k)} \to \xi_n \qquad \text{for} \quad n = 1, 2, \ldots .$$

First it follows from

$$d(x_k, x) = \sum_{n=1}^{\infty} \frac{1}{2^n} \frac{|\xi_n^{(k)} - \xi_n|}{1 + |\xi_n^{(k)} - \xi_n|} \to 0$$

that a fortiori

$$\frac{|\xi_n^{(k)} - \xi_n|}{1 + |\xi_n^{(k)} - \xi_n|} \to 0$$

and thus also $\lim_{k\to\infty} |\xi_n^{(k)} - \xi_n| = 0$. Now assume conversely that (x_k) tends componentwise to x. After having chosen an arbitrary $\varepsilon > 0$ determine $n_0 = n_0(\varepsilon)$ so that

$$\sum_{n=n_0+1}^{\infty} \frac{1}{2^n} \leqq \frac{\varepsilon}{2}.$$

Then for $k = 1, 2, \ldots$ one has a fortiori

$$\sum_{n=n_0+1}^{\infty} \frac{1}{2^n} \frac{|\xi_n^{(k)} - \xi_n|}{1 + |\xi_n^{(k)} - \xi_n|} \leqq \frac{\varepsilon}{2}.$$

Furthermore, because of the hypothesis of componentwise convergence, we can determine $k_0 = k_0(\varepsilon)$ so that

$$\sum_{n=1}^{n_0} \frac{1}{2^n} \frac{|\xi_n^{(k)} - \xi_n|}{1 + |\xi_n^{(k)} - \xi_n|} \leqq \frac{\varepsilon}{2} \qquad \text{for} \quad k \geqq k_0.$$

From the last two estimates it follows immediately that $d(x_k, x) \leqq \varepsilon$ whenever $k \geqq k_0$, i.e., that (x_k) converges to x in the sense of the metric of (s). ■

It is well known that a sequence of real or complex numbers converges if and only if it is a Cauchy sequence, i.e., if it satisfies the Cauchy convergence criterion. The concept of a Cauchy sequence can be carried over immediately to metric spaces:

A sequence (x_n) in a metric space is called a *Cauchy sequence* if for every $\varepsilon > 0$ there exists an $n_0 = n_0(\varepsilon)$ so that $d(x_n, x_m) \leqq \varepsilon$ for every $n, m \geqq n_0$.

A convergent sequence is always a Cauchy sequence. In fact, if $\lim x_n = x$, then to any $\varepsilon > 0$ there exists n_0 such that $d(x_n, x) \leqq \varepsilon/2$ for $n \geqq n_0$. Thus if $n, m \geqq n_0$, the estimate

$$d(x_n, x_m) \leqq d(x_n, x) + d(x, x_m) = d(x_n, x) + d(x_m, x) \leqq \frac{\varepsilon}{2} + \frac{\varepsilon}{2} = \varepsilon$$

shows that (x_n) is indeed a Cauchy sequence. ■

However, a Cauchy sequence is not always convergent. Consider for instance the open interval $E := (0, 1)$ and define on E the usual distance by $d(x, y) := |x - y|$, then the sequence $(1/(n + 1))$ is obviously a Cauchy sequence in the metric space E, but has no limit in E, i.e., it is not convergent. In many investigations the occurrence of non-convergent Cauchy sequences is very disturbing, so that one must restrict oneself to so-called complete spaces, which we now define:

A metric space E is *complete* if every Cauchy sequence in E converges to an element of E. *Thus* Cauchy's convergence criterion (*'a sequence converges if and only if it is a Cauchy sequence'*) *is valid exactly in complete spaces.*

The spaces $l^p(n)$ are all complete. Indeed, if (x_k) is a Cauchy sequence in $l^p(n)$, where $x_k = (\xi_1^{(k)}, \ldots, \xi_n^{(k)})$, then one sees immediately that each of the n sequences of components $(\xi_\nu^{(k)})$, $\nu = 1, \ldots, n$ is a Cauchy sequence in $\mathbf{K}$ and so converges to a scalar ξ_ν. The sequence (x_k) converges thus componentwise, hence by the result obtained above, also in the sense of the metric of $l^p(n)$ to $x := (\xi_1, \ldots, \xi_n) \in l^p(n)$. ■

Since also in (s) metric convergence is equivalent to componentwise convergence, the above arguments show that (s) *too is complete.*

Also the spaces $B(T)$—in particular l^∞—and $C[a, b]$ are complete. Let (x_n) be a Cauchy sequence in $B(T)$. To every $\varepsilon > 0$ there exists $n_0 = n_0(\varepsilon)$ so that $d(x_n, x_m) = \sup_{t \in T} |x_n(t) - x_m(t)| \leqq \varepsilon$ for $n, m \geqq n_0$ and so

$$|x_n(t) - x_m(t)| \leqq \varepsilon \qquad \text{for} \quad n, m \geqq n_0 \qquad \text{and all} \quad t \in T.$$

Thus $(x_n(t))$ is a Cauchy sequence for every $t \in T$. By $x(t) := \lim x_n(t)$ one can therefore define a function x on T for which

$$(1.11) \qquad |x_n(t) - x(t)| \leqq \varepsilon \qquad \text{for} \quad n \geqq n_0 \qquad \text{and all} \quad t \in T.$$

Because of $|x(t)| \leqq |x(t) - x_{n_0}(t)| + |x_{n_0}(t)| \leqq \varepsilon + \sup_{t \in T} |x_{n_0}(t)|$ the function x is bounded on T, hence belongs to $B(T)$. It follows now from (1.11) that (x_n) converges to x uniformly on T, thus also in the sense of the metric on $B(T)$. The proof of the completeness of $C[a, b]$ goes similarly, one only has to use the well-known theorem that a uniformly convergent sequence of continuous functions has a continuous limit. ■

Every convergent sequence of numbers is bounded. In order to carry over this theorem to metric spaces, we need a concept of boundedness. If E is a metric space we call, just like in analysis, the set $K_r(x_0) := \{x \in E: d(x, x_0) < r\}$ the *open*, $K_r[x_0] := \{x \in E: d(x, x_0) \leqq r\}$ the *closed ball* with radius $r > 0$ and center x_0. The distinction between an open and a closed ball is often inessential; if we speak simply of a ball, then it may be open or closed. With the help of the triangle inequality one sees immediately that a subset of E which lies entirely in a ball around x_0, is also contained entirely in an appropriate ball around any other center x_1. This fact makes it possible to call a subset (or a sequence) of E *bounded* if it lies in a ball. Every Cauchy sequence (x_n), in particular every convergent sequence, is bounded. Indeed, for $\varepsilon = 1$ there exists an m such that $d(x_n, x_m) \leqq 1$ for $n \geqq m$. Then for any n we have obviously

$$d(x_n, x_m) \leqq \sum_{\mu=1}^{m-1} d(x_\mu, x_m) + 1,$$

so that (x_n) lies in a ball around x_m. ■

In a metric space the so-called *quadrangle-inequality*

$$|d(x, y) - d(u, v)| \leqq d(x, u) + d(y, v) \tag{1.12}$$

holds. Indeed, applying twice the triangle inequality, one obtains $d(u, v) \leqq d(u, x) + d(x, y) + d(y, v)$, hence

$$d(u, v) - d(x, y) \leqq d(x, u) + d(y, v).$$

Similarly, one has also

$$d(x, y) - d(u, v) \leqq d(x, u) + d(y, v).$$

(1.12) follows immediately from the last two inequalities. ■

Distance is a continuous *function, i.e.,* $x_n \to x$ *and* $y_n \to y$ *imply that* $d(x_n, y_n) \to d(x, y)$. To see this one only has to set $u = x_n$, $v = y_n$ in (1.12). The concept of continuity will be defined in generality and investigated more closely in §7.

The concepts introduced in this section are very few; they consist essentially of those of a metric space, convergence and completeness. Furthermore we studied some examples of complete metric spaces. In the next two sections we will show that this material is already sufficient to obtain interesting and nontrivial results.

Exercises

*1. Every subsequence of a sequence converging to x converges again to x.

*2. If a Cauchy sequence has a subsequence converging to x, then the sequence itself converges to x.

3. In a space E with the discrete metric, a sequence converges if and only if it is constant from a certain term on. A ball in E contains either only its center or it coincides with the whole space.

4. Besides the canonical maximum-metric, one can introduce a further metric on $C[a, b]$ by

$$d_1(x, y) := \int_a^b |x(t) - y(t)|\,dt$$

(justify carefully why (M1) holds). With this metric $C[a, b]$ is not complete. The same holds if we define distance by

$$d_2(x, y) := \left(\int_a^b |x(t) - y(t)|^2\,dt\right)^{1/2}$$

(use Minkowski's inequality to prove (M3)).

5. If on E two metrics d_1 and d_2 are defined, we say that d_1 is *stronger* than d_2 if $d_1(x_n, x) \to 0$ implies that $d_2(x_n, x) \to 0$, i.e., if convergence with respect to d_1 entails the convergence with respect to d_2 to the same limit; in this case one also says that d_2 is *weaker* than d_1. If d_1 is at the same time both stronger and weaker then d_2 (i.e., if convergence with respect to d_1 is equivalent to convergence with respect to d_2), then the two metrics are said to be *equivalent*. Show the following:

(a) The discrete metric is the strongest metric on E (use Exercise 3).

(b) d_1 is stronger than d_2 if and only if for all $x \in E$ the following condition is fulfilled: Every open ball around x with respect to d_2 contains an open ball around x with respect to d_1.

(c) On every metric space (E, d) a metric equivalent to d can be defined by

$$d_1(x, y) := \frac{d(x, y)}{1 + d(x, y)}$$

(cf. Exercise 14).

*6. If $(E_1, d_1), \ldots, (E_n, d_n)$ are metric spaces, then the cartesian product $E = E_1 \times \cdots \times E_n$ becomes a metric space if we define distance by

$$d(x, y) := \sum_{k=1}^{n} d_k(x_k, y_k);$$

here $x = (x_1, \ldots, x_n)$, $y = (y_1, \ldots, y_n)$. Convergence in E is equivalent to componentwise convergence. E is complete if and only if each E_k is complete. Define other metrics on E which are equivalent to d (cf. the spaces $l^p(n)$).

*7. If $(E_1, d_1), (E_2, d_2), \ldots$ are metric spaces, then the cartesian product $E = \prod_{k=1}^{\infty} E_k$ becomes a metric space if we define distance by

$$d(x, y) := \sum_{k=1}^{\infty} \frac{1}{2^k} \frac{d_k(x_k, y_k)}{1 + d_k(x_k, y_k)};$$

here $x = (x_1, x_2, \ldots)$, $y = (y_1, y_2, \ldots)$. Convergence in E is equivalent to componentwise convergence. E is complete if and only if every E_k is complete (cf. the space (s)).

*8. If the convergent sequence (x_n) lies in the closed ball $K_r[z]$, then its limit also lies there.

9. One has

$$\lim_{p \to \infty} \left(\sum_{\nu=1}^{n} |\xi_\nu - \eta_\nu|^p \right)^{1/p} = \max_{\nu=1}^{n} |\xi_\nu - \eta_\nu|.$$

*10. Like in classical analysis, a subset M of a metric space E is said to be *open* if around each point of M there exists a ball which lies entirely in M. We say that M is *closed* if $E \setminus M$ is open. Show the following:

(a) M is closed if and only if the limit of every convergent sequence in M lies again in M.

(b) An open ball is an open, a closed ball is a closed set.

*11. In a metric space E the empty set $\varnothing$ and the whole space E are open; the intersection of finitely many open sets and the union of an arbitrary number of open sets is open. We have the complementary statements: $\varnothing$ and E are closed; the union of finitely many closed sets and the intersection of an arbitrary number of closed sets is closed.

12. If E is equipped with the discrete metric, then any subset of E is open as well as closed.

13. In a metric space E an *open neighborhood* of x_0 is an open set containing x_0. A *neighborhood* of x_0 is a set which contains a ball with center x_0. It follows from Exercise 10 that a neighborhood of x_0 can also be defined as a subset of E which contains an open neighborhood of x_0. Show the following:

(a) A subset of E is open if and only if it is a neighborhood of each of its points.

(b) (x_n) converges to x if and only if for every neighborhood U of x there exists $n_0 = n_0(U)$ such that $x_n \in U$ for all $n \geqq n_0$ (briefly: if almost all terms x_n lie in any neighborhood of x).

14. The set $\mathfrak{U}(x)$ of all neighborhoods of x in a metric space E (see Exercise 13) is called the *filter of neighborhoods* of x. Let two metrics d_1 and d_2 be introduced on E, let $\mathfrak{U}_1(x)$ and $\mathfrak{U}_2(x)$ be the corresponding filters of neighborhoods, $\mathfrak{D}_1$ and $\mathfrak{D}_2$ the corresponding systems of open sets. Show with the help of Exercise 5 that the following assertions are equivalent:

(a) d_1 is stronger than d_2.

(b) $\mathfrak{U}_1(x) \supset \mathfrak{U}_2(x)$ for every $x \in E$.

(c) $\mathfrak{D}_1 \supset \mathfrak{D}_2$.

It follows that d_1 and d_2 are equivalent if and only if $\mathfrak{U}_1(x) = \mathfrak{U}_2(x)$ for all $x \in E$ or, equivalently, if $\mathfrak{D}_1 = \mathfrak{D}_2$.

*15. The *closure* or *closed hull* of a subset M of the metric space E is the intersection of all closed subsets of E containing M; it is denoted by $\overline{M}$. One has $M \subset \overline{M}$; $\overline{M}$ is closed; and M is closed if and only if $M = \overline{M}$. $\overline{M}$ is the set of all limits of convergent sequences from M.

§2. Banach's fixed-point theorem

An important problem in analysis is the determination of all zeros of a real function f, i.e., the solution of the equation $f(x) = 0$. This can often be achieved by putting the equation in the form

$$x = g(x) \tag{2.1}$$

and then using an *iteration procedure*: one chooses an x_0 from the definition set of g, one sets $x_1 := g(x_0)$, $x_2 := g(x_1)$, generally $x_{n+1} := g(x_n)$ for $n = 0, 1, 2, \ldots$ and one hopes that the sequence (x_n) of *iterates* converges to a solution of (2.1). This is indeed the case when we make certain hypotheses which will be indicated now.

Suppose that the function g maps the interval $[a, b]$ into itself and that it satisfies a *Lipschitz-condition* with a *Lipschitz-constant* $q < 1$, i.e., that there exists a number q with $0 \leqq q < 1$ such that

$$|g(x) - g(y)| \leqq q|x - y| \qquad \text{for all} \quad x, y \text{ in } \quad [a, b]. \tag{2.2}$$

Since g maps the interval $[a, b]$ into itself, for any starting point x_0 in $[a, b]$ the above defined sequence (x_n) of iterates exists. We show now that (x_n) converges to a solution x of the equation (2.1) and that, furthermore, x is the only solution in $[a, b]$ of this equation.

It follows from (2.2) by repeated estimates that $|x_2 - x_1| = |g(x_1) - g(x_0)| \leqq q|x_1 - x_0|$, $|x_3 - x_2| = |g(x_2) - g(x_1)| \leqq q|x_2 - x_1| \leqq q^2|x_1 - x_0|$, and in general

$$|x_{n+1} - x_n| \leqq q^n|x_1 - x_0|.$$

From here one obtains for $k = 1, 2, \ldots$ with the help of the triangle inequality that

$$\begin{aligned} |x_{n+k} - x_n| &\leqq \sum_{\nu=1}^{k} |x_{n+\nu} - x_{n+\nu-1}| \leqq \left(\sum_{\nu=1}^{k} q^{n+\nu-1}\right)|x_1 - x_0| \\ &= q^n \frac{1 - q^k}{1 - q}|x_1 - x_0| \leqq \frac{q^n}{1 - q}|x_1 - x_0|. \end{aligned} \tag{2.3}$$

Because $0 \leqq q < 1$, this estimate shows that (x_n) is a Cauchy sequence in $[a, b]$ and converges therefore to a point $x \in [a, b]$. With this x one has

$$\begin{aligned} |g(x) - x| &\leqq |g(x) - x_{n+1}| + |x_{n+1} - x| = |g(x) - g(x_n)| + |x_{n+1} - x| \\ &\leqq q|x - x_n| + |x_{n+1} - x|, \end{aligned}$$

and since the last term of this inequality tends to 0, we have $|g(x) - x| = 0$, i.e., indeed $g(x) = x$. If for a y in $[a, b]$ we also have $g(y) = y$, then $|x - y| = |g(x) - g(y)| \leqq q|x - y|$ implies $|x - y| = 0$, hence $x = y$. So all the assertions made above are proven. ■

If we replace the interval $[a, b]$ by an arbitrary complete metric space E and g by a map A from E into itself which is *contracting*, i.e., which for a fixed q, $0 \leqq q < 1$, satisfies the *Lipschitz-condition*

$$d(Ax, Ay) \leqq q d(x, y) \quad \text{for all} \quad x, y \quad \text{in} \quad E, \tag{2.4}$$

then we can repeat the above argument word for word; we only have to replace the distance $|u - v|$ in $[a, b]$ by the distance $d(u, v)$ in E. Thus if x_0 is an arbitrary point in E and we define the *sequence of iterates* (x_n) by $x_{n+1} := Ax_n$ $(n = 0, 1, 2, \ldots)$, then we obtain the estimate analogous to (2.3)

$$d(x_{n+k}, x_n) \leqq \frac{q^n}{1-q} d(x_1, x_0), \tag{2.5}$$

from where it follows that (x_n) is a Cauchy sequence in E, which—because of the completeness of E—converges to a point x in E. One sees exactly as above that $Ax = x$ and that x is the only *fixed point* of A in E, i.e., the only point which remains fixed under the map A. From (2.5) we obtain furthermore an error estimate if we let $k \to \infty$ and use the continuity of the metric (§1):

$$d(x, x_n) \leqq \frac{q^n}{1-q} d(x_1, x_0). \tag{2.6}$$

We collect our results in the following theorem, known under the name of *Banach's fixed-point theorem* or the *theorem on contracting maps.*

Theorem 2.1 *If A is a* contracting *self-map of the* complete *metric space E, then the equation $x = Ax$ has* exactly *one solution in E. This solution can be obtained by* iteration: *if one chooses an arbitrary starting point x_0 in E and sets $x_{n+1} := Ax_n$, $n = 0, 1, 2, \ldots$, then (x_n) converges to x. Furthermore the* error estimate (2.6) *is valid.*

If one uses the powers or iterates A^n of A, then $x_n = A^n x_0$; under the hypotheses of the theorem the sequence $(A^n x_0)$ converges to the unique fixed point of A in E. One says that the fixed point of A was found by iteration of A. In the next section we shall be able to convince ourselves of the great possibility of practical applications of the fixed-point theorem. Let us emphasize here only that the theorem not only guarantees the existence of a fixed point but also permits its construction.

Exercises

1. The function g satisfies the Lipschitz condition (2.2) if it is differentiable and $|g'(x)| \leqq q < 1$ in $[a, b]$.

$^+$2. Let $\sum \alpha_n$ be a convergent sequence with non-negative terms and A a map of the complete metric space E into itself, which for every natural number n and for all x, y in E satisfies the condition $d(A^n x, A^n y) \leqq \alpha_n d(x, y)$. Then A has exactly one fixed point x in E, this fixed point is the limit of the sequence of iterates $(A^n x_0)$ for an arbitrary x_0 in E, and one has the error estimate

$$d(x, A^n x_0) \leqq \left(\sum_{\nu=n}^{\infty} \alpha_\nu \right) d(Ax_0, x_0).$$

§3. Some applications of Banach's fixed-point theorem

We consider first systems of n linear equations with n unknowns. Such a system can always be written in the form

$$\xi_i - \sum_{k=1}^{n} \alpha_{ik} \xi_k = \gamma_i \qquad (i = 1, \ldots, n) \tag{3.1}$$

and thus brought into the form of a *fixed-point problem*

$$\xi_i = \sum_{k=1}^{n} \alpha_{ik} \xi_k + \gamma_i \qquad (i = 1, \ldots, n); \tag{3.2}$$

indeed, to solve the system (3.2) means to find a fixed point of the map $A\colon \mathbf{K}^n \to \mathbf{K}^n$ defined by

$$A(\xi_1, \ldots, \xi_n) := \left(\sum_{k=1}^{n} \alpha_{1k} \xi_k + \gamma_1, \ldots, \sum_{k=1}^{n} \alpha_{nk} \xi_k + \gamma_n \right). \tag{3.3}$$

To be able to apply the fixed-point theorem of §2 we have to make first a complete metric space out of $\mathbf{K}^n$. For this purpose we have e.g., the metrics defined by (1.2) and (1.3) which transform $\mathbf{K}^n$ into the metric spaces $l^p(n)$. For the cases $p = 1, 2, \infty$ we want to write down the conditions which tell us that A is contracting. Let $x = (\xi_1, \ldots, \xi_n)$ and $y = (\eta_1, \ldots, \eta_n)$ be two arbitrary points in $\mathbf{K}^n$.

In the case $p = 1$ one has

$$\begin{aligned} d(Ax, Ay) &= \sum_{i=1}^{n} \left| \sum_{k=1}^{n} \alpha_{ik}(\xi_k - \eta_k) \right| \leqq \sum_{i=1}^{n} \sum_{k=1}^{n} |\alpha_{ik}| \cdot |\xi_k - \eta_k| \\ &= \sum_{k=1}^{n} |\xi_k - \eta_k| \sum_{i=1}^{n} |\alpha_{ik}| \end{aligned}$$

and so

$$d(Ax, Ay) \leqq \left(\max_{k=1}^{n} \sum_{i=1}^{n} |\alpha_{ik}| \right) d(x, y). \tag{3.4}$$

In the case $p = 2$ one obtains with the help of the Cauchy–Schwarz inequality

$$d(Ax, Ay) = \sqrt{\sum_{i=1}^{n} \left| \sum_{k=1}^{n} \alpha_{ik}(\xi_k - \eta_k) \right|^2} \leqq \sqrt{\sum_{i=1}^{n} \left(\sum_{k=1}^{n} |\alpha_{ik}|^2 \cdot \sum_{k=1}^{n} |\xi_k - \eta_k|^2 \right)},$$

and so

$$d(Ax, Ay) \leqq \sqrt{\sum_{i,k=1}^{n} |\alpha_{ik}|^2} \, d(x, y). \tag{3.5}$$

Finally, in the case $p = \infty$ we find the estimate

$$d(Ax, Ay) = \max_{i=1}^{n} \left| \sum_{k=1}^{n} \alpha_{ik}(\xi_k - \eta_k) \right| \leqq \max_{i=1}^{n} \left(\sum_{k=1}^{n} |\alpha_{ik}| \cdot \max_{k=1}^{n} |\xi_k - \eta_k| \right),$$

and so

$$d(Ax, Ay) \leqq \left(\max_{i=1}^{n} \sum_{k=1}^{n} |\alpha_{ik}| \right) d(x, y). \tag{3.6}$$

If we take into account that convergence in $l^p(n)$ is equivalent to componentwise convergence (see §1), then we obtain from the fixed-point theorem of §2, with the help of the above estimates, the following statement:

If one of the numbers

$$q_1 := \max_{k=1}^{n} \sum_{i=1}^{n} |\alpha_{ik}|, \qquad q_2 := \sqrt{\sum_{i,k=1}^{n} |\alpha_{ik}|^2}, \qquad q_\infty := \max_{i=1}^{n} \sum_{k=1}^{n} |\alpha_{ik}| \tag{3.7}$$

is less than 1, *then the system of equations* (3.1) *possesses exactly one solution. This solution x can be determined by iteration of A; the sequence of iterates converges componentwise to x.*

The reader can verify with simple examples that the three conditions are independent from each other, i.e., that none of them implies another one. We see that it is indeed useful to consider different metrics on the same set.

So far in our considerations we made use only of the *metric* properties of $\mathbf{K}^n$. But $\mathbf{K}^n$ has also an *algebraic* structure: its elements can be added together (componentwise) and multiplied by numbers from $\mathbf{K}$, and these operations satisfy the known rules of vector algebra. We want to take advantage of this circumstance to describe the solution by iteration of the system (3.1) in a more striking fashion. To this end we define the map $K: \mathbf{K}^n \to \mathbf{K}^n$ by

$$K(\xi_1, \ldots, \xi_n) := \left(\sum_{k=1}^{n} \alpha_{ik} \xi_k, \ldots, \sum_{k=1}^{n} \alpha_{nk} \xi_k \right), \tag{3.8}$$

set $x := (\xi_1, \ldots, \xi_n)$, $g := (\gamma_1, \ldots, \gamma_n)$ and can now write (3.3) in the form

$$Ax = g + Kx.$$

As a matrix map K is *linear*, i.e., for all x, y in $\mathbf{K}^n$ and all α in $\mathbf{K}$ one has

$$K(x + y) = Kx + Ky, \qquad K(\alpha x) = \alpha K(x).$$

If we choose $x_0 = g$, then the iterated vectors $x_{n+1} := Ax_n$ $(n = 0, 1, \ldots)$ can be represented as follows:

$$x_1 = Ax_0 = Ag = g + Kg,$$

$$x_2 = Ax_1 = g + Kx_1 = g + K(g + Kg) = g + Kg + K^2 g,$$

in general

$$x_n = Ax_{n-1} = g + Kg + K^2 g + \cdots + K^n g \quad \text{for} \quad n = 1, 2, \ldots. \tag{3.9}$$

Thus (x_n) is the sequence of partial sums of the series

$$g + Kg + K^2g + \cdots = \sum_{n=0}^{\infty} K^n g. \tag{3.10}$$

If we write the system (3.1) in the form

$$x - Kx = g, \tag{3.11}$$

then we can summarize our result as follows:

If the map K is defined by (3.8) *and one of the numbers q_1, q_2, q_∞ of* (3.7) *is less than* 1, *then the equation* (3.11) *possesses exactly one solution x which one obtains as the sum of the componentwise convergent* Neumann series (3.10).

As the next application of the fixed-point theorem we consider a functional equation which is an exact analogue of (3.1), namely the *Fredholm integral equation* (of the second kind)

$$x(s) - \int_a^b k(s, t)x(t)\mathrm{d}t = g(s). \tag{3.12}$$

For the sake of simplicity we suppose that g is continuous on the interval $[a, b]$ and the *kernel* k on the square $[a, b] \times [a, b]$, furthermore we want to seek only solutions x which are again continuous on $[a, b]$. If we define a map $A: C[a, b] \to C[a, b]$ by

$$(Ax)(s) := g(s) + \int_a^b k(s, t)x(t)\mathrm{d}t, \tag{3.13}$$

then we see that the problem to solve the integral equation (3.12) by a continuous x is equivalent to finding a solution of the equation $x = Ax$, that is, a fixed point of A. We introduce in $C[a, b]$ the canonical maximum-metric (see Example 1.4), which turns $C[a, b]$ into a complete metric space and we obtain with it the estimate

$$\begin{aligned} d(Ax, Ay) &= \max_{a \leqq s \leqq b} \left| \int_a^b k(s, t)[x(t) - y(t)]\mathrm{d}t \right| \\ &\leqq \max_{a \leqq s \leqq b} \int_a^b |k(s, t)| \max_{a \leqq t \leqq b} |x(t) - y(t)|\mathrm{d}t, \end{aligned}$$

and so

$$d(Ax, Ay) \leqq \left(\max_{a \leqq s \leqq b} \int_a^b |k(s, t)|\mathrm{d}t \right) d(x, y), \tag{3.14}$$

an inequality which is completely analogous to (3.6). From the fixed-point theorem of §2 now follows:

If

$$q_\infty := \max_{a \leqq s \leqq b} \int_a^b |k(s, t)|\mathrm{d}t < 1 \tag{3.15}$$

then the Fredholm integral equation (3.12) *possesses exactly one solution x which is continuous in* $[a, b]$. *This solution can be obtained by iteration of A; the sequence of iterates converges in the sense of the maximum-metric, i.e.,* uniformly *on* $[a, b]$.

Similarly as in the preceding example, we want to make use also here of the fact that $C[a, b]$ has not only metric but also algebraic properties: functions from $C[a, b]$ can be added (pointwise) and multiplied by numbers from $\mathbf{K}$. Similarly as in (3.8) we define an *integral transformation* $K: C[a, b] \to C[a, b]$ by

$$(3.16) \qquad (Kx)(s) := \int_a^b k(s, t)x(t)\mathrm{d}t \qquad \text{for} \quad a \leqq s \leqq b,$$

so that we have $Ax = g + Kx$. The map K is again *linear*, i.e., for all x, y in $C[a, b]$ and for all α in $\mathbf{K}$ we have

$$K(x + y) = Kx + Ky, \qquad K(\alpha x) = \alpha Kx.$$

As a consequence we can write the sequence of iterates (x_n) beginning with $x_0 = g$ like in (3.10) as a sequence of partial sums of an infinite series and obtain the following result:

If the integral transformation K with a continuous kernel k is defined by (3.16), *and if condition* (3.15) *is satisfied, then the equation*

$$x - Kx = g, \qquad g \in C[a, b],$$

i.e., the Fredholm integral equation (3.12), *possesses exactly one solution x in* $C[a, b]$ *which can be obtained as the sum of the* uniformly *convergent* Neumann series

$$(3.17) \qquad g + Kg + K^2g + \cdots = \sum_{n=0}^{\infty} K^n g.$$

One is led to Fredholm integral equations among others through the study of boundary value problems; these occur frequently in physics and technology, e.g., in buckling and vibration problems, see [152]; there—as in any other book on integral equations—one can find justifications of the facts which we want to describe briefly now.

By a *linear differential operator of order n* we understand a map L which according to the definition

$$(3.18) \qquad (Lx)(t) := \sum_{\nu=0}^{n} f_\nu(t)x^{(\nu)}(t) \qquad (f_\nu \in C[a, b],\ f_n(t) \neq 0 \text{ on } [a, b])$$

associates with every n times continuously differentiable function x on $[a, b]$ a function $Lx \in C[a, b]$. *Boundary conditions* for such a function are equations of the form

$$(3.19) \qquad R_\mu x := \sum_{\nu=0}^{n-1} [\alpha_{\mu\nu}x^{(\nu)}(a) + \beta_{\mu\nu}x^{(\nu)}(b)] = 0 \qquad (\mu = 1, \ldots, n)$$

with given numbers $\alpha_{\mu\nu}, \beta_{\mu\nu}$. A *boundary value problem* for L is the task to find an n times continuously differentiable function x on $[a, b]$ which for a given $y \in C[a, b]$ satisfies the equations

$$(3.20) \qquad Lx = y \quad \text{and} \quad R_\mu x = 0 \quad \text{for} \quad \mu = 1, \ldots, n.$$

If for a fundamental system $x_1, \ldots, x_n$ of the homogeneous differential equation $Lx = 0$ the determinant

$$(3.21) \qquad |R_\mu x_k| \neq 0,$$

then there exists one and only one *Green's function* G, which is continuous on the square $a \leqq s, t \leqq b$ and with whose help the solution x—which is uniquely determined under the present hypotheses—of the boundary value problem (3.20) can be represented in the form

$$(3.22) \qquad x(s) = \int_a^b G(s, t)y(t)\mathrm{d}t.$$

We consider now the boundary value problem with a parameter λ

$$(3.23) \qquad Lx - \lambda rx = 0, \qquad R_\mu x = 0 \qquad (\mu = 1, \ldots, n),$$

where $r \in C[a, b]$ and the product rx is defined, as is customary, pointwise: $(rx)(t) := r(t)x(t)$. Every value λ, for which (3.23) possesses a non-trivial (i.e., not identically vanishing) solution x, is called an *eigenvalue* (or proper value) of the problem, and x is called an *eigensolution* corresponding to λ. Eigenvalues and eigensolutions, or expressions which depend in a simple way on them, have often an important significance in physics and technology (buckling loads, frequencies of resonance, energy levels, vibration figures, etc.). This fact explains the basic role of the theory of eigenvalues in various branches of science.

Clearly, λ is an eigenvalue of (3.23) if and only if for a fundamental system $x_{\lambda 1}, \ldots, x_{\lambda n}$ of the homogeneous differential equation $Lx - \lambda rx = 0$ the determinant

$$(3.24) \qquad D(\lambda) := |R_\mu x_{\lambda k}|$$

vanishes. *Consequently,* (3.21) *is satisfied if and only if $\lambda = 0$ is not an eigenvalue. In this case there exists,* as we have seen, *a Green's function G of the boundary value problem* (3.20). If the problem

$$(3.25) \qquad Lx - \lambda rx = y, \qquad R_\mu x = 0 \qquad (\mu = 1, \ldots, n)$$

has a solution x for a $y \in C[a, b]$, then because of (3.22) this solution x satisfies the integral equation

$$(3.26) \qquad x(s) = \int_a^b G(s, t)[\lambda r(t)x(t) + y(t)]\mathrm{d}t$$

which with

$$(3.27) \qquad k(s, t) := G(s, t)r(t), \qquad g(s) := \int_a^b G(s, t)y(t)\mathrm{d}t$$

becomes a Fredholm integral equation

$$x(s) - \lambda \int_a^b k(s, t)x(t)dt = g(s) \tag{3.28}$$

with the continuous kernel $\lambda k(s, t)$, Conversely, it can be shown that every continuous solution of (3.28) also solves (3.25). *Thus under the hypothesis* (3.21) *the boundary value problem* (3.25) *can be reduced completely to a Fredholm integral equation with continuous kernel.* This fact is an essential reason why we shall consider extensively such integral equations.

Exercises

1. Let the infinite system of linear equations

$$\xi_i - \sum_{k=1}^{\infty} \alpha_{ik}\xi_k = \gamma_i \qquad (i = 1, 2, \ldots) \tag{3.29}$$

be given. A solution of this system is a sequence $(\xi_1, \xi_2, \ldots)$ for which each series $\sum_{k=1}^{\infty} \alpha_{ik}\xi_k$ converges and has sum $\xi_i - \gamma_i$. Show that if $g := (\gamma_1, \gamma_2, \ldots)$ belongs to l^∞ and $\sup_{i=1}^{\infty} \sum_{k=1}^{\infty} |\alpha_{ik}| < 1$, then (3.29) possesses exactly one solution $x = (\xi_1, \xi_2, \ldots)$ in l^∞. This solution can be obtained with the help of a Neumann series which converges uniformly on **N**.

2. If we introduce on $C[a, b]$ instead of the maximum-metric the metric d_1 or d_2 of §1, Exercise 4, then the self-map A of $C[a, b]$ defined in (3.13) is contracting provided we have

$$q_1 := \max_{a \leqq t \leqq b} \int_a^b |k(s, t)| ds < 1 \quad \text{or} \quad q_2 := \sqrt{\int_a^b\int_a^b |k(s, t)|^2 \, ds \, dt} < 1,$$

respectively (cf. these quantities with the numbers q_1, q_2 in (3.7)). However, one cannot apply Banach's fixed-point theorem.

3. If $\lambda = 0$ is not an eigenvalue of the problem (3.23), and if G is the Green's function of (3.20), then in the case

$$|\lambda| < \left[\max_{a \leqq s \leqq b} \int_a^b |G(s, t)r(t)| dt\right]^{-1},$$

the boundary value problem (3.25) has a unique solution for every $y \in C[a, b]$. The solution is obtained as the sum of a series $g + \lambda K g + \lambda^2 K^2 g + \cdots$ uniformly convergent on $[a, b]$, where K is defined through (3.16), k and g through (3.27).

4. Let the integral transformation $K\colon C[a, b] \to C[a, b]$ with continuous kernel $k(s, t)$ be defined by (3.16). Then

$$(K^2x)(s) = \int_a^b \left[\int_a^b k(s, u)k(u, t)du\right] x(t)dt;$$

thus K^2 is also an integral transformation with kernel

$$k_2(s, t) := \int_a^b k(s, u)k(u, t)\mathrm{d}u.$$

In general, define the 'n times iterated kernels' k_n by

$$k_1(s, t) := k(s, t),$$

$$k_n(s, t) := \int_a^b k(s, u)k_{n-1}(u, t)\mathrm{d}u \qquad \text{for} \quad n \geqq 2$$

and show that K^n is an integral transformation with kernel k_n.

5. We use the notation and results of Exercise 4. If

$$q := \max_{a \leqq s \leqq b} \int_a^b |k(s, t)|\mathrm{d}t < 1,$$

then the solution of the integral equation $x - Kx = y$ can be represented in the form

$$x(s) = y(s) + \sum_{n=1}^{\infty} \int_a^b k_n(s, t)y(t)\mathrm{d}t;$$

the series converges uniformly on $[a, b]$. Setting $\mu := \max_{s,t}|k(s, t)|$ one has $|k_n(s, t)| \leqq \mu q^{n-1}$ for $a \leqq s, t \leqq b$, $n \geqq 1$; thus because of $q < 1$ the series

$$\sum_{n=1}^{\infty} k_n(s, t) =: r(s, t)$$

converges uniformly on the square $a \leqq s, t \leqq b$, hence r is continuous there. It follows that

$$x(s) = y(s) + \int_a^b r(s, t)y(t)\mathrm{d}t \qquad \text{i.e.,} \quad x = y + Ry,$$

where R is the integral transformation with kernel $r(s, t)$. We call $r(s, t)$ the *resolving kernel* and R *the resolving transformation.*

6. The reader will find in [9] further applications of Banach's fixed-point theorem (see also Exercise 2 in §106).

II

Normed spaces

§4. Vector spaces

In the preceding section we made use of the fact that certain sets of functions carry not only a metric but also an algebraic structure. Namely one can define in a natural way the sum $x + y$ and the scalar multiple αx of functions x, y with a common set of definition T and values in $\mathbf{K}$ by

$$(x + y)(t) := x(t) + y(t), \qquad (\alpha x)(t) := \alpha x(t) \qquad \text{for} \quad t \in T, \alpha \in \mathbf{K}.$$

These operations in the set E of all $\mathbf{K}$-valued functions on T satisfy the known rules of vector algebra:

(V1) $x + (y + z) = (x + y) + z$,
(V2) $x + y = y + x$,
(V3) in E there exists a zero element 0 such that $x + 0 = x$ for all $x \in E$,
(V4) to every $x \in E$ there exists an element $-x \in E$ such that $x + (-x) = 0$,
(V5) $\alpha(x + y) = \alpha x + \alpha y$,
(V6) $(\alpha + \beta)x = \alpha x + \beta x$,
(V7) $(\alpha\beta)x = \alpha(\beta x)$,
(V8) $1 \cdot x = x$.

The zero element is of course the function which vanishes identically on T, the element $-x$ inverse to x is $t \mapsto -x(t)$.

In general, we call a non-empty set E a *vector space* or a *linear space* over $\mathbf{K}$, if for any two elements x, y in E and for any α in $\mathbf{K}$ a sum $x + y \in E$ and a product $\alpha x \in E$ is defined, so that the axioms (V1) through (V8) of a vector space hold. The elements of E are called *points* or *vectors*. E is called *real* or *complex* according as $\mathbf{K}$ is the field of real or complex numbers.

The definition of a vector space will be familiar to the reader from linear algebra. He will also know from there that the zero element 0, and for a given x the element $-x$, are determined uniquely, that the relations $0x = 0$, $\alpha 0 = 0$ and $(-1)x = -x$ hold for all x in E and α in $\mathbf{K}$ and that the equation $y + x = z$ has the unique solution $x = z + (-y)$. Instead of $z + (-y)$ we write henceforth more shortly $z - y$.

A non-empty set E of scalar-valued functions on T is already a vector space over **K** *if the (pointwise) addition and multiplication by numbers $\alpha \in \mathbf{K}$ do not lead outside E*; axioms (V1) through (V8) are then automatically satisfied. Such a set will be called a *linear function space*. If $T = \mathbf{N}$, then one speaks most often of a *linear sequence space*.

The following sets are linear function spaces or sequence spaces over **K**; the function values or terms of sequences should all lie in **K**:

1. The set $\mathbf{K}^n$ of all n-tuples $(\xi_1, \ldots, \xi_n)$, in particular **K** itself.
2. The set (s) of all sequences.
3. The set l^p, $1 \leqq p < \infty$, of all sequences $x = (\xi_1, \xi_2, \ldots)$ for which $\sum_{n=1}^{\infty} |\xi_n|^p$ converges. Clearly, with x also every multiple αx lies in l^p. It follows immediately from Minkowski's inequality that addition does not lead outside l^p.
4. The set l^∞ of all bounded sequences.
5. The set (c) of all convergent sequences.
6. The set (c_0) of all sequences converging to zero.
7. The set $B(T)$ of all bounded functions on T. l^∞ is, as we have observed earlier, nothing else but $B(\mathbf{N})$.
8. The set $C[a, b]$ of all continuous functions on $[a, b]$.
9. The set $C^{(n)}[a, b]$, $1 \leqq n \leqq \infty$, of all functions which are n-times continuously differentiable on the interval $[a, b]$.
10. The set $C_0(\mathbf{R})$ of all continuous functions on **R** which 'vanish at infinity', i.e., to every $\varepsilon > 0$ there exists a number $\rho = \rho(\varepsilon, x) > 0$ such that $|x(t)| \leqq \varepsilon$ for $|t| \geqq \rho$.
11. The set $BV[a, b]$ of all functions defined on $[a, b]$ which are of bounded variation there.

With the exception of $\mathbf{K}^n$, we have not distinguished in the notation of the above spaces by symbols like (s), l^p, etc. between the real and the complex space, since such a distinction is most often inessential. *Unless we say expressly something different, our results are valid for the real as well as for the complex case.*

After these examples, which show among others how often linear structures occur in analysis, we start to study now the geometry of linear spaces E over **K**. A non-empty subset F of E is called a (*linear*) *subspace* of E if sums and scalar multiples of elements in F lie again in F; then F is itself a vector space. All linear sequence spaces are subspaces of (s). Every subspace of E contains the zero element 0, and $\{0\}$ itself is a subspace. *The intersection of arbitrary many subspaces of E is again a subspace of E.* In particular, the intersection of all subspaces, which contain a non-empty subset M of E, is a subspace; it is called the subspace *generated* or *spanned* by M or also the *linear hull* of M and will be denoted by $[M]$. The set $[M]$ is the smallest subspace which contains M; its elements are all the (finite) linear combinations $\alpha x + \beta y + \cdots$ of elements $x, y, \ldots$ in M. The linear hull of the finite or countable set $\{x_1, x_2, \ldots\}$ will also be denoted by $[x_1, x_2, \ldots]$.

A *finite* subset $\{x_1, \ldots, x_n\}$ of E is called *linearly independent* if $\alpha_1 x_1 + \cdots + \alpha_n x_n = 0$ implies $\alpha_1 = \cdots = \alpha_n = 0$, otherwise it is called *linearly dependent.* An *infinite* subset M of E is said to be linearly independent if every finite subset of M is linearly independent; otherwise it is again said to be linearly dependent.

The kth *unit vector* $e_k := (0, \ldots, 1, 0, \ldots)$, i.e., the sequence which has 1 at the kth place and 0 everywhere else, lies in all the sequence spaces considered so far: l^p, (c_0), (c) and (s). It is obvious that $\{e_1, e_2, \ldots\}$ is a linearly independent subset of all these spaces. It follows from the identity theorem for polynomials that the set of the functions $t \mapsto t^n$ $(n = 0, 1, 2, \ldots)$ defined on $[a, b]$ is a linearly independent subset of $C[a, b]$.

Of special importance are those linearly independent subsets of E which generate all E. Such a set is called a (*Hamel* or *algebraic*) *basis* of E. With the help of a basis $B = \{x_\lambda : \lambda \in L\}$ every vector x in E can be represented in the form $x = \sum_{\lambda \in L} \alpha_\lambda x_\lambda$, where the coefficients α_λ are determined uniquely and only finitely many α_λ are different from zero. The following result answers the question whether there exists at all a basis in E.

Theorem 4.1. *To every linearly independent subset M of E there exists a basis of E containing M (i.e., M can be extended to a basis). If $E \neq \{0\}$, then it contains a basis.*

The proof depends on Zorn's lemma. The set $\mathfrak{M}$ of all linearly independent subsets of E containing M is not empty (it contains M) and is ordered by set theoretical inclusion. Every totally ordered subset $\mathfrak{B}$ of $\mathfrak{M}$ possesses the upper bound $\bigcup_{V \in \mathfrak{B}} V$ in $\mathfrak{M}$. Consequently, there exists a maximal element B in $\mathfrak{M}$. For every vector x not in B, the set $B \cup \{x\}$ is linearly dependent as a proper superset of B. Thus there exist vectors $y_1, \ldots, y_n$ in B and numbers $\alpha_0, \alpha_1, \ldots, \alpha_n$ which are not all zero, so that $\alpha_0 x + \alpha_1 y_1 + \cdots + \alpha_n y_n = 0$. Because of the linear independence of B, we must have $\alpha_0 \neq 0$. Thus we obtain that x is a linear combination of the vectors $y_1, \ldots, y_n$, i.e., lies in the linear hull $[B]$ of B. Since trivially every $x \in B$ lies in $[B]$, we have $[B] = E$. Thus B is indeed a basis of E containing M. If in E there exists an element $x_0 \neq 0$, then the set $\{x_0\}$ is linearly independent, and can thus be extended to a basis of E according to what we just proved. ■

If $E \neq \{0\}$ has a basis of finitely many, say n, elements, then every other basis of E also consists of n elements, as the reader knows from elementary linear algebra. In this case we say that E has *dimension n* or that it is *n-dimensional*, and we write $\dim E = n$. To the trivial space $\{0\}$ we attribute dimension zero. If E does not possess a finite basis, we set $\dim E = \infty$ and say that E is *infinite-dimensional.* With the exception of $\mathbf{K}^n$, all the function and sequence spaces considered in this section are infinite-dimensional, since they contain infinite-dimensional linearly independent subsets. It can be shown that even in an infinite-dimensional vector space two bases have always the same cardinal (power); this cardinal can then be called the *dimension* of the space. See e.g., [8], §7,4.(4), p. 53.

For subsets M, N of E, vectors x_0 in E and scalars α we set:

$$x_0 \pm M := \{x_0 \pm x : x \in M\},$$
$$M \pm N := \{x \pm y : x \in M, y \in N\},$$
$$\alpha M := \{\alpha x : x \in M\}.$$

Of course the sum of sets can be defined analogously for more than two summands.

The sum $F + G$ of two subspaces F, G of E is again a subspace. This sum is called *direct* and will be denoted by $F \oplus G$ if $F \cap G = \{0\}$. Clearly, this is the case if and only if for every element $z = x + y$ in $F + G$ the *components* $x \in F$, $y \in G$ are determined uniquely. If $F \oplus G = E$, then G is called an (*algebraic*) *complement* of F (in E). In this case the union of a basis of F with a basis of G gives a basis of E. Conversely, if one extends according to Theorem 4.1 a basis of the subspace F, through adjoining to it an appropriate linearly independent set M, to a basis of E, then M generates a complement of F. Thus we have proved the following proposition:

Proposition 4.1. *To every subspace of a vector space there exists at least one (algebraic) complement.*

If $E = F \oplus G_1 = F \oplus G_2$, then it can be shown, applying the above-mentioned general concept of dimension, that G_1 and G_2 have the same (possibly transfinite) dimension. We want to prove it here only in the case that G_1 is finite-dimensional. To avoid trivialities, let us assume also that $G_1 \neq \{0\}$. If now $\{y_1, \ldots, y_n\}$ is a basis of G_1 and if $z_1, \ldots, z_m$ are vectors from G_2, then $z_\mu = x_\mu + \sum_{\nu=1}^n \alpha_{\mu\nu} y_\nu$ with $x_\mu \in F$. If $m > n$, then the m vectors $(\alpha_{\mu 1}, \ldots, \alpha_{\mu n})$, $\mu = 1, \ldots, m$, in the n-dimensional space $\mathbf{K}^n$ are linearly dependent, so there exist numbers $\beta_1, \ldots, \beta_m$ so that $\sum_{\mu=1}^m |\beta_\mu| \neq 0$ and $\sum_{\mu=1}^m \beta_\mu(\alpha_{\mu 1}, \ldots, \alpha_{\mu n}) = 0$, i.e., $\sum_{\mu=1}^m \beta_\mu \alpha_{\mu\nu} = 0$ for $\nu = 1, \ldots, n$. It follows that

$$\begin{aligned}\sum_{\mu=1}^m \beta_\mu z_\mu &= \sum_{\mu=1}^m \beta_\mu\left(x_\mu + \sum_{\nu=1}^n \alpha_{\mu\nu} y_\nu\right) \\ &= \sum_{\nu=1}^n \left(\sum_{\mu=1}^m \beta_\mu \alpha_{\mu\nu}\right) y_\nu + \sum_{\mu=1}^m \beta_\mu x_\mu = \sum_{\mu=1}^m \beta_\mu x_\mu\end{aligned}$$

lies in $G_2 \cap F = \{0\}$, so that the vectors $z_1, \ldots, z_m$ are linearly dependent: more than n elements from G_2 are always linearly dependent. Thus $\dim G_2 \leqq \dim G_1$, and since by symmetry the opposite inequality also holds, we have proved our assertion. We want to state it in the form of a proposition:

Proposition 4.2. *Two complements of a given subspace are either both infinite-dimensional, or they have the same finite dimension.*

This proposition gives us the possibility to define the *codimension* of a subspace G of E as the dimension of any complement F of G. If we denote this quantity by $\operatorname{codim}_E G$ or more simply by $\operatorname{codim} G$ if the space E is fixed, then we have

$$\operatorname{codim} G = \infty \qquad \text{if} \quad \dim F = \infty$$

$$\operatorname{codim} G = \dim F \qquad \text{if} \quad 1 \leqq \dim F < \infty$$

$$\operatorname{codim} G = 0 \qquad \text{if } G = E.$$

The codimension of a subspace measures how much this subspace differs from the whole space. A subspace of codimension 1 is called, as in analytic geometry, a *hyperplane* (through 0).

The cartesian product $\prod_{\lambda \in L} E_\lambda$ of a family $(E_\lambda : \lambda \in L)$ of vector spaces over **K** is made into a vector space over **K** by defining the sums and scalar multiples *componentwise*:

$$(x_\lambda) + (y_\lambda) := (x_\lambda + y_\lambda), \qquad \alpha(x_\lambda) := (\alpha x_\lambda).$$

Exercises

1. The unit vectors $e_1, e_2, \ldots$ do not form a basis of (s).

*2. At times it is convenient to multiply the elements of a *complex* vector space only by *real* scalars. *Then E becomes a* real *vector space E_r*. Observe that E_r consists of exactly the same elements as E. E.g., the space $C[a, b]$ of the complex-valued continuous functions on the interval $[a, b]$ is by its very nature a complex vector space; if we only admit real numbers as scalar multipliers, we obtain a real vector space $C_r[a, b]$, which, however, has to be distinguished carefully from the real vector space of all *real-valued* continuous functions on $[a, b]$. Prove the following assertions in the case of a complex vector space E:

(a) If the set $\{x_\lambda : \lambda \in L\}$ is linearly independent in E, then the sets $\{x_\lambda : \lambda \in L\}$ and $\{x_\lambda : \lambda \in L\} \cup \{ix_\lambda : \lambda \in L\}$ are linearly independent in E_r.

(b) If $\{x_\lambda : \lambda \in L\}$ is a basis of E, then $\{x_\lambda : \lambda \in L\} \cup \{ix_\lambda : \lambda \in L\}$ is a basis of E_r.

The reader should make a clear picture for himself of the situation in the case of $\mathbf{C}^n$. The space $\mathbf{C}^n$ has dimension n as a vector space over **C** and dimension $2n$ as a vector space over **R**.

3. A *real* vector space E can be made through *complexification* into a complex vector space E_c in exactly the same way as the field of complex numbers **C** is obtained from the field of real numbers **R**. We indicate the construction and the main facts:

(a) Let E_c be the set of all ordered pairs (x, y), $x, y \in E$. Addition and multiplication by a complex number $\alpha + i\beta$ ($\alpha, \beta \in \mathbf{R}$) are defined as follows: $(x_1, y_1) + (x_2, y_2) := (x_1 + x_2, y_1 + y_2)$, $(\alpha + i\beta)(x, y) := (\alpha x - \beta y, \beta x + \alpha y)$. In this way E_c becomes a vector space over **C**.

(b) $E_0 := \{(x, 0) : x \in E\} \subset E_c$ is a vector space over $\mathbf{R}$, the map $x \mapsto (x, 0)$ is an isomorphism of the spaces E and E_0. For this reason one identifies the elements x and $(x, 0)$ and obtains E as a subset of E_c.

(c) Because of $(x, y) = (x, 0) + (0, y) = (x, 0) + i(y, 0)$, the elements (x, y) of E_c can be written in the form $(x, y) = x + iy$ and one has $(x_1 + iy_1) + (x_2 + iy_2) = (x_1 + x_2) + i(y_1 + y_2)$, $(\alpha + i\beta)(x + iy) = (\alpha x - \beta y) + i(\beta x + \alpha y)$ for $\alpha, \beta \in \mathbf{R}$. The vectors x in E can be found again in E_c as the elements $x = x + i0$.

(d) If the set $\{x_\iota : \iota \in J\} \subset E$ is linearly independent in E, then it is also linearly independent in E_c.

(e) A basis $\{x_\iota : \iota \in J\}$ of E is also a basis of E_c. In particular $\dim E = \dim E_c$.

*4. By the *sum* $F = \sum_{\lambda \in L} F_\lambda$ of a family $(F_\lambda : \lambda \in L)$ of subspaces of a vector space E we understand the set of all elements $x = \sum_{\lambda \in L} x_\lambda$, where x_λ is in F_λ and only finitely many x_λ are different from zero. F is the linear hull of $\bigcup_{\lambda \in L} F_\lambda$. We say that F is the *direct sum* of the F_λ, and we write $F = \bigoplus_{\lambda \in L} F_\lambda$, if for every $x = \sum x_\lambda$ from F the components $x_\lambda \in F_\lambda$ are uniquely determined. Show that the following statements are equivalent:

(a) $F = \bigoplus_{\lambda \in L} F_\lambda$.

(b) From $\sum_{\lambda \in L} x_\lambda = 0$, $x_\lambda \in F_\lambda$ it follows that $x_\lambda = 0$ for all $\lambda \in L$.

(c) $F_\lambda \cap \sum_{\substack{\mu \in L \\ \mu \neq \lambda}} F_\mu = \{0\}$ for all $\lambda \in L$.

§5. Linear maps

We have encountered linear maps already in §3 when solving systems of linear equations and Fredholm integral equations. Now that we dispose of the concept of a vector space, we can give the general definition of a linear map:

The map $A: E \to F$ from the vector space E over $\mathbf{K}$ into the vector space F over $\mathbf{K}$ is said to be *linear* if for all x, y in E and all $\alpha \in \mathbf{K}$ we have

$$A(x + y) = Ax + Ay, \qquad A(\alpha x) = \alpha Ax.$$

By the *linear image* of a set we understand its image under a linear map.

Linear maps are also called *linear transformations*, *linear operators* (or simply *operators*) or *homomorphisms*. An *endomorphism* of the vector space E is a linear self-map of E. We denote by $\mathscr{S}(E, F)$ the set of all linear maps from E into F, $\mathscr{S}(E) := \mathscr{S}(E, E)$ is the set of all endomorphisms of E.

$\mathscr{S}(E, F)$ becomes a vector space (over the common field of scalars of E and F), if we define sums $A + B$ and multiples αA pointwise:

$$(A + B)x := Ax + Bx, \qquad (\alpha A)x := \alpha(Ax) \qquad \text{for all} \quad x \in E.$$

The zero element of this space is the zero map 0 which associates with every element in E the zero element of F.

The composition $B \circ A$ of the linear maps $A: E \to F$, $B: F \to G$ is a linear map from E into G. We denote it more shortly by BA and call it usually the *product* of B by A.

In case the following products exist, they satisfy these rules:

$$A(BC) = (AB)C,$$
$$A(B + C) = AB + AC,$$
$$(A + B)C = AC + BC,$$
$$\alpha(AB) = (\alpha A)B = A(\alpha B).$$

Products are in particular always defined if the factors lie in $\mathscr{S}(E)$. It follows that $\mathscr{S}(E)$ is an 'algebra' over the scalar field $\mathbf{K}$, since a set R is called an *algebra* over $\mathbf{K}$ if R is a vector space over $\mathbf{K}$ and if for any two elements a, b in R a product $ab \in R$ is defined in such a way that the following rules hold:

$$a(bc) = (ab)c,$$
$$a(b + c) = ab + ac,$$
$$(a + b)c = ac + bc,$$
$$\alpha(bc) = (\alpha a)b = a(\alpha b).$$

The algebra R is said to be *commutative* if one always has $ab = ba$. An element e in R such that $ae = ea = a$ for all $a \in R$ is called a *unit element* (*identity*) of R. In an algebra there exists at most one unit element. The algebra $\mathscr{S}(E)$ is in general not commutative but possesses a unit element, namely the *identity map I* defined by $Ix := x$ for all $x \in E$. Sometimes we write I_E instead of I to emphasize that I operates on E. We recall that the powers or iterates of an endomorphism A are defined successively by $A^0 := I$, $A^n := AA^{n-1}$ $(n = 1, 2, \ldots)$.

The importance of linear maps (and so also of linear spaces) is based among others on the fact that many problems concerning equations are *linear problems* in the following sense: Given a linear map $A: E \to F$, determine for which 'right-hand sides' $y \in F$ does the equation

$$Ax = y \tag{5.1}$$

have a solution (i.e., describe the image of A) and check whether the solution, if it exists, is determined uniquely; if not, describe the set of solutions. The most satisfying situation occurs when our equation has exactly one solution for each $y \in F$, i.e., when A is bijective.

In §3 we wrote the system of linear equations (3.1) and the Fredholm integral equation (3.12) with the help of an endomorphism K of $\mathbf{K}^n$ or $C[a, b]$, respectively, in the form

$$(I - K)x = y$$

and gave sufficient conditions for the bijectivity of the endomorphism $I - K$; moreover, we could also construct the solution. Further examples for linear problems are given by the boundary value problems of the form (3.20). Using the notations and hypotheses from there, if we set $E := \{x \in C^{(n)}[a, b]: R_\mu x = 0$ for $\mu = 1, \ldots, n\}$ and define the linear map $A: E \to C[a, b]$ by $Ax := Lx$, then the boundary value problem (3.20) is precisely the task to solve the equation $Ax = y$.

In this section we shall collect some simple concepts and facts which serve to investigate 'operator equations' of the form (5.1), and which in particular are concerned with the questions whether (5.1) has a unique solution or has a solution for every $y \in F$, i.e., with the injectivity or surjectivity of A, respectively.

We first remark that

$$\text{the } \textit{image } A(E) := \{Ax \colon x \in E\}$$

and

$$\text{the } \textit{nullspace } N(A) := \{x \in E \colon Ax = 0\}$$

of A are linear spaces, in particular $0 \in N(A)$, i.e., $A0 = 0$. If the equation (5.1) has the two solutions x_0, x_1, then $A(x_0 - x_1) = Ax_0 - Ax_1 = y - y = 0$, i.e., $x_1 - x_0 \in N(A)$ and so $x_1 \in x_0 + N(A)$. Conversely, if x_1 is contained in $x_0 + N(A)$, then clearly $Ax_1 = Ax_0 = y$. With this we have a first assertion on the set of solutions of (5.1):

Proposition 5.1. *The set of all solutions of* (5.1) *can be represented with the help of one solution x_0 in the form $x_0 + N(A)$. The map A is* injective *if and only if $N(A) = \{0\}$, i.e., if $Ax = 0$ implies $x = 0$.*

If A is injective, the *inverse map* (see introduction) defined on $A(E)$ is denoted by A^{-1} and we call it shortly the *inverse* of A.

It is easy to see that A^{-1} is linear, hence lies in $\mathscr{L}(A(E), E)$. From here and from the discussion of inverse maps in the introduction we immediately obtain the following:

Proposition 5.2. *The linear map $A \colon E \to F$ is* bijective *if and only if there exist linear maps B and C from F into E such that*

$$BA = I_E \quad \textit{and} \quad AC = I_F.$$

In this case $B = C = A^{-1}$.

A bijective linear map $A \colon E \to F$ is also called an *isomorphism* (of the spaces E and F) and we say that E and F are *isomorphic*. This symmetric way of expression is justified because together with A also A^{-1} is an isomorphism.

The product BA of two bijective linear maps $A \colon E \to F$ and $B \colon F \to G$ is again bijective and has as inverse $(BA)^{-1} = A^{-1}B^{-1}$.

A particularly important class of endomorphisms consists of the so-called *projectors*. If $E = F \oplus G$, i.e., if for every $x \in E$ one has a decomposition $x = y + z$ with uniquely determined components $y \in F$, $z \in G$, then one can define a map $P \colon E \to F$ by $Px := y$. The map P is linear and is called projector because, taking over the language of geometry, P projects the space E along (or parallel to) G onto F. Then $I - P$ projects E along F onto G. We have clearly

$$P(E) = F, \qquad N(P) = G \quad \text{and} \quad P^2 = P.$$

If we call an endomorphism A which satisfies $A^2 = A$ *idempotent*, then the last equation asserts that *projectors are idempotent. This property is characteristic for projectors.* Indeed, if P is an idempotent endomorphism of E, then because of $x = Px + (I - P)x$, the space E is the sum of the two subspaces $P(E)$ and $(I - P)(E)$, and this sum is even direct since from $z \in P(E) \cap (I - P)(E)$, i.e., $z = Px = (I - P)y$ it follows that $z = Px = P^2x = P(I - P)y = Py - P^2y = Py - Py = 0$. It can be seen immediately that P is the projector of E onto $P(E)$ along $(I - P)(E)$.

Because of Proposition 4.1, to every subspace F of E there exists a projector which projects E onto F. The image of a projector P of E is given by

$$P(E) = \{x \in E: Px = x\},$$

i.e., it is that subspace on which P acts as the identity. Indeed, if $x \in P(E)$ then $x = Py$ and so $Px = P^2y = Py = x$, and conversely from $x = Px$ it follows trivially that x lies in $P(E)$. With the help of projectors we can describe algebraically the discrepancy of a linear map $A: E \to F$ from bijectivity. Let P be a projector of E onto $N(A)$ and Q_0 a projector of F onto $A(E)$; let the corresponding decompositions of E and F be

$$E = N(A) \oplus U, \qquad F = A(E) \oplus V.$$

We now define a linear map $A_0: U \to A(E)$ by $A_0x := Ax$ for $x \in U$. The map A_0 is clearly surjective, and since $A_0x = 0$ implies that $x \in N(A) \cap U = \{0\}$, it is also injective, hence $A_0^{-1}: A(E) \to U$ exists. Thus $B := A_0^{-1}Q_0$ is a linear map from F into E. If we represent an arbitrary $x \in E$ in the form $x = y + z$ with $y \in N(A)$, $z \in U$, then we obtain $BAx = A_0^{-1}Q_0A(y + z) = A_0^{-1}Q_0Az = A_0^{-1}Az = z = x - y = x - Px$, hence $BA = I_E - P$. Similarly one gets the equation $AB = Q_0$. With the projector $Q := I_F - Q_0$, complementary to Q_0, which projects F along $A(E)$ onto V, we have $AB = I_F - Q$. Thus we proved the following proposition:

Proposition 5.3. *If $A: E \to F$ is linear, P a projector of E onto $N(A)$ and Q a projector of F along $A(E)$, then there exists a linear map $B: F \to E$ such that the equations*

$$BA = I_E - P, \qquad AB = I_F - Q$$

are valid.

Comparing this proposition with Proposition 5.2, one sees in what way one can describe with the help of projectors the deviation of a linear map from bijectivity. Ultimately this depends on the fact that the deviation from injectivity is given by the nullspace and the deviation from surjectivity by a complementary subspace of the image space.

The linear subspace F of E is said to be *invariant* under $A \in \mathscr{S}(E)$ if $A(F) \subset F$; in this case the restriction of A to F is a self-map of F. We say that the pair (F, G)

of subspaces *reduces* A if $E = F \oplus G$ and both F and G are invariant under A. The reader will prove easily the following proposition:

Proposition 5.4. *Let P be a projector of E onto F parallel to G. The subspace F is* invariant *under A if and only if $AP = PAP$ holds. A is* reduced *by (F, G) if and only if $AP = PA$ holds.*

Exercises

1. Let $A: E \to F$ be a linear map. Show:

(a) If $\{Ax_\lambda : \lambda \in L\} \subset A(E)$ is linearly independent, then also $\{x_\lambda : \lambda \in L\} \subset E$ is linearly independent.

(b) If A is injective and $\{x_\lambda : \lambda \in L\} \subset E$ is linearly independent, then $\{Ax_\lambda : \lambda \in L\}$ is also linearly independent.

2. Let E and F be two vector spaces having the same finite dimension and let $A \in \mathscr{S}(E, F)$. With the help of exercise 1 show that the map A is bijective if it is either injective or surjective (cf. Exercise 3).

3. Construct an endomorphism of (s) which is injective but not surjective (cf. Exercise 2).

*4. Let E and F be two vector spaces over $\mathbf{K}$ and let E be finite-dimensional. E and F are isomorphic if and only if $\dim E = \dim F$.

*5. Let P be a projector of E and $M \subset E$. Then $P(M) \subset M + N(P)$.

*6. To $A \in \mathscr{S}(E, F)$ there exists a $B \in \mathscr{S}(F, E)$ with $ABA = A$.

§6. Normed spaces

When we discussed in §3 systems of linear equations and Fredholm integral equations, we encountered the linear structure of the spaces $l^p(n)$ and $C[a, b]$. This linear structure is closely related to the canonical metric structure of $l^p(n)$ and $C[a, b]$. We first define some useful concepts.

The metric d on the vector space E is said to be *translation-invariant* if

$$d(x, y) = d(x + z, y + z) \qquad \text{for all} \quad x, y, z \text{ in } \quad E.$$

If in this case we set $|x| := d(x, 0)$, then $d(x, y) = d(x - y, y - y) = d(x - y, 0) = |x - y|$, i.e., the distance of two points of E can be described with the help of the 'absolute value' $|x|$, just like the distance of two points of $\mathbf{K}$ with the help of the absolute value given there. From the axioms (M1) through (M3) in §1 we obtain immediately the following properties of absolute value:

(A1) $|x| \geqq 0$, where $|x| = 0$ holds exactly when $x = 0$,

(A2) $|-x| = |x|$,

(A3) $|x + y| \leqq |x| + |y|$ (*triangle inequality*).

Conversely, if a vector space E is a *valued space*, i.e., if with every x in E a real number $|x|$, its *absolute value*, is associated in such a way that the axioms (A1) through (A3) hold, then

$$d(x, y) := |x - y| \tag{6.1}$$

defines a translation-invariant metric on E, the *canonical metric* of a valued space. Thus *the valued spaces are exactly the vector spaces with a translation-invariant metric.*

A sequence (x_n) in a valued space converges to x, if $|x_n - x| \to 0$; it is a Cauchy sequence if for every $\varepsilon > 0$ there exists an $n_0 = n_0(\varepsilon)$ such that $|x_n - x_m| \leqq \varepsilon$ for $n, m \geqq n_0$. A valued space is *complete* if it is complete as a metric space.

The canonical metrics of the vector spaces $l^p(n)$, l^∞, $B(T)$, $C[a, b]$ and (s) are all translation-invariant (cf. Examples 1.1 through 1.5), so these spaces are valued spaces with the absolute values

$$|x| := \left(\sum_{\nu=1}^{n} |\xi_\nu|^p\right)^{1/p} \quad \text{for} \quad x = (\xi_1, \ldots, \xi_n) \in l^p(n), \qquad 1 \leqq p < \infty,$$

$$|x| := \max_{\nu=1}^{n} |\xi_\nu| \quad \text{for} \quad x = (\xi_1, \ldots, \xi_n) \in l^\infty(n),$$

$$|x| := \sup_{\nu=1}^{\infty} |\xi_\nu| \quad \text{for} \quad x = (\xi_1, \xi_2, \ldots) \in l^\infty,$$

$$|x| := \sup_{t \in T} |x(t)| \quad \text{for} \quad x : t \mapsto x(t) \quad \text{in} \quad B(T),$$

$$|x| := \max_{t \in [a,b]} |x(t)| \quad \text{for} \quad x : t \mapsto x(t) \quad \text{in} \quad C[a, b],$$

$$|x| := \sum_{\nu=1}^{\infty} \frac{1}{2^\nu} \frac{|\xi_\nu|}{1 + |\xi_\nu|} \quad \text{for} \quad x = (\xi_1, \xi_2, \ldots) \in (s).$$

In §1 we have seen furthermore that these spaces are complete.

The discrete metric (see Example 1.6) is translation-invariant on every vector space; it is generated by the *discrete absolute value*

$$|x| := \begin{cases} 1 & \text{for} \quad x \neq 0 \\ 0 & \text{for} \quad x = 0. \end{cases} \tag{6.2}$$

The reader will notice immediately that the first five items on our list have a further tie to the linear structure. Indeed, beyond (A2), even $|\alpha x| = |\alpha||x|$ holds for all $\alpha \in \mathbf{K}$. On the other hand, the absolute value on (s) and the discrete absolute value do not have this property.

We call an absolute value on a vector space E a *norm* if instead of (A2) even the equation $|\alpha x| = |\alpha||x|$ is valid for all $\alpha \in \mathbf{K}$ and all $x \in E$; the space E is then called a *normed vector space* or a *normed linear space* or simply a *normed space.* In such a space the absolute value $|x|$ will always be denoted by $\|x\|$ and we call this number the norm of x. For the sake of clarity we collect again the three norm axioms:

(N1) $\|x\| \geqq 0$, where $\|x\| = 0$ holds exactly if $x = 0$,
(N2) $\|\alpha x\| = |\alpha| \, \|x\|$,
(N3) $\|x + y\| \leqq \|x\| + \|y\|$ *(triangle inequality).*

The *canonical metric* of a normed space is according to (6.1) given by $d(x, y) := \|x - y\|$. The space is called a *Banach space* if it is complete with respect to this metric.

The spaces $l^p(n)$, $1 \leqq p \leqq \infty$, l^∞, $B(T)$, $C[a, b]$ are with the absolute values indicated above, which are norms, Banach spaces. The absolute value on (s) and the discrete absolute value are not norms.

We want to give a few more examples of Banach spaces; at the end all the examples will be listed in a table. The following spaces were already introduced in §4 as vector spaces.

The vector space (c) of all convergent sequences of numbers is a linear subspace of l^∞; it is therefore reasonable to equip the elements $x = (\xi_n)$ of (c) with the norm which they already have as elements of l^∞, i.e., to set

$$\|x\| := \sup_{n=1}^{\infty} |\xi_n|.$$

With this norm (c) is complete: indeed, if the elements $x_k = (\xi_n^{(k)})$ form a Cauchy sequence in (c), hence also in l^∞, then there exists an $x = (\xi_n)$ in l^∞ to which the sequence (x_k) converges in the sense of the metric of l^∞:

$$\sup_{n=1}^{\infty} |\xi_n^{(k)} - \xi_n| \to 0 \qquad \text{as} \quad k \to \infty.$$

Having chosen $\varepsilon > 0$ we can determine a $k_0 = k_0(\varepsilon)$ so that the following holds:

(6.3) $$|\xi_n^{(k)} - \xi_n| \leqq \varepsilon \qquad \text{for} \quad k \geqq k_0 \qquad \text{and} \quad n = 1, 2, \ldots.$$

Since $(\xi_n^{(k)})$ lies in (c), the limit $\xi^{(k)} := \lim_{n\to\infty} \xi_n^{(k)}$ exists and there exists an $n_0 = n_0(\varepsilon)$ so that

(6.4) $$|\xi_n^{(k_0)} - \xi^{(k_0)}| \leqq \varepsilon \qquad \text{for} \quad n \geqq n_0.$$

From (6.3) and (6.4) we obtain for $m, n \geqq n_0$ the estimate

$$|\xi_m - \xi_n| \leqq |\xi_m - \xi_m^{(k_0)}| + |\xi_m^{(k_0)} - \xi^{(k_0)}| + |\xi^{(k_0)} - \xi_n^{(k_0)}| + |\xi_n^{(k_0)} - \xi_n| \leqq 4\varepsilon;$$

thus the sequence $x = (\xi_1, \xi_2, \ldots)$ is a Cauchy sequence and therefore convergent, hence x lies in (c), i.e., the Cauchy sequence (x_k) has a limit in (c). ■

We add a remark to this proof. If we set $\xi := \lim \xi_n$ and let $n \to \infty$ in (6.3), then we obtain $|\xi^{(k)} - \xi| \leqq \varepsilon$ for $k \geqq k_0$, hence $\lim_{k\to\infty} \xi^{(k)} = \xi = \lim_{n\to\infty} \xi_n$, or expressed even more forcefully: If the sequence $x_k = (\xi_n^{(k)})$ converges in (c), then

(6.5) $$\lim_{k\to\infty} \lim_{n\to\infty} \xi_n^{(k)} = \lim_{n\to\infty} \lim_{k\to\infty} \xi_n^{(k)}.$$

This is the well-known theorem on the interchange of limits of double sequences in the case of uniform convergence; we know that convergence in the sense of the supremum-metric of (c) is equivalent to uniform convergence on **N**.

The vector space (c_0) of all sequences converging to zero is a linear subspace of (c); it is again reasonable to give to its elements $x = (\xi_n)$ the same norm that they already have as elements of (c), i.e., the norm

$$\|x\| := \sup_{n=1}^{\infty} |\xi_n|.$$

If the elements $x_k = (\xi_n^{(k)})$ form a Cauchy sequence in (c_0), hence also in the Banach space (c), then there exists an $x = (\xi_n)$ in (c) to which the sequence (x_k) converges in (c). It follows immediately from (6.5) that $\lim \xi_n = 0$, hence $x \in (c_0)$ and so (c_0) is a Banach space.

The vector space l^p, $1 \leqq p < \infty$, which consists of all number sequences $x = (\xi_n)$ with $\sum_{n=1}^{\infty} |\xi_n|^p < \infty$ is a linear subspace of (c_0); following the procedure in our last two examples we are tempted to equip l^p with the supremum-norm coming from (c_0). Unfortunately, l^p will not be complete for this norm (cf. Exercise 2). However, Minkowski's inequality suggests the definition

$$\|x\| := \left(\sum_{n=1}^{\infty} |\xi_n|^p \right)^{1/p}.$$

Indeed, Minkowski's inequality is just the triangle inequality for this norm. We now prove the completeness of l^p. Let the elements $x_k = (\xi_n^{(k)})$ form a Cauchy sequence in l^p and $\varepsilon > 0$ be given arbitrarily. Then there exists a $k_0 = k_0(\varepsilon)$ so that

$$\text{(6.6)} \qquad \|x_k - x_l\| = \left(\sum_{n=1}^{\infty} |\xi_n^{(k)} - \xi_n^{(l)}|^p \right)^{1/p} \leqq \varepsilon \qquad \text{for} \quad k, l \geqq k_0.$$

A fortiori $|\xi_n^{(k)} - \xi_n^{(l)}| \leqq \varepsilon$ for $k, l \geqq k_0$ and $n = 1, 2, \ldots$, i.e., every sequence of components $(\xi_n^{(1)}, \xi_n^{(2)}, \ldots)$ is a Cauchy sequence and has thus a limit ξ_n. From (6.6) we obtain for each m the inequality

$$\left(\sum_{n=1}^{m} |\xi_n^{(k)} - \xi_n^{(l)}|^p \right)^{1/p} \leqq \varepsilon \qquad \text{for} \quad k, l \geqq k_0.$$

If we let $k \to \infty$, then

$$\left(\sum_{n=1}^{m} |\xi_n - \xi_n^{(l)}|^p \right)^{1/p} \leqq \varepsilon \qquad \text{for} \quad l \geqq k_0$$

and $m = 1, 2, \ldots$, and therefore also

$$\text{(6.7)} \qquad \left(\sum_{n=1}^{\infty} |\xi_n - \xi_n^{(l)}|^p \right)^{1/p} \leqq \varepsilon \qquad \text{for} \quad l \geqq k_0.$$

From here it follows that for $l \geqq k_0$ the sequence $(\xi_1 - \xi_1^{(l)}, \xi_2 - \xi_2^{(l)}, \ldots)$ lies in l^p; since l^p is a vector space, also the sequence $x := (\xi_1, \xi_2, \ldots) = (\xi_1 - \xi_1^{(l)}, \xi_2 - \xi_2^{(l)}, \ldots) + (\xi_1^{(l)}, \xi_2^{(l)}, \ldots)$ belongs to l^p, and inequality (6.7) can now be written in the form $\|x - x_l\| \leqq \varepsilon$ for $l \geqq k_0$, in which it expresses precisely the fact that (x_l) converges to x. Thus l^p is indeed a Banach space. ■

In the vector space $C^{(n)}[a, b]$ of the n times continuously differentiable functions on $[a, b]$ one wishes to introduce a norm in such a fashion that convergence $x_k \to x$ in the sense of the norm is equivalent to the *uniform* convergence $x_k^{(\nu)}(t) \to x^{(\nu)}(t)$ for $t \in [a, b]$, $\nu = 0, 1, \ldots, n$. This can be done for $n < \infty$ as follows. We define the function $p_\nu: C^{(n)}[a, b] \to \mathbf{R}$ by

$$(6.8) \qquad p_\nu(x) := \max_{a \leqq t \leqq b} |x^{(\nu)}(t)| \qquad (\nu = 0, 1, \ldots, n).$$

p_ν is not a norm on $C^{(n)}[a, b]$ if $\nu \geqq 1$ since from $p_\nu(x) = 0$ it does not follow that $x = 0$ (all other properties of a norm are, however, satisfied, so that p_ν is called a *seminorm*), but if we define

$$\|x\| := \sum_{\nu=0}^{n} p_\nu(x)$$

we do get a norm on $C^{(n)}[a, b]$. It is now easy to see that this norm fulfills the wish expressed above and that $C^{(n)}[a, b]$ is complete. The space $C^{(\infty)}[a, b]$ will be considered in Exercise 1.

In the vector space $C_0(\mathbf{R})$ of all continuous functions which 'vanish at infinity' one can introduce a norm by $\|x\| = \max_{t \in \mathbf{R}} |x(t)|$. As in the case $C[a, b]$, one sees easily that $x_k \to x$ in $C_0(\mathbf{R})$ is equivalent to the *uniform* convergence $x_k(t) \to x(t)$ for all $t \in \mathbf{R}$, and that $C_0(\mathbf{R})$ is complete.

For an element x of the vector space $BV[a, b]$ of all functions of bounded variation on $[a, b]$ the *total variation*

$$V(x) := \sup_Z \sum_{\nu=1}^{n} |x(t_\nu) - x(t_{\nu-1})|$$

is finite; here Z runs through all partitions $a = t_0 < t_1 < \cdots < t_n = b$ of the interval $[a, b]$. If we take into account that such a function is constant if and only if $V(x)$ vanishes, we can easily see that

$$\|x\| := |x(a)| + V(x)$$

defines a norm on $BV[a, b]$.

From the estimate

$$|x(t)| - |x(a)| \leqq |x(t) - x(a)| \leqq |x(t) - x(a)| + |x(b) - x(t)| \leqq V(x)$$

valid for all $x \in BV[a, b]$ and all $t \in [a, b]$ it follows that

$$(6.9) \qquad \sup_{a \leqq t \leqq b} |x(t)| \leqq |x(a)| + V(x);$$

in particular $BV[a, b]$ is a linear subspace of $B[a, b]$. The $BV[a, b]$-norm does not coincide, however, with the $B[a, b]$-norm; because of (6.9) we have only, in self-explanatory notation, the relation

$$(6.10) \qquad \|x\|_B \leqq \|x\|_{BV}.$$

$BV[a, b]$ is complete: A Cauchy sequence (x_k) in $BV[a, b]$ is because of (6.10) also a Cauchy sequence in $B[a, b]$, consequently there exists an $x \in B[a, b]$ such that $x_k(t) \to x(t)$ for all $t \in [a, b]$. Furthermore to every $\varepsilon > 0$ there exists $k_0 = k_0(\varepsilon)$ so that for every partition Z of $[a, b]$ and for all $k, l \geqq k_0$ the estimate

$$|x_k(a) - x_l(a)| + \sum_{\nu=1}^{n} |x_k(t_\nu) - x_l(t_\nu) - [x_k(t_{\nu-1}) - x_l(t_{\nu-1})]| \leqq \|x_k - x_l\| \leqq \varepsilon$$

holds. If we let $k \to \infty$ and observe that the inequality one obtains is valid for every partition Z, then it follows that

$$|x(a) - x_l(a)| + V(x - x_l) \leqq \varepsilon \qquad \text{for} \quad l \geqq k_0. \tag{6.11}$$

In particular $x - x_{k_0}$ and so also $x = (x - x_{k_0}) + x_{k_0}$ are of bounded variation. (6.11) now states exactly that (x_l) converges in the metric of $BV[a, b]$ to x, which proves our assertion. ■

With the supremum-norm coming from $B[a, b]$ the space $BV[a, b]$ is not complete (see Exercise 3); for this reason we have not used it.

We list the Banach spaces defined so far in Table 1 and emphasize again that *these spaces will be equipped* always *with the norms indicated there—the* canonical norms—*unless the contrary is explicitly stated.* Further examples of normed spaces will be found in the Exercises of this section.

A linear subspace F of a normed space E becomes a normed space if we equip its elements with the norm that they already have as elements of E; this norm on F is called the norm *induced* by E (or by the norm of E) and F is then said to be a *subspace* of E. Thus, e.g., (c_0) is a subspace of (c) and (c) a subspace of l^∞; on the other hand, though $BV[a, b]$ is a *linear* subspace of $B[a, b]$, however, it is not a *subspace* because the norm of $BV[a, b]$ is not induced by the norm of $B[a, b]$. The subspaces of valued spaces are defined correspondingly.

We have seen at the beginning of this section that the translation-invariance of the metric given by the norm (which yields the triangle inequality (N3)) and its homogeneity property (N2) establish a close connection between the linear and the metric structure of a normed space. This interplay of the two structures has among others the following consequences:

Proposition 6.1. *In a normed space addition, multiplication by a scalar and the norm are* continuous, *i.e., from $x_n \to x$, $y_n \to y$ and $\alpha_n \to \alpha$ it follows that*

$$x_n + y_n \to x + y,$$

$$\alpha_n x_n \to \alpha x,$$

$$\|x_n\| \to \|x\|.$$

The proof of the first two assertions follows from the estimates

$$\|(x_n + y_n) - (x + y)\| = \|(x_n - x) + (y_n - y)\| \leqq \|x_n - x\| + \|y_n - y\|,$$

$$\|\alpha_n x_n - \alpha x\| = \|\alpha_n(x_n - x) + (\alpha_n - \alpha)x\| \leqq |\alpha_n| \cdot \|x_n - x\| + |\alpha_n - \alpha| \cdot \|x\|.$$

Table 1. Banach spaces

Notation	Definition	Canonical norm
$l^p(n)$, $1 \leqq p \leqq \infty$	Set $\mathbf{K}^n$ of all n-tuples $x = (\xi_1, \ldots, \xi_n)$	$\|x\| = \left(\sum_{\nu=1}^{n} \|\xi_\nu\|^p\right)^{1/p}$ if $1 \leqq p < \infty$ $\|x\| = \max_{\nu=1}^{n} \|\xi_\nu\|$ if $p = \infty$
l^p, $1 \leqq p < \infty$	Set of all sequences $x = (\xi_\nu)$ with $\sum_{\nu=1}^{\infty} \|\xi_\nu\|^p < \infty$	$\|x\| = \left(\sum_{\nu=1}^{\infty} \|\xi_\nu\|^p\right)^{1/p}$
l^∞	Set of all bounded sequences $x = (\xi_\nu)$	$\|x\| = \sup_{\nu=1}^{\infty} \|\xi_\nu\|$
(c)	Set of all convergent sequences $x = (\xi_\nu)$	$\|x\| = \sup_{\nu=1}^{\infty} \|\xi_\nu\|$
(c_0)	Set of all sequences $x = (\xi_\nu)$ converging to zero	$\|x\| = \sup_{\nu=1}^{\infty} \|\xi_\nu\|$
$B(T)$	Set of all bounded functions $x: t \mapsto x(t)$ on T	$\|x\| = \sup_{t \in T} \|x(t)\|$
$C[a, b]$	Set of all continuous functions $x: t \mapsto x(t)$ on $[a, b]$	$\|x\| = \max_{a \leqq t \leqq b} \|x(t)\|$
$C^{(n)}[a, b]$, $n = 1, 2, \ldots$	Set of all n times continuously differentiable functions $x: t \mapsto x(t)$ on $[a, b]$	$\|x\| = \sum_{\nu=0}^{n} \max_{a \leqq t \leqq b} \|x^{(\nu)}(t)\|$
$C_0(\mathbf{R})$	Set of all functions $x: t \mapsto x(t)$ continuous on $\mathbf{R}$ which vanish at infinity	$\|x\| = \max_{t \in \mathbf{R}} \|x(t)\|$
$BV[a, b]$	Set of all functions $x: t \mapsto x(t)$ of bounded variation on $[a, b]$	$\|x\| = \|x(a)\| + V(x)$, $V(x)$ the total variation of x

The continuity of the norm follows from the continuity of the metric (see end of §1). ■

In §1 we defined bounded sets in metric spaces. According to this definition a set in a normed space is bounded if and only if it lies in a ball around 0, i.e., if for a certain number r the estimate $\|x\| \leqq r$ holds for all its elements x.

If we set $y = v = 0$ in the quadrangle inequality (1.12) and we use the letter y instead of u, we get the important inequality

$$|\|x\| - \|y\|| \leqq \|x - y\|. \tag{6.12}$$

Exercises

1. In the vector space $C^{(\infty)}[a, b]$ of all infinitely differentiable functions on $[a, b]$, we can define, with the help of the seminorms p_ν introduced in (6.8), the absolute value

$$|x| := \sum_{\nu=0}^{\infty} \frac{1}{2^\nu} \frac{p_\nu(x)}{1 + p_\nu(x)}.$$

$x_k \to x$ is equivalent to $x_k^{(\nu)}(t) \to x^{(\nu)}(t)$ uniformly on $[a, b]$, $\nu = 0, 1, 2, \ldots$. $C^{(\infty)}[a, b]$ is complete. *Hint*: cf. Example 1.5.

2. If one introduces on l^p, $1 \leqq p < \infty$, the supremum-norm induced by l^∞, then l^p is not complete. *Hint*: Use a sequence of elements x_n of the form

$$(\xi_1^{(1)}, \ldots, \xi_n^{(n)}, 0, 0, \ldots).$$

3. $BV[a, b]$ is not complete with the norm induced by $B[a, b]$.

4. Use the Banach spaces l^1, l^2 to state theorems concerning the solution of infinite systems of linear equations of the form $\xi_i - \sum_{k=1}^{\infty} \alpha_{ik} \xi_k = \gamma_i, i = 1, 2, \ldots$ (cf. §3. Exercise 1).

5. The following spaces with the given norms are Banach spaces:

(a) The space (bv) of all sequences $x = (\xi_n)$ of bounded variation $V(x) := \sum_{n=1}^{\infty} |\xi_{n+1} - \xi_n|$ with $\|x\| := |\xi_1| + V(x)$.

(b) The space (bts) of all sequences $x = (\xi_n)$ with bounded partial sums, i.e., all sequences for which $\|x\| := \sup_{n=1}^{\infty} |\sum_{\nu=1}^{n} \xi_\nu| < \infty$.

(c) The space (kr) of all sequences $x = (\xi_n)$ with convergent series $\sum_{n=1}^{\infty} \xi_n$; $\|x\| := \sup_{n=1}^{\infty} |\sum_{\nu=1}^{n} \xi_\nu|$.

*6. If $E_1, \ldots, E_n$ are valued spaces over $\mathbf{K}$ with absolute values $|\cdot|_1, \ldots, |\cdot|_n$, then for the elements $x = (x_1, \ldots, x_n)$ of the vector space $E := E_1 \times \cdots \times E_n$ an absolute value is defined by $|x| := \sum_{k=1}^{n} |x_k|_k$. If the $|\cdot|_k$ are norms, then $|\cdot|$ is also a norm. Convergence in E is equivalent to componentwise convergence. E is complete if and only if each E_k is complete (cf. Exercise 6 in §1).

*7. If $E_1, E_2, \ldots$ are valued spaces over $\mathbf{K}$ with absolute values $|\cdot|_1, |\cdot|_2, \ldots$ then for the elements $x = (x_1, x_2, \ldots)$ of the vector space $E := \prod_{k=1}^{\infty} E_k$ an absolute value is introduced by

$$|x| := \sum_{k=1}^{\infty} \frac{1}{2^k} \frac{|x_k|_k}{1 + |x_k|_k}.$$

Discuss convergence and completeness (cf. Exercise 7 in §1).

8. For a discussion of the important spaces $L^p(a, b)$ $(1 \leqq p < \infty)$ we refer the reader to the textbooks on the theory of measure and integral; see e.g. [181]. We only mention the following: $L^p(a, b)$ consists of all functions

$t \mapsto x(t) \in \mathbf{K}$ which are defined almost everywhere and measurable in $[a, b]$ and for which $\int_a^b |x(t)|^p \, dt$ exists in the sense of Lebesgue; one agrees furthermore to identify two functions which coincide almost everywhere on $[a, b]$ (thus strictly speaking $L^p(a, b)$ does not consist of functions but of equivalence classes of functions). With an obvious definition of addition and multiplication by a scalar $L^p(a, b)$ becomes a vector space, and with

$$\|x\| := \left(\int_a^b |x(t)|^p \, dt\right)^{1/p}$$

a Banach space. If one introduces the L^p-norm on $C[a, b]$, one obtains an incomplete normed space (cf. Exercise 4 in §1).

$^+$9. The closure of a linear subspace F of the normed space E is again a linear subspace of E.

§7. Continuous linear maps

If we want to apply Banach's fixed-point theorem of §2 to a self-map A of the metric space E, we first have to check whether for all x, y in E an estimate of the form

(7.1) $$d(Ax, Ay) \leqq q d(x, y)$$

holds with a certain constant q, and then whether—or under what additional conditions on A—this constant is less than 1. This is how we proceeded in §3, see for instance the estimates (3.4) through (3.6) and (3.14). A self-map A of E or, more generally, a map A from a metric space E into another metric space F, which satisfies (7.1) with whatever constant q is said to be *of bounded dilation.* For maps with bounded dilation $x_n \to x$ clearly implies $Ax_n \to Ax$, so they are continuous if we take over word for word the concept of continuity of classical analysis:

A map A of the metric space E into a second metric space F is said to be *continuous at the point* $x \in E$ if from $x_n \to x$ it follows that $Ax_n \to Ax$. If A is continuous at every point of E, then we say that A is *continuous on* E, or shortly *continuous.* By the *continuous image* of a set we understand its image under a continuous map.

A linear map from the normed space E into the normed space F is, because of $d(Ax, Ay) = \|Ax - Ay\| = \|A(x - y)\|$, then and only then of bounded dilation if with a constant $q \geqq 0$ the estimate

(7.2) $$\|Ax\| \leqq q\|x\| \quad \text{for all} \quad x \text{ in } E$$

holds. Such maps are called shortly *bounded*, and we can now say that bounded linear maps are continuous. The converse of this is also true. Namely, if the linear map A is unbounded, then for every natural number n there exists an x_n such that $\|Ax_n\| > n\|x_n\|$. For $y_n := x_n/\|x_n\|$ we have

(7.3) $$\|y_n\| = 1 \quad \text{and} \quad \|Ay_n\| \to \infty$$

and consequently for $z_n := y_n/\|Ay_n\|$:

$$z_n \to 0, \qquad \|Az_n\| = 1.$$

If A were continuous, then $z_n \to 0$ would imply $Az_n \to A0 = 0$ and thus also $\|Az_n\| \to 0$. Because of $\|Az_n\| = 1$ this is impossible, hence A is not continuous. We state this result as:

Proposition 7.1. *A linear map from a normed space E into a normed space F is* continuous *if and only if it is* bounded.

We denote by $\mathscr{L}(E, F)$ the set of all continuous linear maps from E into F, and by $\mathscr{L}(E) := \mathscr{L}(E, E)$ the set of all continuous endomorphisms of E. The smallest number q which one can choose in (7.2) is called the *norm* of A and is denoted by $\|A\|$. Thus

$$\|Ax\| \leqq \|A\| \, \|x\| \qquad \text{for all} \quad x \in E \tag{7.4}$$

and

$$\|A\| = \sup_{x \neq 0} \frac{\|Ax\|}{\|x\|} = \sup_{\|x\| \leqq 1} \|Ax\| = \sup_{\|x\| = 1} \|Ax\|. \tag{7.5}$$

Obviously $\|0\| = 0$ and $\|I\| = 1$.

If we equip the maps with their norms, then $\mathscr{L}(E, F)$ is a normed space. We have namely

Proposition 7.2. *If A and B are in $\mathscr{L}(E, F)$ then so are $A + B$ and αA, and we have*

$$\|A + B\| \leqq \|A\| + \|B\|, \qquad \|\alpha A\| = |\alpha| \, \|A\|.$$

Furthermore $\|A\| \geqq 0$, and $\|A\| = 0$ is valid only for the zero map $A = 0$. Thus $\mathscr{L}(E, F)$ with the norm defined by (7.5) *is a* normed space.

We only prove the assertion concerning the sum. We have $\|(A + B)x\| = \|Ax + Bx\| \leqq \|Ax\| + \|Bx\| \leqq (\|A\| + \|B\|)\|x\|$, so $A + B$ is bounded and thus also continuous. Furthermore $\|A + B\| \leqq \|A\| + \|B\|$. ■

For products we have the following proposition whose simple proof can be left to the reader.

Proposition 7.3. *If $A \in \mathscr{L}(E, F)$, $B \in \mathscr{L}(F, G)$, then $BA \in \mathscr{L}(E, G)$ and*

$$\|BA\| \leqq \|B\| \, \|A\|; \tag{7.6}$$

in particular, $\mathscr{L}(E)$ is an algebra and for an A in $\mathscr{L}(E)$ we have

$$\|A^n\| \leqq \|A\|^n \qquad (n = 0, 1, 2, \ldots). \tag{7.7}$$

Since $\mathscr{L}(E, F)$ is on one hand a normed space and on the other a set of maps, we have in a natural way two concepts of convergence in $\mathscr{L}(E, F)$, which we are going to distinguish through notation and terminology:

1. $A_n \to A$ means *pointwise convergence*, i.e., $A_n x \to Ax$ for all $x \in E$.
2. $A_n \Rightarrow A$ means *convergence in the sense of the norm*, also called uniform convergence, that is $\|A_n - A\| \to 0$.

Because of $\|A_n x - Ax\| = \|(A_n - A)x\| \leqq \|A_n - A\| \|x\|$, *the pointwise convergence follows from uniform convergence*; the converse is, however, false (see Exercise 8).

By Proposition 6.1 in the normed space $\mathscr{L}(E, F)$ addition, multiplication by scalars and the norm are continuous, i., from $A_n \Rightarrow A$, $B_n \Rightarrow B$ and $\alpha_n \to \alpha$ it follows that

$$A_n + B_n \Rightarrow A + B,$$

$$\alpha_n A_n \Rightarrow \alpha A,$$

$$\|A_n\| \Rightarrow \|A\|;$$

in particular, $(\|A_n\|)$ is bounded. The multiplication of maps is also continuous, more precisely: If A_n and A are in $\mathscr{L}(E, F)$, B_n and B in $\mathscr{L}(F, G)$ and if $A_n \Rightarrow A$, $B_n \Rightarrow B$, then

(7.8) $$B_n A_n \Rightarrow BA.$$

Indeed, from (7.6) one obtains the estimate $\|B_n A_n - BA\| = \|B_n(A_n - A) + (B_n - B)A\| \leqq \|B_n\| \|A_n - A\| + \|B_n - B\| \|A\|$ from where the assertion follows immediately because of the boundedness of $(\|B_n\|)$.

Addition and multiplication by scalars satisfy similar limit relations with respect to *pointwise* convergence: From $A_n \to A$, $B_n \to B$ and $\alpha_n \to \alpha$ it follows that

$$A_n + B_n \to A + B,$$

$$\alpha_n A_n \to \alpha A.$$

In contrast, we cannot conclude any more that $\|A_n\| \to \|A\|$; see, however, Proposition 33.1 for a weaker assertion in this direction. Also (7.8) is not valid without any further ado for pointwise instead of uniform convergence; we refer to the discussion following Proposition 33.1.

The following proposition answers the question when the normed space $\mathscr{L}(E, F)$ is complete:

Proposition 7.4. *If E is a normed space and F a Banach space, then $\mathscr{L}(E, F)$ is also a Banach space.*

For the proof, let (A_n) be a Cauchy sequence in $\mathscr{L}(E, F)$, i.e., to an arbitrary $\varepsilon > 0$ let there exist a $n_0 = n_0(\varepsilon)$ so that $\|A_n - A_m\| \leqq \varepsilon$ for $n, m \geqq n_0$. For every $x \in E$ and all $n, m \geqq n_0$ one has then

(7.9) $$\|A_n x - A_m x\| = \|(A_n - A_m)x\| \leqq \|A_n - A_m\| \|x\| \leqq \varepsilon \|x\|,$$

i.e., $(A_n x)$ is a Cauchy sequence in the Banach space F and has therefore a limit $\lim A_n x$. The map $A: E \to F$ defined by $Ax := \lim A_n x$ is obviously linear. As $m \to \infty$ it follows from (7.9) with the help of Proposition 6.1 that

$$\|A_n x - Ax\| \leqq \varepsilon \|x\|$$

for $n \geqq n_0$ and all $x \in E$. From here it results that $A_{n_0} - A$ and thus also $A = (A - A_{n_0}) + A_{n_0}$ is continuous and $\|A_n - A\| \leqq \varepsilon$ for $n \geqq n_0$. Hence the sequence (A_n) converges uniformly to $A \in \mathscr{L}(E, F)$. ■

We emphasized in §5 that numerous questions in mathematics lead to the problem to solve an equation of the form

$$Ax = y, \tag{7.10}$$

with a linear map $A: E \to F$, or at least to discuss its solvability. If E and F are normed spaces and A is injective, one will try, moreover, to answer the question *whether the solution depends* continuously *on the right-hand side*, i.e., whether the solutions x_n of the equations $Ax = y_n$ converge to a solution of the equation $Ax = y$ when $y_n \to y$. Assuming that our equation can be solved at all for the right-hand-side y, this will certainly be the case when the inverse A^{-1} of A is continuous on its space of definition $A(E)$, since then $x_n = A^{-1}y_n$ converges to $A^{-1}y =: x$ and $Ax = y$. If now A has a continuous inverse, and if the right-hand side y of the equation (7.10) is in some sense 'difficult', then we can try to find an approximate solution of the equation, replacing y by 'simple' right-hand sides y_n which converge to y; the solutions x_n of $Ax = y_n$ will then for sufficiently large n be arbitrarily good approximate solutions of (7.10). For this reason we shall repeatedly consider the question whether the inverse of a given linear map is continuous. A precise criterion for this is given by:

Proposition 7.5. *The linear map $A: E \to F$ of the normed spaces E and F has a* continuous *inverse A^{-1} defined on $A(E)$ if and only if for an appropriate constant $m > 0$ the estimate $m\|x\| \leqq \|Ax\|$ holds for all $x \in E$.*

If we first assume that the estimate is valid, then $Ax = 0$ implies $x = 0$, so A possesses an inverse A^{-1} defined on $A(E)$ by Proposition 5.1. Then for an arbitrary element $y = Ax$ in $A(E)$ one has $m\|A^{-1}y\| = m\|x\| \leqq \|Ax\| = \|y\|$, hence $\|A^{-1}y\| \leqq (1/m)\|y\|$, i.e., A^{-1} is bounded and so continuous. The proof of the converse is left to the reader. ■

Let two norms $\|\cdot\|_1$, $\|\cdot\|_2$ be defined on a vector space E, which make E into the normed spaces E_1 and E_2, respectively. In accordance with the concepts introduced in §1, Exercise 5, we say that $\|\cdot\|_1$ is *stronger* than $\|\cdot\|_2$ if from $\|x_n - x\|_1 \to 0$ it follows that $\|x_n - x\|_2 \to 0$. This is obviously equivalent with saying that the identity map I_E of E as a map from E_1 into E_2 is continuous, i.e., that there exists a $\mu > 0$ such that $\|x\|_2 = \|I_E x\|_2 \leqq \mu\|x\|_1$ holds for all $x \in E$. If we call two norms *equivalent* if each of them is stronger than the other, we obtain immediately the following:

Proposition 7.6. *A norm $\|\cdot\|_1$ on a vector space E is* stronger *than a second norm $\|\cdot\|_2$ on E if and only if with a constant $\mu > 0$ the estimate*

$$\|x\|_2 \leqq \mu\|x\|_1 \qquad \textit{for all} \quad x \in E$$

holds. Thus the two norms are equivalent *if and only if there exist two positive numbers* γ_1 *and* γ_2 *so that*

$$\gamma_1 \leqq \frac{\|x\|_1}{\|x\|_2} \leqq \gamma_2 \qquad \textit{holds for all} \quad x \neq 0 \qquad \textit{in} \quad E.$$

Since convergence in $l^p(n)$ is equivalent to componentwise convergence, the $l^p(n)$-norms on $\mathbf{K}^n$ are equivalent for all $p \geqq 1$.

Exercises

*1. Let $A \in \mathscr{L}(E, F)$, and A_0 be the restriction of A to the subspace E_0 of E. Show:

(a) One has $\|A_0\| \leqq \|A\|$.

(b) If $\|Ax\| \leqq q\|x\|$ on E and $\|Ax_0\| = q\|x_0\|$ for a certain $x_0 \neq 0$, then $\|A\| = q$.

2. Let the endomorphism A of $\mathbf{K}^n$ be defined with the aid of the matrix (α_{ik}) by

$$A(\xi_1, \ldots, \xi_n) := \left(\sum_{k=1}^{n} \alpha_{1k}\xi_k, \ldots, \sum_{k=1}^{n} \alpha_{nk}\xi_k\right).$$

Show the following:

(a) If $\mathbf{K}^n$ is equipped with the norm $\|(\xi_1, \ldots, \xi_n)\| = \max_{k=1}^n |\xi_k|$, then $\|A\| = \max_{i=1}^n \sum_{k=1}^n |\alpha_{ik}|$.

(b) If $\mathbf{K}^n$ is equipped with the norm $\|(\xi_1, \ldots, \xi_n)\| = \sum_{k=1}^n |\xi_k|$, then $\|A\| = \max_{k=1}^n \sum_{i=1}^n |\alpha_{ik}|$.

(c) If $\mathbf{K}^n$ is equipped with the norm $\|(\xi_1, \ldots, \xi_n)\| = (\sum_{k=1}^n |\xi_k|^2)^{1/2}$, then $\|A\| \leqq (\sum_{i,k=1}^n |\alpha_{ik}|^2)^{1/2}$.

3. Every linear map from $l^p(n)$ into a normed space is continuous.

$^+$4. The norm of a continuous projector $\neq 0$ in a normed space is $\geqq 1$.

5. A linear map A from a normed space E into a normed space F is bounded if and only if it transforms every bounded set into a bounded set.

6. The map A from Exercise 5 is bounded if and only if there exists in E a ball around 0 whose image is bounded.

7. Let A_n and A be elements of $\mathscr{L}(E, F)$. The sequence (A_n) converges to A uniformly if and only if $(A_n x)$ converges to Ax uniformly on every bounded subset M of E, i.e., when to every such M and every $\varepsilon > 0$ there exists $n_0 = n_0(\varepsilon, M)$ so that $\|A_n x - Ax\| \leqq \varepsilon$ for all $n \geqq n_0$ and all x in M.

8. Define on l^1 for each natural number n a continuous endomorphism A_n by $A_n(\xi_1, \xi_2, \ldots) := (\xi_1, \xi_2, \ldots, \xi_n, 0, 0, \ldots)$. Show that A_n converges pointwise but not uniformly to I.

9. Let A be a map from the metric space E into the metric space F. Show that the following assertions are equivalent (for the concepts used see Exercise 13 in §1):

(a) A is continuous at the point $x_0 \in E$.

(b) To every $\varepsilon > 0$ there exists a $\delta > 0$ so that $d(Ax, Ax_0) \leqq \varepsilon$ whenever $d(x, x_0) \leqq \delta$.

(c) To every ball V with center Ax_0 there exists a neighborhood U of x_0 so that $A(U) \subset V$.

(d) To every neighborhood W of Ax_0 there exists a neighborhood U of x_0 so that $A(U) \subset W$.

(e) The preimage $A^{-1}(V)$ of every ball V with center Ax_0 is a neighborhood of x_0.

(f) The preimage $A^{-1}(W)$ of every neighborhood W of Ax_0 is a neighborhood of x_0.

*10. Under the hypotheses of Exercise 9 the map $A: E \to F$ is continuous (on E) if and only if the preimage $A^{-1}(M)$ of every open set $M \subset F$ is open in E, or equivalently if the preimage $A^{-1}(N)$ of every closed set $N \subset F$ is closed in E.

11. If E is equipped with the discrete metric and F is an arbitrary metric space, then every map $A: E \to F$ is continuous.

§8. The Neumann series

In §3, at the investigation of systems of linear equations and of Fredholm integral equations of the second kind, we met equations of the form $x - Kx = y$ with a linear map K. In certain cases we could solve the equations with the help of Neumann series. The concepts and theorems we have listed so far make it possible to obtain again the results from there in the abstract framework of normed spaces, and thereby to unify them. For this purpose we define an *infinite series* $\sum_{\nu=0}^{\infty} x_\nu$ with elements x_ν from a normed space E, exactly like in classical analysis, as the sequence of its partial sums $s_n := x_0 + x_1 + \cdots + x_n$. The series is said to be *convergent* with sum s, in symbols $\sum_{\nu=0}^{\infty} x_\nu = s$, if $s_n \to s$. It is called a *Cauchy series*, if (s_n) is a Cauchy sequence, i.e., if to every $\varepsilon > 0$ there exists an $n_0 = n_0(\varepsilon)$ so that for $n > m \geqq n_0$ one has

$$\|s_n - s_m\| = \|x_{m+1} + \cdots + x_n\| \leqq \varepsilon.$$

A convergent series is a Cauchy series, hence the sequence of its terms tends to 0. *In a Banach space every Cauchy series converges.*

Lemma 8.1. *If* $\sum_{\nu=0}^{\infty} \|x_\nu\|$ *converges, then the series* $\sum_{\nu=0}^{\infty} x_\nu$ *is a Cauchy series. Thus it* converges *if the space is* complete, *and in this case the generalized triangle inequality*

$$\left\| \sum_{\nu=0}^{\infty} x_\nu \right\| \leqq \sum_{\nu=0}^{\infty} \|x_\nu\| \tag{8.1}$$

holds.

Because of $\|\sum_{\nu=m}^{n} x_\nu\| \leqq \sum_{\nu=m}^{n} \|x_\nu\|$ it follows that $\sum x_\nu$ is a Cauchy series. The generalized triangle inequality follows, because of the continuity of the norm, from the elementary triangle inequality $\|\sum_{\nu=0}^{n} x_\nu\| \leqq \sum_{\nu=0}^{n} \|x_\nu\|$ as $n \to \infty$, provided that $\sum x_\nu$ converges. ■

Let now K be a continuous endomorphism of the Banach space E. We want to study the equation

$$x - Kx = y \quad \text{or} \quad (I - K)x = y. \tag{8.2}$$

If we bring it to the form $x = y + Kx$ and define the (nonlinear) self-map A of E by $Ax := y + Kx$, then (8.2) becomes the fixed-point equation $Ax = x$. Because of

$$\begin{aligned} d(Ax_1, Ax_2) &= \|Ax_1 - Ax_2\| = \|Kx_1 - Kx_2\| = \|K(x_1 - x_2)\| \\ &\leqq \|K\| \, \|x_1 - x_2\| = \|K\| d(x_1, x_2), \end{aligned}$$

according to Theorem 2.1 this equation possesses in the case $q := \|K\| < 1$ exactly one solution x in E, i.e., equation (8.2) has for every y in E exactly one solution x in E, or in other words: The inverse $(I - K)^{-1}$ exists on the whole space E. If x_0 is arbitrary in E and $x_n := Ax_{n-1}$ $(n = 1, 2, \ldots)$, then x_n converges to x. If we choose $x_0 = y$, then we obtain (cf. (3.9))

$$x_n = y + Ky + K^2 y + \cdots + K^n y \to x,$$

i.e.,

$$x = (I - K)^{-1} y = \sum_{n=0}^{\infty} K^n y \tag{8.3}$$

or

$$(I - K)^{-1} = \sum_{n=0}^{\infty} K^n, \tag{8.4}$$

if convergence of this series is understood in the sense of pointwise convergence of its partial sums. The series (8.3) and (8.4) are called *Neumann series.*

However, the Neumann series in (8.4) converges not only pointwise but even uniformly. Indeed, from the error estimate (2.6) with $x_0 = y$, $x_1 = y + Ky$ it follows that

$$\left\| \left((I - K)^{-1} - \sum_{\nu=0}^{n} K^\nu \right) y \right\| = \|x - x_n\| \leqq \frac{\|K\|^n}{1 - \|K\|} \|Ky\| \leqq \frac{\|K\|^{n+1}}{1 - \|K\|} \|y\|,$$

and this inequality shows that the linear map $(I - K)^{-1} - \sum_{\nu=0}^{n} K^\nu$ is bounded—hence $(I - K)^{-1}$ is also bounded—and that we have the estimate

$$\left\| (I - K)^{-1} - \sum_{\nu=0}^{n} K^\nu \right\| \leqq \frac{\|K\|^{n+1}}{1 - \|K\|}. \tag{8.5}$$

Because of $\|K\| < 1$ the partial sums $\sum_{\nu=0}^{n} K^\nu$ converge indeed uniformly to $(I - K)^{-1}$, i.e., the expansion (8.4) holds, as asserted, also in the sense of uniform convergence. With the aid of Lemma 8.1 it further follows that

$$\|(I - K)^{-1}\| = \left\| \sum_{n=0}^{\infty} K^n \right\| \leqq \sum_{n=0}^{\infty} \|K^n\| \leqq \sum_{n=0}^{\infty} \|K\|^n = \frac{1}{1 - \|K\|}.$$

We state these important results:

Theorem 8.1. *Let K be a continuous endomorphism of the Banach space E such that $\|K\| < 1$. Then the inverse $(I - K)^{-1}$ exists on E, is continuous with*

$$\|(I - K)^{-1}\| \leqq \frac{1}{1 - \|K\|}, \tag{8.6}$$

and can be expanded into the uniformly *convergent* Neumann series

$$(I - K)^{-1} = \sum_{n=0}^{\infty} K^n; \tag{8.7}$$

an error estimate is given by (8.5).

The expansion (8.7) is obviously analogous to the geometric series

$$(1 - q)^{-1} = \sum_{n=0}^{\infty} q^n \qquad \text{for} \quad |q| < 1. \tag{8.8}$$

We want to give now a new proof for (8.7) which goes similarly as the proof of (8.8). Under the hypothesis of Theorem 8.1 it follows from $\|\sum_{\nu=m}^{n} K^\nu\| \leqq \sum_{\nu=m}^{n} \|K^\nu\| \leqq \sum_{\nu=m}^{n} \|K\|^\nu$ that $\sum_{\nu=0}^{\infty} K^\nu$ is a Cauchy series in $\mathscr{L}(E)$. Since $\mathscr{L}(E)$ is complete by Proposition 7.4, there exists an S in $\mathscr{L}(E)$ with $S = \sum_{\nu=0}^{\infty} K^\nu$. Because of the continuity of multiplication (see (7.8)), it follows that $SK = KS = \sum_{n=0}^{\infty} K^{n+1} = S - I$, hence

$$(I - K)S = S - KS = I, \qquad S(I - K) = S - SK = I.$$

Proposition 5.2 now implies that $I - K$ is bijective and has S as its inverse. ■

This new proof shows in particular that $(I - K)^{-1}$ always exists on E and is continuous, if the Neumann series converges at all. Now one can prove exactly as in the scalar case the *root test*, i.e., the assertion that *the series* $\sum x_n$ *with terms* x_n *in a Banach space converges or diverges according as* $\alpha :=$ lim sup$\sqrt[n]{\|x_n\|} < 1$ *or* > 1; in the case $\alpha = 1$ the test does not give any information concerning convergence or divergence. The convergence behavior of the Neumann series is thus determined by the quantity lim sup$\|K^n\|^{1/n}$. We first prove that here lim sup can be replaced by lim; Proposition 8.2 will then be obvious.

Proposition 8.1. *For every continuous endomorphism A of a normed space* $\lim_{n\to\infty} \|A^n\|^{1/n}$ *exists and is* $\leqq \|A^k\|^{1/k}$ *for* $k = 1, 2, \ldots$.

Proof. For $\alpha_n := \|A^n\|$ we have $0 \leqq \alpha_{n+m} \leqq \alpha_n \alpha_m$; hence the sequence $(\alpha_n^{1/n})$ converges to its greatest lower bound (see [183], Section I, Problem 98). ■

Proposition 8.2. *If K is a continuous endomorphism of the Banach space E, then $I - K$ has a continuous inverse on E whenever the Neumann series* $\sum_{n=0}^{\infty} K^n$

converges (uniformly); in this case the expansion (8.7) *is valid. The Neumann series converges or diverges according as* $\lim \|K^n\|^{1/n} < 1$ *or* >1.

We show by an example that the convergence condition of the last proposition is much weaker than the earlier condition $\|K\| < 1$. We define the *Volterra integral transformation* K on $C[a, b]$ by

$$(Kx)(s) := \int_a^s k(s, t)x(t)\mathrm{d}t, \tag{8.9}$$

where the kernel k is continuous in the triangle $a \leqq t \leqq s \leqq b$. This transformation is a linear and continuous self-map of $C[a, b]$ and the *Volterra integral equation*

$$x(s) - \int_a^s k(s, t)x(t)\mathrm{d}t = y(s) \tag{8.10}$$

can be written with its help in the form

$$(I - K)x = y;$$

here y is supposed to be in $C[a, b]$ and we ask only for solutions x lying in $C[a, b]$ equipped with the canonical maximum-norm.

We examine now the iterates K^n. Setting $\mu := \max_{a \leqq t \leqq s \leqq b} |k(s, t)|$ we get

$$|(Kx)(s)| = \left| \int_a^s k(s, t)x(t)\mathrm{d}t \right| \leqq \mu\|x\|(s - a),$$

$$|(K^2x)(s)| = \left| \int_a^s k(s, t)(Kx)(t)\mathrm{d}t \right| \leqq \int_a^s \mu \cdot \mu\|x\|(t - a)\mathrm{d}t = \mu^2\|x\| \frac{(s - a)^2}{2},$$

$$|(K^3x)(s)| = \left| \int_a^s k(s, t)(K^2x)(t)\mathrm{d}t \right| \leqq \int_a^s \mu \cdot \mu^2\|x\| \frac{(t - a)^2}{2} \mathrm{d}t = \mu^3\|x\| \frac{(s - a)^3}{3!},$$

in general

$$|(K^nx)(s)| \leqq \mu^n\|x\| \frac{(s - a)^n}{n!} \qquad (n = 1, 2, \ldots).$$

Thus

$$\|K^nx\| = \max_{a \leqq s \leqq b} |(K^nx)(s)| \leqq \mu^n \frac{(b - a)^n}{n!} \|x\|,$$

hence

$$\|K^n\| \leqq \mu^n \frac{(b - a)^n}{n!}.$$

Because of $\sqrt[n]{n!} \to \infty$ we obtain

$$\lim \|K^n\|^{1/n} = 0, \tag{8.11}$$

hence $(I - K)^{-1}$ exists on $C[a, b]$, i.e., *the Volterra integral equation* (8.10) *has for any continuous right-hand side y exactly one continuous solution, however large* $\|K\|$ *is.*

Exercises

1. Let the linear and continuous self-map of the Banach space $l^\infty(n)$ be given by $K(\xi_1, \dots, \xi_n) := (2\xi_1, \dots, 2\xi_n)$. The corresponding Neumann series diverges, nevertheless the inverse $(I - K)^{-1}$ exists on $l^\infty(n)$ and is continuous there.

2. The finite-dimensional analogue to a Volterra integral transformation of $C[a, b]$ is a self-map K of $l^\infty(n)$ by means of a triangular matrix (α_{ik}) which has only zeros above the main diagonal, so that the components η_i of the image vector $(\eta_1, \dots, \eta_n) := K(\xi_1, \dots, \xi_n)$ are given by

$$\eta_i := \sum_{k=1}^{i} \alpha_{ik} \xi_k \qquad (i = 1, \dots, n).$$

The Neumann series does not converge always (see Exercise 1), and the inverse $(I - K)^{-1}$ does not always exist either. It exists on $l^\infty(n)$ if and only if all entries on the main diagonal $\alpha_{ii} \neq 1$ (observe that the determinant of a triangular matrix is the product of its entries on the main diagonal).

3. For the integral transformation $K\colon C[0, 1] \to C[0, 1]$ defined by

$$(Kx)(s) := \int_0^1 s \cdot x(t) \mathrm{d}t, \qquad 0 \leqq s \leqq 1$$

one has

$$\|K\| = 1 \quad \text{but} \quad \lim \|K^n\|^{1/n} = \tfrac{1}{2}$$

($C[0, 1]$ is equipped with the maximum-norm). Consequently, the Neumann series $\sum_{n=0}^{\infty} K^n y$ converges for every $y \in C[0, 1]$ and yields a continuous solution of the integral equation $x(s) - \int_0^1 sx(t)\mathrm{d}t = y(s)$. This solution can be given explicitly.

4. If we introduce on $C[0, 1]$ the analogue of an l^1-norm by $\|x\|_1 := \int_0^1 |x(t)| \mathrm{d}t$, then the integral transformation K of Exercise 3 has norm $\|K\|_1 = \frac{1}{2}$. However, Theorem 8.1 cannot be applied, since $C[0, 1]$ is not complete with this norm (see §1, Exercise 4).

$^+$5. For $A \in \mathscr{L}(E)$ (E a normed space) the following assertions are equivalent:

(a) $A^n \Rightarrow 0$.
(b) There exists an $m \in \mathbf{N}$ such that $\|A^m\| < 1$.
(c) $\sum_{n=0}^{\infty} \|A^n\|$ converges.

$^+$6. The series $\sum_{\nu=1}^{\infty} x_\nu$ in a Banach space E is *absolutely convergent* if $\sum_{\nu=1}^{\infty} \|x_\nu\|$ converges, it is *unconditionally convergent* if all its rearrangements converge to the same vector x. Absolute convergence implies convergence. In finite-dimensional spaces a series is absolutely convergent if and only if it is unconditionally convergent. In [52] it is shown that this theorem is false in the case $\dim E = \infty$, in fact it is characteristic for finite-dimensional spaces.

§9. Normed algebras

In the second proof for the expansion (8.7) in §8 it was unimportant that K is a linear map. It was based only on formal properties of $\mathscr{L}(E)$ which can also be found in other sets and which we make explicit in the following important definition:

An algebra R is called a *normed algebra* if it is a normed vector space and the inequality

$$\|ab\| \leqq \|a\| \, \|b\| \tag{9.1}$$

holds for products.

In a normed algebra all algebraic operations and the norm are continuous, i.e., from $a_n \to a$, $b_n \to b$, $\alpha_n \to \alpha$ it follows that

$$\begin{aligned} a_n + b_n &\to a + b, \\ \alpha_n a_n &\to \alpha a, \\ a_n b_n &\to ab, \\ \|a_n\| &\to \|a\|; \end{aligned} \tag{9.2}$$

the continuity of the linear operations (addition, multiplication by a scalar) and of the norm are guaranteed by Proposition 6.1, the continuity of multiplication follows from (9.1) just like (7.8) followed from (7.6).

For powers a^n we have, according to (9.1),

$$\|a^n\| \leqq \|a\|^n \qquad \text{for} \quad n = 1, 2, \ldots .$$

If a—not necessarily normed—algebra R has a unit element e and $ab = e$, then a is called a *left-inverse* of b and b a *right-inverse* of a; we also say that b is *left-invertible* and a is *right-invertible*. If a is left- as well as right-invertible, i.e., if there exist elements b and c with $ba = ac = e$, then a is said to be *invertible* or *regular*; in this case $b = c$ and this uniquely determined element b is called the *inverse* of a and denoted by a^{-1}. The unit element is invertible, and *the set of invertible elements forms obviously a group with respect to multiplication*. If a and b are invertible, then $(ab)^{-1} = b^{-1}a^{-1}$.

In an algebra R with unit element e we set $a^0 := e$. If R is normed and $\neq \{0\}$, then $\|e\| = \|e^2\| \leqq \|e\|^2$ immediately implies $\|e\| \geqq 1$.

In the literature a normed algebra, which as a normed space is complete, is usually called a *Banach algebra*. For the sake of simplicity we shall use this word only for complete normed algebras which have a unit element e with $\|e\| = 1$. Thus a Banach algebra consists of at least two elements.

$\mathscr{S}(E)$ is an algebra, $\mathscr{L}(E)$ a normed algebra and by Proposition 7.4 even a Banach algebra, *provided that E is a Banach space* $\neq \{0\}$.

Observe that in the algebra $\mathscr{S}(E)$ the word 'inverse' has two meanings, an operator-theoretical one and one from the theory of algebras:

If $A \in \mathscr{S}(E)$ has an inverse in the sense of the theory of algebras, i.e., if there exists a $B \in \mathscr{S}(E)$ so that $BA = AB = I$, then by Proposition 5.2 the map A is in particular injective, thus has an inverse as a map, namely the inverse map A^{-1}, and we have $B = A^{-1}$: the inverse as a map coincides with the inverse in the sense of algebra-theory. If we only know that A has an inverse as a map—the inverse map—then A does not have necessarily an inverse in the algebra $\mathscr{S}(E)$; such an inverse exists according to Proposition 5.2 only if A is also surjective. Expressed differently: *The inverse map A^{-1} is the inverse of A in the sense of the algebra $\mathscr{S}(E)$ if and only if A^{-1} is defined on the whole space E.* If we speak of the inverse A^{-1} of a map $A \in \mathscr{S}(E)$, we mean, according to the convention made in §5, the inverse map, even if it is not defined on all of E.

An analysis of the second proof in §8 for the Neumann expansion (8.7) shows immediately that it only uses properties of $\mathscr{L}(E)$ which are present in every Banach algebra; the same holds for Propositions 8.1 and 8.2. Therefore we can state without a new proof the following two propositions:

Proposition 9.1. *For every element x of a normed algebra the limit*

$$\lim_{n\to\infty} \|x^n\|^{1/n} \quad \textit{exists and is} \quad \leqq \|x^k\|^{1/k} \quad \textit{for} \quad k = 1, 2, \ldots. \tag{9.3}$$

Proposition 9.2. *The element $e - x$ in a Banach algebra E has an inverse in E whenever the Neumann series $\sum_{n=0}^{\infty} x^n$ converges; in this case*

$$(e - x)^{-1} = \sum_{n=0}^{\infty} x^n. \tag{9.4}$$

The Neumann series converges or diverges according as $\lim \sqrt[n]{\|x^n\|} < 1$ or > 1; *in particular, it converges for all x with* $\|x\| < 1$.

We want to discuss now the question whether the invertibility of an element x_0 of a Banach algebra is perturbed if x_0 undergoes a small change. If x lies sufficiently close to x_0, more precisely: if

$$\|x - x_0\| < \frac{1}{\|x_0^{-1}\|}, \tag{9.5}$$

then it follows from

$$x = x_0 - (x_0 - x) = x_0[e - x_0^{-1}(x_0 - x)]$$

that x is invertible; indeed, in the product on the right-hand side the first factor is invertible according to hypothesis, and the second factor has an inverse according to Proposition 9.2 because $\|x_0^{-1}(x_0 - x)\| \leqq \|x_0^{-1}\| \, \|x_0 - x\| < 1$. The inverse x^{-1} is given by

$$x^{-1} = [e - x_0^{-1}(x_0 - x)]^{-1} x_0^{-1} = x_0^{-1} + \sum_{n=1}^{\infty} [x_0^{-1}(x_0 - x)]^n x_0^{-1},$$

from where we obtain the estimate

$$\|x^{-1} - x_0^{-1}\| \leqq \frac{\|x_0 - x\|}{1 - \|x_0^{-1}\| \, \|x_0 - x\|} \|x_0^{-1}\|^2. \tag{9.6}$$

It follows that the inverse x^{-1} depends continuously on x: *If $x_n \to x_0$ then x_n^{-1} converges to x_0^{-1}.*

We recall the concept of an open set (see §1, Exercise 10): A subset M of a metric space is said to be *open* if around each point of M there exists a ball which lies entirely in M. With the help of this concept we can summarize our results as follows:

Proposition 9.3. *The group of invertible elements of a Banach algebra is open, and the inverse x^{-1} is a continuous function of x. Quantitatively: if x satisfies condition* (9.5), *then x possesses an inverse and this inverse satisfies the inequality* (9.6).

This proposition can be of importance for the solution of equations of the form $Ax = y$, where A is a continuous linear self-map of the Banach space E. In §7 we replaced a 'difficult' right-hand side y by 'simple' right-hand sides y_n which converged to y. One can, however, think of replacing a 'difficult' transformation A by a 'simple' transformation lying close to A and to obtain approximate solutions for the original problem by solving the modified problem. If we combine the two procedures, we obtain with the help of the preceding proposition easily the following result: *If A has a continuous inverse on E, if the continuous linear maps $A_n \Rightarrow A$ and the right-hand sides $y_n \to y$, then the equations $A_n x_n = y_n$ have unique solutions x_n at least for n greater than some n_0, and the sequence of solutions x_n converges to the solution of the equation $Ax = y$.*

As an example we want to consider the Fredholm integral equation

$$x(s) - \int_a^b k(s, t)x(t)dt = y(s) \qquad \text{or} \quad (I - K)x = y \tag{9.7}$$

in $C[a, b]$, where the kernel k should again be continuous and the integral transformation k is defined as usually. According to the Weierstrass approximation theorem, there exists a sequence of polynomials $p_n(s, t) := \sum_{i,k} \alpha_{ik}^{(n)} s^i t^k$ in two variables which converges uniformly on $[a, b] \times [a, b]$ to $k(s, t)$, i.e.,

$$\max_{a \leqq s, t \leqq b} |p_n(s, t) - k(s, t)| \to 0 \qquad \text{as} \quad n \to \infty. \tag{9.8}$$

If we define the continuous endomorphisms P_n of $C[a, b]$ by

$$(P_n x)(s) := \int_a^b p_n(s, t)x(t)\mathrm{d}t,$$

then

$$\|(P_n - K)x\| = \max_s \left| \int_a^b [p_n(s, t) - k(s, t)]x(t)\mathrm{d}t \right|$$
$$\leqq (b - a)\max_{s,t} |p_n(s, t) - k(s, t)| \, \|x\|,$$

hence $\|P_n - K\| \leqq (b - a)\max_{s,t}|p_n(s, t) - k(s, t)|$. Because of (9.8) we have

(9.9) $$P_n \Rightarrow K \quad \text{and thus also} \quad I - P_n \Rightarrow I - K.$$

If we already know that $(I - K)^{-1}$ *exists* and is *continuous* on $C[a, b]$, then it follows from the remark after Proposition 9.3 that for sufficiently large n the Fredholm integral equation with *polynomial kernel*

(9.10) $$(I - P_n)x = y$$

is solvable by an x_n belonging to $C[a, b]$, and that the sequence (x_n) converges in $C[a, b]$—i.e., uniformly on $[a, b]$—to the solution of (9.7).

A polynomial kernel $p(s, t)$ can obviously be written in the form

$$p(s, t) = \sum_{i=1}^{m} u_i(s)v_i(t), \qquad u_i \quad \text{and} \quad v_i \quad \text{in} \quad C[a, b];$$

the corresponding integral transformation P is then given by

$$(Px)(s) = \int_a^b \sum_{i=1}^{m} u_i(s)v_i(t)x(t)\mathrm{d}t = \sum_{i=1}^{m} \left(\int_a^b v_i(t)x(t)\mathrm{d}t \right) u_i(s)$$

or

(9.11) $$Px = \sum_{i=1}^{m} \int_a^b v_i(t)x(t)\mathrm{d}t \cdot u_i.$$

The image space of P is contained in the finite-dimensional linear hull $[u_1, \ldots, u_m]$ of the functions $u_1, \ldots, u_m$ and is therefore itself finite-dimensional. Linear operators with finite-dimensional image spaces are called *finite-dimensional operators* or *operators of finite rank*. For such operators, the equation $(I - P)x = y$ can be solved, as we shall see in §17, by purely algebraic methods. If we anticipate this result, we can say that the solution x_n of (9.10) can be determined in an elementary fashion and $\lim x_n$ is the solution of (9.7).

We return again to Proposition 9.3. A subset M of the metric space E is said to be *closed* if its complement in E is open. *M is closed if and only if the limit of every convergent sequence in M lies again in M* (§1, Exercise 10). From Proposition 9.3 we now obtain:

Proposition 9.4. *The set of non-invertible elements in a Banach algebra is closed; thus the limit of a sequence of non-invertible elements is itself non-invertible.*

A sequence of invertible elements can also have a non-invertible limit, as the sequence $((1/n)e)$ shows.

Exercises

1. The Banach space $C[a, b]$ becomes a commutative Banach algebra if we define the product xy pointwise: $(xy)(t) := x(t)y(t)$.

2. The Banach space l^1 becomes a commutative Banach algebra if the product xy is defined by the *convolution* of the sequences $x = (\xi_0, \xi_1, \ldots)$, $y = (\eta_0, \eta_1, \ldots)$ thus:

$$xy := (\xi_0\eta_0, \xi_0\eta_1 + \xi_1\eta_0, \ldots, \xi_0\eta_n + \xi_1\eta_{n-1} + \cdots + \xi_n\eta_0, \ldots);$$

it is useful to start indexing of the sequences by 0 to make the connection with the Cauchy multiplication of power series easily. The properties of l^1 which have to be proved can be obtained readily from the theory of power series.

3. With the aid of Exercise 2 and Proposition 9.2 prove the following well-known

Theorem: If the power series $\sum_{n=0}^{\infty} \alpha_n \zeta^n$ has a convergence radius different from zero and $\alpha_0 \neq 0$, then in a certain neighborhood of 0 the reciprocal $(\sum_{n=0}^{\infty} \alpha_n \zeta^n)^{-1}$ can be expanded into a power series $\sum_{n=0}^{\infty} \beta_n \zeta^n$.

*4. The set L of left-invertible elements of a Banach algebra E is open and contains with any two elements their product (L is a multiplicative semi-group). The same holds for the set of right-invertible elements. *Hint*: If y is a left-inverse of x, then $y(x + z) = e + yz$ is invertible for all z with sufficiently small norm.

§10. Finite-dimensional normed spaces

In the next sections we shall frequently need some properties of finite-dimensional normed spaces. We list therefore these facts briefly here and begin with the fundamental

Lemma 10.1. *Given finitely many* linearly independent *elements* $x_1, \ldots, x_n$ *in a normed vector space, there exists* $\mu > 0$ *so that*

$$|\alpha_1| + \cdots + |\alpha_n| \leqq \mu\|\alpha_1 x_1 + \cdots + \alpha_n x_n\| \tag{10.1}$$

for all numbers $\alpha_1, \ldots, \alpha_n$.

For the proof, we first consider only linear combinations $\alpha_1 x_1 + \cdots + \alpha_n x_n$ with $|\alpha_1| + \cdots + |\alpha_n| = 1$ and set

$$\gamma := \inf_{\sum|\alpha_\nu| = 1} \|\alpha_1 x_1 + \cdots + \alpha_n x_n\|.$$

By the definition of the greatest lower bound, there exists a *minimising sequence*, i.e., a sequence of elements

$$y_k := \sum_{\nu=1}^{n} \alpha_\nu^{(k)} x_\nu \quad \text{with} \quad \sum_{\nu=1}^{n} |\alpha_\nu^{(k)}| = 1 \quad \text{and} \quad \|y_k\| \to \gamma.$$

Since $|\alpha_\nu^{(k)}| \leqq 1$, there exists by the Bolzano–Weierstrass theorem an increasing sequence of integers $k_1, k_2, \ldots$, such that $\alpha_\nu^{(k_l)} \to \beta_\nu$ ($\nu = 1, \ldots, n$). Obviously $|\beta_1| + \cdots + |\beta_n| = 1$, hence by the linear independence of $x_1, \ldots, x_n$ one has $x := \beta_1 x_1 + \cdots + \beta_n x_n \neq 0$ and so $\|x\| > 0$. From the continuity of the operations and of the norm (see Proposition 6.1) it follows furthermore that

$$\|y_{k_l}\| = \left\| \sum_{\nu=1}^{n} \alpha_\nu^{(k_l)} x_\nu \right\| \to \left\| \sum_{\nu=1}^{n} \beta_\nu x_\nu \right\| = \|x\|.$$

Since on the other hand $\|y_{k_l}\| \to \gamma$, we have $\|x\| = \gamma > 0$, thus also $\mu := 1/\gamma > 0$ and by the definition of γ

$$(10.2) \qquad 1 \leqq \mu \|\alpha_1 x_1 + \cdots + \alpha_n x_n\| \qquad \text{for} \quad |\alpha_1| + \cdots + |\alpha_n| = 1.$$

If now the $\alpha_1, \ldots, \alpha_n$ are arbitrary (but not all equal to zero, to exclude the trivial case), then we obtain (10.1) immediately if we replace in (10.2) the coefficients α_ν by

$$\frac{\alpha_\nu}{|\alpha_1| + \cdots + |\alpha_n|}. \qquad \blacksquare$$

The lemma just proved is the source from which we derive the next five propositions.

Proposition 10.1. *In a finite-dimensional normed space convergence is equivalent to* componentwise *convergence, i.e., if $\{x_1, \ldots, x_n\}$ is a basis of the space then $y_k := \sum_{\nu=1}^{n} \alpha_\nu^{(k)} x_\nu$ converges to $y := \sum_{\nu=1}^{n} \alpha_\nu x_\nu$ if and only if $\alpha_\nu^{(k)} \to \alpha_\nu$ for $\nu = 1, \ldots, n$.*

Because of the continuity of the operations (Proposition 6.1) it is clear that componentwise convergence implies convergence in the sense of the norm. Conversely, it follows from the above lemma that $\sum_{\nu=1}^{n} |\alpha_\nu^{(k)} - \alpha_\nu| \leqq \mu \|y_k - y\|$. $\blacksquare$

This proposition immediately implies:

Proposition 10.2. *All norms on a* finite-dimensional *vector space are equivalent.*

This result goes far beyond the statement we could make at the end of §7 that all the $l^p(n)$-norms on $\mathbf{K}^n$ are equivalent.

Proposition 10.3. *Every linear map from a* finite-dimensional *normed space into an arbitrary normed space is* continuous.

Indeed, if $\{x_1, \ldots, x_n\}$ is a basis of the first space, A the given map and if $y_k := \sum_{\nu=1}^{n} \alpha_\nu^{(k)} x_\nu$ converges to $y := \sum_{\nu=1}^{n} \alpha_\nu x_\nu$, then $\alpha_\nu^{(k)}$ converges to α_ν (Proposition 10.1), and because of the continuity of the operations it follows that

$$Ay_k = \sum_{\nu=1}^{n} \alpha_\nu^{(k)} Ax_\nu \to \sum_{\nu=1}^{n} \alpha_\nu Ax_\nu = A\left(\sum_{\nu=1}^{n} \alpha_\nu x_\nu\right) = Ay. \qquad \blacksquare$$

On infinite-dimensional normed spaces there do exist discontinuous linear maps, see Exercise 1.

Proposition 10.4. *In a* finite-dimensional *normed space the Bolzano-Weierstrass theorem holds, i.e., every* bounded *sequence contains a* convergent *subsequence.*

Proof. Let $\{x_1, \ldots, x_n\}$ be a basis of the space, $y_k := \sum_{\nu=1}^{n} \alpha_\nu^{(k)} x_\nu$ and $\|y_k\| \leqq \gamma$ for $k = 1, 2, \ldots$. According to Lemma 10.1 for an appropriate $\mu > 0$ we have

$$\sum_{\nu=1}^{n} |\alpha_\nu^{(k)}| \leqq \mu \left\| \sum_{\nu=1}^{n} \alpha_\nu^{(k)} x_\nu \right\| = \mu \|y_k\| \leqq \mu\gamma,$$

hence each numerical sequence $(\alpha_\nu^{(k)})$, $\nu = 1, \ldots, n$, is bounded. By the theorem of Bolzano–Weierstrass for number sequences, there exists an increasing sequence of integers $k_1, k_2, \ldots$ such that $\alpha_\nu^{(k_l)} \to \beta_\nu$ $(\nu = 1, \ldots, n)$. Consequently $y_{k_l} = \sum_{\nu=1}^{n} \alpha_\nu^{(k_l)} x_\nu$ converges to $\sum_{\nu=1}^{n} \beta_\nu x_\nu$, thus (y_{k_l}) is a convergent subsequence of (y_k). ■

Proposition 10.5. *Every* finite-dimensional *normed space is* complete; *hence every finite-dimensional subspace of a normed space is closed.*

Indeed, if (y_n) is a Cauchy-sequence in a finite-dimensional space, then (y_k) is bounded (see §1), hence by Proposition 10.4 contains a convergent subsequence. Its limit is also the limit of (y_k). The closedness of finite-dimensional subspaces is now clear. ■

Let us return to Proposition 10.4. In infinite-dimensional normed spaces the Bolzano–Weierstrass theorem need not hold any more, see Exercise 2. It is reasonable to inquire whether among all normed spaces the finite-dimensional ones are distinguished by the validity of this important theorem. The answer to this question will be affirmative. The basis of our investigation is the following:

Lemma 10.2 (Lemma of F. Riesz). *If F is a proper* closed *subspace of the normed space E, then to every η with $0 < \eta < 1$ there exists a vector x_η in E with*

$$\|x_\eta\| = 1 \quad \textit{and} \quad \|x - x_\eta\| \geqq \eta \qquad \textit{for all} \quad x \in F.$$

Proof. There exists a vector y in E which does not lie in F. Let $d := \inf_{x \in F} \|x - y\|$ and (x_n) be a minimising sequence in F, i.e., $\|x_n - y\| \to d$. It follows that $d > 0$ (since otherwise ($d = 0$) we would have $x_n \to y$ and because of the closedness of F the vector y would lie in F in contradiction to its choice). From here and from the hypothesis $0 < \eta < 1$ it follows that $d/\eta > d$, thus there exists $z \in F$ with $0 < \|z - y\| \leqq d/\eta$. If we set $\gamma := 1/\|z - y\|$ and $x_\eta := \gamma(y - z)$, then $\|x_\eta\| = 1$ and for all $x \in F$ we obtain because of $\gamma \geqq \eta/d$ and $(1/\gamma)x + z \in F$ that

$$\|x - x_\eta\| = \|x - \gamma(y - z)\| = \|(x + \gamma z) - \gamma y\| = \gamma \left\| \left(\frac{1}{\gamma} x + z \right) - y \right\|$$

$$\geqq \frac{\eta}{d} \cdot d = \eta. \qquad \blacksquare$$

We are now in the position to characterize the finiteness of dimension—an *algebraic* property—of a normed space by a *metric* property—the validity of the Bolzano-Weierstrass theorem. The deep reason that this is possible is the fact that the algebraic and the metric structure of a normed space do not stand unrelated side by side but are intertwined through continuity.

Theorem 10.1. *The Bolzano–Weierstrass theorem is valid in a normed space, i.e., every bounded sequence contains a convergent subsequence, if and only if the space has finite dimension.*

In one direction the assertion is nothing but Proposition 10.4. Conversely, suppose that the normed space E is infinite-dimensional, so that no finite-dimensional subspace of E coincides with E. Let $x_1 \in E$, $\|x_1\| = 1$ and $F_1 := [x_1]$ the linear hull of x_1. As a finite-dimensional subspace, F_1 is distinct from E and by Proposition 10.5 it is closed. According to the lemma of F. Riesz, there exists an $x_2 \in E$ with $\|x_2\| = 1$ and $\|x - x_2\| \geqq \frac{1}{2}$ for all $x \in F_1$. Let $F_2 := [x_1, x_2]$ be the linear hull of x_1, x_2. Again $F_2 \neq E$ and F_2 is closed, hence there exists $x_3 \in E$ such that $\|x_3\| = 1$ and $\|x - x_3\| \geqq \frac{1}{2}$ for all $x \in F_2$. One continues in this way, more precisely: if $x_1, \ldots, x_n$ have been constructed, then $F_n := [x_1, \ldots, x_n] \neq E$ and F_n is closed, hence there exists $x_{n+1} \in E$ such that $\|x_{n+1}\| = 1$ and $\|x - x_{n+1}\| \geqq \frac{1}{2}$ for all $x \in F_n$. The sequence (x_n) is bounded since $\|x_n\| = 1$ but contains no convergent subsequence, not even a Cauchy subsequence because for $n \neq m$ one has $\|x_n - x_m\| \geqq \frac{1}{2}$. $\blacksquare$

Because of the fundamental importance of the Bolzano–Weierstrass theorem we distinguish in the theory of metric spaces those sets in which it is valid by a special name: A subset M of a metric space is *relatively compact*—or *relatively sequentially compact*—if every sequence in M contains a convergent subsequence. Its limit does not have to lie in M; if, however, it always belongs to M, we say that M is *compact* or *sequentially compact*.

A compact set M is always closed. Indeed, if the sequence (x_n) from M converges to x, then because of compactness there exists a subsequence (x_{n_k}) which converges to an element y in M. Since $y = x$, the point x lies in M. ■

It follows immediately that *a set is compact if and only if it is relatively compact and closed.*

A relatively compact, and a fortiori a compact set is bounded. If M were unbounded and y arbitrary in M, there would exist for every natural number n a point x_n in M such that $d(x_n, y) \geqq n$. On the other hand, because M is relatively compact, there would exist a subsequence (x_{n_k}) with $x_{n_k} \to x$. Because of the continuity of distance (§1) it would follow that $d(x_{n_k}, y) \to d(x, y)$ which contradicts $d(x_{n_k}, y) \geqq n_k$. ■

A set U in a normed space is called a *neighborhood of zero* if it contains a ball around 0 (cf. §1, Exercise 13). With this terminology, the above observations concerning compact and relatively compact sets, Exercise 10b in §1 and Theorem 10.1, the reader should have no difficulty in proving the following proposition.

Proposition 10.6. *For a normed space E the following assertions are equivalent:*

(a) *E is finite-dimensional.*
(b) *Every closed and bounded subset of E is compact.*
(c) *The closed unit ball $K_1[0]$ is compact.*
(d) *Every bounded subset of E is relatively compact.*
(e) *There exists in E a relatively compact neighborhood of zero.*

Exercises

1. Define the linear map $A: l^2 \to l^2$ by $Ax := (\sum_{\nu=1}^{\infty} (\xi_\nu/\nu), 0, 0, \ldots)$ and show that A is continuous (*Hint*: Cauchy–Schwarz inequality). If we introduce, however, on $l^2 \subset l^\infty$ the l^∞-norm $\|x\|_\infty = \sup|\xi_\nu|$, then A is not continuous (consider the sequence of elements $x_n := (1, 1, \ldots, 1, 0, 0, \ldots)$—the first n terms $=1$, all others $=0$).

2. From the bounded sequence of unit vectors $e_n := (0, \ldots, 0, 1, 0, \ldots)$ in l^2—the nth term $=1$, all others $=0$—no Cauchy subsequence can be extracted.

3. A distance problem: Let F be a proper subspace of the normed space E and $\delta > 0$. Does there exist a $y \in E$ so that $\|x - y\| \geqq \delta$ for all $x \in F$?

Show on an appropriate subspace of l^2 that the answer is negative if F is not closed. For a closed F it is, however, affirmative.

*4. Let M be a non-empty subset of the metric space E; with the metric taken from E also M is a metric space. If M is complete as a metric space, then it is closed as a subset of E. If E itself is complete, then M is complete if and only if it is closed.

5. Let E, F be metric spaces, $A: E \to F$ a continuous map and $M \subset E$ compact. Then $A(M)$ too is compact, briefly: the continuous image of a compact set is compact.

6. A real-valued continuous function on a compact metric space is bounded and assumes its greatest lower bound and its least upper bound (use Exercise 5).

7. Let T be a compact metric space and $C(T)$ the set of all continuous functions $x: T \to \mathbf{K}$. With the usual pointwise definition of addition and multiplication by scalars $C(T)$ is a vector space over $\mathbf{K}$. Defining the norm by $\|x\| := \max_{t \in T} |x(t)|$ (cf. Exercise 6) $C(T)$ becomes a Banach space.

8. A compact metric space is complete.

We make an observation before the remaining exercises. A relatively compact subset of a metric space is bounded and in finite-dimensional normed spaces, according to Proposition 10.6, exactly the bounded sets are relatively compact. In the following Exercises we shall state precise conditions which have to accompany boundedness in order to make subsets of certain Banach spaces relatively compact. To show that the conditions are sufficient, choose by the diagonal process from a given sequence a subsequence which at first converges componentwise (for the diagonal process see the proof of Proposition 13.1).

9. $M \subset l^p$, $1 \leqq p < \infty$, is relatively compact if and only if M is bounded and $\sum_{\nu=n}^{\infty} |\xi_\nu|^p \to 0$ as $n \to \infty$, uniformly for all $(\xi_1, \xi_2, \ldots)$ in M.

10. $M \subset (c_0)$ is relatively compact if and only if M is bounded and $\sup_{\nu \geqq n} |\xi_\nu| \to 0$ as $n \to \infty$, uniformly for all $(\xi_1, \xi_2, \ldots)$ in M.

11. $M \subset (c)$ is relatively compact if and only if M is bounded and $\sup_{\nu, \mu \geqq n} |\xi_\nu - \xi_\mu| \to 0$ as $n \to \infty$, uniformly for all $(\xi_1, \xi_2, \ldots)$ in M.

§11. The Neumann series in non-complete normed spaces

The investigations in §8 concerning the solution of the equation

$$(I - K)x = y \tag{11.1}$$

with the aid of the Neumann series $\sum_{n=0}^{\infty} K^n y$ were always performed in *complete* normed spaces. In this section we want to investigate how far we can liberate ourselves from this sometimes obnoxious condition of completeness. The first question will be whether the Neumann series—if it converges at all for a certain y—yields a solution of equation (11.1). The following completely elementary but fundamental proposition gives an affirmative answer to this question.

Proposition 11.1. *If K is a continuous endomorphism of the normed, not necessarily complete space E and if the Neumann series*

$$\sum_{n=0}^{\infty} K^n y \tag{11.2}$$

converges for a certain y in E to a vector x in E, then

$$(I - K)x = y.$$

Because of the continuity of K it follows from $x = \sum_{n=0}^{\infty} K^n y$ that $Kx = \sum_{n=0}^{\infty} K^{n+1} y = \sum_{n=1}^{\infty} K^n y = x - y$, i.e., indeed $x - Kx = y$. ■

Thus the Neumann series (11.2) can converge only for vectors y from the image space $(I - K)(E)$. Under certain hypotheses concerning K convergence will take place for all these vectors. The following proposition presents a condition of this kind.

Proposition 11.2. *For the map K from Proposition* 11.1 *the following assertions are equivalent:*

(a) $K^n \to 0$ *as* $n \to \infty$.

(b) $I - K$ *is injective, and the Neumann series* $\sum_{n=0}^{\infty} K^n y$ *converges for all vectors y from the image space of* $I - K$.

If (a) holds, then from $(I - K)x = 0$ it first follows that $x = Kx$, and then successively $x = K^n x$ for $n = 1, 2, \ldots$, hence $x = 0$. Thus $I - K$ is injective. If y is from $(I - K)(E)$, then there exists a vector x—and, according to what we have just proved, exactly one x—such that $x - Kx = y$, i.e., $x = y + Kx$. It follows that $x = y + K(y + Kx) = y + Ky + K^2x$, and in general $x = y + Ky + \cdots + K^{n-1}y + K^n x$ for all natural numbers n. Because of $K^n x \to 0$ we have $x = \sum_{n=0}^{\infty} K^n y$. Now suppose that (b) is valid, and let x be an arbitrary vector in E. Then $y = x - Kx$ lies in $(I - K)(E)$ and we see as above that $x = y + Ky + \cdots + K^{n-1}y + K^n x$. Since the series $\sum_{n=0}^{\infty} K^n y$ converges by hypothesis, its sum solves equation (11.1) and since by the injectivity of $I - K$ the only such solution is the x we started out with, it follows that $y + Ky + \cdots + K^{n-1}y \to x$ and so $K^n x \to 0$: thus (K^n) converges indeed pointwise to 0. ■

Towards the end of §7 we discoursed on the special importance which the continuity of the inverse has in applications. Therefore we do not want to omit the discussion of the continuity of $(I - K)^{-1}$, and prove in this direction the following:

Proposition 11.3. *If for the map K of Proposition* 11.1 *we have* $K^n \Rightarrow 0$, *then the inverse* $(I - K)^{-1}$ *is* continuous *on its space of definition* $(I - K)(E)$ (cf. §8, Exercise 5).

Proof. $(I - K)^{-1}$ exists by Proposition 11.2 since $K^n \Rightarrow 0$ implies a fortiori that $K^n \to 0$. If m is a natural number such that $\|K^m\| < 1$—such an m exists according to hypothesis—then with $\varepsilon := 1 - \|K^m\| > 0$ for every $x \in E$ we have

$$\|(I - K^m)x\| \geqq \|x\| - \|K^m x\| \geqq \|x\| - \|K^m\|\,\|x\| = \varepsilon\|x\|; \tag{11.3}$$

for the first estimate we used (6.12). If $(I - K)^{-1}$ were not continuous on $(I - K)(E)$, we would have by Proposition 7.5 a sequence (x_n) with $(1/n)\|x_n\| > \|(I - K)x_n\|$. For $y_n := x_n/\|x_n\|$ we would thus have

$$\|y_n\| = 1 \quad \text{and} \quad (I - K)y_n \to 0. \tag{11.4}$$

Because of $I - K^m = (I + K + \cdots + K^{m-1})(I - K)$ it would follow that $(I - K^m)y_n \to 0$ as $n \to \infty$, in contradiction to (11.3) according to which $\|(I - K^m)y_n\| \geqq \varepsilon\|y_n\| = \varepsilon > 0$ for all n. Thus $(I - K)^{-1}$ is indeed continuous. ■

When applying Proposition 11.1, one can often verify without any effort that for a certain y the Neumann series (11.2) is a Cauchy series. It will be, in general, more difficult to show that this Cauchy series converges. This leads to the question whether one can determine classes of operators K for which the *convergence* of the Neumann series follows already from its being *Cauchy*. We shall see below that this is the case for operators with complete image space, for finite-dimensional operators, and for compact operators—a concept yet to be defined.

Proposition 11.4. *If the image space $K(E)$ of the map K from Proposition* 11.1 *is complete, then the Neumann series* (11.2) *converges whenever it is a Cauchy series. Its sum solves equation* (11.1).

We only have to modify slightly the proof of Proposition 11.1. If (11.2) is a Cauchy series, i.e., if the sequence of partial sums $s_n := y + Ky + \cdots K^n y$ is a Cauchy sequence, then obviously also the elements $Ks_n = Ky + K^2 y + \cdots + K^{n+1}y$ form a Cauchy sequence in the complete subspace $K(E)$. Consequently the sequence (Ks_n) converges, i.e., the series $\sum_{n=1}^{\infty} K^n y$, and with it also the Neumann series (11.2). The last assertion of the proposition follows from Proposition 11.1. ■

Proposition 11.4 can be invoked in particular if K is a continuous operator of finite rank, since then by Proposition 10.5 its finite-dimensional image space $K(E)$ is complete. The next proposition shows that under weaker hypotheses we can prove even more.

Proposition 11.5. *Let K be a finite-dimensional endomorphism of the vector space E. If the Neumann series* (11.2) *is a Cauchy series with respect to some norm $\|\cdot\|_0$ on E, then it converges with respect to* every *norm $\|\cdot\|$ on E to a solution of the equation* (11.1)*—even if it is not continuous with respect to this norm. If* (11.2) *is a Cauchy series with respect to the norm $\|\cdot\|_0$ for every y in E, then the inverse $(I - K)^{-1}$ exists on E and is continuous for any norm on E with respect to which K itself is continuous.*

Proof. We set $F := K(E)$ and denote by $(E, \|\cdot\|_0)$ the space E equipped with the norm $\|\cdot\|_0$. We define $(E, \|\cdot\|)$, $(F, \|\cdot\|_0)$ and $(F, \|\cdot\|)$ analogously. The restriction K_F of K to F is obviously a linear self-map of F. Since F is finite-dimensional, K_F must be continuous for every norm of F according to Proposition 10.3. Assume now that the Neumann series (11.2) is a Cauchy series with respect to the norm $\|\cdot\|_0$, i.e., that the sequence of partial sums $s_n := y + Ky + \cdots$

$+ K^n y$ is a Cauchy sequence. Because of the completeness of $(F, \|\cdot\|_0)$ it follows, as in the proof of the preceding proposition, that the sequence (Ks_n) converges to some $z \in F$. Since by Proposition 10.2 all norms on F are equivalent, we have $Ks_n \to z$ also in $(F, \|\cdot\|)$. If we set $x := y + \sum_{n=1}^{\infty} K^n y = y + z$, then x is the sum of (11.2) in $(E, \|\cdot\|)$ and from the continuity of K_F on $(F, \|\cdot\|)$ it follows that

$$Kx = Ky + K_F\left(\sum_{n=1}^{\infty} K^n y\right) = Ky + \sum_{n=1}^{\infty} K_F K^n y = Ky + \sum_{n=1}^{\infty} K^{n+1} y = x - y,$$

so that indeed $x - Kx = y$.

Now let (11.2) be a Cauchy series for every y in $(E, \|\cdot\|_0)$. Then by what we have just proved $I - K$ is surjective, furthermore obviously $K^n \to 0$. Similarly as at the beginning of the proof of Proposition 11.2 it follows that $I - K$ is also injective, hence bijective (we cannot apply Proposition 11.2 directly because there K was supposed to be continuous). Now let K be also continuous on $(E, \|\cdot\|)$ but assume that $(I - K)^{-1}$ is not continuous there. As it was shown in the proof of Proposition 11.3, there exists then a sequence (y_n) in $(E, \|\cdot\|)$ for which (11.4) holds. Because of

$$|1 - \|Ky_n\|| = |\|y_n\| - \|Ky_n\|| \leqq \|y_n - Ky_n\|$$

it follows that $\|Ky_n\| \to 1$. Thus (Ky_n) is a bounded sequence in $(F, \|\cdot\|)$. By Proposition 10.4 it contains a convergent subsequence (Ky_{n_m}):

$$Ky_{n_m} \to z, \qquad \|z\| = 1.$$

With the help of (11.4) it follows that

$$y_{n_m} = (I - K)y_{n_m} + Ky_{n_m} \to z.$$

Because we assumed that K is continuous with respect to the norm $\|\cdot\|$, we have $Ky_{n_m} \to Kz$ and so $(I - K)y_{n_m} \to (I - K)z$. Going back again to (11.4) we get $(I - K)z = 0$ and so $z = 0$ (since $I - K$ is injective) in contradiction to $\|z\| = 1$. Thus $(I - K)^{-1}$ must be continuous on $(E, \|\cdot\|)$. ■

Exercise 1 of §10 shows that a finite-dimensional operator can very well be continuous with respect to one norm but discontinuous with respect to another one.

The above proof of the continuity of $(I - K)^{-1}$ used, besides the continuity of K, only the following property: If (x_n) is a bounded sequence, then (Kx_n) has a convergent subsequence. Every linear map K of a normed space E into itself—or into another normed space F—with this property is said to be *compact* or *completely continuous. A compact transformation is always continuous*; otherwise there would exist a sequence (x_n) with $\|x_n\| = 1$ and $\|Kx_n\| \to \infty$, thus in spite of the boundedness of (x_n), the sequence (Kx_n) could not have a convergent subsequence. Thus we can state the following proposition:

Proposition 11.6. *For a compact self-map K of a normed space the inverse $(I - K)^{-1}$ is continuous provided it exists at all.*

We shall see in the second next section examples of compact operators. For our present purposes it is decisive that compactness of an operator guarantees the convergence of certain sequences, even in non-complete spaces. Let us suppose for instance that the Neumann series (11.2) with a compact self-map K of the normed space E is a Cauchy series, i.e., that the sequence of the partial sums $s_n := \sum_{\nu=0}^{n} K^\nu y$ is a Cauchy sequence. Then the sequence of the $Ks_n = \sum_{\nu=1}^{n+1} K^\nu y$ is also a Cauchy sequence. (s_n) is bounded, hence (Ks_n) contains a convergent subsequence; but then the Cauchy sequence (Ks_n), i.e., the series $\sum_{n=1}^{\infty} K^n y$ must also converge. This obviously implies that the series (11.2) converges, its sum is according to Proposition 11.1 a solution of the equation (11.1). Let us state this result:

Proposition 11.7. *If K is a compact self-map of a normed space, then the Neumann series* (11.2) *converges to a solution of equation* (11.1) *whenever it is a Cauchy series.*

We can follow another way to obtain a solution for equation (11.1)—or at least for an equation closely related to it—from a Cauchy series (11.2). We only have to think of the procedure with which one makes solvable the equation $x^2 - 2 = 0$, which does not have a solution in the set $\mathbf{Q}$ of rational numbers: One enlarges the incomplete metric space $\mathbf{Q}$ to the complete metric space $\mathbf{R}$, extends the function $x \mapsto x^2$ to $\mathbf{R}$ (i.e., defines the squares of real numbers $\tilde{x}$), and shows that the 'extended equation' $\tilde{x}^2 - 2 = 0$ has solutions in $\mathbf{R}$. The space $\mathbf{R}$ is a particularly economical extension of $\mathbf{Q}$: Every $\tilde{x}$ in $\mathbf{R}$ is the limit of a sequence (x_n) in $\mathbf{Q}$. We say therefore that $\mathbf{Q}$ is dense in $\mathbf{R}$. In the next section we shall deal with the question whether one can in a similar way enlarge a non-complete normed space E to a complete normed space $\tilde{E}$, extend equation (11.1) to $\tilde{E}$, and solve it there with the aid of the Cauchy series (11.2) (see Proposition 12.4).

Exercises

1. Consider Exercises 3, 4 in §8 in the light of Proposition 11.5.

2. If K is a finite-dimensional self-map of the vector space E over $\mathbf{K}$ and $\{x_1, \ldots, x_r\}$ is a basis of $K(E)$, then there exist r linear maps $f_\rho: E \to \mathbf{K}$ with which

$$Kx = \sum_{\rho=1}^{r} f_\rho(x)x_\rho \qquad \text{for all} \quad x \text{ in } \quad E.$$

Thus we have $K^n x = \sum_{\rho=1}^{r} f_\rho(K^{n-1}x)x_\rho$ for $n = 1, 2, \ldots$. Use this observation to prove the convergence statement of Proposition 11.5 under the sole application of Proposition 10.1.

*3. The zero transformation 0 is compact on all normed spaces, the identity transformation I only on finite-dimensional ones (thus there exist continuous transformations which are not compact).

$^+$4. The linear map $K: E \to F$ (E, F normed) is compact if and only if one of the following equivalent assertions is true:

(a) $\{Kx: x \in E, \|x\| \leqq 1\}$, i.e., the image of the closed unit ball in E, is relatively compact.

(b) There exists a neighborhood U of zero in E whose image $K(U)$ is relatively compact.

*5. Scalar multiples and sums of compact operators are compact. The product of a compact operator with a continuous operator is compact, independently of the order of the factors (observe that the continuous image of a bounded sequence is again bounded).

*6. A compact operator on a infinite-dimensional normed space does not have a continuous inverse. *Hint*: Exercises 3 and 5.

§12. The completion of metric and normed spaces

We investigate first the completion of general metric spaces and use the method of the *Cantor fundamental sequences*, with the help of which one extends the incomplete space **Q** of the rational numbers to the complete space **R** of real numbers, as is well known. To describe this method adequately, we need the concept of an isometry.

A map A from the metric space (E_1, d_1) into the metric space (E_2, d_2) is said to be *isometric* or an *isometry* if it preserves distances, i.e., if for two points x, y in E_1 one has $d_1(x, y) = d_2(Ax, Ay)$.

An isometry is obviously injective and continuous; the inverse map $A^{-1}: A(E_1) \to E_1$ is again isometric. Therefore we may give the following symmetric definition:

Two metric spaces are said to be *isometric* if there exists an isometric map from one onto the other.

From the point of view of a metric theory, isometric spaces differ only by the names of their elements and can therefore be identified.

A subset M of the metric space E is *dense* in E if every point in E is the limit of a sequence in M, or equivalently, if for an arbitrary $\varepsilon > 0$ and for every $x \in E$ there exists a $y \in M$ with $d(x, y) \leqq \varepsilon$.

A non-empty subset M of a metric space E is itself a metric space if one assigns to any two elements of M the distance they already have as elements of E. One then calls M a *subspace* of E and says that the metric of M is *induced* by the metric of E, or simply by E (cf. the definition of a subspace of a normed space in §6 and Exercise 4 in §10).

Let now E be an incomplete metric space. We call two Cauchy sequences (x_n), (y_n) from E equivalent, and write $(x_n) \sim (y_n)$, when $d(x_n, y_n) \to 0$. This relation in the set F of all Cauchy sequences from E is reflexive, symmetric and transitive, i.e., an equivalence relation. It decomposes F into classes $\hat{x}, \hat{y}, \ldots$ of equivalent Cauchy sequences. Let $\hat{E}$ be the set of all these equivalence classes.

It follows from the quadrangle inequality that for two Cauchy sequences (x_n), (y_n) the limit $\lim d(x_n, y_n)$ always exists and that it does not change if we pass to equivalent Cauchy sequences. Therefore if $\hat{x}$, $\hat{y}$ are equivalence classes with representatives (x_n), (y_n), the definition $d(\hat{x}, \hat{y}) := \lim d(x_n, y_n)$ makes sense. It is easy to see that d is a metric on $\hat{E}$.

We now indicate a subspace of $\hat{E}$ which is isometric with E. For this we observe that an equivalence class $\hat{x}$ can contain at most one constant sequence $(x, x, x, \ldots)$. Let E_0 be the set of all equivalence classes which contain a constant sequence. The map $A: E \to E_0$ which associates with each $x \in E$ the equivalence class $\hat{x} \in \hat{E}$ containing $(x, x, x, \ldots)$ is obviously an isometry of the spaces E and E_0.

E_0 is dense in $\hat{E}$: To a representative (x_n) of $\hat{x} \in \hat{E}$ and to a preassigned $\varepsilon > 0$ there exists an $m \in \mathbf{N}$ so that $d(x_n, x_m) \leqq \varepsilon$ for $n \geqq m$. The equivalence class $\hat{y}$ of the sequence $(x_m, x_m, \ldots)$ lies in E_0 and satisfies $d(\hat{y}, \hat{x}) = \lim_{n\to\infty} d(x_m, x_n) \leqq \varepsilon$.

We show now that $\hat{E}$ is complete. Let $(\hat{x}_n)$ be a Cauchy sequence in $\hat{E}$. Since E_0 is dense in $\hat{E}$, to each natural number n there corresponds a $\hat{y}_n \in E_0$ so that $d(\hat{x}_n, \hat{y}_n) \leqq 1/n$. Because of

$$d(\hat{y}_n, \hat{y}_m) \leqq d(\hat{y}_n, \hat{x}_n) + d(\hat{x}_n, \hat{x}_m) + d(\hat{x}_m, \hat{y}_m) \leqq \frac{1}{n} + d(\hat{x}_n, \hat{x}_m) + \frac{1}{m},$$

the $\hat{y}_n$ form a Cauchy sequence. Let $(y_n, y_n, \ldots)$ be the constant sequence in $\hat{y}_n$, i.e., $Ay_n = \hat{y}_n$ with the above defined isometry A. Then $d(y_n, y_m) = d(Ay_n, Ay_m) = d(\hat{y}_n, \hat{y}_m)$, and thus $(y_1, y_2, \ldots)$ is a Cauchy sequence in E. If we denote its equivalence class by $\hat{y}$, then

$$d(\hat{x}_n, \hat{y}) \leqq d(\hat{x}_n, \hat{y}_n) + d(\hat{y}_n, \hat{y}) \leqq \frac{1}{n} + d(\hat{y}_n, \hat{y}) = \frac{1}{n} + \lim_{m\to\infty} d(y_n, y_m),$$

hence $\lim d(\hat{x}_n, \hat{y}) = 0$. The Cauchy sequence $(\hat{x}_n)$ has therefore the limit $\hat{y}$.

Finally, we imbed E into $\hat{E}$ with the aid of the isometry A. This means: If $\hat{x}$ is in E_0, and $A^{-1}\hat{x} = x$, i.e., $(x, x, \ldots)$ is the constant sequence in $\hat{x}$, then we remove $\hat{x}$ from $\hat{E}$ and replace it by x. Thus we obtain a new set $\tilde{E} = E \cup (\hat{E} \setminus E_0)$. On it we define a metric d_1 so that the situation on $\hat{E}$ shall be conserved:

$d_1(\hat{x}, \hat{y}) := d(\hat{x}, \hat{y})$	for $\hat{x}, \hat{y}$ in $\hat{E}\setminus E_0$.
$d_1(x, y) := d(x, y) = d(Ax, Ay)$	for x, y in E.
$d_1(x, \hat{y}) := d(Ax, \hat{y})$	for x in E and $\hat{y}$ in $\hat{E}\setminus E_0$.

Now $(\tilde{E}, d_1)$ is a complete metric space which contains (E, d) as a dense subspace.

Assume that $(\bar{E}, d_2)$ is a second complete metric space which contains (E, d) as a dense subspace. We show that $(\bar{E}, d_2)$ and $(\tilde{E}, d_1)$ are isometric. Let $\bar{x}$ be an arbitrary element from $\bar{E}$. Then there exists a sequence (x_n) in E so that $d_2(x_n, \bar{x}) \to 0$. Since (x_n) is a Cauchy sequence in E, there exists an $\tilde{x}$ in $\tilde{E}$ so that $d_1(x_n, \tilde{x}) \to 0$. The point $\tilde{x}$ is independent of the particular choice of the approximating sequence (x_n). Indeed, if (y_n) is another sequence in E which converges to $\bar{x}$ in $\bar{E}$, then it follows from

$$d_1(y_n, \tilde{x}) \leqq d(y_n, x_n) + d_1(x_n, \tilde{x}) \leqq d_2(y_n, \bar{x}) + d_2(\bar{x}, x_n) + d_1(x_n, \tilde{x})$$

that also $d_1(y_n, \tilde{x}) \to 0$. We associate with each $\bar{x}$ in $\bar{E}$ the element $\tilde{x}$ in $\tilde{E}$ determined in this way. This map $B: \bar{E} \to \tilde{E}$ is obviously surjective, and leaves the elements of E fixed. It follows further from the continuity of the metric that B is an isometry; indeed, if $\bar{x}$, $\bar{y}$ are two points in $\bar{E}$ and (x_n), (y_n) approximating sequences in E whose limits $\tilde{x}$, $\tilde{y}$ in $\tilde{E}$ are given by $\tilde{x} = B\bar{x}$ and $\tilde{y} = B\bar{y}$, then

$$d_2(\bar{x}, \bar{y}) = \lim d_2(x_n, y_n) = \lim d(x_n, y_n) = \lim d_1(x_n, y_n) = d_1(\tilde{x}, \tilde{y}).$$

We summarise now these results:

Proposition 12.1. *To every non-complete metric space E there exists a complete metric space $\tilde{E}$, determined uniquely up to an isometry, in which E is dense and which induces on E the original metric of E. We call $\tilde{E}$ the* completion *or the* complete hull *of E.*

Let now E be a non-complete *normed* space. Let us denote the metric of its completion $\tilde{E}$ by d, thus for x, y in E we have $d(x, y) = \|x - y\|$. We want to make out of $\tilde{E}$ first a vector space and then a normed space. For x, y in $\tilde{E}$ there exist sequences (x_n), (y_n) in E which converge to x and y, respectively. From $\|(x_n + y_n) - (x_m + y_m)\| \leqq \|x_n - x_m\| + \|y_n - y_m\|$ we deduce that $(x_n + y_n)$ is a Cauchy sequence, and has therefore a limit z in $\tilde{E}$. It is easy to see that z depends only on x and y and not on the choice of the approximating sequences (x_n), (y_n). This observation justifies the definition $x + y := \lim(x_n + y_n)$. If x and y lie in E, then this sum coincides with the sum already defined in E: set $x_n = x$ and $y_n = y$ for all n. In a similar way the multiplication by a scalar already defined on E can be extended to $\tilde{E}$ by $\alpha x := \lim(\alpha x_n)$.

The reader should check that $\tilde{E}$ is now a vector space. Next we introduce a norm on $\tilde{E}$ by $\|x\| := d(x, 0)$, which is clearly an extension of the norm which exists already on E. It is trivial that $\|x\| \geqq 0$ and that $\|x\| = 0$ if and only if $x = 0$. The verification of the other properties of a norm will be based on the continuity of the metric (see §1) from which it follows in particular that $x_n \to x$ implies $d(x_n, 0) \to d(x, 0)$, i.e., $\|x_n\| \to \|x\|$. If the sequences (x_n), (y_n) from E converge to x, y from $\tilde{E}$, then $\|x_n + y_n\| \leqq \|x_n\| + \|y_n\|$ implies for $n \to \infty$ the triangle inequality $\|x + y\| \leqq \|x\| + \|y\|$; the equality $\|\alpha x\| = |\alpha|\,\|x\|$ is proved completely analogously. Finally, we have $d(x, y) = \lim d(x_n, y_n) = \lim\|x_n - y_n\| = \|x - y\|$, i.e., the metric d on $\tilde{E}$ derives from the norm introduced above. In summary, we say that E can be completed into a Banach space $\tilde{E}$. If $\bar{E}$ is another Banach space which is a completion of E, then the map B defined in the proof of Proposition 12.1 is not only an isometry from $\bar{E}$ onto $\tilde{E}$—so that in particular x and Bx always have the same norm—but it is also a linear transformation, i.e., an 'isomorphism' of normed spaces. The precise definition of this important concept is as follows:

The linear map T from the normed space F onto the normed space G is an *isomorphism of normed spaces* or an *isometric isomorphism*, if

$$\|Tx\| = \|x\| \qquad \text{for all} \quad x \text{ in } \quad F.$$

An isomorphism of normed spaces T is because of

$$\|Tx - Ty\| = \|T(x - y)\| = \|x - y\|$$

an isometry, hence injective, thus an (algebraic) isomorphism of the vector spaces F, G.

$T^{-1}: G \to F$ is again an isomorphism of normed spaces; thus one is justified to say with a symmetric locution that two normed spaces are *isomorphic as normed spaces* (or *isometrically isomorphic*) if there exists a linear map which is an isomorphism of normed spaces from one space onto the other.

From the point of view of the theory of normed spaces, isometrically isomorphic spaces differ from each other only in the names of their elements and can therefore be identified.

The result of our investigations above can be described as follows:

Proposition 12.2. *To every non-complete normed space E there exists a Banach space $\tilde{E}$, determined uniquely up to an isomorphism of normed spaced, such that E is a subspace dense in $\tilde{E}$. We call $\tilde{E}$ the* completion *or the* complete hull *of E.*

The next proposition asserts that continuous linear maps between normed spaces can be uniquely extended to their completions with preservation of the norm.

Proposition 12.3. *Let E, F be two normed spaces and $\tilde{E}$, $\tilde{F}$ their completions. If A is a continuous linear map from E into F, then there exists exactly one continuous linear map $\tilde{A}$ from $\tilde{E}$ into $\tilde{F}$ with $\tilde{A}x = Ax$ for all $x \in E$. This extension preserves the norm, i.e., $\|\tilde{A}\| = \|A\|$.*

Proof. If x is in $\tilde{E}$ and (x_n) a sequence from E converging to x, then (Ax_n) is a Cauchy sequence because of $\|Ax_n - Ax_m\| = \|A(x_n - x_m)\| \leqq \|A\|\,\|x_n - x_m\|$, hence converges to an element y of $\tilde{F}$. It is easy to see that y depends only on x and not on the approximating sequence (x_n). The expression $\tilde{A}x := \lim Ax_n$ is thus well defined and yields a map $\tilde{A}: \tilde{E} \to \tilde{F}$ which is obviously linear and an extension of A. From $\|Ax_n\| \leqq \|A\|\,\|x_n\|$ the estimate $\|\tilde{A}x\| \leqq \|A\|\,\|x\|$ follows as $n \to \infty$; thus $\tilde{A}$ is bounded and $\|\tilde{A}\| \leqq \|A\|$. Sincc conversely also $\|A\| \leqq \|\tilde{A}\|$ (see §7, Exercise 1), we must have $\|\tilde{A}\| = \|A\|$. Finally, if $\bar{A}: \tilde{E} \to \tilde{F}$ is a second continuous and linear extension of A to $\tilde{E}$, then $\bar{A}x = \lim Ax_n$, hence $\bar{A} = \tilde{A}$. ∎

It does not present any difficulties any more to answer the question which we posed at the end of §11 concerning the solution of the equation $x - Kx = y$. From the last two propositions and Proposition 11.1 we have immediately:

Proposition 12.4. *Let K be a continuous endomorphism of the non-complete normed space E and $\tilde{K}$ the continuous and linear extension of K to the completion*

$\tilde{E}$ of E. If the Neumann series $\sum_{n=0}^{\infty} K^n y$ is a Cauchy series for a y in E, then its sum $\tilde{x} \in \tilde{E}$ satisfies the equation $\tilde{x} - \tilde{K}\tilde{x} = y$.

Exercises

1. The isometric image of a complete metric space is again complete (completeness is preserved under isometries).

2. The image of a Cauchy sequence under a continuous map does not need to be a Cauchy sequence.

3. The map A from the metric space E into the metric space F is said to be *uniformly continuous* if for every $\varepsilon > 0$ there exists a $\delta > 0$ such that $d(x, y) \leqq \delta$ implies $d(Ax, Ay) \leqq \varepsilon$. If A is *uniformly* continuous, then the image (Ax_n) of a Cauchy sequence (x_n) in E is a Cauchy sequence in F.

4. Let $A: E \to F$ be a map of the metric spaces E, F. If A is uniformly continuous and bijective, and if the inverse map A^{-1} is continuous, then, together with F, also E is complete (use Exercise 3).

5. A map with bounded dilation is uniformly continuous (see Exercise 3).

6. A continuous map from a compact metric space into a second metric space is uniformly continuous (see Exercise 3).

+7. Every metric space (E, d) is isometric to a subset of the Banach space $B(E)$. *Hint*: Let a be a fixed element of E. For every $x \in E$ define the function $f_x: E \to \mathbf{R}$ by

$$f_x(t) := d(x, t) - d(a, t) \qquad (t \in E)$$

and show successively with the help of the quadrangle inequality:

(a) $|f_x(t)| \leqq d(x, a)$ for all $t \in E$; thus $f_x \in B(E)$.

(b) For all $x, y \in E$ we have $\| f_x - f_y \| = d(x, y)$; here $\| \cdot \|$ denotes of course the supremum-norm on $B(E)$. It follows that the map $x \mapsto f_x$ from (E, d) onto the metric space $\{f_x : x \in E\} \subset B(E)$, equipped with the distance $\|f_x - f_y\|$ induced by $B(E)$, is an isometry.

8. With the help of Exercise 7 give a second proof of Proposition 12.1 concerning completion. *Hint*: Exercise 15 of §1. Observe that a closed subset of a complete metric space is itself complete.

Remark. This second proof uses the completeness of $B(E)$ and so ultimately the completeness of $\mathbf{R}$. The proof given in the main text does not do this; it even given us the possibility to construct $\mathbf{R}$ as the completion of $\mathbf{Q}$ (a procedure which goes back to Georg Cantor).

§13. Compact operators

We recall at the outset the definition of compact operators in §11:

A linear map K from a normed space E into a normed space F is said to be *compact* or *completely continuous* if the image (Kx_n) of every bounded sequence (x_n) from E contains a convergent subsequence.

We observed in §11 (and the exercises that followed it) that compact operators are continuous and that scalar multiples and sums of compact operators are compact. The product of a compact operator with a continuous one is compact, whatever the order of the factors. The identity transformation I is compact only on finite-dimensional spaces, and a compact operator can have a continuous inverse only if its space of definition is finite-dimensional.

We denote by $\mathscr{K}(E, F)$ the set of all compact operators $K: E \to F$; the set of all compact self-maps of E is denoted by $\mathscr{K}(E) := \mathscr{K}(E, E)$. By the above results *$\mathscr{K}(E, F)$ is a linear subspace of the vector space $\mathscr{L}(E, F)$, and $\mathscr{K}(E)$ is even a (two-sided) 'ideal' in the algebra $\mathscr{L}(E)$*—as is well known, a subset M of an algebra R is a (two-sided) *ideal* if it is a linear subspace of R and for every x in R and m in M the products xm and mx lie in M.

The next proposition shows that $\mathscr{K}(E, F)$ is even a closed subspace of $\mathscr{L}(E, F)$ if F is complete; thus *if $E \neq \{0\}$ is complete, then $\mathscr{K}(E)$ is a closed ideal in the Banach algebra $\mathscr{L}(E)$.*

Proposition 13.1. *If the sequence of compact maps K_n from a normed space E into a Banach space F converges uniformly to K, then K is compact.*

For the proof, let (x_i) be a bounded sequence in E: $\|x_i\| \leqq \gamma$. Then there exists a subsequence (x_{1i}) of (x_i) so that $(K_1 x_{1i})$ converges, next a subsequence (x_{2i}) of (x_{1i}) so that $(K_2 x_{2i})$ converges, etc. The diagonal elements $y_i := x_{ii}$ form a sequence which from a certain index on is a subsequence of every one of the sequences $(x_{k1}, x_{k2}, \ldots)$, and therefore the sequence $(K_n y_i)$ converges for every operator K_n. Choose now an arbitrary $\varepsilon > 0$ and determine an n_0 so that $\|K_{n_0} - K\| \leqq \varepsilon$. If we fix an i_0 so that for $i, k \geqq i_0$ one has $\|K_{n_0} y_i - K_{n_0} y_k\| \leqq \varepsilon$, then for these subscripts i, k

$$\begin{aligned}\|Ky_i - Ky_k\| &\leqq \|Ky_i - K_{n_0} y_i\| + \|K_{n_0} y_i - K_{n_0} y_k\| + \|K_{n_0} y_k - Ky_k\| \\ &\leqq \varepsilon\|y_i\| + \varepsilon + \varepsilon\|y_k\| \leqq (2\gamma + 1)\varepsilon.\end{aligned}$$

Thus (Ky_i) is a Cauchy sequence in the Banach space F and so a convergent subsequence of (Kx_i). ■

A continuous operator K of finite rank is compact. Indeed, because of $\|Kx\| \leqq \|K\|\,\|x\|$, the image of a bounded sequence is a bounded sequence in the finite-dimensional image-space of K, and contains therefore by Theorem 10.1 a convergent subsequence. If we also take into account Proposition 13.1, we may state:

Proposition 13.2. *A continuous operator of finite rank is compact, and a uniformly convergent sequence of such operators $K_n: E \to F$ has always a compact limit operator if F is complete.*

The question whether conversely every compact operator $K: E \to F$ (F complete) is the limit of a uniformly convergent sequence of continuous, finite-dimensional operators $K_n: E \to F$ has a negative answer (cf. [53], [48]). A totally different formulation of this approximation problem is given in [46].

Based on Proposition 13.2 we now give two important examples of compact operators.

Example 13.1. The integral transformation $K: C[a, b] \to C[a, b]$ with continuous kernel k

$$(Kx)(s) := \int_a^b k(s, t)x(t)\mathrm{d}t \tag{13.1}$$

is compact. Indeed, in §9 (see from (9.7) on) we observed that K is the uniform limit of a sequence of continuous operators of finite rank.

Example 13.2. Let (α_{ik}) be an infinite matrix with

$$\sum_{i,k=1}^{\infty} |\alpha_{ik}|^2 < \infty. \tag{13.2}$$

In the first place, it follows from the Cauchy–Schwarz inequality that for every $x = (\xi_1, \xi_2, \ldots)$ in l^2 the series $\sum_{k=1}^{\infty} \alpha_{ik}\xi_k$ $(i = 1, 2, \ldots)$ converge, and then that

$$\sum_{i=1}^{\infty} \left| \sum_{k=1}^{\infty} \alpha_{ik}\xi_k \right|^2 \leqq \sum_{i=1}^{\infty} \sum_{k=1}^{\infty} |\alpha_{ik}|^2 \sum_{k=1}^{\infty} |\xi_k|^2.$$

Thus the matrix transformation K, defined by

$$Kx = K(\xi_1, \xi_2, \ldots) := \left(\sum_{k=1}^{\infty} \alpha_{1k}\xi_k, \sum_{k=1}^{\infty} \alpha_{2k}\xi_k, \ldots \right), \tag{13.3}$$

maps l^2 into itself, is obviously linear, and because of

$$\|Kx\| \leqq \left(\sum_{i,k=1}^{\infty} |\alpha_{ik}|^2 \right)^{1/2} \|x\|$$

also continuous. If we define the continuous operators $K_n: l^2 \to l^2$ of finite rank by

$$K_n x = K_n(\xi_1, \xi_2, \ldots) := \left(\sum_{k=1}^{\infty} \alpha_{1k}\xi_k, \ldots, \sum_{k=1}^{\infty} \alpha_{nk}\xi_k, 0, 0, \ldots \right),$$

then

$$\|K - K_n\| \leqq \left(\sum_{i=n+1}^{\infty} \sum_{k=1}^{\infty} |\alpha_{ik}|^2 \right)^{1/2},$$

and since the right-hand side of this inequality tends to 0 as $n \to \infty$, we have $K_n \Rightarrow K$. *The matrix transformation* $K: l^2 \to l^2$ *defined by* (13.3) *is compact provided that* (13.2) *is satisfied.*

By Proposition 8.2 the solution of the Fredholm integral equation

$$x(s) - \int_a^b k(s, t)x(t)\mathrm{d}t = y(s) \tag{13.4}$$

with continuous kernel k depends on the right-hand side y continuously in the sense of the maximum-norm if, for the corresponding integral transformation K, defined by (13.1), the norm $\|K\|$, or only the limit $\lim \|K^n\|^{1/n}$, is less than 1. Based on Proposition 11.6 and example 13.1 we can state the much more general result that *this continuous dependence is already present whenever* (13.4) *has for* every $y \in C[a, b]$ *a* unique *solution* $x \in C[a, b]$.

Because of Proposition 5.1 the deviation from unique solvability of the integral equation (13.4) can be measured by the 'size' of the nullspace of $I - K$, e.g., by dim $N(I - K)$. The following proposition shows that this deviation is at any rate finite.

Proposition 13.3. *For a compact self-map K of a normed space the nullspace $N(I - K)$ has* finite *dimension.*

Indeed, if (x_n) is a bounded sequence in $N(I - K)$, then we can select a convergent subsequence from (Kx_n), and because of $Kx_n = x_n$, also from (x_n). By Theorem 10.1 the dimension *of* $N(I - K)$ is therefore finite. ■

If for a certain y the function x_0 is a solution of the integral equation (13.4), then there are finitely many solutions $x_1, \ldots, x_n$ of the corresponding homogeneous integral equation

$$x(s) - \int_a^b k(s, t)x(t)\mathrm{d}t = 0,$$

so that all solutions of (13.4), and only these, can be written in the form $x_0 + \alpha_1 x_1 + \cdots + \alpha_n x_n$ with arbitrary coefficients α_ν.

To finish, we cast a glance at *finite-dimensional operators.* We denote the set of all such operators $K: E \to F$, where E and F are vector spaces, not necessarily normed, by $\mathscr{E}(E, F)$; instead of $\mathscr{E}(E, E)$ we write more shortly $\mathscr{E}(E)$. If E and F are normed, let $\mathscr{F}(E, F)$ be the set of continuous, finite-dimensional maps from E into F, and $\mathscr{F}(E) := \mathscr{F}(E, E)$. Since scalar multiples and sums of finite-dimensional operators, and their products with arbitrary operators, are obviously always again finite-dimensional, we may state that $\mathscr{E}(E, F)$ *and* $\mathscr{F}(E, F)$ *are vector spaces,* and that $\mathscr{E}(E)$ *and* $\mathscr{F}(E)$ *are ideals in the algebra* $\mathscr{S}(E)$ *and* $\mathscr{L}(E)$, *respectively.*

Let $K: E \to F$ be finite-dimensional and $\{y_1, \ldots, y_n\}$ a basis of $K(E)$; if $K = 0$, we set $n = 1$ and take for y_1 any non-zero element of F. For every $x \in E$ we can represent Kx in the form

$$Kx = \sum_{\nu=1}^{n} f_\nu(x) y_\nu \tag{13.5}$$

with scalar coefficients $f_\nu(x)$. From $K(x + y) = K(x) + K(y)$ it follows that $\sum f_\nu(x + y)y_\nu = \sum [f_\nu(x) + f_\nu(y)]y_\nu$ and, comparing coefficients, $f_\nu(x + y) = f_\nu(x) + f_\nu(y)$; similarly we obtain $f_\nu(\alpha x) = \alpha f_\nu(x)$. The maps $x \mapsto f_\nu(x)$ from E into $\mathbf{K}$ are thus linear. Linear maps from a vector space into its field of scalars are called *linear functionals* or *linear forms*. The set of all linear forms on E, i.e., the vector space $\mathscr{S}(E, \mathbf{K})$ will be denoted by E^* and called the *algebraic dual space* or briefly the *algebraic dual* of E.

A finite-dimensional operator $K: E \to F$ can thus always be represented in the form (13.5) with vectors $y_1, \ldots, y_n$ from F and linear forms $f_1, \ldots, f_n$ from E^*, and conversely if the y_ν and f_ν have this meaning, then (13.5) always defines a finite-dimensional operator $K: E \to F$. If, furthermore, $\{y_1, \ldots, y_n\}$ is a basis of the image space $K(E)$, then the coefficient functionals $f_1, \ldots, f_n$ are linearly independent, as one sees easily.

Because of Proposition 13.2 we are mainly interested in *continuous* operators of finite rank. We characterize them by the following:

Proposition 13.4. *If the finite-dimensional linear map* $K: E \to F$ *(*E, F *normed spaces) is given by* (13.5) *with linearly independent vectors* $y_1, \ldots, y_n$, *then* K *is continuous—i.e., compact—if and only if all coefficient functionals are continuous.*

For the proof we only have to observe that for any sequence (x_i) in E the assertion $Kx_i \to Kx$ is equivalent to $f_\nu(x_i) \to f_\nu(x)$ for $\nu = 1, \ldots, n$ (Proposition 10.1). ■

The vector space $\mathscr{L}(E, \mathbf{K})$ of all *continuous* linear forms on a normed space E is denoted by E' and is called the *topological dual space* or simply the *dual* of E. A linear form f on E lies in E' if and only if with a certain $q > 0$ the estimate $|f(x)| \leqq q\|x\|$ is valid for all $x \in E$. *The space* E', *equipped with the norm*

$$\|f\| = \sup_{\|x\| = 1} |f(x)|$$

of a linear map, is a Banach space (Proposition 7.4).

Exercises

1. A family F of continuous functions on $[a, b]$ is said to be *equicontinuous* if to every $\varepsilon > 0$ there exists a $\delta > 0$ such that for every $x \in F$ and for any two points t_1, t_2 from $[a, b]$ with $|t_1 - t_2| \leqq \delta$ one has $|x(t_1) - x(t_2)| \leqq \varepsilon$ (observe that δ should not depend on x). The *theorem of Arzelà–Ascoli* (cf. [185], p. 144) asserts that a subset of $C[a, b]$ is relatively compact if it is (norm-) bounded and equicontinuous. Use this theorem to give a new proof for the compactness of the integral transformation K in Example 13.1. *Hint*: Show with the help of the continuity of the kernel k that the image $K(M)$ of a bounded set M is an equicontinuous family of functions.

2. Apply Exercise 9 of §10 to supply a new proof for the compactness of the matrix transformation K in Example 13.2.

3. Show with the help of an appropriate diagonal matrix that hypothesis (13.2) is not necessary for the compactness of the matrix transformation K in Example 13.2.

4. Every linear map from a finite-dimensional normed space into a second normed space is compact.

5. Let E, F be non-complete normed spaces, $\tilde{E}$, $\tilde{F}$ their completions. If $K: E \to F$ is compact, then also the continuous linear extension $\tilde{K}: \tilde{E} \to \tilde{F}$ of K to $\tilde{E}$ is compact, and $\tilde{K}(\tilde{E}) \subset F$. If $F = E$ and $\tilde{I}$ the identity transformation on $\tilde{E}$, then $N(I - K) = N(\tilde{I} - \tilde{K})$ and $(I - K)(E) = (\tilde{I} - \tilde{K})(\tilde{E}) \cap E$.

6. If E is a non-complete normed space and $\tilde{E}$ its completion, then the duals E' and $(\tilde{E})'$ are isomorphic as normed spaces.

+7. Let E and F be Banach spaces. The linear map $A: E \to F$ is said to be *nuclear* if it can be represented in the form

$$Ax = \sum_{\nu=1}^{\infty} f_\nu(x) y_\nu \quad \text{with} \quad f_\nu \in E',\ y_\nu \in F,\ \sum_{\nu=1}^{\infty} \|f_\nu\| \, \|y_\nu\| < \infty.$$

Let $\mathcal{N}(E, F)$ be the set of all nuclear maps from E into F, and let $\mathcal{N}(E) := \mathcal{N}(E, E)$. Show:

(a) $A \in \mathcal{N}(E, F)$ can always be represented in the form

$$Ax = \sum_{\nu=1}^{\infty} \alpha_\nu g_\nu(x) z_\nu \quad \text{with} \quad \alpha_\nu \geqq 0,\ \sum_{\nu=1}^{\infty} \alpha_\nu < \infty,\ \|g_\nu\| = \|z_\nu\| = 1.$$

(b) $\mathcal{N}(E, F)$ is a linear subspace of $\mathcal{L}(E, F)$.

(c) The product of a nuclear operator with a continuous one is nuclear, whatever the order of the factors.

(d) $\mathcal{N}(E)$ is a (two-sided) ideal of $\mathcal{L}(E)$.

(e) $\mathcal{F}(E, F) \subset \mathcal{N}(E, F) \subset \mathcal{K}(E, F)$. (For the proof of the second inclusion use (a) and a diagonal process in order to extract from a bounded sequence $(x_n) \subset E$ a subsequence (x_{n_k}) so that $(g_\nu(x_{n_k}))$ is a Cauchy sequence for every ν).

III

Bilinear systems and conjugate operators

§14. Bilinear systems

A finite-dimensional operator $K: E \to F$ can always be represented in the form

$$Kx = \sum_{\nu=1}^{n} f_\nu(x) y_\nu$$

by means of vectors $y_1, \ldots, y_n$ from F and linear forms $f_1, \ldots, f_n$ from E^*; see (13.5). In almost all cases which occur in practice the coefficient functionals are generated in a simple fashion by elements of a vector space E^+. Let us consider for instance a degenerate kernel

$$k(s, t) := \sum_{\nu=1}^{n} x_\nu(s) x_\nu^+(t)$$

with functions x_ν, x_ν^+ which are continuous on $[a, b]$, and the corresponding integral operator $K: C[a, b] \to C[a, b]$ defined by

$$(Kx)(s) := \int_a^b k(s, t) x(t) \mathrm{d}t = \sum_{\nu=1}^{n} \left(\int_a^b x(t) x_\nu^+(t) \mathrm{d}t \right) \cdot x_\nu(s).$$

The coefficient functionals

$$(14.1) \qquad f_\nu(x) := \int_a^b x(t) x_\nu^+(t) \mathrm{d}t$$

are 'generated' by the elements $x_1^+, \ldots, x_n^+$ of $C[a, b]$. Of course, one can take for E^+ also the vector space $[x_1^+, \ldots, x_n^+]$; in other cases—e.g., if one has to do with polynomial kernels as in (9.10)—one will take for E^+ the vector space of all polynomials. If against our assumption the x_ν^+ are not continuous, only integrable, then we can choose E^+ to consist of all functions integrable on $[a, b]$.

As a second example consider the operator $K: \mathbf{K}^n \to \mathbf{K}^n$ which by means of the matrix $(\alpha_{\nu\mu})$ and the unit vectors $x_1, \ldots, x_n$ is given by

$$Kx = K(\xi_1, \ldots, \xi_n) := \left(\sum_{\mu=1}^{n} \alpha_{1\mu}\xi_\mu, \ldots, \sum_{\mu=1}^{n} \alpha_{n\mu}\xi_\mu \right) = \sum_{\nu=1}^{n} \left(\sum_{\mu=1}^{n} \alpha_{\nu\mu}\xi_\mu \right) x_\nu .$$

Here

$$f_\nu(x) = \sum_{\mu=1}^{n} \alpha_{\nu\mu}\xi_\mu, \tag{14.2}$$

hence the f_ν are 'generated' by the vectors $x_\nu^+ := (\alpha_{\nu 1}, \ldots, \alpha_{\nu n})$ from $E^+ := \mathbf{K}^n$. Here also one can take of course for E^+ the linear hull $[x_1^+, \ldots, x_n^+]$ of the vectors x_ν^+; this space will possibly be smaller than $\mathbf{K}^n$.

We can describe the common trait of the two examples as follows. Let two vector spaces E, E^+ over the scalar field $\mathbf{K}$ be given and let a scalar $\langle x, x^+ \rangle$ be associated with each pair of vectors $x \in E$, $x^+ \in E^+$—in our examples we have

$$\langle x, x^+ \rangle = \int_a^b x(t)x^+(t)\,\mathrm{d}t \quad \text{and} \quad \langle x, x^+ \rangle = \sum_{\mu=1}^{n} \xi_\mu \xi_\mu^+ . \tag{14.3}$$

This scalar-valued function $(x, x^+) \mapsto \langle x, x^+ \rangle$ on $E \times E^+$ is *bilinear* or a *bilinear form*, i.e., it is linear in both variables:

$$\langle x + y, x^+ \rangle = \langle x, x^+ \rangle + \langle y, x^+ \rangle, \qquad \langle \alpha x, x^+ \rangle = \alpha \langle x, x^+ \rangle$$
$$\langle x, x^+ + y^+ \rangle = \langle x, x^+ \rangle + \langle x, y^+ \rangle, \qquad \langle x, \alpha x^+ \rangle = \alpha \langle x, x^+ \rangle .$$

By means of such a bilinear form and appropriate vectors x_ν in E, x_ν^+ in E^+, the integral or matrix operator considered above can then be represented in the form

$$Kx = \sum_{\nu=1}^{n} \langle x, x_\nu^+ \rangle x_\nu . \tag{14.4}$$

If a bilinear form is defined on $E \times E^+$, then we call the pair (E, E^+) of vector spaces a *bilinear system* with respect to this bilinear form; the last addition is usually omitted, we speak simply of the bilinear system (E, E^+) and denote always by $\langle x, x^+ \rangle$ the value of the bilinear form which, according to definition, exists on $E \times E^+$.

The concept of a bilinear system, in particular in its narrower form of a dual system, will turn out to be one of the ruling concepts of functional analysis. We want to consider some more examples.

Example 14.1. Every pair (E, E^+) of vector spaces is a bilinear system with respect to the trivial bilinear form $\langle x, x^+ \rangle := 0$ for all (x, x^+).

Example 14.2. If r is the smaller one of the natural numbers m, n and if $(\alpha_1, \ldots, \alpha_r)$ is chosen arbitrarily, then $(\mathbf{K}^m, \mathbf{K}^n)$ is a bilinear system with respect to the bilinear form

(14.5) $$\langle x, x^+ \rangle := \sum_{\nu=1}^{r} \alpha_\nu \xi_\nu \xi_\nu^+ .$$

For $r = m = n$ and $\alpha_1 = \cdots = \alpha_n = 1$ we obtain precisely the second bilinear form in (14.3).

Example 14.3. Let $P[a, b]$ be the vector space of all polynomials on $[a, b]$ and $w \in C[a, b]$. Then $(C[a, b], P[a, b])$ and $(C[a, b], C[a, b])$ are bilinear systems with respect to the bilinear form

(14.6) $$\langle x, x^+ \rangle := \int_a^b w(t)x(t)x^+(t)\mathrm{d}t;$$

for $w(t) \equiv 1$ on $[a, b]$ we obtain precisely the first bilinear form in (14.3).

Example 14.4. $(C[a, b], BV[a, b])$ is a bilinear system with respect to the bilinear form

(14.7) $$\langle x, x^+ \rangle := \int_a^b x(t)\mathrm{d}x^+(t).$$

Example 14.5. Let $1 < p$, $q < \infty$ and $1/p + 1/q = 1$. Then (l^p, l^q) is a bilinear system with respect to the bilinear form

(14.8) $$\langle x, x^+ \rangle := \sum_{\nu=1}^{\infty} \xi_\nu \xi_\nu^+ .$$

The convergence of the series follows from the Hölder inequality.

Example 14.6. Let E be a vector space and E^+ an arbitrary *subspace* of E^*, the space of all linear forms on E. Then (E, E^+) becomes through the *canonical bilinear form*

(14.9) $$\langle x, x^+ \rangle := x^+(x)$$

a bilinear system.

If x^* is a linear form on E, then henceforth $\langle x, x^* \rangle$ will mean as in (14.9) *the value of x^* at the point x*—even if we do not speak explicitly of a bilinear system. The boundedness of a linear form x^* on a normed space E is then expressed by the inequality $|\langle x, x^* \rangle| \leqq \|x\| \|x^*\|$ for all x in E.

The concept of a bilinear system is symmetric: If (E, E^+) is a bilinear system with respect to the bilinear form $\langle x, x^+ \rangle$, then (E^+, E) is a bilinear system with respect to the bilinear form

(14.10) $$\langle x^+, x \rangle := \langle x, x^+ \rangle;$$

this is the *canonical bilinear form* for (E^+, E); *we shall always use it except when the contrary is explicitly stated.*

Example 14.6 is particularly important for our purposes. In connection with (13.5) it shows that for every finite-dimensional operator $K: E \to F$ there always exists a bilinear system (E, E^+), e.g., the bilinear system (E, E^*), so that with appropriate vectors $y_1, \ldots, y_n$ from F and $x_1^+, \ldots, x_n^+$ from E^+ the representation

(14.11) $$Kx = \sum_{\nu=1}^{n} \langle x, x_\nu^+ \rangle y_\nu$$

is valid for every $x \in E$. Conversely, an operator defined by (14.11) has obviously finite rank. *If $K \neq 0$, then the vectors $x_1^+, \ldots, x_n^+$, as well as the vectors $y_1, \ldots, y_n$, can be thought of as being linearly independent* (see Exercise 3).

Exercises

1. If (E, E^+) is a given bilinear system, then in general not every finite-dimensional operator $K: E \to F$ will be representable in the form (14.11) by means of (E, E^+).

2. The set of all self-maps of E which can be written in the form (14.11) by means of a given bilinear system (E, E^+) is an algebra.

*3. Show that if $K \neq 0$, then the vectors $x_1^+, \ldots, x_n^+$ and the vectors $y_1, \ldots, y_n$ can be chosen linearly independent. *Hint*: If for instance $y_n = \alpha_1 y_1 + \cdots + \alpha_{n-1} y_{n-1}$, then

$$Kx = \sum_{\nu=1}^{n-1} \langle x, x_\nu^+ + \alpha_\nu x_n^+ \rangle y_\nu;$$

continuing in this way reduce the number of terms in the sum until the remaining y_ν are linearly independent. One can reduce further in a similar fashion, if the vectors from E^+ occurring in $\langle \cdot, \cdot \rangle$ are still linearly dependent.

$^+$4. Assume that the *normed* spaces E, E^+ form a bilinear system with the bilinear form $\langle x, x^+ \rangle$. The bilinear form is said to be *continuous* if $\langle x_n, x_n^+ \rangle \to \langle x, x^+ \rangle$ whenever $x_n \to x, x_n^+ \to x^+$; it is called *bounded* if there exists a constant $\gamma > 0$ so that $|\langle x, x^+ \rangle| \leq \gamma \|x\| \, \|x^+\|$ for all $x \in E, x^+ \in E^+$. Show that a bilinear form is continuous if and only if it is bounded.

§15. Dual systems

If (E, E^+) is a bilinear system, then every x^+ generates by means of

(15.1) $$f_{x^+}(x) := \langle x, x^+ \rangle$$

a linear form f_{x^+} on E.

The correspondence $x^+ \mapsto f_{x^+}$ is obviously a linear map $A: E^+ \to E^*$. *Different* elements of E^+ generate *different* linear forms on E if and only if

$Ax^+ = 0$, i.e., $\langle x^+, x \rangle = 0$ for all $x \in E$, implies that $x^+ = 0$. In this case E^+ is isomorphic to the subspace $A(E^+)$ of E^*, and can be identified with it. In a completely analogous way every x in E generates by means of

(15.2) $$F_x(x^+) := \langle x, x^+ \rangle$$

a linear form F_x on E^+, and the correspondence $x \mapsto F_x$ is an isomorphic map from E into $(E^+)^*$ if and only if from $\langle x, x^+ \rangle = 0$ for all $x^+ \in E^+$ it follows that x vanishes.

We call a bilinear system (E, E^+):

a *left dual system* if from $\langle x, x^+ \rangle = 0$ for all x in E it follows that $x^+ = 0$,

a *right dual system* if from $\langle x, x^+ \rangle = 0$ for all x^+ in E^+ it follows that $x = 0$,

a *dual system* if it is both a left and a right dual system.

Using a self-explanatory language we can say that (E, E^+) is a left or a right dual system if the 'left' space E or the 'right' space E^+, respectively, contains 'many' elements, or is 'large', cf. also Proposition 15.1.

If we make the above-described identifications, then *in the case of a left dual system every $x^+ \in E^+$ is a linear form on E* with the values

(15.3) $$x^+(x) := \langle x, x^+ \rangle, \qquad x \in E.$$

In the case of a right dual system every $x \in E$ is a linear form on E^+; it is defined by

(15.4) $$x(x^+) := \langle x, x^+ \rangle, \qquad x^+ \in E^+.$$

Observe that we identify E^+ with a subspace of E^* and E with a subspace of $(E^+)^*$ only through the *canonical imbeddings*

(15.5) $$x^+ \mapsto f_{x^+} \quad \text{and} \quad x \mapsto F_x,$$

and not with the help of any other isomorphism which might possibly exist. In other words: *to consider x^+ as a linear form on E and x as a linear form on E^+ means always to use definitions* (15.3) *and* (15.4), *respectively.*

For the proof of the important Proposition 15.1, and for some other purposes, we need:

Lemma 15.1. *If $f_1, \ldots, f_n$ are linear forms on E and if for a further linear form f on E we have*

(15.6) $$\bigcap_{\nu=1}^{n} N(f_\nu) \subset N(f),$$

i.e., if $f_1(x) = \cdots = f_n(x) = 0$ implies that $f(x) = 0$, then f is a linear combination of $f_1, \ldots, f_n$.

Without restricting the generality, we may assume that the $f_1, \ldots, f_n$ are linearly independent, and we use mathematical induction. Let first $n = 1$. Since $f_1 \neq 0$, there exists an x_1 with $f_1(x_1) \neq 0$. Obviously, for every $x \in E$ the vector $y_x := x - (f_1(x)/f_1(x_1))x_1$ lies in $N(f_1)$, hence x can be represented in the form

$$(15.7) \qquad x = \frac{f_1(x)}{f_1(x_1)} x_1 + y_x \qquad \text{with} \quad y_x \in N(f_1).$$

Since by hypothesis $f(y_x)$ vanishes, we have $f(x) = (f(x_1)/f_1(x_1))f_1(x)$, i.e., $f = \alpha f_1$.

Now let us assume that the Lemma is already proven for $n - 1$ linearly independent linear forms. Then $\bigcap_{\nu=1, \nu \neq \mu}^{n} N(f_\nu)$ is not a subset of $N(f_\mu)$, since otherwise by the induction hypothesis f_μ would be a linear combination of the other linear forms, in contradiction to the assumed linear independence of $f_1, \ldots, f_n$. Thus for every μ there exists an x_μ with $f_\nu(x_\mu) = 0$ for $\nu \neq \mu$ and $f_\mu(x_\mu) \neq 0$; for reasons of homogeneity we may even assume that $f_\mu(x_\mu) = 1$. For every x the vector $y_x := x - \sum_{\nu=1}^{n} f_\nu(x)x_\nu$ lies obviously in $\bigcap_{\nu=1}^{n} N(f_\nu)$, hence by (15.6) also in $N(f)$, so that $f(x) = \sum_{\nu=1}^{n} f(x_\nu)f_\nu(x)$, i.e.,

$$f = \sum_{\nu=1}^{n} f(x_\nu) f_\nu. \qquad \blacksquare$$

Proposition 15.1. *The bilinear system* (E, E^+) *is a* left *dual system if and only if for finitely many linearly independent vectors* $x_1^+, \ldots, x_n^+$ *in* E^+ *there always exist vectors* $x_1, \ldots, x_n$ *in* E *such that*

$$(15.8) \qquad \langle x_i, x_k^+ \rangle = \delta_{ik} \qquad \textit{for} \quad i, k = 1, \ldots, n;$$

it is a right *dual system if and only if for finitely many linearly independent vectors* $x_1, \ldots, x_n$ *in* E *there always exist vectors* $x_1^+, \ldots, x_n^+$ *in* E^+ *so that* (15.8) *is valid. The elements* $x_1, \ldots, x_n$ *and* $x_1^+, \ldots, x_n^+$*, respectively, so determined are linearly independent.*

We have to give the proof only for a left dual system since (E, E^+) is a right dual system if and only if (E^+, E) is a left dual system. Thus let (E, E^+) be a left system. Then the linearly independent vectors $x_1^+, \ldots, x_n^+$ from E^+ are also linearly independent linear forms on E according to definition (15.3). The elements $x_1, \ldots, x_n$ which satisfy (15.8) can be constructed, because of Lemma 15.1, just as it was already done in the proof of the Lemma. They are linearly independent, since from $\alpha_1 x_1 + \cdots + \alpha_n x_n = 0$ it follows because of (15.8) that

$$\alpha_k = \langle \alpha_1 x_1 + \cdots + \alpha_k x_k + \cdots + \alpha_n x_n, x_k^+ \rangle = \langle 0, x_k^+ \rangle = 0$$

for $k = 1, \ldots, n$. Conversely, let (15.8) be valid, to put it briefly. Then in particular for every $x^+ \neq 0$ in E^+ there exists an x in E such that $\langle x, x^+ \rangle = 1$, so (E, E^+) is a left dual system. $\blacksquare$

(E, E^+) is a left dual system if E^+ is any subspace of E^*, since a linear form is the zero form exactly if it vanishes identically. But (E, E^+) does not need to be a right dual system, as one can easily see by taking 'small' spaces E^+. However, E^* itself contains sufficiently many linear forms to make a right dual system out of (E, E^*), as it will result from the following theorem.

Theorem 15.1. *If $\{x_\lambda: \lambda \in L\}$ is a basis of the non-trivial vector space E—there always exists one by Theorem 4.1—, then there exists exactly one linear form f on E which assumes at the points x_λ arbitrarily prescribed values α_λ.*

For the proof we represent every x in E in the form $\sum_{\lambda \in L} \xi_\lambda x_\lambda$ with uniquely determined coefficients ξ_λ, of which at most finitely many are $\neq 0$, and define f by $f(x) := \sum_{\lambda \in L} \alpha_\lambda \xi_\lambda$. Obviously f is the only linear form which satisfies the requirement. ■

Proposition 15.2. *(E, E^*) is a dual system.*

It is sufficient to show that for every $x_0 \neq 0$ in E there exists a linear form f on E so that $f(x_0) \neq 0$. For this purpose one extends according to Theorem 4.1 the set $\{x_0\}$ to a basis of E and one knows by Theorem 15.1 that there is an $f \in E^*$ with $f(x_0) = 1$. ■

On the basis of this theorem *we can consider E as a subspace of the vector space $E^{**} := (E^*)^*$ of all linear forms on E^**: the element x from E becomes through

$$x(f) := f(x) \qquad \text{for} \quad f \in E^*$$

a linear form on E^* (see (15.4)). And finally, Proposition 15.2 in connection with (14.11) shows that *to every finite-dimensional operator $K: E \to F$ there always exists a* dual system (E, E^+), *e.g., the system (E, E^*), with which one has*

$$\text{(15.9)} \qquad Kx = \sum_{\nu=1}^{n} \langle x, x_\nu^+ \rangle y_\nu$$

for all $x \in E$; in case $K \neq 0$ one can take the vectors $x_1^+, \ldots, x_n^+$ in E^+ as well as the vectors $y_1, \ldots, y_n$ in F *linearly independent*, as we already observed in connection with (14.11).

What is interesting for the applications, to which we shall return, is just that frequently for the representation of K we do not have to take the not very handy system (E, E^*) but we can use other systems (E, E^+) where the elements of E are *explicitly* known, e.g., the system $(C[a, b], C[a, b])$ in the case of the integral transformation with degenerate continuous kernel discussed in §14. For certain normed spaces E, the elements of E', i.e., the *continuous* linear forms, can be precisely described, so that in these cases it might be favorable to use the system (E, E'). We shall see in §28 that (E, E') is a dual system; this important result will be much more laborious to prove than Proposition 15.2.

Exercises

1. Check which bilinear systems in the Examples 14.1 through 14.5 are left or right dual systems, respectively.

2. Let $E = l^\infty$ or $=(s)$. Construct a dual system (E, E^+).

3. Given the linearly independent functions $x_1, \ldots, x_n$ continuous on $[a, b]$ there exist functions $y_1, \ldots, y_n$ with the same properties so that

$$\int_a^b x_i(t)y_k(t)\mathrm{d}t = \delta_{ik}$$

for $i, k = 1, \ldots, n$.

4. If $f_1, \ldots, f_n$ are linearly independent linear forms on E, then the system $f_k(x) = \xi_k$ $(k = 1, \ldots, n)$ of equations has a solution x in E for any right-hand side.

5. If the finite-dimensional operator $K: E \to F$ is represented in the form (15.9) with linearly independent vectors $x_1^+, \ldots, x_n^+$ with the help of a dual system, or only a left dual system (E, E^+), then $K(E) = [y_1, \ldots, y_n]$ (see Lemma 25.1).

$^+$6. Whenever (E, E^+) is a dual system—or only a right dual system—and $x_1, \ldots, x_n$ are linearly independent vectors in E, $y_1, \ldots, y_n$ arbitrary vectors in F, there exists a finite-dimensional operator $K: E \to F$ of the form (15.9) so that $Kx_\nu = y_\nu$ for $\nu = 1, \ldots, n$.

*7. For every linear form $f \neq 0$ on E the nullspace $N(f)$ is a hyperplane through 0. (*Hint*: see (15.7)). Conversely, if H is a hyperplane through 0 in E, there exists a linear form $f \neq 0$ on E so that $H = N(f)$.

$^+$8. If the linear forms $f_1, \ldots, f_n$ in E^* are linearly independent, then codim $\bigcap_{\nu=1}^n N(f_\nu) = n$. (*Hint*: consider the vector y_x towards the end of the proof of Lemma 15.1.) Conversely, if F is a subspace of E of codimension n, there exist n linearly independent forms $f_1, \ldots, f_n$ on E with which

$$F = \bigcap_{\nu=1}^{n} N(f_\nu).$$

*9. Let $\{x_\lambda : \lambda \in L\}$ be a basis of the vector space E over $\mathbf{K}$ and let f run through E^*. The correspondence $f \mapsto (f(x_\lambda): \lambda \in L)$ is an isomorphism of E^* onto the product space $\prod_{\lambda \in L} \mathbf{K}_\lambda$, where $\mathbf{K}_\lambda := \mathbf{K}$.

10. $\mathbf{K}^n$ is isomorphic to its algebraic dual $(\mathbf{K}^n)^*$.

11. We know that in the sense of the canonical imbedding $x \mapsto F_x$ (see (15.5)) we have $E \subset E^{**}$. Show that $E = E^{**}$ if and only if E is finite-dimensional.

$^+$12. If f_0 is a linear form on the proper subspace F of E, then there exists a linear form f on E with $f_0(x) = f(x)$ for all x in F (*Extension theorem* for linear forms). *Hint*: extend a basis of F to a basis of E.

*13. (E, E^+) is a left dual system if from $\langle x, x^+ \rangle = \langle x, y^+ \rangle$ for all $x \in E$ it follows that $x^+ = y^+$; on the other hand, (E, E^+) is a right dual system if $\langle x, x^+ \rangle = \langle y, x^+ \rangle$ for all $x^+ \in E$ implies that $x = y$.

+14. A linear form f on the algebra E is said to be *multiplicative if* $f(xy) = f(x)f(y)$ for all $x, y \in E$. Show that two multiplicative linear forms on E coincide if and only if they have the same nullspaces. *Hint*: Lemma 15.1.

§16. Conjugate operators

The importance of bilinear and dual systems is not based only on the fact that finite-dimensional operators can be represented with their help; more importantly they play a significant role in the study of operator equations $Ax = y$. As a preparation we recall a known theorem from the theory of systems of linear equations which anyway will be proved in a much more general form (see Proposition 16.2):

The system of equations

$$\sum_{k=1}^{n} \alpha_{ik} \xi_k = \eta_i \qquad (i = 1, \ldots, n) \tag{16.1}$$

is solvable if and only if

$$\sum_{i=1}^{n} \eta_i \xi_i^+ = 0 \tag{16.2}$$

for all solutions $(\xi_1^+, \ldots, \xi_n^+)$ *of the* transposed homogeneous system

$$\sum_{i=1}^{n} \alpha_{ik} \xi_i^+ = 0 \qquad (k = 1, \ldots, n). \tag{16.3}$$

The matrix

$$A^+ := \begin{pmatrix} \alpha_{11} & \alpha_{21} & \cdots & \alpha_{n1} \\ \vdots & \vdots & & \vdots \\ \alpha_{1n} & \alpha_{2n} & \cdots & \alpha_{nn} \end{pmatrix}$$

of the system (16.3) is obtained from the matrix

$$A := \begin{pmatrix} \alpha_{11} & \alpha_{12} & \cdots & \alpha_{1n} \\ \vdots & \vdots & & \vdots \\ \alpha_{n1} & \alpha_{n2} & \cdots & \alpha_{nn} \end{pmatrix}$$

through reflexion in the main diagonal. We can, however, describe this 'transposition' of A in a way which makes sense for arbitrary operators, not only matrix operators. If we consider namely the bilinear system $(\mathbf{K}^n, \mathbf{K}^n)$ with the bilinear form $\langle x, x^+ \rangle := \sum_{i=1}^{n} \xi_i \xi_i^+$, then for all x and x^+ we have

$$\langle Ax, x^+ \rangle = \sum_{i=1}^{n} \left(\sum_{k=1}^{n} \alpha_{ik} \xi_k \right) \xi_i^+ = \sum_{k=1}^{n} \left(\sum_{i=1}^{n} \alpha_{ik} \xi_i^+ \right) \xi_k = \langle x, A^+ x^+ \rangle; \tag{16.4}$$

it is also easy to see that A^+ is uniquely determined by this relation. The above solvability theorem can now be stated more shortly as follows:

The equation $Ax = y$ is solvable if and only if $\langle y, x^+\rangle = 0$ for all solutions x^+ of the equation $A^+x^+ = 0$.

In general, if A is an endomorphism of an *arbitrary* vector space E, then we shall try to describe the solvability of the equation $Ax = y$ in a similar way. Thus one will pick a bilinear system (E, E^+), check whether to A there exists a linear self-map A^+ of E^+ so that we have $\langle Ax, x^+\rangle = \langle x, A^+x^+\rangle$ as in (16.4), and whether with this operator A^+ there exists a solvability theorem like the one above. This program is easy to carry out if we use the dual system (E, E^*). For any fixed x^* in E^* the map $x \mapsto \langle Ax, x^*\rangle$ is again a linear form on E, which we denote by A^*x^* and for which by definition

$$\langle Ax, x^*\rangle = (A^*x^*)(x) = \langle x, A^*x^*\rangle$$

holds. A^* is a linear self-map of E^*; indeed, for all x in E and all x^*, y^* in E^* we have

$$\begin{aligned}\langle x, A^*(x^* + y^*)\rangle &= \langle Ax, x^* + y^*\rangle = \langle Ax, x^*\rangle + \langle Ax, y^*\rangle \\ &= \langle x, A^*y^*\rangle + \langle x, A^*y^*\rangle = \langle x, A^*x^* + A^*y^*\rangle\end{aligned}$$

and since (E, E^*) is a left dual system, it follows that $A^*(x^* + y^*) = A^*x^* + A^*y^*$. One sees in an analogous way that $A^*(\alpha x^*) = \alpha A^*x^*$.

A^* is uniquely determined by the equation

$$\langle Ax, x^*\rangle = \langle x, A^*x^*\rangle \text{ for all } x \text{ in } E \text{ and all } x^* \text{ in } E^*. \tag{16.5}$$

Indeed, if for some self-map B of E we have identically $\langle Ax, x^*\rangle = \langle x, Bx^*\rangle$, then also $\langle x, A^*x^*\rangle = \langle x, Bx^*\rangle$, and, since (E, E^*) is a left dual system, $A^*x^* = Bx^*$ for all x^* in E^*, i.e., $B = A^*$.

The linear self-map A^* of E^* uniquely determined by (16.5) is called the *algebraic dual map* (transformation, operator) of $A: E \to E$.

We can indeed state now a solvability theorem for the equation $Ax = y$ with the help of the algebraic dual transformation, which coincides completely with the one formulated above. For its proof we need the following proposition on the richness of E^*.

Proposition 16.1. *If F is a proper subspace of E and if y does not lie in F, then there exists a linear form f on E which vanishes at all points of F and assumes the value 1 at y.*

Indeed, if $\{x_\lambda : \lambda \in L\}$ is a basis of F, then $\{x_\lambda : \lambda \in L\} \cup \{y\}$ is a linearly independent set and thus can be extended by Theorem 4.1 to a basis of E. According to Theorem 15.1 there exists a linear form f on E with $f(x_\lambda) = 0$ for all $\lambda \in L$ and $f(y) = 1$. f fulfills the requirements. ■

Proposition 16.2. *For every endomorphism A of the vector space E the equation*

$$Ax = y, \qquad y \in E \tag{16.6}$$

is solvable in E if and only if the following holds:

(16.7) $\langle y, x^* \rangle = 0$ *for all solutions x^* of the equation $A^*x^* = 0$;*

here $\langle x, x^* \rangle$ is the canonical bilinear form of the dual system (E, E^*).

Proof. If (16.6) has a solution, say $x = x_0$, and if $A^*x^* = 0$, then $\langle y, x^* \rangle = \langle Ax_0, x^* \rangle = \langle x_0, A^*x^* \rangle = \langle x_0, 0 \rangle = 0$; thus (16.7) is a necessary condition for solvability. Now suppose that (16.7) holds for some y in E. If (16.6) were not solvable, i.e., if y did not lie in the subspace $A(E)$, then, according to Proposition 16.1, there would exist an x^* in E^* such that

$$\langle x, A^*x^* \rangle = \langle Ax, x^* \rangle = 0 \quad \text{for all} \quad x \text{ in } \quad E, \qquad \text{but} \quad \langle y, x^* \rangle = 1.$$

Since (E, E^*) is a left dual system, this would imply $A^*x^* = 0$. Because of (16.7) we would have $\langle y, x^* \rangle = 0$ in contradiction to $\langle y, x^* \rangle = 1$. Therefore (16.6) must be solvable. ∎

This proposition is very satisfactory from a theoretical point of view, but from a practical one it is almost worthless. In order to be able to apply it, we need to control the linear forms on E. One is certainly in the position to do this if one knows a basis of E (see Theorem 15.1 and Exercise 9 in §15). But such bases can be constructed only in a few cases; their existence was proved non-constructively with the help of Zorn's lemma, and unfortunately there exists no constructive proof. Therefore we should not be satisfied with Proposition 16.2 but should go on probing whether similar propositions can be proved using other bilinear systems (E, E^+), while naturally such systems will be of particular interest to us, where the elements of E^+ are *known explicitly*. The following definition of the conjugate operator is the basis of such investigations:

If (E, E^+) is a bilinear system, and if for the linear operator $A: E \to E$ there exists a linear operator $A^+: E^+ \to E^+$ such that

(16.8) $\langle Ax, x^+ \rangle = \langle x, A^+x^+ \rangle$ for all x in E and all x^+ in E^+,

then A is said to be *conjugable*, and A^+ is called an *operator conjugate to A* or a *conjugate of A*.

For the sake of greater precision we shall say occasionally that A is E^+-conjugable and call A^+ an E^+-conjugate operator to A (an E^+-conjugate of A).

If A possesses an E^+-conjugate A^+, then A is E-conjugate to A^+, where we have provided the bilinear system (E^+, E) with its canonical bilinear form (14.10). Furthermore one sees, exactly as when we considered the algebraic dual transformation, that *in the case of a* left dual system (E, E^+) *there exists at most one conjugate operator A^+ to A and that the* linearity *of A^+ does not have to be postulated explicitly*; it follows already from (16.8). *Thus in a* dual system *A and A^+ determine each other uniquely.*

If E^+ is a subspace of E^*, then a linear operator A: $E \to E$ possesses at most one E^+-conjugate A^+, completely characterized by (16.8), since in this case (E, E^+) is a left dual system.

Example 16.1. We have seen above that A^* is the only E^*-conjugate of $A \in \mathscr{S}(E)$.

Example 16.2. Let A be a *continuous* endomorphism of the normed space E. If A' denotes the restriction of A^* to E', then we have for all $x \in E$ and x' in E'

$$|(A'x')(x)| = |\langle x, A'x'\rangle| = |\langle Ax, x'\rangle| = |x'(Ax)| \leqq \|x'\| \, \|Ax\| \leqq \|x'\| \, \|A\| \, \|x\|,$$

hence the linear form $A'x'$ is *again continuous*, i.e., A' maps the dual E' into itself, and is thus the (uniquely determined) E'-conjugate of A. Furthermore it follows that $\|A'x'\| \leqq \|A\| \, \|x'\|$, hence also A' is continuous and

$$\|A'\| \leqq \|A\|. \tag{16.9}$$

We shall see in §29 that we have even $\|A'\| = \|A\|$, and in §41 we shall find out that *only* continuous endomorphisms of E are conjugable with respect to the dual E'.

The continuous endomorphism A' of E', which for a continuous A is uniquely determined by

$$\langle Ax, x'\rangle = \langle x, A'x'\rangle \qquad \text{for all} \quad x \in E \qquad \text{and all} \quad x' \in E', \tag{16.10}$$

is called briefly the *operator dual* to A.

Example 16.3. A finite-dimensional endomorphism K of E, which with respect to a bilinear system (E, E^+) can be represented in the form

$$Kx = \sum_{\nu=1}^{n} \langle x, x_\nu^+\rangle x_\nu \qquad (x_\nu \in E, \, x_\nu^+ \in E^+), \tag{16.11}$$

is E^+-conjugable; the finite-dimensional operator $K^+ : E^+ \to E^+$, given by

$$K^+x^+ := \sum_{\nu=1}^{n} \langle x^+, x_\nu\rangle x_\nu^+ = \sum_{\nu=1}^{n} \langle x_\nu, x^+\rangle x_\nu^+ \tag{16.12}$$

is obviously conjugate to K.

Example 16.4. Let $k(s, t)$ be continuous on $a \leqq s, \, t \leqq b$ and define, as usual, the integral operator $K : C[a, b] \to C[a, b]$ belonging to this kernel by

$$(Kx)(s) := \int_a^b k(s, t)x(t)dt. \tag{16.13}$$

With the bilinear form

$$\langle x, x^+\rangle := \int_a^b x(t)x^+(t)dt \tag{16.14}$$

$(C[a, b], C[a, b])$ becomes a dual system, hence the conjugate K^+ is uniquely determined, if it exists. Because of

$$(16.15) \qquad \langle Kx, x^+\rangle = \int_a^b \left(\int_a^b k(s, t)x(t)dt\right)x^+(s)ds$$

$$= \int_a^b \left(x(t) \cdot \int_a^b k(s, t)x^+(s)ds\right)dt = \langle x, K^+x^+\rangle,$$

where, after an interchange of variables, $K^+ : C[a, b] \to C[a, b]$ is defined by

$$(16.16) \qquad (K^+x^+)(s) = \int_a^b k(t, s)x^+(t)dt,$$

the operator K is indeed conjugable. The reader should observe the analogy with the transpose matrix (see (16.4)).

Example 16.5. If (E, E^+) is an arbitrary bilinear system, then the identity map I^+ of E^+ is conjugate to the identity map I of E, and the zero map 0^+ of E^+ to the zero map 0 of E.

Example 16.6. If (E, E^+) is the trivial bilinear system of Example 14.1, then every endomorphism of E^+ is conjugate to every endomorphism of E.

If, given a bilinear system (E, E^+), the operators A^+, B^+ are conjugate to A, B, respectively, then obviously $A^+ + B^+$ is conjugate to $A + B$, αA^+ to αA and B^+A^+ to AB, so that the conjugable operators form an algebra. An equation of the form $(AB)^+ = B^+A^+$ is, however, meaningless, since $(AB)^+$ is not necessarily uniquely determined unless (E, E^+) is a left dual system. In this case if A is furthermore bijective and if its inverse is conjugable, then it follows from $AA^{-1} = A^{-1}A = I$ immediately that

$$(A^{-1})^+A^+ = A^+(A^{-1})^+ = I^+;$$

thus A^+ is also bijective by Proposition 5.2 and $(A^+)^{-1} = (A^{-1})^+$. We state these results as:

Proposition 16.3. *If (E, E^+) is a* left dual system*—which is, e.g., always the case if E^+ is a subspace of E^*—then the set of all conjugable operators forms an* algebra, *and the following rules of conjugations are valid*

$$(A + B)^+ = A^+ + B^+$$

$$(\alpha A)^+ = \alpha A^+$$

$$(AB)^+ = B^+A^+.$$

If A is bijective and A^{-1} too is conjugable, then also A^+ is bijective and

$$(A^+)^{-1} = (A^{-1})^+.$$

If (E, E^+) is a left dual system, then E^+ can be imbedded in the algebraic dual E^* of E, and $x^+ \in E^+$ can then be considered as the linear form on E defined by $x^+(x) := \langle x, x^+ \rangle$ (see §15). The following evident proposition is to be understood in the sense of this imbedding.

Proposition 16.4. *If* (E, E^+) *is a* left dual system, *then* $A \in \mathscr{S}(E)$ *is* E^+-*conjugable if and only if the algebraic dual operator* A^* *maps the space* E^+ *into itself. In this case* A^+ *is the restriction of* A^* *onto* E^+.

If a dual system is given, then the conjugable operator of finite rank can be characterized very simply. We have namely the following:

Proposition 16.5. *If* (E, E^+) *is a dual system, then the finite-dimensional endomorphism* K *of* E *is* E^+-*conjugable if and only if it can be represented in the form* (16.11).

Proof. If the operator K has the representation (16.11), then by Example 16.3 it is surely E^+-conjugable. Let us assume conversely that K is E^+-conjugable. Since the case $K = 0$ is trivial, we may assume that $K \neq 0$. Then K can be represented with the help of a basis $\{x_1, \ldots, x_n\}$ of $K(E)$ and of linear forms $x_1^*, \ldots, x_n^*$ from E^* in the form

$$Kx = \sum_{\nu=1}^{n} \langle x, x_\nu^* \rangle x_\nu .$$

The E^*-conjugate K^* of K is given by

$$K^*x^* = \sum_{\nu=1}^{n} \langle x_\nu, x^* \rangle x_\nu^*$$

(see Example 16.3). Since (E, E^+) is a dual system, for the linearly independent vectors $x_1, \ldots, x_n$ there exist, by Proposition 15.1, elements $y_1^+, \ldots, y_n^+$ of E^+ with $\langle x_\nu, y_\mu^+ \rangle = \delta_{\nu\mu}$. Consequently we have

$$K^*y_\mu^+ = \sum_{\nu=1}^{n} \langle x_\nu, y_\mu^+ \rangle x_\nu^* = x_\mu^* , \qquad \mu = 1, \ldots, n.$$

But since K^* maps E^+ into itself (Proposition 16.4), we obtain from this equation that all the x_μ^* lie already in E^+. Thus K is indeed representable in the form (16.11). ■

If a bilinear system (E, E^+) and an endomorphism A of E are given, then one cannot expect a solvability criterion of the kind of Proposition 16.2 even if A is conjugable. If e.g., A is continuous on the normed space E and A' the dual transformation from Example 16.2, then it is easy to see that such a criterion can be valid only if the image space $A(E)$ is *closed*. We shall see in §29 that in this case the criterion does indeed hold. The problem will therefore consist in finding, for a given bilinear system, classes of operators for which a solvability theorem of the kind we described can be proved. We shall consider such classes of operators in the next sections.

For the sake of simplicity we considered in this section only *self-maps* of a vector space. It is clear how the concept of a conjugate map has to be formulated for linear transformations $A: E \to F$:

If (E, E^+), (F, F^+) are bilinear systems, and if for the linear operator $A: E \to F$ there exists a linear operator $A^+: F^+ \to E^+$ such that

$$(16.17) \quad \langle Ax, y^+ \rangle = \langle x, A^+ y^+ \rangle \qquad \text{for all } x \text{ in } E \text{ and all } y^+ \text{ in } F^+,$$

then we say that A is *conjugable* and A^+ is called an *operator conjugate to A* or a *conjugate of A*.

If (E, E^+) *is a* left dual system, *then there can exist at most one conjugate operator* A^+ *to A, whose* linearity *does not have to be postulated explicitly* because it is guaranteed by (16.17).

It should be clear from Examples 16.1 and 16.2 how the algebraic dual of $A: E \to F$, and in case of a continuous A, the dual operator, are to be defined. The reader will easily provide the necessary modifications to Proposition 16.3, see Exercise 7.

Exercises

1. Let E be the vector space of all sequences $x = (\xi_1, \xi_2, \ldots)$, where only finitely many components are $\neq 0$, let $E^+ := l^2$ and $\langle x, x^+ \rangle := \sum_{n=1}^{\infty} \xi_n \xi_n^+$. With this bilinear form (E, E^+) is a dual system. Define $A \in \mathscr{S}(E)$ by $Ax := (\xi_1, 2\xi_2, 3\xi_3, \ldots)$ and show that A is not conjugable.

2. Under the hypotheses and with the notations of Exercise 1 let $B \in \mathscr{S}(E)$ be defined by $Bx := (\xi_1, \frac{1}{2}\xi_2, \frac{1}{3}\xi_3, \ldots)$. B is conjugable and bijective. However, $B^{-1} = A$ is not conjugable (see Proposition 16.3, cf. also Exercise 3).

3. $(\mathbf{K}^2, \mathbf{K}^3)$ with the bilinear form $\langle x, x^+ \rangle := \xi_1 \xi_1^+ + \xi_2 \xi_2^+$ is a right but not a left dual system. Every endomorphism A of $\mathbf{K}^2$ is conjugable and has infinitely many conjugates. Among these one can find, if A is bijective, bijective ones as well as not bijective ones (see Proposition 16.3, cf. also Exercise 2). *Hint*: Use the matrix representation of A.

$^+$4. If (E, E^+) is a bilinear system, then for every conjugable endomorphism A of E the conjugate operator A^+ is uniquely determined if and only if (E, E^+) is a left dual system. *Hint*: If (E, E^+) is not a left dual system, then

$$F^+ := \{x^+ \in E^+ : \langle x, x^+ \rangle = 0 \qquad \text{for all } x \text{ in } E\}$$

is a non-trivial subspace of E^+. The identity map I^+, as well as every projector of E^+ parallel to F^+, are conjugate to I.

5. If (E, E^+) is a bilinear system, then not every endomorphism of E^+ is necessarily conjugate to some endomorphism of E. *Hint*: Use the bilinear system from Exercise 3.

6. Let (E, E^+) be a bilinear system, A an endomorphism of E and A^+ conjugate to A. If the equation $Ax = y$ is solvable, then $\langle y, x^+ \rangle = 0$ for all x^+ in $N(A^+)$. This necessary condition is also sufficient if and only if for every $z \notin A(E)$ there exists an $x^+ \in N(A^+)$ such that $\langle z, x^+ \rangle \neq 0$. In particular, in the case of a left dual system it is sufficient if and only if for every $z \notin A(E)$ there exists an $x^+ \in E$ so that $\langle Ax, x^+ \rangle = 0$ for all $x \in E$ but $\langle z, x^+ \rangle \neq 0$ (cf. the proof of Proposition 16.2). These postulates of existence are postulates concerning the richness of E^+.

*7. Let (E, E^+) be a left dual system, (F, F^+) a bilinear system. The set of all linear maps $A: E \to F$ which possess conjugate transformations $A^+: F^+ \to E^+$ is a vector space, and in this vector space the following rules of conjugation are valid: $(A + B)^+ = A^+ + B^+$, $(\alpha A)^+ = \alpha A^+$. If (G, G^+) is a third bilinear system and if the linear maps $A: E \to F$ and $B: F \to G$ are conjugable, then also $BA: E \to G$ is conjugable and $(BA)^+ = A^+B^+$. Observe that the last equation is valid for every operator B^+ conjugate to B and investigate more closely how conjugates of a given operator differ from each other (use spaces like F^+ in Exercise 4).

$^+$8. Let the dual system $(C[a, b], C[a, b])$ with the bilinear form

$$\langle x, x^+ \rangle := \int_a^b x(t)x^+(t)\mathrm{d}t$$

be given. A finite-dimensional endomorphism of $C[a, b]$ is conjugable if and only if it is an integral transformation with degenerate kernel

$$\sum_{\nu=1}^{n} x_\nu(s)x_\nu^+(t) \qquad (x_\nu, x_\nu^+ \in C[a, b]).$$

§17. The equation $(I - K)x = y$ with finite-dimensional K

We observed in §9 that the Fredholm integral equation

$$(17.1) \quad x(s) - \int_a^b k(s, t)x(t)\mathrm{d}t = y(s) \qquad \text{with continuous kernel} \quad k(s, t)$$

can be solved in the following way if we already know that it is solvable for every $y \in C[a, b]$ and that the solution depends continuously on the right-hand side: One approximates $k(s, t)$ uniformly by a sequence of degenerate kernels and solves the integral equations corresponding to these kernels for the given y; the sequence of these solutions converges then to the solution of (17.1).

Our considerations following Example 13.2 show us that in the above assumptions we do not have to postulate explicitly the continuous dependence of the solution on the right-hand side: *It is enough to suppose that* (17.1) *is uniquely solvable for all* y *in* $C[a, b]$. At the end of §19 we shall see that we can drop even this requirement of unique solvability.

According to these remarks, it will be useful for the invcstigation of the integral equation (17.1) to study equations of the form $(I - K)x = y$ with a *finite-dimensional* operator K. We assume that $K \neq 0$ maps a vector space E into itself and that (E, E^+) is a bilinear system with the help of which K can be represented in the form

$$Kx = \sum_{i=1}^{n} \langle x, x_i^+ \rangle x_i, \tag{17.2}$$

where the vectors $x_1, \ldots, x_n$ from E and the vectors $x_1^+, \ldots, x_n^+$ from E^+ should be linearly independent. Such a bilinear system always exists, see (15.9). The equation

$$(I - K)x = y \tag{17.3}$$

can now be written in the form

$$x - \sum_{i=1}^{n} \langle x, x_i^+ \rangle x_i = y. \tag{17.4}$$

Every solution x of this equation has the form

$$x = y + \sum_{i=1}^{n} \xi_i x_i \tag{17.5}$$

and substituting this solution into (17.4) we get

$$\sum_{i=1}^{n} \left[\xi_i - \langle y, x_i^+ \rangle - \sum_{k=1}^{n} \xi_k \langle x_k, x_i^+ \rangle \right] x_i = 0. \tag{17.6}$$

With

$$\eta_i := \langle y, x_i^+ \rangle, \qquad \alpha_{ik} := \langle x_k, x_i^+ \rangle \tag{17.7}$$

it follows from (17.6) because of the linear independence of $x_1, \ldots, x_n$ that the coefficients $\xi_1, \ldots, \xi_n$ satisfy the system of equations

$$\xi_i - \sum_{k=1}^{n} \alpha_{ik} \xi_k = \eta_i \qquad (i = 1, \ldots, n). \tag{17.8}$$

Conversely, if $(\xi_1, \ldots, \xi_n)$ is a solution of this system with the right-hand side $\eta_i := \langle y, x_i^+ \rangle$ $(i = 1, \ldots, n)$, and if we define x by (17.5), then x solves obviously the equation (17.4). We record this result:

(17.9) $\quad x = y + \sum_{i=1}^{n} \xi_i x_i$ solves (17.4) $\Leftrightarrow$ $(\xi_1, \ldots, \xi_n)$ solves (17.8) with

$$\eta_i := \langle y, x_i^+ \rangle.$$

The solution of equation (17.4) is thus indeed, as it was already indicated in §9, possible by *elementary algebraic* methods.

If we consider the homogeneous problems

$$(17.10)\qquad (I - K)x = x - \sum_{i=1}^{n} \langle x, x_i^+ \rangle x_i = 0,$$

$$(17.11)\qquad \xi_i - \sum_{k=1}^{n} \alpha_{ik}\xi_k = 0 \qquad (i = 1, \ldots, n),$$

then we get from (17.9):

$$(17.12)\quad x = \sum_{i=1}^{n} \xi_i x_i \text{ solves } (17.10) \Leftrightarrow (\xi_1, \ldots, \xi_n) \text{ solves } (17.11).$$

Since vectors of the form $z_\mu = \sum_{i=1}^{n} \xi_i^{(\mu)} x_i$ ($\mu = 1, \ldots, m$) are linearly independent if and only if this is true for the coefficient vectors $(\xi_1^{(\mu)}, \ldots, \xi_n^{(\mu)})$, and since, as known, the maximal number of linearly independent solutions of (17.11) is given by $n - \operatorname{rank}(\delta_{ik} - \alpha_{ik})$, it follows immediately from (17.12) that

$$(17.13)\qquad \dim N(K - I) = n - \operatorname{rank}(\delta_{ik} - \alpha_{ik}).$$

For a deeper investigation of the image space of $I - K$ we avail ourselves, according to our program in §16, of the operator $K^+ : E^+ \to E^+$ which is given by

$$(17.14)\qquad K^+x^+ := \sum_{i=1}^{n} \langle x^+, x_i \rangle x_i^+ \qquad \text{with} \quad \langle x^+, x_i \rangle := \langle x_i, x^+ \rangle,$$

and is conjugate to K (see Example 16.3). Let I^+ be the identity transformation on E^+. Since K^+ is represented with respect to the bilinear system (E^+, E) exactly in the same way as K with respect to (E, E^+), we can apply directly the results obtained for (17.3). To the equation

$$(17.15)\qquad (I^+ - K^+)x^+ = x^+ - \sum_{i=1}^{n} \langle x^+, x_i \rangle x_i^+ = y^+$$

there corresponds, because of $\langle x_k^+, x_i \rangle = \langle x_i, x_k^+ \rangle = \alpha_{ki}$ (cf. (17.7)), the system of equations

$$(17.16)\quad \xi_i^+ - \sum_{k=1}^{n} \alpha_{ki}\xi_k^+ = \eta_i^+ \; (i = 1, \ldots, n) \quad \text{with } \eta_i^+ = \langle y^+, x_i \rangle,$$

to the homogeneous equation

$$(17.17)\qquad (I^+ - K^+)x^+ = x^+ - \sum_{i=1}^{n} \langle x^+, x_i \rangle x_i^+ = 0$$

the homogeneous system

$$(17.18)\qquad \xi_i^+ - \sum_{k=1}^{n} \alpha_{ki}\xi_k^+ = 0 \qquad (i = 1, \ldots, n),$$

and the following assertions are true: Every solution x^+ of (17.15) has the form $x^+ = y^+ + \sum_{i=1}^n \xi_i^+ x_i^+$, and

$$(17.19)\quad x^+ = y^+ + \sum_{i=1}^n \xi_i^+ x_i^+ \text{ solves (17.15)} \Leftrightarrow (\xi_1^+, \dots, \xi_n^+) \text{ solves (17.16)},$$

$$(17.20)\quad x^+ = \sum_{i=1}^n \xi_i^+ x_i^+ \text{ solves (17.17)} \Leftrightarrow (\xi_1^+, \dots, \xi_n^+) \text{ solves (17.18)},$$

$$(17.21)\quad \dim N(I^+ - K^+) = n - \operatorname{rank}(\delta_{ki} - \alpha_{ki}).$$

Since the rank of a matrix coincides with the rank of its transpose, it follows from (17.13) and (17.21) that

$$(17.22)\quad \dim N(I - K) = \dim N(I^+ - K^+).$$

For every y in E and every solution $x^+ = \sum_{i=1}^n \xi_i^+ x_i^+$ of equation (17.17) we have

$$\langle y, x^+ \rangle = \sum_{i=1}^n \xi_i^+ \langle y, x_i^+ \rangle = \sum_{i=1}^n \xi_i^+ \eta_i \qquad \text{with} \quad \eta_i := \langle y, x_i^+ \rangle.$$

If we take this equation into account and apply the criterion for the solvability of systems of linear equations, which we indicated at the beginning of §16, then from (17.9) and (17.20) the assertion

$$(17.23)\quad (I - K)x = y \text{ is solvable} \Leftrightarrow \langle y, x^+ \rangle = 0 \text{ for all } x^+ \text{ in } N(I^+ - K^+)$$

follows. Analogously we have because of (17.19) and (17.10)

$$(17.24)\quad (I^+ - K^+)x^+ = y^+ \text{ is solvable} \Leftrightarrow \langle y^+, x \rangle = 0 \text{ for all } x \text{ in } N(I - K).$$

Herewith we obtained the required description of the image space of $I - K$ and in addition also that of the image space of $I^+ - K^+$. We can even express numerically, by means of the codimension, how strongly these image spaces differ from E and E^+, respectively. If $m := \dim N(I^+ - K^+) = 0$, then because of (17.23) obviously $(I - K)(E) = E$, and so also $\operatorname{codim}(I - K)(E) = 0$. If $m > 0$ and $\{z_1^+, \dots, z_m^+\}$ is a basis of $N(I^+ - K^+)$, then according to (17.23) the vector y lies in $(I - K)(E)$ if and only if $\langle y, z_\mu^+ \rangle = 0$ for all μ. Thus if we want to find elements z which do not lie in $(I - K)(E)$, we must look for such z for which $\langle z, z_\mu^+ \rangle \neq 0$ for at least one μ. If now (E, E^+) is a left dual system, then there exist m linearly independent vectors $z_1, \dots, z_m$ in E with

$$(17.25)\quad \langle z_\nu, z_\mu^+ \rangle = \delta_{\nu\mu} \qquad \text{for} \quad \nu, \mu = 1, \dots, m$$

(Proposition 15.1). If a linear combination $z := \alpha_1 z_1 + \cdots + \alpha_m z_m$ of these vectors lies in $(I - K)(E)$ then according to (17.23) $\alpha_\mu = \langle z, z_\mu^+ \rangle = 0$ for all μ, hence $[z_1, \dots, z_m] \cap (I - K)(E) = \{0\}$. Again one sees with (17.23) that for every vector x in E the vector $y_x := x - \sum_{\nu=1}^m \langle x, z_\nu^+ \rangle z_\nu$ lies in $(I - K)(E)$, since for all μ we have

$$\langle y_x, z_\mu^+ \rangle = \langle x, z_\mu^+ \rangle - \sum_{\nu=1}^m \langle x, z_\nu^+ \rangle \langle z_\nu, z_\mu^+ \rangle = \langle x, z_\mu^+ \rangle - \langle x, z_\mu^+ \rangle = 0.$$

Thus every $x \in E$ can be represented in the form

$$x = \alpha_1 z_1 + \cdots + \alpha_m z_m + y_x \qquad \text{with} \quad y_x \in (I - K)(E).$$

In conclusion we have therefore

$$E = [z_1, \ldots, z_m] \oplus (I - K)(E)$$

and so

$$\text{codim}(I - K)(E) = m = \dim N(I^+ - K^+), \tag{17.26}$$

with which we have expressed numerically the deviation of the image space $(I - K)(E)$ from the whole space E—in the case that (E, E^+) is a left dual system.

But even then, when (E, E^+) is not a left dual system, we can determine elements $z_1, \ldots, z_m$ so that (17.25) holds, and that is sufficient to ensure (17.26). With the vectors $x_1, \ldots, x_n$ from (17.2) we write z_ν tentatively in the form

$$z_\nu = \sum_{\lambda=1}^{n} \beta_{\nu\lambda} x_\lambda \qquad (\nu = 1, \ldots, m),$$

and need to choose the $\beta_{\nu\lambda}$ for a fixed ν so that

$$\langle z_\nu, z_\mu^+ \rangle = \sum_{\lambda=1}^{n} \beta_{\nu\lambda} \langle x_\lambda, z_\mu^+ \rangle = \delta_{\nu\mu} \qquad \text{for} \quad \mu = 1, \ldots, m.$$

This is possible because the m rows of the matrix

$$\begin{pmatrix} \langle x_1, z_1^+ \rangle & \langle x_2, z_1^+ \rangle & \cdots & \langle x_n, z_1^+ \rangle \\ \cdots & \cdots & \cdots & \cdots \\ \langle x_1, z_m^+ \rangle & \langle x_2, z_m^+ \rangle & \cdots & \langle x_n, z_m^+ \rangle \end{pmatrix}$$

of the system are linearly independent, i.e., the matrix has maximal rank. Indeed,

$$\alpha_1(\langle x_1, z_1^+ \rangle, \ldots, \langle x_n, z_1^+ \rangle) + \cdots + \alpha_m(\langle x_1, z_m^+ \rangle, \ldots, \langle x_n, z_m^+ \rangle) = 0$$

implies

$$\left\langle x_\nu, \sum_{\mu=1}^{m} \alpha_\mu z_\mu^+ \right\rangle = \sum_{\mu=1}^{m} \alpha_\mu \langle x_\nu, z_\mu^+ \rangle = 0 \qquad \text{for} \quad \nu = 1, \ldots, n,$$

hence by the definition of K^+ in (17.14) we have

$$K^+\left(\sum_{\mu=1}^{m} \alpha_\mu z_\mu^+ \right) = \sum_{\nu=1}^{n} \left\langle x_\nu, \sum_{\mu=1}^{m} \alpha_\mu z_\mu^+ \right\rangle x_\nu^+ = 0;$$

but since also $(I^+ - K^+)(\sum_{\mu=1}^{m} \alpha_\mu z_\mu^+) = 0$, it follows that $\sum_{\mu=1}^{m} \alpha_\mu z_\mu^+ = 0$ and so $\alpha_1 = \cdots = \alpha_m = 0$.

Thus (17.26) is valid if (E, E^+) is only a bilinear system. For reasons of symmetry we have therefore

$$\operatorname{codim}(I^+ - K^+)(E^+) = \dim N(I - K),$$

and with (17.22) we obtain finally

$$\begin{aligned}\dim N(I - K) &= \operatorname{codim}(I - K)(E) = \dim N(I^+ - K^+)\\ &= \operatorname{codim}(I^+ - K^+)(E^+) < \infty.\end{aligned}$$

Before we summarize these results, we introduce two more abbreviations: For a linear map $A: E \to F$ we set

$$\alpha(A) := \dim N(A) \quad \text{and} \quad \beta(A) := \operatorname{codim} A(E).$$

Proposition 17.1. *Let $K: E \to E$ be an operator of finite rank and let it be represented with the help of a bilinear system (E, E^+) in the form*

$$Kx = \sum_{i=1}^{n} \langle x, x_i^+ \rangle x_i,$$

where the vectors $x_1, \ldots, x_n$ from E and the vectors $x_1^+, \ldots, x_n^+$ from E^+ are linearly independent (such a representation is always *possible, it is sufficient to choose $E^+ = E^*$). Let the operator $K^+: E^+ \to E^+$ be given by*

$$K^+x^+ = \sum_{i=1}^{n} \langle x^+, x_i \rangle x_i^+$$

(see Example 16.3) *and let I^+ be the identity transformation of E^+. Then we have*

$$\alpha(I - K) = \beta(I - K) = \alpha(I^+ - K^+) = \beta(I^+ - K^+) < \infty, \tag{17.27}$$

in particular, the operators $I - K$ and $I^+ - K^+$ are already bijective *if only they are* injective *or* surjective. *Furthermore*:

$$(I - K)x = y \text{ is solvable } \Leftrightarrow \langle y, x^+ \rangle = 0 \text{ for all } x^+ \in N(I^+ - K^+),$$

$$(I^+ - K^+)x^+ = y^+ \text{ is solvable} \Leftrightarrow \langle x, y^+ \rangle = 0 \text{ for all } x \in N(I - K).$$

§18. The equation $(R - S)x = y$ with a bijective R and finite-dimensional S

We consider again the integral equation (17.1) and the corresponding integral operator $K: C[a, b] \to C[a, b]$ which is defined by $K(x)(s) := \int_a^b k(s, t)x(t)\mathrm{d}t$. In the preceding section we have used the uniform approximability of the kernel $k(s, t)$ by degenerate kernels to develop a method for the solution of (17.1); now we ask whether the solvability solution of integral equations with degenerate kernel, described by Proposition 17.1, carries over to equation (17.1) by means of this approximability. From the consideration of convergence following (9.7) we obtain that for K there exists a finite-dimensional, continuous self-map S of $C[a, b]$ such that $\|K - S\| < 1$; by Theorem 8.1 the map $R := I - (K - S)$ is bijective and because of $I - K = I - (K - S) - S = R - S$, the integral equation (17.1) is changed into an equation of the kind named in the title. Therefore we examine in general the behavior of solutions of the equation

$$(R - S)x = y,$$

where R is a *bijective* and S a *finite-dimensional* linear map of an arbitrary vector space E. To abbreviate, we set $A := R - S$, then we have

(18.1) $\qquad A = R(I - R^{-1}S) \qquad$ with a finite-dimensional $R^{-1}S$,

hence $N(A) = N(I - R^{-1}S)$ and thus

(18.2) $$\alpha(A) = \alpha(I - R^{-1}S).$$

Also the equation

(18.3) $$\beta(A) = \beta(I - R^{-1}S)$$

is easy to prove: It is trivial if $m := \beta(I - R^{-1}S)$ vanishes; if $m \neq 0$ then there exist m linearly independent elements $z_1, \ldots, z_m$ in E (observe that m is finite according to Proposition 17.1), which generate a subspace $[z_1, \ldots, z_m]$ complementary to $(I - R^{-1}S)(E)$. It follows from the bijectivity of R that

$$[Rz_1, \ldots, Rz_m]$$

is an m-dimensional complement of $A(E)$, hence (18.3) is valid also in this case. From (18.2), (18.3) the important assertion

(18.4) $$\alpha(A) = \beta(A) < \infty$$

follows with the help of (17.27), which tells us in particular that the operator A is already bijective if only it is injective or surjective.

We want to study now the image space of A with the help of an operator conjugate to A. In order to be able to apply the conjugation rules of Proposition 16.3, we consider a *left dual system* (E, E^+) with the bilinear form $\langle x, x^+ \rangle$ and make, furthermore, the following *assumptions*:

(a) let S have a representation $Sx = \sum_{i=1}^{n} \langle x, x_i^+ \rangle x_i$,
(b) let R and R^{-1} be E^+-conjugable.

There exists at least one left dual system (E, E^+) with respect to which (a) and (b) are fulfilled; it is enough to choose $E^+ = E^*$ (cf. Exercise 2 in §16 for the conjugability of R^{-1}). Since S is E^+-conjugable (see Example 16.3) by Proposition 16.3 also $A = R - S$ is E^+-conjugable and $A^+ = R^+ - S^+$, furthermore by the same proposition R^+ is bijective and $(R^+)^{-1} = (R^{-1})^+$. Because of (a) we have $R^{-1}Sx = \sum_{i=1}^{n} \langle x, x_i^+ \rangle R^{-1}x_i$, so that the solvability criterion of Proposition 17.1 can be applied to the equation $(I - R^{-1}S)x = y$.

With its help we have the following chain of assertions:

$$
\begin{aligned}
Ax = y \text{ is solvable} &\Leftrightarrow (I - R^{-1}S)x = R^{-1}y \text{ is solvable}\\
&\Leftrightarrow \langle R^{-1}y, z^+\rangle = 0 \text{ for all } z^+ \text{ with } (I^+ - (R^{-1}S)^+)z^+ = 0\\
&\Leftrightarrow \langle R^{-1}y, R^+x^+\rangle = 0 \text{ for all } x^+ \text{ with } (R^+ - S^+)x^+\\
&\qquad = (I^+ - S^+(R^+)^{-1})R^+x^+ = (I^+ - (R^{-1}S)^+)R^+x^+ = 0\\
&\Leftrightarrow \langle y, x^+\rangle = 0 = \langle y, (R^{-1})^+R^+x^+\rangle\\
&\qquad\qquad = 0 \text{ for all } x^+ \in N(A^+).
\end{aligned}
$$

Thus we have for the equation $Ax = y$ the solvability criterion of Proposition 17.1. One sees in an entirely similar way that the equation $A^+x^+ = y^+$ is solvable if and only if $\langle x, y^+\rangle = 0$ for all $x \in N(A)$.

Since, finally, $A^+ = R^+ - S^+$ has exactly the same form as $A = R - S$ (namely a bijective operator minus a finite-dimensional operator) (18.4) is valid for also A^+ instead of A:

$$\alpha(A^+) = \beta(A^+) < \infty, \tag{18.5}$$

furthermore, because of the bijectivity of R^+

$$
\begin{aligned}
A^+(E^+) &= (R(I - R^{-1}S))^+(E^+) = (I^+ - (R^{-1}S)^+)R^+(E^+)\\
&= (I^+ - (R^{-1}S)^+)(E^+);
\end{aligned}
$$

with Proposition 17.1 and (18.2) we obtain from here the equation $\beta(A^+) = \beta(I^+ - (R^{-1}S)^+) = \alpha(I - R^{-1}S) = \alpha(A)$, and from here again it follows with the help of (18.4) and (18.5) that

$$\beta(A) = \alpha(A) = \beta(A^+) = \alpha(A^+) < \infty,$$

i.e., precisely assertion (17.27) for A instead of $I - K$.

We summarize what we have proved so far:

Theorem 18.1. *Let the endomorphism A of E have the form $A = R - S$, where R is a* bijective *and S a* finite-dimensional *operator. Let (E, E^+) be a* left dual system *with respect to which S can be represented in the form*

$$Sx = \sum_{i=1}^{n} \langle x, x_i^+\rangle x_i \tag{18.6}$$

and both R and R^{-1} are conjugable (there always exists such a left dual system, it is sufficient to choose $E^+ = E^*$). Then A is conjugable and

$$\alpha(A) = \beta(A) = \alpha(A^+) = \beta(A^+) < \infty, \tag{18.7}$$

in particular, A and A^+ are already bijective *if only* one *of them is* injective *or* surjective. *Furthermore, the following holds:*

$$
\begin{aligned}
Ax = y \textit{ is solvable} &\Leftrightarrow \langle y, x^+\rangle = 0 \textit{ for all } x^+ \in N(A^+),\\
A^+x^+ = y^+ \textit{ is solvable} &\Leftrightarrow \langle x, y^+\rangle = 0 \textit{ for all } x \in N(A).
\end{aligned}
$$

We shall see in §20 how one can get rid of the hypothesis that (E, E^+) is a left dual system.

We apply now the above theorem to the case of a *normed* space E with the dual E'. According to Example 16.2 every continuous endomorphism of E is conjugable with respect to the left dual system (E, E'), furthermore every continuous operator of finite rank on E can be represented in the form (18.6) with x_i^+ in E' (Proposition 13.4). With these remarks we get from Theorem 18.1:

Proposition 18.1. *Let E be a normed space, and let the linear self-map A of E have the form $A = R - S$, where R and R^{-1} are in $\mathscr{L}(E)$ and S in $\mathscr{F}(E)$. Then the assertions of Theorem* 18.1 *are valid for A, where we have to set $E^+ = E'$ and $A^+ = A'$.*

Exercise

Let K be a compact self-map of the normed space E. If there exists an $S \in \mathscr{F}(E)$ such that $\|K - S\| < 1$, then for $A = I - K$ the assertions of Theorem 18.1 are valid with the left dual system (E, E').

§19. The Fredholm integral equation with continuous kernel

We started §18 with the question concerning the behavior of the solutions of the Fredholm integral equation

$$x(s) - \int_a^b k(s, t)x(t)dt = y(s) \quad \text{or} \quad (I - K)x = y, \tag{19.1}$$

where the kernel $k(s, t)$ is continuous on the square $[a, b] \times [a, b]$ and the function y is continuous on the interval $[a, b]$; we look for solutions x in $C[a, b]$. We return now to this question. Since we are concerned with the study of continuous operators on the normed space $C[a, b]$, it would seem reasonable to apply Proposition 18.1. For this we would have, however, to know the dual of $C[a, b]$, which we will have at our disposal only in §35 after rather laborious considerations. But Example 16.4 suggests to use the dual system $(C[a, b], C[a, b])$ with the bilinear form

$$\langle x, x^+ \rangle := \int_a^b x(t)x^+(t)dt$$

for our investigations—and this we shall do because the integral operator K with continuous kernel $k(s, t)$ is conjugable with respect to this dual system, and the conjugate operator K^+ is again an integral operator with the kernel $k(t, s)$.

According to the considerations concerning approximation after (9.7), there exists to the integral operator K a (finite-dimensional) integral operator S with degenerate kernel $\sum_{i=1}^n x_i(s)x_i^+(t)$ such that

$$Sx = \sum_{i=1}^n \langle x, x_i^+ \rangle x_i \tag{19.2}$$

and

$$(19.3)\qquad \|K - S\| \leqq (b-a)\max_{s,t}\left|k(s,t) - \sum_{i=1}^{n} x_i(s)x_i^+(t)\right| < 1.$$

By Theorem 8.1 the map $R := I - (K - S)$ is bijective, and $I - K = I - (K - S) - S = R - S$ has the form of the operators considered in Theorem 18.1. We have to check whether the further hypotheses of this theorem are fulfilled.

As we already observed above, K^+ exists, and with it also

$$R^+ = (I - (K - S))^+ = I^+ - K^+ + S^+.$$

We now investigate whether also R^{-1} is conjugable. Setting $H := K - S$ we obtain, because of $\|H\| < 1$, from Theorem 8.1

$$R^{-1} = (I - H)^{-1} = \sum_{\nu=0}^{\infty} H^\nu.$$

Since, furthermore, $H^+ = K^+ - S^+$ is an integral operator with the kernel $k(t,s) - \sum_{i=1}^{n} x_i(t)x_i^+(s)$, according to (19.3) the estimate

$$\|H^+\| \leqq (b-a)\max_{t,s}\left|k(t,s) - \sum_{i=1}^{n} x_i(t)x_i^+(s)\right| < 1$$

is valid, hence—again by Theorem 8.1—$R^+ = I^+ - H^+$ is bijective and

$$(R^+)^{-1} = (I^+ - H^+)^{-1} = \sum_{\nu=0}^{\infty} (H^+)^\nu.$$

If we set $S_k := \sum_{\nu=0}^{k} H^\nu$, then by Proposition 16.3 we have $S_k^+ = \sum_{\nu=0}^{k} (H^+)^\nu$ and, furthermore, we have the limits

$$S_k \Rightarrow R^{-1} \quad\text{and}\quad S_k^+ \Rightarrow (R^+)^{-1}.$$

Therefore from $\langle S_k x, x^+\rangle = \langle x, S_k^+ x^+\rangle$ the equation $\langle R^{-1}x, x^+\rangle = \langle x, (R^+)^{-1}x^+\rangle$ follows for all x, x^+ in $C[a,b]$ as $k \to \infty$. But this means that R^{-1} is conjugable. Thus all hypotheses of Theorem 18.1 are satisfied and we can state the following theorem, called thc *Fredholm alternative*:

Theorem 19.1. *Let the function $k(s,t)$ be continuous in $a \leqq s, t \leqq b$. Then for the integral equations*

$$\text{(I)}\qquad x(s) - \int_a^b k(s,t)x(t)\,dt = y(s),$$

$$\text{(I}^+\text{)}\qquad x^+(s) - \int_a^b k(t,s)x^+(t)\,dt = y^+(s)$$

in $C[a,b]$ the following alternative *holds*:

Either both *equations have for all y, y^+ a* unique *solution, or* none *of the equations is solvable for* every *right-hand side; in the second case, the solution, if it exists at* *all, is not determined uniquely, i.e., the homogeneous equations*

$$\text{(H)} \qquad x(s) - \int_a^b k(s, t)x(t)dt = 0,$$

$$\text{(H}^+\text{)} \qquad x^+(s) - \int_a^b k(t, s)x^+(t)dt = 0$$

have non-trivial *solutions; the maximal numbers of linearly independent solutions of these homogeneous equations are finite and equal. Finally, the following solvability criterion is valid:*

$$\text{(I) } \textit{is solvable} \Leftrightarrow \int_a^b y(t)x^+(t)dt = 0 \textit{ for all solutions } x^+ \textit{ of } (H^+),$$

$$\text{(I}^+\text{) } \textit{is solvable} \Leftrightarrow \int_a^b y^+(t)x(t)dt = 0 \textit{ for all solutions } x \textit{ of } (H).$$

At the beginning of §17 we described a method for the solution of integral equation (I), which can always be used when (I) has a *unique* solution for *all* $y \in C[a, b]$: One approximates the kernel $k(s, t)$ uniformly by a sequence of degenerate kernels and solves the integral equations corresponding to these kernels for the given right-hand side by the algebraic method of §17; the sequence of solutions so obtained will then converge to the solution of (I).

Because of Theorem 19.1 we can say: *The method of solution just described can be applied if only* (I) *is* solvable *for all* $y \in C[a, b]$ *or if* (H) *has only the* trivial *solution.*

§20. Quotient spaces

Theorem 18.1 was based on the hypothesis that we have a left dual system. If we have initially only a bilinear system with respect to which—with the notations of the theorem just named—S can be represented in the form (18.6) and both R and R^{-1} are conjugable, then, as we shall see in the present section, one can always obtain easily a left dual system which satisfies the requirements of our theorem.

In order to keep the following formulas as simple as possible, we denote exceptionally the bilinear system by (E, F) instead of (E, E^+); let the corresponding bilinear form be $\langle x, y\rangle$. Decisive is the observation that in the representation (18.6) of the operator S, which we write now in the form $Sx = \sum_{i=1}^n \langle x, y_i\rangle x_i$, the y_i can be clearly replaced by any $z_i \in F$ which satisfies

$$\langle x, y_i\rangle = \langle x, z_i\rangle \qquad \text{for all} \quad x \in E.$$

If we call two elements y, z in F *equivalued*, in symbol: $y \sim z$, if $\langle x, y\rangle = \langle x, z\rangle$ for any $x \in E$, then we can say briefly: In the above representation of S every y_i may be replaced by an equivalued element z_i. Clearly $y \sim z$ holds if and only if $y - z$ lies in the linear subspace

$$\text{(20.1)} \qquad N := \{u \in F : \langle x, u\rangle = 0 \text{ for all } x \in E\}.$$

It follows from the linearity properties of N alone that the relation $x \sim y$ is reflexive, symmetric and transitive, i.e., an *equivalence relation.* It partitions F into pairwise disjoint classes of equivalent elements (*equivalence classes* or *residue classes*). These equivalence classes are precisely the sets of the form $y + N$. We denote the set of all equivalence classes by F/N, the equivalence class of y (i.e., the equivalence class which contains y) by $\hat{y}$; thus $\hat{y} = y + N$, and we have $\hat{y} = \hat{z}$ if and only if $y - z$ lies in N. Since by definition $\langle x, y\rangle$ does not change its value when y runs through a residue class, we can define a scalar-valued function on F/N by

$$(20.2) \qquad [x, \hat{y}] := \langle x, z\rangle \qquad \text{with an arbitrary} \quad z \in \hat{y}.$$

If we have $[x, \hat{y}] = [x, \hat{z}]$ for all $x \in E$, then obviously $\hat{y} = \hat{z}$. If we succeed in making a vector space out of F/N so that $[x, \hat{y}]$ becomes a bilinear form on $E \times (F/N)$, then $(E, F/N)$ is even a left dual system. For this purpose we have to define the sum $\hat{y} + \hat{z}$ so that for all $x \in E$ the following equation is valid:

$$[x, \hat{y} + \hat{z}] = [x, \hat{y}] + [x, \hat{z}] = \langle x, y\rangle + \langle x, z\rangle = \langle x, y + z\rangle = [x, \widehat{y + z}],$$

hence the sum must be defined by

$$(20.3) \qquad \hat{y} + \hat{z} = \widehat{y + z} \qquad \text{with} \quad y \in \hat{y}, z \in \hat{z}.$$

The choice of representatives plays no role here: if also $y_1 \in \hat{y}$ and $z_1 \in \hat{z}$, i.e., $y_1 = y + u$ and $z_1 = z + v$ with $u, v \in N$, then $(y_1 + z_1) - (y + z) = u + v \in N$, and so $\widehat{y_1 + z_1} = \widehat{y + z}$. One sees in an entirely analogous way that we must define the product $\alpha\hat{y}$ by the equation

$$(20.4) \qquad \alpha\hat{y} := \widehat{\alpha y} \qquad \text{with} \quad y \in \hat{y}$$

if we want to have $[x, \alpha\hat{y}] = \alpha[x, \hat{y}]$. This definition too is independent of the choice of the representative: if also $y_1 \in \hat{y}$, i.e., $y_1 = y + u$ with $u \in N$, then $\alpha y_1 - \alpha y = \alpha(y_1 - y) = \alpha u \in N$, and so $\widehat{\alpha y_1} = \widehat{\alpha y}$. Observe that the independence of the definitions (20.3) and (20.4) from the choice of representatives is based on the fact that N is a linear subspace of F. With the addition and scalar multiplication so introduced F/N becomes indeed a vector space (over the scalar field of F); the zero element $\hat{0}$ of F/N is the equivalence class N. By (20.2) a bilinear form is defined on $E \times (F/N)$ for which $(E, F/N)$ is indeed a left dual system. We shall see in a moment that this left dual system satisfies the requirements, but first we want to extract the essence of our considerations which is independent from bilinear systems. The reader will remember that the partitioning into equivalence classes of the space F and the introduction of a structure of vector space by the definitions (20.3) and (20.4) were based alone on the linearity properties of N—not, however, on the definition (20.1) of N. Consequently we may state the following proposition:

Proposition 20.1. *If N is an arbitrary linear subspace of the vector space F, then the set $F/N = \{\hat{y} = y + N; y \in F\}$ of equivalence classes becomes a vector space over the field of scalars of F by the definitions*

$$\hat{y} + \hat{z} := \widehat{y + z}, \qquad \alpha\hat{y} := \widehat{\alpha y} \qquad \text{with} \quad y \in \hat{y}, z \in \hat{z}.$$

The vector space F/N is called the *quotient space* of F modulo N. The map $h: F \to F/N$ defined by $h(y) := \hat{y}$ is linear and surjective; it is called the *canonical homomorphism* from F onto F/N.

We now return to the initial situation: (E, F) is a bilinear system with the bilinear form $\langle x, y\rangle$, N is given by (20.1), and $(E, F/N)$ is a left dual system with respect to the bilinear form defined in (20.2). Let us assume that $A \in \mathscr{S}(E)$ is F-conjugable and let $B \in \mathscr{S}(F)$ be an operator conjugate to A, that is

$$\langle Ax, y\rangle = \langle x, By\rangle \qquad \text{for all} \quad x \in E \qquad \text{and all} \quad y \in F.$$

For $u \in N$ we have $\langle x, Bu\rangle = \langle Ax, u\rangle = 0$ for every $x \in E$, hence Bu also lies in N. Thus if y_1, y_2 are two representatives of the equivalence class $\hat{y}$, i.e., if $y_2 = y_1 + u$ with a $u \in N$, then $\widehat{By_2} = \widehat{B(y_1 + u)} = \widehat{By_1 + Bu} = \widehat{By_1} + \widehat{Bu} = \widehat{By_1}$. Therefore we can define uniquely a map $\hat{B}: F/N \to F/N$ by

$$\hat{B}\hat{y} := \widehat{By} \qquad \text{with} \quad y \in \hat{y}.$$

We have $[Ax, \hat{y}] = \langle Ax, y\rangle = \langle x, By\rangle = [x, \widehat{By}] = [x, \hat{B}\hat{y}]$ for all $x \in E$ and all $\hat{y} \in F/N$; thus A is conjugable also with respect to the left dual system $(E, F/N)$, and $\hat{B}$ is the operator conjugate to A (for the unique determination and the linearity of $\hat{B}$ see the end of §16).

We now modify the assumptions in Theorem 18.1: We suppose that (E, E^+) is not a left dual system but only a bilinear system. With

$$N := \{x^+ \in E^+ : \langle x, x^+\rangle = 0 \qquad \text{for all} \quad x \in E\}$$

we then construct the left dual system $(E, E^+/N)$ with the bilinear form defined by $[x, \widehat{x^+}] := \langle x, x^+\rangle$, $x^+ \in \widehat{x^+}$. According to the above results, and using the notation of Theorem 18.1, the operators R and R^{-1} are conjugable with respect to this left dual system, and because of (18.6) also S is representable in the form $Sx = \sum_{i=1}^{n} [x, \widehat{x_i^+}]x_i$. It follows that *the assertions of Theorem* 18.1 *are valid if one replaces the E^+-conjugate operator A^+ by the corresponding (E^+/N)-conjugate operator $\widehat{A^+}$ and uses in the criterion for solvability the bilinear form* $[x, \widehat{x^+}]$ *instead of* $\langle x, x^+\rangle$. This means that also in this case the behavior of solutions of the equation $Ax = y$ can be described exactly.

The construction of a quotient space can be used, among others, to associate with a linear operator $A: E \to F$ a map which possesses the essential properties of A and which is furthermore injective. This is the *canonical injection*

$$\hat{A}: E/N(A) \to F$$

which is uniquely defined by $\hat{A}\hat{x} := Ax$, $x \in \hat{x}$. It is easy to see that it is linear and that its image space coincides with $A(E)$. To pass from A to $\hat{A}$ means to collect together in a class, or to identify, those elements of E which are mapped by A onto the same vector of F.

From the given properties of the canonical injection one obtains immediately the following *homomorphism theorem*:

Proposition 20.2. *If the map $A: E \to F$ is linear, then the vector spaces $E/N(A)$ and $A(E)$ are* isomorphic.

Exercises

1. Obtain from a bilinear system, by the method of constructing a quotient space, a right dual system and a dual system.

*2. If $A: E \to F$ is linear and $E = N(A) \oplus U$ (cf. the proof of Proposition 5.3), then the vector spaces $E/N(A)$ and U are isomorphic.

*3. If $E = F \oplus G$, then the vector spaces E/F and G are isomorphic. Therefore codim $F = \dim E/F$. *Hint*: Apply Exercise 2 to the canonical homomorphism $h: E \to E/F$ and use Exercise 4 in §5. Observe that hereby Proposition 4.2 is proved again.

*4. If f is a non-zero linear form on the vector space E over $\mathbf{K}$, then $\dim E/N(f) = 1$, and there exists an $x_0 \in E$ so that $E = [x_0] \oplus N(f)$ (see Exercise 3).

§21. The quotient norm

If $A: E \to F$ is a continuous linear map of the normed spaces E and F, then one will ask whether one can introduce a norm on the quotient space $E/N(A)$ so that also the canonical injection $\hat{A}: E/N(A) \to F$ is continuous. Since for all representatives x of the residue class $\hat{x}$ the estimate $\|\hat{A}\hat{x}\| = \|Ax\| \leqq \|A\| \|x\|$ is valid, we also have the inequality $\|\hat{A}\hat{x}\| \leqq \|A\| \cdot \inf_{x \in \hat{x}} \|x\|$. From it we see that $\hat{A}$ is certainly continuous if $\|\hat{x}\| := \inf_{x \in \hat{x}} \|x\|$ does define a norm on $E/N(A)$. The following proposition will show that this is indeed the case (observe that because of the continuity of A the nullspace $N(A)$ is closed).

Proposition 21.1. *If F is a* closed *subspace of the normed space E, and if for every equivalence class $\hat{x}$ in E/F one sets*

$$\|\hat{x}\| := \inf_{x \in \hat{x}} \|x\|,$$

then a norm, the so-called quotient norm, *is defined hereby on the quotient space E/F. The normed space E/F is complete if E itself is complete.*

From the properties of a norm we only prove that $\|\hat{x}\| = 0$ implies $\hat{x} = 0$. From $\|\hat{x}\| = 0$ it follows that there exists a sequence of vectors $y_n \in \hat{x}$ so that $\|y_n\| \to 0$, hence also $y_n \to 0$. Because of the closedness of F also the residue class $\hat{x} = x + F$ is closed as a subset of E, hence the limit 0 of the sequence (y_n) from $\hat{x}$ lies in $\hat{x}$, and so indeed $\hat{x} = 0$.

Now let E be complete and $(\hat{x}_n)$ a Cauchy sequence in E/F. Then there exists a sequence $n_1 < n_2 < n_3 < \cdots$ of subscripts such that $\|\hat{x}_n - \hat{x}_{n_k}\| < 1/2^k$ for $n \geqq n_k$ and in particular

$$\|\hat{x}_{n_{k+1}} - \hat{x}_{n_k}\| < \frac{1}{2^k} \quad \text{for} \quad k = 1, 2, \ldots.$$

Thus there exists in each equivalence class $\hat{x}_{n_{k+1}} - \hat{x}_{n_k}$ a representative y_k with $\|y_k\| \leqq 1/2^k$, so that, according to Lemma 8.1, the series $\sum_{k=1}^{\infty} y_k$ converges to an element y of the complete space E. Because of $\|(\hat{y}_1 + \cdots + \hat{y}_k) - \hat{y}\| \leqq \|(y_1 + \cdots + y_k) - y\|$ we have a fortiori

$$\hat{y}_1 + \cdots + \hat{y}_k = (\hat{x}_{n_2} - \hat{x}_{n_1}) + \cdots + (\hat{x}_{n_{k+1}} - \hat{x}_{n_k}) = \hat{x}_{n_{k+1}} - \hat{x}_{n_1} \to \hat{y},$$

and so $\hat{x}_{n_k} \to \hat{y} + \hat{x}_{n_1}$. Therefore the whole Cauchy sequence $(\hat{x}_n)$ converges to $\hat{y} + \hat{x}_{n_1}$ (see Exercise 2 in §1). ■

The canonical homomorphism $h: E \to E/F$ is continuous because of $\|h(x)\| = \|\hat{x}\| \leqq \|x\|$, but it has still a further important property which we have to define first:

A map $A: E \to F$ between metric spaces E, F is said to be *open*, if the image of every open set is an open subset of the subspace $A(E)$.

We now show the openness of the canonical homomorphism $h: E \to E/F$. We observe first that

$$K_r(\hat{0}) \subset h(K_r(0)); \tag{21.1}$$

to $\hat{x} \in K_r(\hat{0})$ there exists namely an $x \in \hat{x}$ with $\|x\| < r$, so that $\hat{x} = h(x) \in h(K_r(0))$. Let now $G \subset E$ be open and $\hat{y} := h(x_0)$, $x_0 \in G$, any vector from $h(G)$. To x_0 there exists a ball $K_r(x_0)$ lying entirely in G. It follows then from (21.1) that

$$K_r(\hat{y}) = \hat{y} + K_r(\hat{0}) \subset \hat{y} + h(K_r(0)) = h(x_0 + K_r(0)) = h(K_r(x_0)) \subset h(G),$$

thus $h(G)$ is indeed open. We state this result:

Proposition 21.2. *If F is a closed subspace of the normed space E, then the canonical homomorphism $h: E \to E/F$ is* continuous *and* open.

In §20 we introduced the canonical injection with the intention to associate with a given linear map A a second map $\hat{A}$, which possesses all essential properties of A and which is, furthermore, injective. The following proposition is to be considered in the light of this intention.

Proposition 21.3. *The linear map $A: E \to F$ with closed nullspace $N(A)$ is continuous or open if and only if the corresponding canonical injection*

$$\hat{A}: E/N(A) \to F$$

is continuous or open, respectively. In the case of continuity we have $\|\hat{A}\| = \|A\|$.

Proof. If A is continuous, then also $\hat{A}$ is continuous and $\|\hat{A}\| \leqq \|A\|$: this is how we defined the quotient norm at the beginning of this section. If conversely $\hat{A}$ is continuous and $\hat{x}$ is the equivalence class of x, then $\|Ax\| = \|\hat{A}\hat{x}\| \leqq \|\hat{A}\|\,\|\hat{x}\| \leqq \|\hat{A}\|\,\|x\|$, hence also A is continuous and $\|A\| \leqq \|\hat{A}\|$. From here we get, together with the above inequality for the norm, that $\|\hat{A}\| = \|A\|$. Let us now assume that A is open, and let $\hat{G}$ be an open subset of $E/N(A)$. Since by Proposition 21.2 the canonical homomorphism $h: E \to E/N(A)$ is continuous, the set $G := h^{-1}(\hat{G})$ must be open in E (§7 Exercise 10), and because of $\hat{A}(\hat{G}) = A(G)$, also $\hat{A}(\hat{G})$ is open in $A(E) = \hat{A}(E/N(A))$. Thus $\hat{A}$ is an open map. Now let conversely $\hat{A}$ be open and let G be an open subset of E. Since h is open by Proposition 21.2, also $h(G)$ must be open. It follows that $A(G) = \hat{A}(h(G))$ is open in $A(E) = \hat{A}(E/N(A))$, i.e., A is an open map. ■

Exercises

1. A linear form on a normed space is continuous if and only if its nullspace is closed. Use Exercise 4 in §20.

$^+$2. A finite-dimensional map of normed spaces is continuous if and only if its nullspace is closed (cf. Exercise 1).

*3. Let F be a subspace of finite codimension of the normed space E. There exists a continuous projector P with $P(E) = F$ if and only if F is closed. *Hint*: Project E along F onto a finite-dimensional complementary subspace of F and use Exercise 2.

4. The operator $K: E \to F$ is compact if and only if its canonical injection $\hat{K}$ is compact.

*5. An injective map $A: E \to F$ of the metric spaces E, F is open if and only if $A^{-1}: A(E) \to E$ is continuous.

6. Let the linear map $A: E \to F$ of the normed spaces E, F be surjective and open. Then for every sequence (y_n) in F which converges to $y_0 = Ax_0$ there exists a sequence (x_n) in E which converges to x_0 and for which $Ax_n = y_n$ $(n = 1, 2, \ldots)$.

*7. Let F be a subspace of the metric space E (thus F is a metric space with the metric induced by E). Show first on a simple example ($E := \mathbf{R}^2$, $F := \mathbf{R} \times \{0\}$), that an open subset of the subspace F does not need to be an open subset of the space E, and then prove the proposition: $M \subset F$ is open in F if and only if there exists a set G open in E so that $M = G \cap F$. *Hint*: Balls in F are related to balls in E in the following way: $\{x \in F: d(x, x_0) < r\} = \{y \in E: d(y, x_0) < r\} \cap F$.

§22. Quotient algebras

To a linear subspace M of the algebra R we can construct the quotient space R/M. If M is even a (two-sided) ideal, then by the definition

$$\hat{x}\hat{y} := \widehat{xy} \quad \text{with} \quad x \in \hat{x},\ y \in \hat{y}$$

of a product one can introduce a multiplication in R/M. The product is independent of the choice of the representatives of the equivalence classes; indeed, if also $x_1 \in \hat{x}$ and $y_1 \in \hat{y}$ then $x_1 = x + u$, $y_1 = y + v$ with u, v from M, hence

$$x_1y_1 = xy + xv + uy + uv = xy + w \qquad \text{with} \quad w = xv + uy + uv \in M$$

and so $\widehat{x_1y_1} = \widehat{xy}$. With this multiplication R/M becomes an algebra (over the field of scalars of R), the *quotient algebra* of R modulo M. The canonical homomorphism $h: R \to R/M$, which associates with each element x from R its residue class $h(x) = \hat{x}$, is then a *homomorphism of algebras*, i.e., h is linear and furthermore $h(xy) = h(x)h(y)$. If R has a unit element e, then $\hat{e}$ is the unit element of the quotient algebra. If y is a right inverse of x, i.e., if $xy = e$, then $\hat{y}$ is a right inverse of $\hat{x}$; correspondingly for left inverse and inverse elements.

The set $\mathscr{E}(E)$ of all finite-dimensional endomorphisms of the vector space E is an ideal in $\mathscr{S}(E)$; the set $\mathscr{F}(E)$ of all continuous finite-dimensional and the set $\mathscr{K}(E)$ of all compact endomorphisms of the normed space E are ideals in $\mathscr{L}(E)$ (see Exercise 5 in §11). Therefore $\mathscr{S}(E)/\mathscr{E}(E)$, $\mathscr{L}(E)/\mathscr{F}(E)$ and $\mathscr{L}(E)/\mathscr{K}(E)$ are examples of quotient algebras. The unit element of these algebras is the residue class $\hat{I}$ of the identity map I of E.

If R is a *normed* algebra and M a *closed* ideal in R, then the quotient norm of Proposition 21.1 is even a norm of algebras, i.e., it satisfies $\|\hat{x}\hat{y}\| \leqq \|\hat{x}\| \, \|\hat{y}\|$. *The quotient algebra is in this case a normed algebra.* If R is a Banach algebra and $M \neq R$, then we see easily with the help of Proposition 21.1 that also R/M is a Banach algebra.

If $E \neq \{0\}$ is a Banach space, then $\mathscr{L}(E)$ is a Banach algebra and $\mathscr{K}(E)$ a closed ideal in $\mathscr{L}(E)$ (see Propositions 7.4 and 13.1), thus *$\mathscr{L}(E)/\mathscr{K}(E)$ is a Banach algebra whenever $\mathscr{L}(E) \neq \mathscr{K}(E)$, i.e., when E is infinite-dimensional* (see Exercise 3 in §11).

Exercises

*1. The closure $\overline{M}$ of an ideal M in the normed algebra R is again an ideal in R.

2. If E is an infinite-dimensional Banach space, then $\mathscr{L}(E)/\overline{\mathscr{F}(E)}$ is a Banach algebra (use Exercise 1).

IV

Fredholm operators

§23. Operators with finite deficiency

In §18 we had considered operators $A = R - S$ on a vector space E with a bijective R and a finite-dimensional S. If we denote by $\hat{T}$ the residue class of the operator T in the quotient algebra $\hat{\mathscr{S}} := \mathscr{S}(E)/\mathscr{E}(E)$, then $\hat{A} = \hat{R} - \hat{S} = \hat{R}$, and since R possesses by Proposition 5.2 an inverse (in the sense of the theory of algebras) in $\mathscr{S}(E)$, also $\hat{R}$, and with it $\hat{A}$, is invertible in $\hat{\mathscr{S}}$. We can make a corresponding statement for a larger class of operators. If, namely, the dimension $\alpha(A)$ of the nullspace of A as well as the codimension $\beta(A)$ of the image space of A are finite, then it follows immediately from Proposition 5.3 that, with the operator B indicated there, the equation $\hat{B}\hat{A} = \hat{A}\hat{B} = \hat{I}$ holds, because the projectors P and Q in Proposition 5.3 have now finite rank. Thus $\hat{A}$ is also in this case an invertible element of $\hat{\mathscr{S}}$. Conversely, if the residue class $\hat{A}$ of an operator A is invertible, that is, if there exist operators B and C such that $\hat{B}\hat{A} = \hat{A}\hat{C} = \hat{I}$, then with certain finite-dimensional operators K_1, K_2 the equations

$$(23.1) \qquad BA = I - K_1, \qquad AC = I - K_2$$

are valid. From here follow the inclusions

$$N(A) \subset N(BA) = N(I - K_1), \qquad A(E) \supset (AC)(E) = (I - K_2)(E)$$

and with them the estimates $\alpha(A) \leqq \alpha(I - K_1)$, $\beta(A) \leqq \beta(I - K_2)$. Thus because of Proposition 17.1 both dimensions $\alpha(A)$, $\beta(A)$ are finite. We summarize:

Proposition 23.1. *The operator $A \in \mathscr{S}(E)$ possesses an invertible residue class $\hat{A} \in \mathscr{S}(E)/\mathscr{E}(E)$, i.e., there exist operators B, C in $\mathscr{S}(E)$ and K_1, K_2 in $\mathscr{E}(E)$ which satisfy* (23.1), *if and only if $\alpha(A)$ and $\beta(A)$ are* finite.

Since $\alpha(A)$ is called the *nulldeficiency*, $\beta(A)$ the *image-deficiency* of A, one says shortly that A has *finite deficiency* if both $\alpha(A)$ and $\beta(A)$ are finite. According to the last proposition, exactly the operators with finite deficiency have invertible residue classes in $\mathscr{S}(E)/\mathscr{E}(E)$. We denote by $\Delta(E)$ the set of endomorphisms of E with finite deficiency.

We have seen in §18 that Fredholm integral equations with continuous kernels lead to operators with equal finite deficiencies. F. Noether [147] was the first to notice that so-called *singular integral equations* generate operators whose deficiencies are still finite but different; cf. also [115].

If we take into account that the invertible elements of an algebra form a group and that the factors of an invertible product ab are either both invertible or both non-invertible, then we obtain from Proposition 23.1 immediately the following assertion about the structure of $\Delta(E)$:

Proposition 23.2. *The product of operators with finite deficiency has finite deficiency, thus $\Delta(E)$ is a* semi-group. *If the operator AB has finite deficiency, then either both factors or none have finite deficiency. The sum of an operator with finite deficiency and of a finite-dimensional operator has finite deficiency.*

With every operator A of finite deficiency one associates its *index*:

$$\text{ind}(A) := \alpha(A) - \beta(A).$$

The basic assertion concerning the index is made by the following *index theorem*:

Theorem 23.1. *For endomorphisms A, B of finite deficiency we have*

$$\text{ind}(AB) = \text{ind}\, A + \text{ind}\, B. \tag{23.2}$$

We begin the proof with the observation that AB has finite deficiency by Proposition 23.2, and that the complementary spaces which follow exist because of Proposition 4.1. To

$$E_1 := B(E) \cap N(A) \tag{23.3}$$

we determine subspaces E_2, E_3 and E_4 of E such that

$$B(E) = E_1 \oplus E_2, \tag{23.4}$$

$$N(A) = E_3 \oplus E_1 \tag{23.5}$$

and

$$E = \overbrace{E_3 \oplus E_1}^{N(A)} \!\!\!\!\!\!\!\!\underbrace{ \oplus E_2}_{B(E)} \oplus E_4 \tag{23.6}$$

(see Exercise 4 of §4 for the definition of the direct sum of arbitrarily many subspaces). It follows from the last decomposition that

$$\begin{aligned} A(E) = A(E_2 \oplus E_4) &= A(E_2) \oplus A(E_4) = A(E_1 \oplus E_2) \oplus A(E_4) \\ &= (AB)(E) \oplus A(E_4). \end{aligned} \tag{23.7}$$

Furthermore let F be a subspace of E with

$$N(AB) = N(B) \oplus F. \tag{23.8}$$

The restriction of B to F is injective, hence F is isomorphic to its image $B(F) = B(F \oplus N(B)) = B(N(AB)) = B(E) \cap N(A) = E_1$. By Exercise 4 of §5 we have thus

$$\dim F = \dim E_1. \tag{23.9}$$

For the same reason

$$\dim A(E_4) = \dim E_4. \tag{23.10}$$

From (23.5), (23.6), (23.8) and (23.7) we obtain (in this order), taking into consideration (23.9) and (23.10):

$$\begin{aligned}
\alpha(A) &= \dim E_1 + \dim E_3, \\
\beta(B) &= \dim E_3 + \dim E_4, \\
\alpha(AB) &= \alpha(B) + \dim F = \alpha(B) + \dim E_1, \\
\beta(AB) &= \beta(A) + \dim A(E_4) = \beta(A) + \dim E_4.
\end{aligned}$$

From these four equations it follows that

$$\begin{aligned}
\operatorname{ind}(AB) &= \alpha(AB) - \beta(AB) = \alpha(B) + \dim E_1 - \beta(A) - \dim E_4 \\
&= \alpha(B) + \alpha(A) - \dim E_3 - \beta(A) - \beta(B) + \dim E_3 \\
&= \alpha(A) - \beta(A) + \alpha(B) - \beta(B) = \operatorname{ind}(A) + \operatorname{ind}(B). \qquad \blacksquare
\end{aligned}$$

The next proposition is a *stability* result concerning the index; it says that the index of an operator A with finite deficiency does not change if one adds to A arbitrary finite-dimensional operators (or, as one says, if one 'perturbs' A by operators of this kind). We first state a lemma which follows immediately from Proposition 23.2 and the index theorem:

Lemma 23.1. *If for an operator A of finite deficiency an equation of the form*

$$AB = C \quad \textit{or} \quad BA = C \qquad \textit{with} \quad \operatorname{ind}(C) = 0$$

holds, then also B has finite deficiency and $\operatorname{ind}(B) = -\operatorname{ind}(A)$.

Proposition 23.3. *If A is an endomorphism of finite deficiency of E and S a finite-dimensional one, then* $\operatorname{ind}(A + S) = \operatorname{ind} A$.

Proof. We first observe that $A + S$ has finite deficiency by Proposition 23.2. Because of Proposition 23.1 there exists a $B \in \mathscr{S}(E)$ and an $L \in \mathscr{E}(E)$ so that $BA = I - L$. Since by Proposition 17.1 we have $\operatorname{ind}(I - L) = 0$, it follows with the help of the above lemma that $\operatorname{ind}(B) = -\operatorname{ind}(A)$. Furthermore

$$B(A + S) = BA + BS = I - L + BS = I - L_1 \quad \text{with} \quad L_1 := L - BS \in \mathscr{E}(E).$$

From here we obtain as above

$$\operatorname{ind}(A + S) = -\operatorname{ind}(B) = \operatorname{ind}(A). \qquad \blacksquare$$

Finally we present another result which permits in certain cases to ascertain whether an operator has finite deficiency.

Proposition 23.4. *Let K be an endomorphism of E and F a vector space between $K(E)$ and E; let $\tilde{I}$, $\tilde{K}$ be the restrictions of I, K to F. Then the following assertions are true:*

(a) *$N(I - K) = N(\tilde{I} - \tilde{K})$, hence also $\alpha(I - K) = \alpha(\tilde{I} - \tilde{K})$.*

(b) *From $F = G \oplus (\tilde{I} - \tilde{K})(F)$ it follows that $E = G \oplus (I - K)(E)$; in particular $\beta(I - K) = \beta(\tilde{I} - \tilde{K})$.*

(c) *$I - K$ has finite deficiency if and only if $\tilde{I} - \tilde{K}$ does.*

Proof. (a) is trivial because $N(I - K) \subset K(E) \subset F$. Let now the decomposition of F indicated in (b) hold. We first show that $G \cap (I - K)(E) = \{0\}$. If $y \in G$ and at the same time $y = (I - K)x$, then $y + Kx = x$, and since the left-hand side lies in F, we also have $x \in F$ and so $y = (\tilde{I} - \tilde{K})x$; but then by hypothesis y must vanish. Let now x be chosen arbitrarily in E and set $z := (I - K)x$. Then $x = Kx + z = g + (I - K)y + z$ with $g \in G$ and $y \in F$, hence $x = g + (I - K)(y + x)$. Thus (b) is also proved. (c) follows immediately from (a) and (b). ∎

Exercises

1. If A has finite deficiency, then so do A^n ($n = 0, 1, 2, \ldots$). Conversely, together with a power A^n ($n \geqq 1$) also A has finite deficiency.

2. Construct on (s) an endomorphism A with finite deficiency such that $\operatorname{ind}(A) \neq 0$.

3. Extend the concept of an operator with finite deficiency to linear maps $A: E \to F$, characterize these maps by equations of the form (23.1) and prove an index theorem for them.

*4. Let $\Delta_\alpha(E) := \{A \in \mathscr{S}(E): \alpha(A) < \infty\}$, $\Delta_\beta(E) := \{A \in \mathscr{S}(E): \beta(A) < \infty\}$. Show the following:

(a) $A \in \Delta_\alpha(E) \Leftrightarrow$ there exist $B \in \mathscr{S}(E)$ and $K \in \mathscr{E}(E)$ such that $BA = I - K \Leftrightarrow$ the residue class $\hat{A} \in \mathscr{S}(E)/\mathscr{E}(E)$ of A is left-invertible.

(b) $A \in \Delta_\beta(E) \Leftrightarrow$ there exist $C \in \mathscr{S}(E)$ and $K \in \mathscr{E}(E)$ such that $AC = I - K$ $\Leftrightarrow$ the residue class $\hat{A} \in \mathscr{S}(E)/\mathscr{E}(E)$ of A is right-invertible.

(c) $\Delta_\alpha(E)$ and $\Delta_\beta(E)$ are (multiplicative) semi-groups.

§24. Fredholm operators on normed spaces

If A is a *continuous* endomorphism with finite deficiency of the normed space E, then it would be desirable to be able to choose the operators B, C, K_1 and K_2 in the characterizing equations (23.1) also *continuous*, so that we do not have to go outside the algebra $\mathscr{L}(E)$. In order to study this situation more closely, let us return to the proof of Proposition 5.3 from which we obtained the equations (23.1). With the notation of this proof, we shall assume first that there exists a continuous projector P from E onto $N(A)$, and an equally continuous projector

Q_0 from E onto $A(E)$. The operator $B := A_0^{-1}Q_0$ is certainly continuous if A_0^{-1} is continuous, i.e., if A_0 is open (see Exercise 5 in §21). But A_0 is open if A itself is open; indeed, if $P_0 = I - P$ is the (continuous) projector onto U along $N(A)$ and if $M \subset U$ is open in the subspace U, then by Exercise 10 in §7 also $P_0^{-1}(M) = \{x + y : x \in M, y \in N(A)\}$ is open, hence, because of the assumed openness of A, the image $A(P_0^{-1}(M)) = \{Ax : x \in M\} = A_0(M)$ is open in $A(E) = A_0(U)$ and thus A_0 is an open map. In summary, we can state: If $A \in \mathscr{L}(E)$ is open and if there exists a continuous projector P from E onto $N(A)$ and a continuous projector Q_0 from E onto $A(E)$—hence also a continuous projector $Q := I - Q_0$ of E along $A(E)$—then there exists a $B \in \mathscr{L}(E)$ such that

$$BA = I - P, \qquad AB = I - Q. \tag{24.1}$$

For an A with finite deficiency the projectors P, Q are of finite rank, so that in this case all the operators which figure in (23.1) can indeed be chosen to be continuous.

If we multiply the first equation in (24.1) from the left by A, then we obtain

$$ABA = A, \tag{24.2}$$

independently of the dimension of the nullspace $N(A)$. If we call a subspace F of E *continuously projectable* if there exists a continuous projector P with $P(E) = F$, and furthermore if we call an operator $A \in \mathscr{L}(E)$ *relatively regular* if (24.2) holds with some $B \in \mathscr{L}(E)$, then we can now say the following: An open operator $A \in \mathscr{L}(E)$ with continuously projectable nullspace and image space is relatively regular. The converse of this assertion is also valid. Indeed, if A is relatively regular, i.e., if (24.2) holds, then

$$(AB)^2 = ABAB = AB \quad \text{and} \quad (BA)^2 = BABA = BA,$$

hence as idempotent operators AB and BA are continuous projectors. From $A(E) = (ABA)(E) \subset (AB)(E) \subset A(E)$ it follows that $(AB)(E) = A(E)$, hence the image space of A is continuously projectable. From $N(A) \subset N(BA) \subset N(ABA) = N(A)$ one obtains $N(BA) = N(A)$, hence $(I - BA)(E) = N(A)$, and thus also the nullspace of A is continuously projectable. Finally, to show that A is open, we prove the identity

$$A(G) = B^{-1}(G + N(A)) \cap A(E) \qquad \text{for every} \quad G \subset E. \tag{24.3}$$

If we take into consideration that the projector AB operates on $A(E)$ as the identity, then we obtain

$$\begin{aligned} B^{-1}(G + N(A)) \cap A(E) &= (AB)[B^{-1}(G + N(A)) \cap A(E)] \\ &\subset A(G + N(A)) = A(G). \end{aligned} \tag{24.4}$$

On the other hand, since also BA is a projector, we can represent every $g \in G$ as a sum $g = x + y$ with $x \in N(BA)$, $y \in (BA)(E)$; it follows that $BAg = BAy = y = g - x$, hence $(BA)(G) \subset G + N(BA) \subset G + N(ABA) = G + N(A)$ and so $A(G) \subset B^{-1}(G + N(A)) \cap A(E)$. From this inclusion and from (24.4) we obtain the asserted identity (24.3). If now G is an open subset of E, then for every $x \in E$

obviously $G + x$ is open, hence also the sets $G + N(A) = \bigcup_{x \in N(A)} (G + x)$ and $B^{-1}(G + N(A))$ are open in E (§1 Exercise 11, §7 Exercise 10). It follows, with the aid of (24.3), that $A(G)$ is an open subset of the subspace $A(E)$ (see Exercise 7 in §21). We summarize:

Proposition 24.1. *A continuous endomorphism of a normed space is relatively regular if and only if it is open and its nullspace and image space are continuously projectable.*

By virtue of this proposition we can formulate as follows the result found above concerning operators with finite deficiency: For a relatively regular operator $A \in \mathscr{L}(E)$ with finite deficiency there exist continuous endomorphisms B, C and continuous, finite-dimensional endomorphisms K_1, K_2 such that

$$(24.5) \qquad BA = I - K_1, \qquad AC = I - K_2;$$

one can even choose $B = C$.

We call a continuous endomorphism of a normed space E a *Fredholm operator* if it is relatively regular and has finite deficiency. We denote the set of all Fredholm operators on E by $\Phi(E)$. For a Fredholm operator A on E the equations (24.5) hold, hence its residue class $\hat{A}$ in the quotient algebra $\mathscr{L}(E)/\mathscr{F}(E)$ is invertible (observe that $\mathscr{L}(E)/\mathscr{F}(E)$ is in general not normed, since the ideal $\mathscr{F}(E)$ of continuous operators of finite rank is in general not closed).

The question arises now whether conversely a continuous endomorphism A, for which the equations (24.5) hold (i.e., whose residue class $\hat{A}$ is invertible in $\mathscr{L}(E)/\mathscr{F}(E)$), is a Fredholm operator. We shall treat this question in the next section in a much more general form and answer it positively. Right now we want to turn again to the concept of continuous projectability.

If $E = F \oplus G$ and if the projector P onto F along G is continuous, then we call G a *topologically complementary space* (or a *topological complement*) of F; of course F is then a topologically complementary space of G. *It is precisely the continuously projectable subspaces which have topological complements.* Such subspaces are necessarily *closed* because they are nullspaces of continuous projectors. However, closed subspaces are not necessarily continuously projectable; cf. [62]. But Exercise 3 in §21 shows that a closed subspace F is continuously projectable if it has a finite codimension; every algebraic complement of F is then a topological complement. This assertion brings up the question whether also the simplest subspaces, the finite-dimensional ones, are continuously projectable. If $x_1, \ldots, x_n$ is a basis of the finite-dimensional subspace F of E and P a continuous projector from E onto F, then P has according to Proposition 13.4 the representation

$$(24.6) \qquad Px = \sum_{k=1}^{n} \langle x, x_k' \rangle x_k \qquad \text{with appropriate} \quad x_k' \in E'.$$

Since P operates on $P(E) = F$ as the identity, we have

$$Px_i = \sum_{k=1}^{n} \langle x_i, x'_k \rangle x_k = x_i,$$

hence

$$\langle x_i, x'_k \rangle = \delta_{ik} \quad \text{for} \quad i, k = 1, \ldots, n. \tag{24.7}$$

Conversely, if to $x_1, \ldots, x_n$ there exist continuous linear forms $x'_1, \ldots, x'_n$ so that (24.7) is valid, and one defines P by (24.6), then $Px_i = x_i$ for $1 \leqq i \leqq n$, from where $P^2 = P$ and $P(E) = F$ follow: P is a continuous projector from E onto F. Because of Proposition 15.1 we can thus say: Every finite-dimensional subspace of E is continuously projectable if and only if (E, E') is a right dual system. In §28 we shall give, with the help of the fundamental Hahn–Banach theorem, a positive answer to the question whether (E, E') is a right dual system, and thus a dual system. Anticipating this theorem we can summarize our results as follows:

Proposition 24.2. *Finite-dimensional and closed finite-codimensional subspaces of a normed space are* continuously *projectable; every algebraic complement of a closed subspace of finite codimension is even a topological complement.*

With the aid of Proposition 24.1 from here we obtain immediately:

Proposition 24.3. *A continuous endomorphism of a normed space is a* Fredholm operator *if and only if it is open, has finite deficiency and a closed image space.*

Exercises

1. $A \in \mathscr{L}(E, F)$ is called *relatively regular* if there exists a $B \in \mathscr{L}(F, E)$ with $ABA = A$. Show that A is relatively regular if and only if A is open and has a continuously projectable nullspace and image space.
2. $A \in \mathscr{L}(E, F)$ is called a *Fredholm operator* if it is relatively regular and has finite deficiency. To a Fredholm operator A there exists a $B \in \mathscr{L}(F, E)$ and $K_1 \in \mathscr{F}(E)$, $K_2 \in \mathscr{F}(F)$ so that $BA = I_E - K_1$, $AB = I_F - K_2$.
3. An element a of an arbitrary algebra R is called *relatively regular* if for some $b \in R$ the equation $aba = a$ is valid. Show that the following elements of R are always relatively regular (in the last three examples it is assumed that R possesses a unit element e):

   The zero element 0; every a with $a^2 = a$ (idempotent elements); every multiple of a relatively regular element; the unit element e; every a for which there exists an $n \geqq 1$ with $a^n = e$ (elements of finite order); every left- or right-invertible element.
4. If the element a of the algebra R is relatively regular (see Exercise 3), then there exists a $b \in R$ such that $aba = a$ and $bab = b$ (b is said to be *relatively inverse* to a).

5. The element a of the algebra R is relatively regular (see Exercise 3) if and only if $aba - a$ is relatively regular for some $b \in R$.

$^+$6. Let the ideal J of the algebra R consist of only relatively regular elements. Show: $a \in R$ is relatively regular if and only if the residue class $\hat{a}$ of a is relatively regular in R/J. *Hint*: Exercise 5.

§25. Fredholm operators in saturated operator algebras

Fredholm operators on normed spaces are essentially defined purely algebraically as relatively regular operators with finite deficiency. This suggests the idea to abandon completely metric hypotheses for the investigation of Fredholm operators. For this purpose we consider an algebra $\mathscr{A}(E)$ of endomorphisms of a vector space E. We say that $A \in \mathscr{A}(E)$ is *relatively regular* (in $\mathscr{A}(E)$), if there exists $B \in \mathscr{A}(E)$ with $ABA = A$ (cf. Exercise 3 in §24); A is called a *Fredholm operator* (in $\mathscr{A}(E)$) if A is relatively regular in $\mathscr{A}(E)$ and has finite deficiency.

Because of Exercise 6 in §5, every element of the algebra $\mathscr{S}(E)$ is relatively regular in $\mathscr{S}(E)$. Thus *$A \in \mathscr{S}(E)$ is a Fredholm operator in $\mathscr{S}(E)$ if and only if A has finite deficiency*. On the other hand, $A \in \mathscr{L}(E)$ *is*, according to Proposition 24.3, *a Fredholm operator if and only if it is open, has finite deficiency and, furthermore, a closed image space.*

The set of Fredholm operators in $\mathscr{A}(E)$ will be denoted by $\Phi(\mathscr{A}(E))$; in §24 we already introduced the shorter notation $\Phi(E)$ for the important set $\Phi(\mathscr{L}(E))$. Let $\mathscr{F}(\mathscr{A}(E))$ be the (two-sided) ideal of finite-dimensional operators in $\mathscr{A}(E)$; for $\mathscr{F}(\mathscr{L}(E))$ we have written earlier $\mathscr{F}(E)$. In order to develop a theory which is valid for Fredholm operators in $\mathscr{S}(E)$ and in $\mathscr{L}(E)$, and even for continuous Fredholm operators on topological vector spaces (§89), we make some postulates concerning the structure and richness of the ideal $\mathscr{F}(\mathscr{A}(E))$; these are formulated in the following definition:

An algebra of operators $\mathscr{A}(E)$ is said to be *saturated* if it contains the identity transformation I, and if, furthermore, there exists a vector space E^+ with the following property: (E, E^+) is a bilinear system with respect to a bilinear form $\langle x, x^+ \rangle$, every finite-dimensional operator K in $\mathscr{A}(E)$ can be represented in the form

$$Kx = \sum_{i=1}^{n} \langle x, x_i^+ \rangle y_i \tag{25.1}$$

with vectors $y_1, \ldots, y_n$ from E and $x_1^+, \ldots, x_n^+$ from E^+, and conversely every operator K which can be so represented lies in $\mathscr{A}(E)$. One can choose the vectors $y_1, \ldots, y_n$ and $x_1^+, \ldots, x_n^+$ linearly independent (see Exercise 3 in §14).

Occasionally it will be necessary to emphasize the space E^+; then we say more precisely that $\mathscr{A}(E)$ is an E^+*-saturated* algebra. With this terminology *$\mathscr{S}(E)$ is an E^*-saturated, and $\mathscr{L}(E)$ an E'-saturated algebra.* The concept of saturation and most investigations in this and in the following section go back to H. Kroh [108, 109]; see also [107].

Given an E^+-saturated algebra $\mathscr{A}(E)$, there always exists a left dual system $(E, \hat{E}^+)$ so that $\mathscr{A}(E)$ is also $\hat{E}^+$-saturated. Namely if we set

$$N := \{x^+ \in E^+ : \langle x, x^+ \rangle = 0 \text{ for all } x \in E\} \quad \text{and} \quad \hat{E}^+ := E^+/N,$$

then, as we have seen in §20, $(E, \hat{E}^+)$ is a left dual system with the bilinear form defined by $[x, \hat{x}^+] := \langle x, x^+ \rangle$, where $\hat{x}^+$ denotes, as usual, the residue class of x^+ in $\hat{E}^+$, and obviously $\mathscr{A}(E)$ is $\hat{E}^+$-saturated. *Therefore we may always assume that* (E, E^+) *is a* left dual system *when we study* E^+*-saturated algebras* $\mathscr{A}(E)$.

In this section we want to show that an operator in a saturated algebra $\mathscr{A}(E)$ is a Fredholm operator if and only if its equivalence class is invertible in $\mathscr{A}(E)/\mathscr{F}(\mathscr{A}(E))$. For this we need some simple preliminary remarks.

Let first $\mathscr{A}(E)$ be an arbitrary algebra of operators containing I. A subspace F of E is said to be $\mathscr{A}(E)$*-projectable* if there exists a projector P in $\mathscr{A}(E)$ with $P(E) = F$. Then the nullspace $N(P)$ is also $\mathscr{A}(E)$-projectable because of $N(P) = (I - P)(E)$. We say furthermore that $E = F \oplus G$ is an $\mathscr{A}(E)$*-direct sum*, and call G an $\mathscr{A}(E)$*-complement* of F, if there exists a projector $P \in \mathscr{A}(E)$ with $P(E) = F$, $N(P) = G$; obviously also F is then an $\mathscr{A}(E)$-complement of G. The $\mathscr{A}(E)$-projectable subspaces are exactly those which possess $\mathscr{A}(E)$-complements.

Lemma 25.1. *If (E, E^+) is a left dual system, if the vectors $x_1^+, \ldots, x_n^+$ from E^+ are linearly independent, and if the operator K is defined by*

$$Kx := \sum_{i=1}^{n} \langle x, x_i^+ \rangle y_i$$

with elements $y_1, \ldots, y_n$ from E, then $K(E) = [y_1, \ldots, y_n]$.

Clearly $K(E) \subset [y_1, \ldots, y_n]$. To prove the opposite inclusion let $y := \alpha_1 y_1 + \cdots + \alpha_n y_n$ be an arbitrary element from $[y_1, \ldots, y_n]$. By Proposition 15.1 there exist vectors $x_1, \ldots, x_n$ in E with $\langle x_k, x_i^+ \rangle = \delta_{ik}$. If we set $x := \sum_{k=1}^{n} \alpha_k x_k$, then

$$Kx = \sum_{i=1}^{n} \left\langle \sum_{k=1}^{n} \alpha_k x_k, x_i^+ \right\rangle y_i = \sum_{i=1}^{n} \alpha_i y_i = y,$$

hence $y \in K(E)$ and so $[y_1, \ldots, y_n] = K(E)$. ■

Lemma 25.2. *The nullspace of a finite-dimensional operator K in the E^+-saturated algebra of operators $\mathscr{A}(E)$ is $\mathscr{A}(E)$-projectable; furthermore* $\operatorname{codim} N(K) = \dim K(E)$.

For the proof we may assume that (E, E^+) is a left dual system. Let $Kx = \sum_{i=1}^{n} \langle x, x_i^+ \rangle y_i$, where the vectors $x_1^+, \ldots, x_n^+$ from E^+ and $y_1, \ldots, y_n$ from E are linearly independent. By Proposition 15.1 there exist linearly independent elements $x_1, \ldots, x_n$ in E with $\langle x_i, x_k^+ \rangle = \delta_{ik}$. We define $P \in \mathscr{A}(E)$ by $Px := \sum_{i=1}^{n} \langle x, x_i^+ \rangle x_i$. Because of $Px_k = x_k$ we have $P^2 = P$, hence P is a projector.

Obviously $N(P) = N(K)$, hence $(I - P)(E) = N(K)$: the nullspace of K is indeed $\mathscr{A}(E)$-projectable. Furthermore $E = P(E) \oplus N(P) = P(E) \oplus N(K)$, from where with the aid of Lemma 25.1 it follows that codim $N(K) = \dim P(E) = \dim[x_1, \ldots, x_n] = \dim[y_1, \ldots, y_n] = \dim K(E)$. ■

Observe that the assertion concerning the dimension holds for *every* finite-dimensional endomorphism K because K lies always in the E^*-saturated algebra $\mathscr{S}(E)$. It also follows from Proposition 20.2, taking into account Exercise 3 in §20.

Lemma 25.3. *If $\mathscr{A}(E)$ is an E^+-saturated algebra of operators and if the finite-codimensional subspace F of E is $\mathscr{A}(E)$-projectable, then every algebraic complement of F is also an $\mathscr{A}(E)$-complement and, furthermore, every subspace H of E containing F is $\mathscr{A}(E)$-projectable.*

Proof. Without restricting the generality, we may again assume that (E, E^+) is a left dual system. According to hypothesis, there exists a projector P in $\mathscr{A}(E)$ such that $N(P) = F$, hence $E = P(E) \oplus F$ and $\dim P(E) < \infty$. P can be represented in the form $Px = \sum_{i=1}^{n} \langle x, x_i^+ \rangle x_i$, where the vectors $x_1, \ldots, x_n$ from E and the vectors $x_1^+, \ldots, x_n^+$ from E^+ are linearly independent. By Lemma 25.1 we have $P(E) = [x_1, \ldots, x_n]$, hence

$$\text{codim } F = \dim P(E) = n. \tag{25.2}$$

Furthermore we have

$$F = N(P) = \{x \in E : \langle x, x_i^+ \rangle = 0 \text{ for } i = 1, \ldots, n\}. \tag{25.3}$$

Now let G be an algebraic complement of F and $G^+ := [x_1^+, \ldots, x_n^+]$. Then (G, G^+) is a bilinear system with the bilinear form $[y, y^+] := \langle y, y^+ \rangle$ induced by (E, E^+) and because of (25.3) it is even a right dual system. For any basis $\{y_1, \ldots, y_n\}$ of G, which because of (25.2) has n elements, there exist by Proposition 15.1 n linearly independent vectors $y_1^+, \ldots, y_n^+$ from G^+ with $\langle y_i, y_k^+ \rangle = \delta_{ik}$. The map Q, which we define by

$$Qx := \sum_{i=1}^{n} \langle x, y_i^+ \rangle y_i, \tag{25.4}$$

is a projector in $\mathscr{A}(E)$ whose image space, according to Lemma 25.1, is equal to G; furthermore we have

$$N(Q) = \{x \in E : \langle x, y_i^+ \rangle = 0 \text{ for } i = 1, \ldots, n\}. \tag{25.5}$$

Since $\{x_1^+, \ldots, x_n^+\}$ and $\{y_1^+, \ldots, y_n^+\}$ are bases of G^+, it follows from (25.3) and (25.5) that

$$F = N(P) = \{x \in E : \langle x, x^+ \rangle = 0 \text{ for all } x^+ \in G^+\} = N(Q);$$

therefore Q projects E parallel to F onto G, i.e., G is indeed an $\mathscr{A}(E)$-complement of F.

Now we prove the second assertion of the lemma. Given the subspace $H \supset F$, there exist finite-dimensional spaces M and N such that

$$H = F \oplus M \quad \text{and} \quad E = H \oplus N.$$

Let $\{y_1, \ldots, y_m\}$ be a basis of M and $\{y_{m+1}, \ldots, y_n\}$ a basis of N. Then

$$\{y_1, \ldots, y_m, y_{m+1}, \ldots, y_n\}$$

is a basis of the complement $G := M \oplus N$ of F in E. With the map Q defined by (25.4) we can project E parallel to F onto G; we also recall that for the vectors y_i, y_i^+ in (25.4) we have $\langle y_i, y_k^+ \rangle = \delta_{ik}$. The operator defined by $Q_0 x := \sum_{i=m+1}^{n} \langle x, y_i^+ \rangle y_i$ is a projector in $\mathscr{A}(E)$; according to Lemma 25.1 we have $Q_0(E) = [y_{m+1}, \ldots, y_n] = N$ and

$$N(Q_0) = \{x \in E : \langle x, y_i^+ \rangle = 0 \text{ for } i = m+1, \ldots, n\}.$$

If we take into consideration that every vector x of E can be written in the form

$$x = \alpha_1 y_1 + \cdots + \alpha_m y_m + \alpha_{m+1} y_{m+1} + \cdots + \alpha_n y_n + z \qquad \text{with} \quad z \in F$$

and that x belongs to H if and only if $\alpha_{m+1} = \cdots = \alpha_n = 0$, then we see immediately that $N(Q_0) = H$. Thus Q_0 projects E parallel to H onto N and consequently H is $\mathscr{A}(E)$-projectable by $I - Q_0$. ■

We now arrive at the main theorem of this section.

Theorem 25.1. *The operator A from the E^+-saturated algebra $\mathscr{A}(E)$ is a* Fredholm operator *if and only if there exist operators B, C in $\mathscr{A}(E)$ and K_1, K_2 in $\mathscr{F}(\mathscr{A}(E))$ so that the equations*

$$BA = I - K_1, \qquad AC = I - K_2 \tag{25.6}$$

hold, i.e., if and only if the residue class $\hat{A}$ of A is invertible in $\mathscr{A}(E)/\mathscr{F}(\mathscr{A}(E))$.

Proof. We assume first that A is a Fredholm operator, i.e., that it is relatively regular and has finite deficiency. Then there exists a $B \in \mathscr{A}(E)$ with $ABA = A$, and one sees like in §24 (between (24.2) and (24.3)) that BA and AB are projectors in $\mathscr{A}(E)$ and that E is projected onto $N(A)$ by $K_1 := I - BA$ and onto a complement of $A(E)$ along $A(E)$ by $K_2 := I - AB$. Since dim $N(A)$ and codim $A(E)$ are finite, the operators K_1, K_2 have finite rank. Thus the existence of equations (25.6) is proved. Let us now conversely assume (25.6), and let us suppose again without restricting the generality that (E, E^+) is a left dual system. First it follows by Proposition 23.1 that A has finite deficiency. Furthermore we have $A(E) \supset (AC)(E) = (I - K_2)(E) \supset N(K_2)$, from where we obtain with the aid of Lemmas 25.2 and 25.3 the $\mathscr{A}(E)$-projectability of the image space $A(E)$. Thus there exists also a projector $P \in \mathscr{A}(E)$ which projects E along $A(E)$ onto a complement of $A(E)$; in particular, P and also $K_2 - P$ are finite-dimensional. Since $\mathscr{A}(E)$ is saturated, we can represent $K_2 - P$ in the form

$$(K_2 - P)x = \sum_{i=1}^{n} \langle x, x_i^+ \rangle y_i \tag{25.7}$$

with linearly independent vectors $x_1^+, \ldots, x_n^+$ from E^+. By Lemma 25.1 we have therefore

$$(25.8) \qquad (K_2 - P)(E) = [y_1, \ldots, y_n].$$

Because of the second equation in (25.6) we have

$$(25.9) \qquad AC + (K_2 - P) = I - P,$$

and since $I - P$ projects the space E onto $A(E)$, we obtain with the help of (25.8) that $[y_1, \ldots, y_n] = (K_2 - P)(E) = (I - P - AC)(E) \subset A(E)$. Since thus every y_i lies in $A(E)$, there exist z_i with $y_i = Az_i$. The operator K defined by

$$Kx := \sum_{i=1}^{n} \langle x, x_i^+ \rangle z_i$$

lies in $\mathscr{A}(E)$ because this algebra is saturated, and so, with the aid of (25.7), it follows that

$$AKx = \sum_{i=1}^{n} \langle x, x_i^+ \rangle Az_i = \sum_{i=1}^{n} \langle x, x_i^+ \rangle y_i = (K_2 - P)x \qquad \text{for all} \quad x \in E,$$

i.e., $K_2 - P = AK$. From here we get, using (25.9),

$$I - P = AC + (K_2 - P) = AC + AK = A(C + K)$$

and therefore

$$A(C + K)A = (I - P)A = A - PA = A$$

(PA vanishes because of $N(P) = A(E)$). Thus A is relatively regular. With this everything is proved. ■

We want to state separately an observation made in the first part of the proof:

Proposition 25.1. *The nullspace and the image space of a relatively regular endomorphism A belonging to an operator algebra $\mathscr{A}(E)$ containing I are $\mathscr{A}(E)$-projectable. More precisely: There exists $B \in \mathscr{A}(E)$ so that E is projected by $I - BA$ onto $N(A)$ and by AB onto $A(E)$.*

The next proposition specializes Theorem 25.1 to the E'-saturated algebra $\mathscr{L}(E)$.

Proposition 25.2. *A continuous endomorphism of the normed space E is a Fredholm operator if and only if its residue class in $\mathscr{L}(E)/\mathscr{F}(E)$ is invertible.*

Since the invertible elements of an algebra form a multiplicative group, we obtain from Theorem 25.1 immediately the following analogue of Proposition 23.2:

Proposition 25.3. *The following assertions are valid concerning the set $\Phi := \Phi(\mathscr{A}(E))$ of Fredholm operators in the saturated algebra $\mathscr{A}(E)$:*

(a) *Φ is a multiplicative semi-group.*
(b) *If the product AB lies in Φ, then either both factors lie in Φ or none.*
(c) *With A also $A + K$ lies in Φ for every $K \in \mathscr{F}(\mathscr{A}(E))$.*

The finite-dimensional operators in an E^+-saturated algebra are constructed with the help of the space E^+. Thus one will obtain 'many' operators of finite rank if E^+ is 'large'. The following proposition makes this indication precise.

Proposition 25.4. *If $\mathscr{A}(E)$ is an E^+-saturated algebra and (E, E^+) a* right dual system, *then the following assertions are valid:*

(a) *Given the linearly independent vectors $x_1, \ldots, x_n$ and the arbitrary vectors $y_1, \ldots, y_n$, there exists a finite-dimensional operator K in $\mathscr{A}(E)$ with $Kx_i = y_i$ for $i = 1, \ldots, n$.*
(b) *Every finite-dimensional subspace of E is $\mathscr{A}(E)$-projectable.*
(c) *Every finite-dimensional operator in $\mathscr{A}(E)$ is relatively regular.*

In order to prove (a) we determine according to Proposition 15.1 vectors $x_1^+, \ldots, x_n^+$ in E^+ with $\langle x_i, x_k^+ \rangle = \delta_{ik}$. Then the operator K defined by

$$Kx := \sum_{k=1}^{n} \langle x, x_k^+ \rangle y_k \tag{25.10}$$

clearly satisfies the requirement. To prove (b) let $\{x_1, \ldots, x_n\}$ be a basis of the finite-dimensional subspace F. If we replace in (25.10) every y_k by x_k, then K projects the space E onto F (we have already made such a consideration immediately before Proposition 24.2). Now we prove (c). Let $\{x_1, \ldots, x_n\}$ be a basis of the image space of $A \in \mathscr{F}(\mathscr{A}(E))$. Then there exist vectors $y_1, \ldots, y_n$ with $Ay_i = x_i$ and by (a) an operator $B \in \mathscr{F}(\mathscr{A}(E))$ such that $Bx_i = y_i$. Then $ABx_i = Ay_i = x_i$, and since Ax has for every x the form $Ax = \sum_{i=1}^n \alpha_i(x)x_i$, it follows that

$$ABAx = \sum_{i=1}^{n} \alpha_i(x) ABx_i = \sum_{i=1}^{n} \alpha_i(x)x_i = Ax,$$

and so $ABA = A$. ■

Between the saturation of an algebra and the conjugability of the operators belonging to it there are close and important connections. First we prove:

Proposition 25.5. *If (E, E^+) is a* left dual system *and $\mathscr{A}(E)$ an E^+-saturated algebra of operators, then every $A \in \mathscr{A}(E)$ is E^+-conjugable.*

Proof. Let A^* be the operator algebraically dual to A and x^+ an arbitrary element of E^+. Because of Proposition 16.4 we only need to show that A^*x^+ lies in E^+. Since this is trivial in the case $A^*x^+ = 0$, we may assume that $A^*x^+ \neq 0$.

With a vector $y \neq 0$ from E we define the operator K by $Kx := \langle x, x^+ \rangle y$. Since $\mathscr{A}(E)$ is E^+-saturated, K lies in $\mathscr{A}(E)$. Then also the finite-dimensional operator KA belongs to $\mathscr{A}(E)$. We can represent KA in the form

$$KAx = \langle Ax, x^+ \rangle y = \langle x, A^* x^+ \rangle y, \tag{25.11}$$

and because $\mathscr{A}(E)$ is E^+-saturated, also in the form

$$KAx = \sum_{\nu=1}^{n} \langle x, x_\nu^+ \rangle x_\nu \qquad (x_\nu \in E,\ x_\nu^+ \in E^+); \tag{25.12}$$

the elements $x_1^+, \ldots, x_n^+$ may be chosen linearly independent. It now follows from Lemma 25.1 that

$$(KA)(E) = [y] = [x_1, \ldots, x_n],$$

consequently $x_\nu = \alpha_\nu y$ $(\nu = 1, \ldots, n)$. With this (25.12) changes into the equation

$$KAx = \sum_{\nu=1}^{n} \langle x, x_\nu^+ \rangle \alpha_\nu y = \langle x, y^+ \rangle y \qquad \text{with} \quad y^+ := \sum_{\nu=1}^{n} \alpha_\nu x_\nu^+ \in E^+.$$

If we invoke (25.11) we see (since $y \neq 0$) that we must have

$$\langle x, A^* x^+ \rangle = \langle x, y^+ \rangle \qquad \text{for all} \quad x \in E.$$

Because (E, E^+) is a left dual system, it follows that the element $A^* x^+$ coincides with y^+ and thus belongs indeed to E^+. ■

From the last proposition in combination with Proposition 16.5 we immediately obtain:

Proposition 25.6. *If (E, E^+) is a dual system, then the operator algebra $\mathscr{A}(E)$ is E^+-saturated if and only if the following conditions are fulfilled:*

(a) *I lies in $\mathscr{A}(E)$,*
(b) *every $A \in \mathscr{A}(E)$ is E^+-conjugable,*
(c) *every E^+-conjugable endomorphism of finite rank belongs to $\mathscr{A}(E)$.*

Thus in case of a dual system (E, E^+) the algebra $\mathscr{A}_\sigma(E)$ of all E^+-conjugable endomorphisms of E is the largest E^+-saturated algebra of operators on E, more precisely: $\mathscr{A}_\sigma(E)$ is itself E^+-saturated and contains every E^+-saturated algebra of operators $\mathscr{A}(E)$.

Exercises

$^+$1. Let A be an operator from the saturated algebra $\mathscr{A}(E)$. Together with A also A^n $(n = 0, 1, 2, \ldots)$ lies in $\Phi(\mathscr{A}(E))$. Conversely, if a power A^n $(n \geqq 1)$ lies in $\Phi(\mathscr{A}(E))$, then so does A.

2. The only difficulty in the proof of Theorem 25.1 was to show the relative regularity of A when the equations (25.6) hold (in fact only the validity of the second equation was used). Show with the help of Exercise 6 from §24: If $\mathscr{A}(E)$ is an E^+-saturated algebra and (E, E^+) a right dual system, then it follows from the validity of one of the equations (25.6) that A is relatively regular; cf. Exercise 8.

3. For a finite-dimensional operator K from the E^+-saturated algebra $\mathscr{A}(E)$ the following assertions are equivalent:

(a) K is relatively regular.

(b) The image space $K(E)$ is $\mathscr{A}(E)$-projectable.

(c) There exists a subspace G^+ of E^+ so that $(K(E), G^+)$, with the bilinear form induced by (E, E^+), is a dual system.

$^+$4. Let (E, E^+) be a left dual system and $\mathscr{A}(E)$ an E^+-saturated algebra. Show: Every finite-dimensional operator from $\mathscr{A}(E)$ is relatively regular if and only if either $E^+ = \{0\}$ or (E, E^+) is a dual system. *Hint*: Exercise 3.

5. Let K be a finite-dimensional operator from the saturated algebra $\mathscr{A}(E)$. Obviously $E = K^0(E) \supset K(E) \supset K^2(E) \supset \cdots$, and since $\dim K(E)$ is finite, there exists an exponent q such that $K^q(E) = K^{q+1}(E)$. Show that K^q is relatively regular. *Hint*: Use Exercise 3 and the fact that K^q maps the subspace $K^q(E)$ onto itself.

$^+$6. Assume that the algebra $\mathscr{A}(E)$ is saturated and let $\hat{A}$ denote the residue class of $A \in \mathscr{A}(E)$ in $\hat{\mathscr{A}} := \mathscr{A}(E)/\mathscr{F}(\mathscr{A}(E))$. Show that $i(\hat{A}) := \operatorname{ind}(A)$ defines a homomorphic map i from the multiplicative group $\mathscr{G}$ of all invertible elements of $\hat{\mathscr{A}}$ into the additive group of integers. The kernel of this homomorphism is an invariant subgroup of $\mathscr{G}$ and consists of the residue classes of Fredholm operators with index 0.

$^+$7. A (two-sided) ideal $\mathscr{J}$ in the algebra $\mathscr{A}(E)$ is called a *Φ-ideal if* $\mathscr{J} \supset \mathscr{F}(\mathscr{A}(E))$ and $I - K \in \Phi(\mathscr{A}(E))$ for all $K \in \mathscr{J}$. Let $\mathscr{A}(E)$ be saturated and $\mathscr{J}$ a Φ-ideal in $\mathscr{A}(E)$. Let the residue class of $A \in \mathscr{A}(E)$ in $\mathscr{A}(E)/\mathscr{J}$ be denoted by $\bar{A}$. Show: $A \in \Phi(\mathscr{A}(E)) \Leftrightarrow \bar{A}$ is invertible.

$^+$8. An algebra of operators $\mathscr{A}(E)$ is said to be *normal* if it contains I and if all $K \in \mathscr{F}(\mathscr{A}(E))$ are relatively regular (cf. Exercise 4). For a normal $\mathscr{A}(E)$ show that $A \in \mathscr{A}(E)$ is a Fredholm operator if and only if the residue class $\hat{A}$ of A in $\mathscr{A}(E)/\mathscr{F}(\mathscr{A}(E))$ is invertible.

$^+$9. Let $\mathscr{A}(E)$ be an algebra of operators and $\hat{\mathscr{A}} := \mathscr{A}(E)/\mathscr{F}(\mathscr{A}(E))$. We say that $A \in \mathscr{A}(E)$ is an *Atkinson operator* (in $\mathscr{A}(E)$) if at least one of its deficiencies $\alpha(A)$, $\beta(A)$ is finite and A is relatively regular (in $\mathscr{A}(E)$), cf. [92]. Let $\mathrm{A}(\mathscr{A}(E))$ be the set of all Atkinson operators in $\mathscr{A}(E)$, $\mathrm{A}_\alpha := \{A \in \mathrm{A}(\mathscr{A}(E)) : \alpha(A) < \infty\}$, $\mathrm{A}_\beta := \{A \in \mathrm{A}(\mathscr{A}(E)) : \beta(A) < \infty\}$. We shall denote henceforth these sets more briefly by $\mathrm{A}, \mathrm{A}_\alpha, \mathrm{A}_\beta$. For $A \in \mathrm{A}$ the index is defined by

$$\operatorname{ind}(A) := \begin{cases} \alpha(A) - \beta(A) & \text{if} \quad A \in \mathrm{A}_\alpha \cap \mathrm{A}_\beta = \Phi(\mathscr{A}(E)), \\ +\infty & \text{if} \quad \alpha(A) = +\infty, \\ -\infty & \text{if} \quad \beta(A) = +\infty. \end{cases}$$

Let $\hat{A}$ be the residue class of $A \in \mathscr{A}(E)$ in $\hat{\mathscr{A}}$. Show that if $\mathscr{A}(E)$ is normal, then the following assertions are valid:

(a) $A \in \mathrm{A}_\alpha \Leftrightarrow \hat{A}$ is left-invertible.

(b) $A \in \mathrm{A}_\beta \Leftrightarrow \hat{A}$ is right-invertible.

(c) $\mathrm{A}_\alpha, \mathrm{A}_\beta$ are (multiplicative) semi-groups.

(d) $AB \in \mathrm{A}_\alpha \Rightarrow B \in \mathrm{A}_\alpha$; $AB \in \mathrm{A}_\beta \Rightarrow A \in \mathrm{A}_\beta$.

(e) $A \in \mathrm{A} \Rightarrow A + K \in \mathrm{A}$ for all $K \in \mathscr{F}(\mathscr{A}(E))$ and $\operatorname{ind}(A + K) = \operatorname{ind}(A)$.

(f) $\hat{A}$ can be replaced in (a) and (b) by the residue class $\bar{A}$ of A modulo an arbitrary Φ-ideal in $\mathscr{A}(E)$ (cf. Exercise 7). *Hint*: Exercise 6 in §24 and Proposition 25.1.

§26. Representation theorems for Fredholm operators

We want to show in this section that one obtains all Fredholm operators in a saturated algebra $\mathscr{A}(E)$ if one adds finite-dimensional endomorphisms to left- or right-invertible operators. Invertibility is to be understood here in the sense of the theory of algebras: $A \in \mathscr{A}(E)$ is e.g., left-invertible in $\mathscr{A}(E)$ (or has a left-inverse in $\mathscr{A}(E)$), if there exists a $B \in \mathscr{A}(E)$ with $BA = I$.

For our investigations we need the following proposition which can be proved without difficulty with the help of Proposition 25.1.

Proposition 26.1. *An operator A from an algebra $\mathscr{A}(E)$ containing I has a left-inverse (right-inverse, inverse) in $\mathscr{A}(E)$ if and only if A is relatively regular and injective (surjective, bijective).*

Observe that this proposition, in connection with Proposition 24.1, gives a precise answer to the question: when does a continuous endomorphism on the normed space E have a one- or two-sided inverse in $\mathscr{L}(E)$; cf. Exercise 4.

Proposition 26.2. *The operator A from the E^+-saturated algebra $\mathscr{A}(E)$ is a Fredholm operator with* $\operatorname{ind}(A) \leqq 0$ $(\geqq 0, =0)$ *if and only if A has the form*

$$A = R + K \tag{26.1}$$

where $R \in \mathscr{A}(E)$ has finite deficiency and is left-invertible (right-invertible, invertible) in $\mathscr{A}(E)$, and K is finite-dimensional.

For the proof we may assume that (E, E^+) is a left dual system. Let first A be a Fredholm operator and

$$m := \alpha(A), \qquad n := \beta(A), \qquad p := \min(m, n).$$

By Proposition 25.1 there exists a projector P in $\mathscr{A}(E)$ such that

$$P(E) = N(A); \tag{26.2}$$

since $\mathscr{A}(E)$ is saturated, we can represent it in the form

$$Px = \sum_{i=1}^{m} \langle x, x_i^+ \rangle x_i, \tag{26.3}$$

where the vectors $x_1, \ldots, x_m$ from E and the vectors $x_1^+, \ldots, x_m^+$ from E^+ are linearly independent. Observe that because of Lemma 25.1 $\{x_1, \ldots, x_m\}$ is a basis of $P(E)$, hence $Px_k = x_k$ and so

$$(26.4) \qquad \langle x_i, x_k^+ \rangle = \delta_{ik}.$$

With the aid of a basis $\{y_1, \ldots, y_n\}$ of a complement F of $A(E)$ in E we define $K \in \mathscr{F}(\mathscr{A}(E))$ by

$$(26.5) \qquad Kx := \sum_{i=1}^{p} \langle x, x_i^+ \rangle y_i.$$

By Proposition 25.3 the operator

$$R := A - K$$

is a Fredholm operator. We now show that it is injective or surjective or bijective according as $\operatorname{ind}(A) \leqq 0$ or $\geqq 0$ or $=0$, respectively, and because of Proposition 26.1 we will have proved by this one direction of our proposition. Let first $\operatorname{ind}(A) \leqq 0$, i.e., $p = m$. Obviously $K(E) \subset F$ and so

$$(26.6) \qquad A(E) \cap K(E) = \{0\}.$$

It follows that if $Rx = 0$, i.e., $Ax = Kx$, then

$$Ax = 0 \quad \text{and} \quad Kx = 0.$$

By (26.5) we have then $\langle x, x_i^+ \rangle = 0$ for $i = 1, \ldots, m$, hence by (26.3) $Px = 0$. Therefore x lies in $N(A) \cap N(P)$, and because of (26.2) in $P(E) \cap N(P) = \{0\}$, thus we see that R is injective. Now assume that $\operatorname{ind}(A) \geqq 0$, i.e., $p = n$. Because of Lemma 25.1 we have now

$$(26.7) \qquad K(E) = [y_1, \ldots, y_n] = F,$$

furthermore, because of (26.4), for every $x \in E$ we have

$$(26.8) \qquad KPx = \sum_{i=1}^{n} \langle Px, x_i^+ \rangle y_i$$
$$= \sum_{i=1}^{n} \left\langle \sum_{k=1}^{m} \langle x, x_k^+ \rangle x_k, x_i^+ \right\rangle y_i = \sum_{i=1}^{n} \langle x, x_i^+ \rangle y_i = Kx.$$

If we represent now an arbitrary $z \in E$ in the form

$$z = Au_1 - Ku_2$$

(this is possible because of (26.7)) and set

$$v_1 := u_1 - Pu_1, \qquad v_2 := Pu_2,$$

then it follows from (26.8) and (26.2) that

$$Kv_1 = 0 \quad \text{and} \quad Av_2 = 0,$$
$$Kv_2 = Ku_2 \quad \text{and} \quad Av_1 = Au_1,$$

respectively, and so

$$R(v_1 + v_2) = (A - K)(v_1 + v_2) = Au_1 - Ku_2 = z,$$

thus R is surjective. Finally, if $\text{ind}(A) = 0$, i.e., $p = m = n$, then, by what we have just proved, R is bijective.

Let now conversely the representation (26.1) be given and suppose for instance that R has a left inverse in $\mathscr{A}(E)$. Then by Proposition 26.1 it is relatively regular and injective. Since it was assumed to have finite deficiency, it must therefore be a Fredholm operator with $\text{ind}(R) = -\beta(R) \leqq 0$. From Propositions 25.3 and 23.3 it follows that A is a Fredholm operator and $\text{ind}(A) \leqq 0$. One proceeds analogously if R has a right inverse or an inverse in $\mathscr{A}(E)$. ■

Exercises

$^+$1. Let $\mathscr{A}(E)$ be a saturated algebra and $J := \{\text{ind}(A): A \in \Phi(\mathscr{A}(E))\}$ the set of values of the index function. To every $n \in J$ choose a Fredholm operator A_n with $\text{ind}(A_n) = n$. Show that every Fredholm operator $A \in \mathscr{A}(E)$ with $\text{ind}(A) = n$ can be represented in the form $A = A_n R + K$, where $R \in \mathscr{A}(E)$ has an inverse in $\mathscr{A}(E)$ and $K \in \mathscr{A}(E)$ is finite-dimensional. Conversely, every operator A which is so representable is a Fredholm operator in $\mathscr{A}(E)$ with $\text{ind}(A) = n$. *Hint*: Use Exercise 6 in §25 and the partition of a group into cosets modulo an invariant subgroup.

2. If E is a Banach space, then $\overline{\mathscr{F}(E)}$ is a closed Φ-ideal in $\mathscr{L}(E)$ (see Exercise 7 in §25) and $\text{ind}(I - K) = 0$ for all $K \in \overline{\mathscr{F}(E)}$.

3. If E is a Banach space, then the set $\Phi(E)$ of continuous Fredholm operators is open in $\mathscr{L}(E)$. *Hint*: Use Exercise 7 in §25 and Exercise 2.

$^+$4. For $A \in \mathscr{L}(E)$, E a normed space, there exists a left inverse $B \in \mathscr{L}(E)$ if and only if A is injective and open and $A(E)$ is continuously projectable; there exists a right inverse $C \in \mathscr{L}(E)$ if and only if A is surjective and open and $N(A)$ is continuously projectable; finally, there exists a two-sided inverse $D \in \mathscr{L}(E)$ if and only if A is bijective and open.

§27. The equation $Ax = y$ with a Fredholm operator A

In this section we want to develop, similarly as in §18, a solvability theory for the equation named in the title, with the help of the conjugate operator A^+. In order to ensure the uniqueness of A^+ we shall assume that the bilinear systems (E, E^+) used will be left dual systems. We have seen earlier that this is not an essential hypothesis (cf. §20 and §25). It guarantees furthermore that every operator in an E^+-saturated algebra $\mathscr{A}(E)$ is automatically E^+-conjugable (s. Proposition 25.5). We shall make frequent use of this fact without always mentioning it explicitly.

Proposition 27.1. *Let (E, E^+) be a left dual system, $\mathscr{A}(E)$ an E^+-saturated algebra of operators and A a Fredholm operator in $\mathscr{A}(E)$. Denote, as usual, by A^+ the endomorphism of E^+ conjugate to A. Then the following criterion of solvability is valid: The equation*

$$Ax = y \tag{27.1}$$

is solvable if and only if

(27.2) $$\langle y, x^+\rangle = 0 \quad \text{for all} \quad x^+ \in N(A^+);$$

furthermore we have

(27.3) $$\beta(A) = \alpha(A^+).$$

Proof. If (27.1) is solvable and $x^+ \in N(A^+)$, then $\langle y, x^+\rangle = \langle Ax, x^+\rangle = \langle x, A^+x^+\rangle = \langle x, 0\rangle = 0$, so that (27.2) is valid. Conversely, let (27.2) be satisfied and suppose that (27.1) is not solvable, i.e., that y is not in $A(E)$. Because of Proposition 25.1, there exists a decomposition $E = A(E) \otimes F$ and a projector P in $\mathscr{A}(E)$ such that

$$P(E) = F, \qquad N(P) = A(E).$$

Since $\mathscr{A}(E)$ is saturated, we can represent P in the form

(27.4) $$Px = \sum_{i=1}^{n} \langle x, x_i^+\rangle x_i$$

with linearly independent vectors $x_1, \ldots, x_n$; thus we have

(27.5) $$A(E) = N(P) = \{z \in E : \langle z, x_i^+\rangle = 0 \text{ for } i = 1, \ldots, n\}.$$

From here it follows on the one hand that $\langle x, A^+x_i^+\rangle = \langle Ax, x_i^+\rangle = 0$ for all x in E, hence

(27.6) $$x_i^+ \in N(A^+) \qquad \text{for} \quad i = 1, \ldots, n$$

and so, because of the hypothesis (27.2),

(27.7) $$\langle y, x_i^+\rangle = 0 \qquad \text{for} \quad i = 1, \ldots, n;$$

and on the other hand it follows from (27.5), because of the hypothesis '$y \notin A(E)$' that $\langle y, x_i^+\rangle \neq 0$ for at least one i. This contradiction to (27.7) shows that y must lie in $A(E)$. Now we prove (27.3). For this we assume that in (27.4) also the vectors $x_1^+, \ldots, x_n^+$ are chosen to be linearly independent; by Lemma 25.1 we have then $F = P(E) = [x_1, \ldots, x_n]$, and so $\beta(A) = n$. With the aid of (27.6) it follows from here that $\beta(A) \leqq \alpha(A^+)$. In order to prove the opposite inequality, let $y_1^+, \ldots, y_m^+$ be linearly independent vectors from $N(A^+)$. By Proposition 15.1 there exist linearly independent elements $y_1, \ldots, y_m$ in E with $\langle y_i, y_k^+\rangle = \delta_{ik}$. If a linear combination $y := \alpha_1 y_1 + \cdots + \alpha_m y_m$ of these elements lies in $A(E)$, i.e., if $y = Ax$, then we have

$$\begin{aligned} \alpha_k &= \langle \alpha_1 y_1 + \cdots + \alpha_m y_m, y_k^+\rangle = \langle Ax, y_k^+\rangle = \langle x, A^+ y_k\rangle \\ &= 0 \qquad \text{for} \quad k = 1, \ldots, m, \end{aligned}$$

hence $y = 0$. Therefore $[y_1, \ldots, y_m] \cap A(E) = \{0\}$ and so $m \leqq \beta(A)$ and thus also $\alpha(A^+) \leqq \beta(A)$. ■

Corollary. *If under the hypotheses of Proposition* 27.1 *the vectors* $x_1, \ldots, x_n$ *and* $x_1^+, \ldots, x_n^+$ *in* (27.4) *are linearly independent, then* $x_1^+, \ldots, x_n^+$ *is a basis of* $N(A^+)$.

We want to examine now A^+ more closely, and for this we make some preparations. Since the algebraic dual operator of a finite-dimensional endomorphism is itself again finite-dimensional, as (15.9) and Example 16.3 show, we obtain from Proposition 16.4 immediately:

Lemma 27.1. *If* (E, E^+) *is a left dual system and* K *a finite-dimensional,* E^+*-conjugable endomorphism of* E, *then* K^+ *is also finite-dimensional.*

Now we want to agree on one more notation. If $\mathscr{A}(E)$ is an algebra of E^+-conjugable operators—and (E, E^+) again a left dual system—then denote by $\mathscr{A}^+(E^+)$ the set of all operators of E^+ which are conjugate to endomorphisms from $\mathscr{A}(E)$:

$$\mathscr{A}^+(E^+) := \{A^+ : A \in \mathscr{A}(E)\}.$$

Because of Proposition 16.3 $\mathscr{A}^+(E^+)$ is an algebra of operators it contains the identity transformation I^+ of E^+ whenever $\mathscr{A}(E)$ contains the identity transformation I of E. If E is a normed space with the (normed) dual E', then $\mathscr{L}'(E') := \{A' : A \in \mathscr{L}(E)\}$ is a subalgebra of the algebra $\mathscr{L}(E')$ of all continuous endomorphisms of the Banach space E'; the continuity of the dual operator A' was asserted in Example 16.2.

Proposition 27.2. *If* (E, E^+) *is a dual system and* $\mathscr{A}(E)$ *an* E^+*-saturated algebra, then* $\mathscr{A}^+(E^+)$ *is an* E*-saturated algebra and*

$$\mathscr{F}(\mathscr{A}^+(E^+)) := \{K^+ : K \in \mathscr{F}(\mathscr{A}(E))\}.$$

Proof. Every finite-dimensional endomorphism $\bar{K}$ of E^+ of the form $\bar{K}x^+ = \sum_{i=1}^n \langle x^+, x_i\rangle x_i^+$ lies in $\mathscr{A}^+(E^+)$ since it is conjugate to the operator K from $\mathscr{A}(E)$ defined by $Kx := \sum_{i=1}^n \langle x, x_i^+\rangle x_i$ (see Example 16.3). Now let A^+ be an arbitrary operator of finite rank from $\mathscr{A}^+(E^+)$ conjugate to $A \in \mathscr{A}(E)$. Then A is E-conjugate to A^+, hence by Lemma 27.1 finite-dimensional and so representable in the form $Ax = \sum_{i=1}^n \langle x, x_i^+\rangle x_i$; by Example 16.3 already mentioned, we have $A^+x^+ = \sum_{i=1}^n \langle x^+, x_i\rangle x_i^+$, i.e., A^+ has the required representation. ■

If now under the hypotheses of the last proposition A is a Fredholm operator in $\mathscr{A}(E)$, then it follows from Theorem 25.1 by conjugation of the equations (25.6) that A^+ is a Fredholm operator in the E-saturated algebra $\mathscr{A}^+(E^+)$. If we apply Proposition 27.1 to the Fredholm operator A^+ and to its E-conjugate operator A—all this in the framework of the left dual system (E^+, E) and the algebra $\mathscr{A}^+(E^+)$—then we obtain the following theorem on solvability:

Theorem 27.1. *If* (E, E^+) *is a* dual system *and* $\mathscr{A}(E)$ *an* E^+*-saturated algebra, then for every Fredholm operator* A *in* $\mathscr{A}(E)$ *the conjugate operator* A^+ *is a*

Fredholm operator in $\mathscr{A}^+(E^+)$*—and thus also in every operator algebra* $\mathscr{B}(E^+) \supset \mathscr{A}^+(E^+)$*—furthermore we have the equations*

$$(27.8) \quad \alpha(A) = \beta(A^+), \qquad \beta(A) = \alpha(A^+) \qquad \textit{and so} \quad \operatorname{ind}(A) = -\operatorname{ind}(A^+).$$

Finally:

$$(27.9) \qquad Ax = y \textit{ is solvable } \Leftrightarrow \langle y, x^+ \rangle = 0 \textit{ for all } x^+ \in N(A^+),$$

$$(27.10) \quad A^+x^+ = y^+ \textit{ is solvable} \Leftrightarrow \langle x, y^+ \rangle = 0 \textit{ for all } x \in N(A).$$

If we want to apply this theorem to Fredholm operators in $\mathscr{L}(E)$, then the question arises again, as in §24, whether a normed space E together with its dual E' forms a dual system (E, E'). We shall attack this question in the next section and answer it positively. If we anticipate this result for a moment, we can state the following proposition:

Proposition 27.3. *If* E *is a* normed *space and* A *a Fredholm operator in* $\mathscr{L}(E)$, *then the dual operator* A' *is a Fredholm operator in* $\mathscr{L}(E')$, *and the equations* (27.8), *as well as the solvability criteria* (27.9) *and* (27.10) *are valid—of course with* A' *in the place of* A^+.

Fredholm operators on Banach spaces and on topological vector spaces will be studied in §37 and §89, respectively; see also §39 and §51.

Exercises

1. The assertions of Theorem 27.1 are valid for every endomorphism A with finite deficiency of E if $E^+ = E^*$ and $A^+ = A^*$.
2. Under the hypotheses of Theorem 27.1 the equations $\beta(A^+) = \beta(A^*)$ and $\alpha(A^+) = \alpha(A^*)$ are valid.
3. Under the hypotheses of Theorem 27.1 A is a Fredholm operator in $\mathscr{A}(E)$ if and only if A^+ is a Fredholm operator in $\mathscr{A}^+(E^+)$.

V

Four principles of functional analysis and some applications

§28. The extension principle of Hahn–Banach

In the course of our investigations we have encountered several times the question whether a normed space E, together with its dual E', forms a *dual system* (E, E') with respect to the canonical bilinear form defined by $\langle x, x' \rangle := x'(x)$. Since the bilinear system (E, E') is a left dual system, we only need to examine the question whether it is a right dual system, i.e., whether to every $x_0 \neq 0$ in E there exists a continuous linear form f for which, after suitable normalization, we have $f(x_0) = 1$. In the case of a positive answer the restriction f_0 of f to $[x_0] = \{\alpha x_0 : \alpha \in \mathbf{K}\}$ is a continuous linear form with the values $f_0(x) = f_0(\alpha x_0) = \alpha f_0(x_0) = \alpha$. If we define conversely a linear form f_0 on $[x_0]$ by $f_0(\alpha x_0) = \alpha$, then $f_0(x_0) = 1$ and f_0 is continuous by Proposition 10.3. If we can always, i.e., for any initial vector $x_0 \neq 0$, extend this continuous linear form f_0 to a continuous linear form f on the whole space E, then $f(x_0) \neq 0$; hence (E, E') is a dual system.

The extension theorem of Hahn–Banach ensures that continuous linear forms which are defined on arbitrary subspaces, not only on finite-dimensional ones, can always be extended to continuous linear forms on the entire space E and implies thus that (E, E') is a dual system. We prove it immediately in the form in which we shall need it later when studying locally convex vector spaces. Here the concept of a semi-norm will enter, which we have already mentioned in §6. We recall it once more:

The map p from a vector space into its scalar field is called a *semi-norm* if it has the following properties:

(HN1) $p(x) \geqq 0$,

(HN2) $p(\alpha x) = |\alpha| p(x)$,

(HN3) $p(x + y) \leqq p(x) + p(y)$.

For a semi-norm p one has obviously $p(0) = 0$, and p is a norm if and only if p vanishes *only* at the origin.

The following *extension theorem of Hahn–Banach* is one of the fundamental principles of functional analysis.

Principle 28.1. *Let E be a vector space over* $\mathbf{K}$, *p a semi-norm on E and F a linear subspace of E. If the linear form f defined on F satisfies the estimate*

$$(28.1) \qquad |f(x)| \leqq p(x) \quad \textit{for all} \quad x \in F,$$

then there exists a linear form g on E with the following properties:

$$(28.2) \qquad g(x) = f(x) \quad \textit{for all} \quad x \in F,$$

$$(28.3) \qquad |g(x)| \leqq p(x) \quad \textit{for all} \quad x \in E.$$

Thus g extends f to the entire space E conserving the semi-norm estimate (28.1).

The proof will be divided into two main sections.

(I) Let E be *real*, i.e., $\mathbf{K} = \mathbf{R}$. In this case for a linear form h on the subspace H of E from an estimate of the form

$$(28.4) \qquad |h(x)| \leqq p(x) \quad \text{for all} \quad x \in H$$

the estimate

$$(28.5) \qquad h(x) \leqq p(x) \quad \text{for all} \quad x \in H$$

follows trivially. Conversely, if (28.5) is valid, then $-h(x) = h(-x) \leqq p(-x) = p(x)$ for all $x \in H$, so (28.4) holds. The two estimates (28.4) and (28.5) are thus equivalent. We may therefore replace the hypothesis (28.1) by the assumption

$$(28.6) \qquad f(x) \leqq p(x) \quad \text{for all} \quad x \in F$$

and need to prove instead of (28.3) only the estimate

$$(28.7) \qquad g(x) \leqq p(x) \quad \text{for all} \quad x \in E.$$

We first show in partial step (Ia) that f can be extended in the required way to a 'small' superspace of F.

(Ia) Let $x_0 \notin F$ and H be the linear hull of $\{x_0\} \cup F$, i.e.,

$$H = \{\alpha x_0 + y : \alpha \in \mathbf{R},\ y \in F\};$$

here the element $\alpha x_0 + y$ determines its components $\alpha \in \mathbf{R}$ and $y \in F$ uniquely. A *linear* extension h of f to H must have, because of $h(x) = h(\alpha x_0 + y) = \alpha h(x_0) + f(y)$, the form

$$(28.8) \qquad h(x) = \alpha \xi_0 + f(y) \quad \text{with} \quad \xi_0 \in \mathbf{R}.$$

Conversely, every h which, with an *arbitrary* $\xi_0 \in \mathbf{R}$, is defined by (28.8), extends f linearly to H. We will be able to extend f to H in the required way, i.e., conserving linearity *and* the estimate (28.6), if there exists a $\xi_0 \in \mathbf{R}$ with

$$(28.9) \qquad \alpha \xi_0 + f(y) \leqq p(\alpha x_0 + y) \quad \text{for all} \quad \alpha \in \mathbf{R} \quad \text{and all} \quad y \in F.$$

For such an ξ_0 we have necessarily

$$\xi_0 \leqq p(x_0 + y) - f(y) \qquad \text{for all} \quad y \in F \tag{28.10}$$

(set $\alpha = 1$), and also

$$\xi_0 \geqq -p(x_0 + y) - f(y) \qquad \text{for all} \quad y \in F \tag{28.11}$$

(replace α by -1 and y by $-y$). Conversely, if ξ_0 satisfies the last two inequalities then it also satisfies (28.9). Indeed, if we replace in (28.10) first y by $(1/\alpha)y$ with $\alpha > 0$, we get

$$\xi_0 \leqq p\left(x_0 + \frac{1}{\alpha}y\right) - f\left(\frac{1}{\alpha}y\right) = \frac{1}{\alpha}p(\alpha x_0 + y) - \frac{1}{\alpha}f(y),$$

hence (28.9) holds for $\alpha > 0$. If we now replace in (28.11) y by $(1/\alpha)y$ with $\alpha < 0$, we see quite similarly that (28.9) holds also for $\alpha < 0$. Finally, (28.9) holds for $\alpha = 0$ by the hypothesis (28.6). We can thus extend f in the required way to H exactly if there exists a ξ_0 which satisfies the estimates (28.10) and (28.11). Such a ξ_0 exists if and only if

$$\sup_{y \in F}\{-p(x_0 + y) - f(y)\} \leqq \inf_{y \in F}\{p(x_0 + y) - f(y)\}$$

holds. But this inequality holds indeed, since for u, v in F we have

$$\begin{aligned} f(u) - f(v) = f(u - v) &\leqq p(u - v) = p((x_0 + u) + (-x_0 - v)) \\ &\leqq p(x_0 + u) + p(x_0 + v) \end{aligned}$$

and so

$$-p(x_0 + v) - f(v) \leqq p(x_0 + u) - f(u).$$

Thus f can be extended to H as required. The extendability to the whole space E will be proved in part (Ib) with the help of Zorn's lemma.

(Ib) Let $\mathfrak{M}$ be the set of all linear forms h with the following properties:

$$h \text{ is defined on the space } D_h, \qquad F \subset D_h \subset E,$$

$$h(x) = f(x) \qquad \text{for all} \quad x \in F,$$

$$h(x) \leqq p(x) \qquad \text{for all} \quad x \in D_h;$$

briefly, $\mathfrak{M}$ is the set of all extensions of the required kind of f to superspaces of F. Since f lies in $\mathfrak{M}$, we have $\mathfrak{M} \neq \emptyset$. In $\mathfrak{M}$ we define an order relation '$\prec$' by agreeing that

$$h_1 \prec h_2 \Leftrightarrow [D_{h_1} \subset D_{h_2} \quad \text{and} \quad h_1(x) = h_2(x) \qquad \text{for all} \quad x \in D_{h_1}].$$

For every totally ordered subset $\mathfrak{K}$ of $\mathfrak{M}$ the union $D := \bigcup_{h \in \mathfrak{K}} D_h$ is obviously a linear subspace of E. On D we define a map h_0 by $h_0(x) := h(x)$ if $x \in D_h$ for some $h \in \mathfrak{K}$. The total order of $\mathfrak{K}$ ensures that h_0 is uniquely defined and linear. Trivially $h_0(x) \leqq p(x)$ for all $x \in D_{h_0} = D$, thus h_0 lies in $\mathfrak{M}$. Furthermore $h \prec h_0$ for all $h \in \mathfrak{K}$, thus h_0 is even an upper bound of $\mathfrak{K}$ in $\mathfrak{M}$. By Zorn's lemma

$\mathfrak{M}$ possesses a maximal element g, i.e., a linear form which does not have a proper extension bounded from above by p. Because of (Ia) we must have $D_g = E$. Thus the proof for the case of a real vector space is concluded.

(II) Let now E be *complex*, i.e., $\mathbf{K} = \mathbf{C}$. For the linear form h on E let

$$h(x) = h_1(x) + ih_2(x)$$

be the decomposition into real and imaginary part. Then

$$h(ix) = h_1(ix) + ih_2(ix) = ih(x) = ih_1(x) - h_2(x)$$

thus $h_1(ix) = -h_2(x)$ and so

$$h(x) = h_1(x) - ih_1(ix). \tag{28.12}$$

h_1 is a linear form on the real space E_r belonging to the space E (see Exercise 2 in §4). Conversely, for every linear form h_1 on E_r a linear form h on E is defined by (28.12). Let us now represent f in the form $f(x) = f_1(x) - if_1(ix)$. For the linear form f_1 on the real space F_r belonging to F we have, because of (28.1), the estimate $|f_1(x)| \leqq |f(x)| \leqq p(x)$ for all $x \in F_r$. By (I) there exists a linear form g_1 on E_r such that

$$g_1(x) = f_1(x) \quad \text{for all} \quad x \in F_r, \qquad |g_1(x)| \leqq p(x) \quad \text{for all} \quad x \in E_r. \tag{28.13}$$

If we define according to (28.12) the linear form g on E by $g(x) := g_1(x) - ig_1(ix)$, then $g(x) = f(x)$ for $x \in F$. Further we obtain with the help of the polar representation $g(x) = \rho e^{i\varphi}$, $\rho \geqq 0$, that $g(e^{-i\varphi}x) = e^{-i\varphi}g(x) = \rho$ is real and thus $g(e^{-i\varphi}x) = g_1(e^{-i\varphi}x)$. By (28.13) we have for all $x \in E$

$$|g(x)| = |e^{-i\varphi}g(x)| = |g(e^{-i\varphi}x)| = |g_1(e^{-i\varphi}x)| \leqq p(e^{-i\varphi}x) = p(x).$$

Thus we have proved that g extends the linear form f in the required way. The proof of Principle 28.1 is now concluded. ■

The following theorem is the *Hahn–Banach extension theorem* for *normed* spaces.

Theorem 28.1. *Given a* continuous *linear form f on the subspace F of the normed space E, there exists a* continuous *linear form g on the whole space E with the following properties*:

$$g(x) = f(x) \quad \textit{for all} \quad x \in F, \qquad \|g\| = \|f\|.$$

Proof. We define on E a semi-norm p by $p(x) := \|f\|\,\|x\|$. Then we have $|f(x)| \leqq p(x)$ for $x \in F$. By the above principle there exists an extension g of f to E which satisfies the estimate $|g(x)| \leqq \|f\|\,\|x\|$ for $x \in E$. Thus g is continuous and $\|g\| \leqq \|f\|$. On the other hand, $|f(x)| = |g(x)| \leqq \|g\|\,\|x\|$ for $x \in F$ and so $\|f\| \leqq \|g\|$, thus altogether $\|f\| = \|g\|$ (because of this equality, g is also called a *norm-preserving* extension of f). ■

On the basis of our remark made at the beginning of this section, we immediately obtain from the above Theorem that a normed space E with its topological dual E' forms a dual system (E, E') with respect to the canonical bilinear form $\langle x, x' \rangle := x'(x)$. In Theorem 28.3 we will be able to make this result more precise. We remind the reader that now Propositions 24.2, 24.3 and 27.3 are completely proved.

Theorem 28.2. *Let F be a subspace of the normed space E and $x_0 \notin F$. If F is closed or if at least*

$$\delta := \inf_{x \in F} \|x - x_0\| > 0,$$

then there exists a continuous linear form f on E with the following properties:

$$f(x) = 0 \quad \textit{for} \quad x \in F, \qquad f(x_0) = \delta, \qquad \|f\| = 1.$$

Proof. Let H be the linear hull of $\{x_0\} \cup F$. The elements of H are the vectors of the form $\alpha x_0 + x$ ($\alpha \in \mathbf{K}$, $x \in F$); here α and x are uniquely determined. We define the linear form h on H by

$$h(\alpha x_0 + x) := \alpha\delta.$$

Clearly

$$h(x) = 0 \quad \text{for} \quad x \in F \quad \text{and} \quad h(x_0) = \delta.$$

h is continuous: For $y := \alpha x_0 + x$ with $\alpha \neq 0$, $x \in F$, we have

$$\|y\| = \left\| -\alpha\left(-\frac{1}{\alpha}x - x_0\right) \right\| = |\alpha| \left\| \left(-\frac{1}{\alpha}x\right) - x_0 \right\| \geqq |\alpha|\delta,$$

hence $\|h(y)\| \leqq \|y\|$, and since this inequality is obviously valid also for $\alpha = 0$, we see that h is continuous and satisfies $\|h\| \leqq 1$. We now show that also $\|h\| \geqq 1$ and therefore $\|h\| = 1$. To every $\varepsilon > 0$ there exists an $x \in F$ with $\|x - x_0\| \leqq \delta + \varepsilon$. With this x we form the vector $z := (x - x_0)/\|x - x_0\|$ for which $\|z\| = 1$, hence

$$\|h\| \geqq |h(z)| = \frac{1}{\|x - x_0\|}\delta \geqq \frac{\delta}{\delta + \varepsilon}.$$

Letting $\varepsilon \to 0$ we get $\|h\| \geqq 1$. The linear form f of our theorem is now obtained by extending h to E according to Theorem 28.1. ∎

If we choose in Theorem 28.2 the closed subspace $F := \{0\}$, then we obtain:

Theorem 28.3. *To every vector $x_0 \neq 0$ of the normed space E there exists a continuous linear form f on E such that $f(x_0) = \|x_0\|$ and $\|f\| = 1$. In particular, (E, E') is a* dual system.

Exercises

1. Compare Principle 28.1 and Theorem 28.1 with Exercise 12 in §15, furthermore Theorem 28.2 with Proposition 16.1.

*2. A map p from the vector space E into the scalar field $\mathbf{R}$ is called a *sublinear functional* on E if $p(\alpha x) = \alpha p(x)$ for $\alpha \geqq 0$ and $p(x + y) \leqq p(x) + p(y)$ hold. Prove the following

Extension theorem: Let E be a real vector space, p a sublinear functional on E, and F a linear subspace of E. If the linear form f defined on F satisfies the inequality $f(x) \leqq p(x)$ for all $x \in F$, then there exists on E a linear form g with the following properties: $g(x) = f(x)$ for all $x \in F$, $g(x) \leqq p(x)$ for all $x \in E$.

*3. A vector x_0 of the normed space E can be approximated arbitrarily closely by linear combinations of elements of the set $M \subset E$ (i.e., $x_0 \in \overline{[M]}$) if and only if every continuous linear form on E, which vanishes on M, also vanishes on x_0.

$^+$4. Let E be a normed space over $\mathbf{K}$ and $\gamma > 0$. There exists a continuous linear form f on E with $\|f\| \leqq \gamma$, which assumes at given points x_n prescribed values $f(x_n) = \alpha_n$ $(n = 1, 2, \ldots)$ if and only if for arbitrary $n \in \mathbf{N}$ and $\beta_\nu \in \mathbf{K}$ the inequality $|\sum_{\nu=1}^n \beta_\nu \alpha_\nu| \leqq \gamma \|\sum_{\nu=1}^n \beta_\nu x_\nu\|$ holds.

§29. Normal solvability

The solvability criteria of Theorem 18.1 and Proposition 27.1 motivate the following definition.

The endomorphism A of the vector space E is said to be *normally solvable with respect to the left dual system* (E, E^+) or briefly E^+*-normally solvable*, if it is E^+-conjugable and the equation

$$Ax = y \tag{29.1}$$

possesses a solution exactly if

$$\langle y, x^+ \rangle = 0 \qquad \text{for all} \quad x^+ \in N(A^+). \tag{29.2}$$

In this case we also say that the equation (29.1) is E^+*-normally solvable*.

In this definition we assume (E, E^+) to be a left dual system to make sure, among others, that the conjugate operator A^+ is uniquely determined. If E is a *normed* space, then we call an endomorphism A of E shortly *normally solvable* if it is continuous and E'-normally solvable.

According to Proposition 16.2 a linear self-map of the vector space E is always E^*-normally solvable; a Fredholm operator on a normed space is always normally solvable (Proposition 27.3).

The definition of normal solvability can be stated more concisely if we use the concept of orthogonal space. If (E, E^+) is a bilinear system and M a non-empty subset of E, then

$$M^\perp := \{x^+ \in E^+ : \langle x, x^+ \rangle = 0 \text{ for all } x \in M\}$$

is called the *space orthogonal* to M in E^+; similarly for $M \subset E^+$ the space orthogonal to M in E is defined by

$$M^\perp := \{x \in E: \langle x, x^+ \rangle = 0 \text{ for all } x^+ \in M\}.$$

$M^\perp$ is a linear subspace of E^+ or E, respectively. Under the above hypotheses on (E, E^+) and on A *the endomorphism A is E^+-normally solvable if and only if*

$$A(E) = N(A^+)^\perp. \tag{29.3}$$

We first make some remarks on orthogonal spaces. We fix a bilinear system (E, E^+) and for $(M^\perp)^\perp$ we write more briefly $M^{\perp\perp}$. The following is trivial:

Lemma 29.1. (a) *If $M \subset N$ then $N^\perp \subset M^\perp$.*
(b) *We have $M \subset M^{\perp\perp}$.*

A set M such that $M = M^{\perp\perp}$ is said to be *orthogonally closed*; such a set is always a linear subspace (of E or of E^+). The spaces E and E^+ are orthogonally closed.

Lemma 29.2. *A linear subspace M of E is orthogonally closed if and only if to every $x_0 \notin M$ there exists an $x^+ \in E^+$ so that $\langle x, x^+ \rangle = 0$ for all $x \in M$ but $\langle x_0, x^+ \rangle \neq 0$. The orthogonally closed subspaces of E^+ can be characterized analogously.*

In the first place, if $M = M^{\perp\perp}$ and $x_0 \notin M$, then also $x_0 \notin M^{\perp\perp}$, so not for all $x^+ \in M^\perp$ is $\langle x_0, x^+ \rangle = 0$; this is precisely the assertion. Conversely, assume that the condition of the lemma is satisfied. If $x_0 \notin M$, then there exists an $x^+ \in M^\perp$ with $\langle x_0, x^+ \rangle \neq 0$, hence $x_0 \notin M^{\perp\perp}$. This means that $M^{\perp\perp} \subset M$; because of Lemma 29.1b we have $M = M^{\perp\perp}$. ■

Lemma 29.3. *Every orthogonal space $M^\perp$ is orthogonally closed.*

Because of Lemma 29.1b we have $M^\perp \subset (M^\perp)^{\perp\perp} =: M^{\perp\perp\perp}$. On the other hand, from $M \subset M^{\perp\perp}$, by assertion (a) of that lemma, the opposite inclusion $M^\perp \supset M^{\perp\perp\perp}$ follows. ■

Proposition 29.1. *If (E, E^+) is a* left *dual system and A an E^+-conjugable endomorphism of E, then A is E^+-normally solvable if and only if its image space $A(E)$ is* orthogonally closed.

If A is E^+-normally solvable, i.e., if (29.3) holds, then $A(E)$ is an orthogonal space and thus, by Lemma 29.3, it is orthogonally closed. If, conversely, $A(E)$ is orthogonally closed, then one obtains the E^+-normal solvability of A by reproducing the proof of Proposition 16.2 with some unessential modifications; Proposition 16.1 used there has to be replaced by Lemma 29.2. ■

The fact whether a subspace is orthogonally closed depends on the choice of the bilinear system which has been fixed. For two particularly important bilinear systems the following proposition determines all orthogonally closed subspaces.

Proposition 29.2. *Every subspace of the vector space E is orthogonally closed with respect to* (E, E^*). *If E is a normed space, then a subspace of E is orthogonally closed with respect to* (E, E') *if and only if it is closed.*

The first assertion and one direction of the second assertion follows from Lemma 29.2, Proposition 16.1 and Theorem 28.2. The other direction is almost trivial: An orthogonally closed subspace of E is the orthogonal space of some subset of E' and as such closed. ■

Orthogonally closed subspaces of E' will be discussed in §41.

From the last two propositions it follows again that an endomorphism of E is always E^*-normally solvable. Furthermore, we obtain the important

Proposition 29.3. *A* continuous *endomorphism of a normed space is normally solvable if and only if its image is* closed.

We conclude this section with a proposition on the norm of the dual of an operator.

Proposition 29.4. *Let E, F be normed spaces with dual spaces E', F'. If $A' \in \mathscr{L}(F', E')$ is the operator dual to $A \in \mathscr{L}(E, F)$, then $\|A'\| = \|A\|$.*

Proof. The inequality $\|A'\| \leqq \|A\|$ is proved as in example 16.2. We now show $\|A\| \leqq \|A'\|$. To every x with $Ax \neq 0$ there exists according to Theorem 28.3 a $y' \in F'$ such that $\langle Ax, y'\rangle = \|Ax\|$ and $\|y'\| = 1$. With this y' we have

$$\|Ax\| = \langle Ax, y'\rangle = \langle x, A'y'\rangle \leqq \|A'y'\|\,\|x\| \leqq \|A'\|\,\|y'\|\,\|x\| = \|A'\|\,\|x\|,$$

and since this inequality trivially also holds when $Ax = 0$, we have $\|A\| \leqq \|A'\|$. ■

Exercises

*1. Under the hypotheses and with the notations of Proposition 29.4 we have

$$\overline{A(E)}^{\perp} = N(A'), \qquad \overline{A(E)} = N(A')^{\perp};$$
$$\overline{A'(F')}^{\perp} = N(A), \qquad \overline{A'(F')} \subset N(A)^{\perp};$$

in the last inclusion we have '=' whenever $\overline{A'(F')}$ is orthogonally closed (the orthogonals are taken with respect to the dual systems (E, E'), (F, F')).

$^+$2. Let E be a normed space and F a closed subspace of E. Then $(E/F)'$ is isomorphic as a normed space to $F^\perp$. *Hint*: Associate with every $f \in (E/F)'$ the linear form $x' \in E'$ defined by $x'(x) := f(\hat{x})$; here $\hat{x}$ is the residue class of x in E/F.

3. Let (E, E^+) be a bilinear system. The intersection of arbitrarily many orthogonally closed subspaces is again orthogonally closed.

§30. The normal solvability of the operators I-K with compact K

Operators of the kind named in the title arise in the study of the Fredholm integral equation

$$x(s) - \int_a^b k(s, t)x(t)dt = y(s) \quad \text{or} \quad (I - K)x = y. \tag{30.1}$$

If the kernel k is continuous, then the integral operator K is a compact self-map of the Banach space $C[a, b]$ (see Example 13.1). In Theorem 19.1 we saw, among others, that equation (30.1) is normally solvable with respect to the dual system $(C[a, b], C[a, b])$ with the bilinear form

$$\langle x, x^+ \rangle := \int_a^b x(t)x^+(t)dt.$$

In this section we study the normal solvability of $I - K$ with respect to the dual system (E, E').

Proposition 30.1. *If K is a compact self-map of the normed space E, then $I - K$ is normally solvable or, equivalently,* (Proposition 29.3): *The image space of $I - K$ is closed.*

Proof. We set $A := I - K$ and show that $y_n := Ax_n \to y$ implies $y \in A(E)$. Let

$$\alpha_n := \inf_{u \in N(A)} \|x_n - u\|;$$

for every n there exists a $u_n \in N(A)$ such that

$$\|x_n - u_n\| \leqq 2\alpha_n.$$

Setting $v_n := x_n - u_n$ we have

$$y_n = Av_n \quad \text{and} \quad \|v_n\| \leqq 2\alpha_n.$$

The sequence (v_n) is bounded: Otherwise it would contain a subsequence—which we will denote again by (v_n)—for which $\|v_n\| \to \infty$. If we set $w_n := v_n/\|v_n\|$, then

$$Aw_n = \frac{Av_n}{\|v_n\|} = \frac{y_n}{\|v_n\|} \to 0. \tag{30.2}$$

Because of $\|w_n\| = 1$, the sequence (Kw_n) contains a convergent subsequence; let us say that $Kw_{n_j} \to z$. It follows from (30.2) that $w_{n_j} = (I - K)w_{n_j} + Kw_{n_j} =$

$Aw_{n_j} + Kw_{n_j} \to z$. From here it follows, if we apply (30.2) once more, that $Az = \lim Aw_{n_j} = 0$, hence $z \in N(A)$. Consequently we have

$$\|w_n - z\| = \left\| \frac{x_n - u_n}{\|v_n\|} - z \right\| = \frac{1}{\|v_n\|} \|x_n - (u_n + \|v_n\| z)\| \geqq \frac{\alpha_n}{\|v_n\|} \geqq \frac{1}{2},$$

in contradiction to $\lim w_{n_j} = z$. Thus the sequence (v_n) is indeed bounded. Therefore (Kv_n) contains a convergent subsequence (Kv_{n_j}), thus also the sequence of the vectors $v_{n_j} = Av_{n_j} + Kv_{n_j} = y_{n_j} + Kv_{n_j}$ converges to some $v \in E$. Consequently $y = \lim y_n = \lim y_{n_j} = \lim Av_{n_j} = Av \in A(E)$. ■

Exercise

Under the hypothesis of Proposition 30.1 the image spaces of all the powers $(I - K)^n$ $(n = 0, 1, 2, \ldots)$ are closed.

§31. The Baire category principle

After we have posed and answered in the last section the question, suggested by Proposition 29.3, whether the image space of a special class of operators is closed, we will, in the next paragraphs, examine this problem in a general way.

If E and F are normed spaces and E is, furthermore, complete, then the image space of an operator $A \in \mathscr{L}(E, F)$ is certainly closed when A is injective and the inverse A^{-1} is continuous on $A(E)$; indeed, if the sequence $y_n := Ax_n$ converges to $y \in F$, then (x_n) is a Cauchy sequence because of $\|x_n - x_m\| = \|A^{-1}(y_n - y_m)\| \leqq \|A^{-1}\| \, \|y_n - y_m\|$, and converges thus to an $x \in E$, so that $y_n = Ax_n \to Ax$. Consequently, $y = Ax$ and lies therefore in $A(E)$. What we just proved, combined with Proposition 21.1, shows that $A(E)$ is already closed if the inverse of the canonical injection $\hat{A}: E/N(A) \to F$ is continuous; indeed, its image space coincides with $A(E)$. $(\hat{A})^{-1}$ is continuous according to Exercise 5 in §21 if and only if $\hat{A}$ is open; because of Proposition 21.3 this is the case exactly when A is open. We state: *$A \in \mathscr{L}(E, F)$ has a closed image space whenever E is complete and A itself is open.*

This result leads us to the question: when is a continuous map *open*. An answer will be given by the open mapping principle in the following section. In order to prove it, we need the Baire category theorem, to which this paragraph is devoted.

First we define the *diameter* $\delta(M)$ of a non-empty subset M of the metric space E by

$$\delta(M) := \sup\{d(x, y): x, y \in M\}$$

and prove the *Cantor intersection theorem*:

Proposition 31.1. *Let a sequence of non-empty closed subsets F_n with $F_1 \supset F_2 \supset \cdots$ and $\delta(F_n) \to 0$ be given in the* complete *metric space E. Then $\bigcap_{n=1}^{\infty} F_n$ contains exactly one point $x \in E$.*

Proof. We choose an arbitrary $\varepsilon > 0$ and determine an n_0 so that $\delta(F_n) \leqq \varepsilon$ for $n \geqq n_0$. If we choose for every n an $x_n \in F_n$, then clearly $d(x_n, x_m) \leqq \varepsilon$ for all $m, n \geqq n_0$, thus (x_n) is a Cauchy sequence. Because of the completeness of E, the sequence (x_n) converges to an $x \in F$. Since for every $k = 1, 2, \ldots$ the subsequence $(x_k, x_{k+1}, \ldots)$ also converges to x and is contained in the closed set F_k, x, too, belongs to F_k and so x lies in $F := \bigcap_{k=1}^{\infty} F_k$. Since for every $y \in F$ we obviously have $d(x, y) \leqq \delta(F_n)$ and since $\delta(F_n)$ tends to zero, $d(x, y)$ must vanish: F contains only the one point x. ■

We now formulate one of the fundamental theorems of functional analysis, the *Baire category theorem*:

Principle 31.1. *If the* complete *metric space E is represented as the union $E = \bigcup_{n=1}^{\infty} F_n$ of countably many closed sets F_n, then at least one F_n contains a closed (and a fortiori an open) ball.*

We make first a preliminary remark: If a closed set $F \subset E$ contains no closed ball, then every closed ball $K_r[x_0]$ contains a closed ball disjoint from F. Indeed, the ball $K_{r/2}[x_0]$ contains certainly a point $x_1 \notin F$, and since F is closed, there exists a positive $r_1 \leqq r/2$ so that $K_{r_1}[x_1] \cap F = \varnothing$. Furthermore $K_{r_1}[x_1] \subset K_r[x_0]$ since for $x \in K_{r_1}[x_1]$ we have $d(x, x_0) \leqq d(x, x_1) + d(x_1, x_0) \leqq r_1 + r/2 \leqq r$. Now let us assume that Baire's theorem is false, i.e., that no F_n contains a ball. Then there exists, according to our preliminary remark, for an arbitrarily chosen closed ball $K^{(0)}$ a closed ball $K^{(1)}$ which is disjoint from F_1; obviously we may assume $\delta(K^{(1)}) \leqq 1$. To $K^{(1)}$ there exists—again according to the preliminary remark—a closed ball $K^{(2)}$ with $K^{(2)} \subset K^{(1)}$, $\delta(K^{(2)}) \leqq \frac{1}{2}$ and $K^{(2)} \cap F_2 = \varnothing$. Proceeding in this manner we obtain a sequence of closed balls $K^{(n)}$ such that $K^{(1)} \supset K^{(2)} \supset \cdots$, $\delta(K^{(n)}) \to 0$ and

$$K^{(n)} \cap F_n = \varnothing \quad \text{for} \quad n = 1, 2, \ldots. \tag{31.1}$$

By Proposition 31.1 the intersection $\bigcap_{n=1}^{\infty} K^{(n)}$ contains (exactly) one point x_0. Because of (31.1) the point x_0 lies in none of the F_n, in contradiction to the assumed representation $E = \bigcup_{n=1}^{\infty} F_n$. Thus Principle 31.1 is proved. ■

Exercises

1. Let J be the set of the irrational numbers in the interval $[0, 1]$. Show that a representation of the form $J = \bigcup_{n=1}^{\infty} F_n$ with closed sets F_n is impossible.
2. There exists no real-valued function on the interval $[0, 1]$ which is continuous at every rational point and discontinuous at every irrational point. *Hint*: Use Exercise 1.

§32. The open mapping principle and the closed graph theorem

The following theorem, called the open mapping principle or the Banach–Schauder theorem, shows that numerous continuous linear maps are open.

Principle 32.1. *Every continuous linear map A from the Banach space E* onto *the Banach space F is open.*

We divide the proof into several steps.

(a) First we make a preliminary remark which we will need in the next step and of which the reader can convince himself immediately. If M is a subset of a normed space and $\alpha \neq 0$, then $\overline{\alpha M} = \alpha \overline{M}$; if, furthermore, M is open, then also the subsets αM and $x + M$, for any element x of the space, are open.

(b) Next we show: If U is an open ball around 0 in E, then $\overline{A(U)}$ contains an open ball around 0 in F.

Let U have radius $2r$ and let $K := K_r(0)$. Obviously $E = \bigcup_{n=1}^{\infty} nK$, hence also $F = A(E) = \bigcup_{n=1}^{\infty} nA(K)$ and a fortiori $F = \bigcup_{n=1}^{\infty} \overline{nA(K)}$. According to the Baire category principle (Principle 31.1), one of the sets $\overline{nA(K)}$, say $\overline{mA(K)}$, contains an open ball S. Because of $\overline{mA(K)} = m\overline{A(K)}$—see (a)—we have $(1/m)S \subset \overline{A(K)}$. Since, furthermore, $U \supset K - K$, and thus also $A(U) \supset A(K) - A(K)$, we obtain the inclusions

$$\overline{A(U)} \supset \overline{A(K) - A(K)} \supset \overline{A(K)} - \overline{A(K)} \supset \frac{1}{m} S - \frac{1}{m} S = \bigcup_{x \in (1/m)S} \left(x - \frac{1}{m} S\right).$$

$(1/m)S - (1/m)S$ is open as the union of the sets $x - (1/m)S$, which are open according to (a) (see Exercise 11 in §1), and contains the origin of F. Hence there exists an open ball around the origin of F contained in $(1/m)S - (1/m)S$ and thus also in $\overline{A(U)}$.

(c) Now we prove a sharpening of (b): If U is an open ball around 0 in E, then already $A(U)$ contains an open ball around 0 in F.

In the proof let E_r and F_r be open balls around 0 with radius r in E or F, respectively. We set $U := E_{2r_0}$ and choose positive numbers r_n such that

$$\sum_{n=1}^{\infty} r_n < r_0. \tag{32.1}$$

According to (b) there exist $\sigma_n > 0$ which satisfy

$$F_{\sigma_n} \subset \overline{A(E_{r_n})} \qquad \text{for} \quad n = 0, 1, 2, \ldots; \tag{32.2}$$

we may obviously assume that $\sigma_n \to 0$. We will show that

$$F_{\sigma_0} \subset A(E_{2r_0}) = A(U), \tag{32.3}$$

which will prove (c). By (32.2) each $y \in F_{\sigma_0}$ lies in $\overline{A(E_{r_0})}$, hence there exists an $x_0 \in E_{r_0}$ such that $\|y - Ax_0\| < \sigma_1$, i.e., $y - Ax_0 \in F_{\sigma_1}$. Again by (32.2) we see that $y - Ax_0 \in \overline{A(E_{r_1})}$, and so there exists $x_1 \in E_{r_1}$ so that $\|y - Ax_0 - Ax_1\| < \sigma_2$, i.e., $y - Ax_0 - Ax_1 \in F_{\sigma_2}$. Continuing in this way we obtain a sequence of vectors $x_n \in E_{r_n}$ with

$$\|y - A(x_0 + x_1 + \cdots + x_n)\| < \sigma_{n+1} \qquad \text{for} \quad n = 0, 1, 2, \ldots. \tag{32.4}$$

From $\|x_n\| < r_n$ it follows because of (32.1) that $\sum_{n=0}^{\infty} \|x_n\|$ converges; by Lemma 8.1 there exists an $x \in E$ with $x = \sum_{n=0}^{\infty} x_n$; we have $\|x\| \leqq \sum_{n=0}^{\infty} \|x_n\| \leqq r_0 + \sum_{n=1}^{\infty} r_n < 2r_0$, thus x lies in $E_{2r_0} = U$. If we now let $n \to \infty$ in (32.4), we get $\|y - Ax\| \leqq 0$, hence $Ax = y$. Thus we have obtained the required assertion (32.3).

(d) Finally, let M be an open subset of E and $y := Ax$, $x \in M$, an arbitrary vector in $A(M)$. There exists an open ball U around 0 in E such that $x + U \subset M$, and because of (c) an open ball V around 0 in F such that $V \subset A(U)$. Consequently we have

$$y + V \subset Ax + A(U) = A(x + U) \subset A(M),$$

the image $A(M)$ of the open set M is thus open in F. ■

Theorem 32.1. *The continuous linear map A from the Banach space E into the Banach space F is open if and only if the image space $A(E)$ is closed.*

Proof. Already at the beginning of §31 we have observed that if A is open, then $A(E)$ is closed. If, conversely, $A(E)$ is closed, thus also complete, then A as a continuous linear map from the Banach space E *onto* the Banach space $A(E)$ is open according to Principle 32.1, i.e., the image $A(M)$ of every open set $M \subset E$ is open in $A(E)$. So A is open also as a map from E into F. ■

In this book we have referred already several times to the importance of the *continuity* of the inverse operator. The following theorem guarantees the continuity of the inverse, whenever it exists, in the case of complete spaces.

Theorem 32.2. *If the continuous linear map A from the Banach space E onto the Banach space F is injective, then the inverse map A^{-1}* is continuous.

By Principle 32.1 the map A is open, thus $A^{-1}: F \to E$ is continuous by Exercise 5 of §21. ■

To formulate the closed graph theorem, we need some preparations.

Let E and F be normed spaces over $\mathbf{K}$. On the vector space $E \times F$ with elements (x, y) $(x \in E, y \in F)$ we introduce a norm by

$$\|(x, y)\| := \|x\| + \|y\|.$$

$E \times F$ equipped with this norm is called the *product space* of the spaces E and F. Convergence in the product space is equivalent to componentwise convergence, i.e., $(x_n, y_n) \to (x, y)$ holds if and only if $x_n \to x$ and $y_n \to y$. $E \times F$ is complete if and only if both spaces E and F are complete (cf. Exercise 6 in §6).

If D is a linear subspace of E and A a linear map from D into F, then

$$G_A := \{(x, Ax)\| : x \in D\}$$

is called the *graph* of A. Obviously the graph is a linear subspace of $E \times F$. We call the map A *closed* if its graph G_A is closed.

It is clear that A *is closed if and only if from* $x_n \in D$, $x_n \to x$ *and* $Ax_n \to y$ *it follows that* x *lies in* D *and* $Ax = y$. The following *closed graph theorem* can often be used to prove continuity.

Theorem 32.3. *A* closed *linear map* A *from the Banach space* E *into the Banach space* F *is* continuous.

Proof. The graph G_A is itself a Banach space as a closed subspace of $E \times F$. The map $P: G_A \to E$ defined by $P(x, Ax) := x$ is linear, continuous and bijective, hence by Theorem 32.2 its inverse P^{-1} is continuous, i.e., if $x_n \to x$, then $(x_n, Ax_n) \to (x, Ax)$ and, in particular, $Ax_n \to Ax$. Thus A is indeed continuous. ■

The theorems of this section have many applications to problems in analysis. Before giving some examples for this, we want to obtain from the Baire category principle another fundamental assertion in the next paragraph.

Exercises

1. Let $E := C[a, b]$ and $D \subset E$ the subspace of all $x \in E$ which have a continuous derivative x' on $[a, b]$. Let the differentiation operator $A: D \to E$ be defined by $Ax := x'$. Show that A is not continuous but it is closed. (To show that A is not continuous consider the sequence of functions $t \mapsto [(t - a)/(b - a)]^n$ on $[a, b]$, $n = 1, 2, \ldots$.)

$^+$2. Let E, F be normed spaces, D a linear subspace of E and $A: D \to F$ a linear map. Show the following:

(a) If A is continuous and D closed, then A is closed.
(b) If A is continuous and closed and F complete, then D is closed.
(c) If D is not closed, then $A := I$ is continuous but not closed.
(d) If A is closed and injective, then A^{-1} is also closed.
(e) If E is complete, A closed and injective, $A(E)$ dense in F and A^{-1} continuous, then $A(E) = F$.
(f) If A is closed and $\alpha \neq 0$, then also αA is closed.
(g) Together with A also $A - \alpha I$ is closed for every α.

$^+$3. With the notation of Exercise 2 the following hold:

(a) D becomes a normed space D_A if we define $\|x\|_A := \|x\| + \|Ax\|$ for $x \in D$.
(b) If E and F is complete and A is closed, then D_A is a Banach space and $A: D_A \to F$ is continuous.

$^+$4. We use again the notation of Exercise 2. If F is a Banach space and the closed operator $A: D \to F$ is bijective, then A^{-1} is continuous.

$^+$5. Let E be a Banach space and let (E, E) be a left dual system with respect to the continuous bilinear form $\langle x, y\rangle$ (*continuity* of the bilinear form means that $x_n \to x$, $y_n \to x$ implies that $\langle x_n, y_n\rangle \to \langle x, y\rangle$). If $A \in \mathscr{S}(E)$ is E-conjugate to itself, i.e., if $\langle Ax, x^+\rangle = \langle x, Ax^+\rangle$ for all x, x^+ in E, then A is continuous.

Hint: Show that A is closed by considering, for an arbitrary $z \in E$, the expression $\langle Ax_n, z\rangle$ in the case $x_n \to x$, $Ax_n \to y$.

$^+$6. Let the Banach space E be the direct sum of the closed subspaces F and G. Then G is a topological complement of F. *Hint*: Show that the projector P of E onto F along G is closed.

$^+$7. Let two norms $\|\cdot\|_1$, $\|\cdot\|_2$ be defined on the vector space E and assume that E is complete with respect to both norms. Show that if $\|\cdot\|_1$ is stronger than $\|\cdot\|_2$, then the two norms are equivalent.

8. All continuous linear forms and all continuous finite-dimensional operators on a normed space are open.

$^+$9. Let E_1, E_2, F be Banach spaces and $A_k: E_k \to F$ $(k = 1, 2)$ continuous linear maps. Assume that the equation $A_1x = A_2y$ has for all $x \in E_1$ exactly one solution $y \in E_2$. Show that the map $A: E_1 \to E_2$ defined by $Ax := y$ is linear and continuous.

$^+$10. Let E be a Banach space, F a normed space, $A \in \mathscr{L}(E, F)$ and $K \in \mathfrak{K}(E, F)$. Show that if $A(E) \subset K(E)$, then A is compact. *Hint*: Let $\hat{K}$ be the canonical injection of K. Show with the help of the closed graph theorem that $B := \hat{K}^{-1}A$ is continuous, and then invoke Exercise 4 of §21.

§33. The principle of uniform boundedness

Each of the two following fundamental theorems is known under the name of *principle of uniform boundedness*.

Principle 33.1. *If the family $(f_\iota: \iota \in J)$ of continuous, real-valued functions f_ι on the* complete *metric space E is* pointwise *bounded from above, then on a certain closed ball $K \subset E$ it is even bounded from above* uniformly, *i.e., with an appropriate constant γ we have*

$$(33.1) \qquad f_\iota(x) \leqq \gamma \quad \textit{for all} \quad \iota \in J \quad \textit{and all} \quad x \in K.$$

Proof. Because of the continuity of the functions f_ι the sets

$$F_n := \{x \in E: f_\iota(x) \leqq n \quad \text{for all} \quad \iota \in J\}$$

are closed for $n = 1, 2, \ldots$. From the pointwise boundedness of the family $(f_\iota: \iota \in J)$ it follows that $E = \bigcup_{n=1}^{\infty} F_n$. By the Baire category theorem (Principle 31.1), a certain F_m contains a closed ball K, so that (33.1) holds with $\gamma := m$. ■

Principle 33.2. *If E is a* complete, *F an arbitrary normed space and if the family $(A_\iota: \iota \in J)$ of continuous linear maps from E into F is* pointwise *bounded* (i.e., if to every $x \in E$ there *exists a number α_x so that $\|A_\iota x\| \leqq \alpha_x$ for all $\iota \in J$*), *then the family of* norms $(\|A_\iota\|: \iota \in J)$ *is bounded.*

Proof. The family $(p_\iota: \iota \in J)$ of continuous, real-valued functions p_ι on E, defined by $p_\iota(x) := \|A_\iota x\|$, is pointwise bounded. By the above principle there exists a closed ball $K := \{x \in E: \|x - x_0\| \leqq r\}$ and a constant γ so that $p_\iota(x) \leqq \gamma$ for all $\iota \in J$ and $x \in K$. For every $y \in E$ with $\|y\| \leqq 1$ the point $x := x_0 + ry$ lies in K, thus we have

$$\|A_\iota y\| = \left\| A_\iota\left(\frac{x - x_0}{r}\right)\right\| \leqq \frac{1}{r}[\|A_\iota x\| + \|A_\iota x_0\|] = \frac{1}{r}[p_\iota(x) + p_\iota(x_0)] \leqq \frac{2\gamma}{r},$$

and thus $\|A_\iota\| = \sup_{\|y\| \leqq 1} \|A_\iota y\| \leqq 2\gamma/r$ for all $\iota \in J$. ■

From the principle of uniform boundedness important consequences follow for pointwise convergent sequences of operators. Let a sequence of linear maps A_n from a normed space E into a normed space F be given. If $(A_n x)$ converges for every $x \in E$ to an element of F, then we can define a map $A: E \to F$ by $Ax := \lim A_n x$; obviously A is linear. If all the A_n are continuous, A does not need to be continuous (see Exercise 1). If, however, E is complete, then—since the pointwise convergent sequence (A_n) is certainly pointwise bounded—according to Principle 33.2 there exists a number β so that $\|A_n\| \leqq \beta$, i.e., $\|A_n x\| \leqq \beta\|x\|$ for $n = 1, 2, \ldots$ and all $x \in E$. For $n \to \infty$ we obtain $\|Ax\| \leqq \beta\|x\|$, thus A is bounded and $\|A\| \leqq \beta$; we clearly even have $\|A\| \leqq \liminf \|A_n\|$. We state these results as a proposition:

Proposition 33.1. *If E is a* complete, *F an arbitrary normed space and if the sequence of continuous linear operators $A_n: E \to F$ converges* pointwise *to the map $A: E \to F$, then A is linear and* continuous, *the sequence of norms $\|A_n\|$ is bounded and $\|A\| \leqq \liminf \|A_n\|$.*

In §7 we made the simple observation that from $A_n \to A$, $B_n \to B$ and $\alpha_n \to \alpha$ it follows that $A_n + B_n \to A + B$ and $\alpha_n A_n \to \alpha A$ (the arrow in $A_n \to A$ etc. means, according to the convention of §7, pointwise convergence); these limit relations are valid also in the case of non-continuous operators A_n, B_n. The question concerning the behavior of the product sequence $(B_n A_n)$ was left open but can be answered now. If E, F, G are normed spaces, if $A_n \in \mathscr{L}(E, F)$, $B_n \in \mathscr{L}(F, G)$ and if $A_n \to A$, $B_n \to B$, then, provided F is complete, the sequence $(\|B_n\|)$ is bounded according to Proposition 33.1 and from the inequality

$$\begin{aligned}\|B_n A_n x - BAx\| &= \|B_n(A_n - A)x + (B_n - B)Ax\| \leqq \|B_n\|\,\|(A_n - A)x\| \\ &\quad + \|(B_n - B)Ax\|\end{aligned}$$

we conclude that $B_n A_n \to BA$.

The following *Theorem of Banach-Steinhaus* is applied in many ways in classical analysis.

Theorem 33.1. *A sequence (A_n) of continuous linear maps from a Banach space E into a Banach space F converges pointwise to a* continuous *linear map $A: E \to F$ if and only if the following conditions* (a) *and* (b) *are both satisfied:*

(a) *The sequence of norms $\|A_n\|$ is bounded;*
(b) *The sequence $(A_n x)$ converges for all elements x of a set M dense in E.*

Proof. (a) is a necessary condition because of Proposition 33.1, the necessity of (b) is trivial. Now assume that conversely (a) and (b) are satisfied and let y be an arbitrary vector of E. We choose $\varepsilon > 0$, set $\gamma := \sup_n \|A_n\|$ and pick an $x \in M$ with $|x - y| \leqq \varepsilon/3\gamma$. Since the sequence $(A_n x)$ converges because of (b), there exists an n_0 so that $\|A_n x - A_m x\| \leqq \varepsilon/3$ for $n, m \geqq n_0$. For these subscripts n, m we have then

$$\|A_n y - A_m y\| \leqq \|A_n y - A_n x\| + \|A_n x - A_m x\| + \|A_m x - A_m y\|$$

$$\leqq \gamma \frac{\varepsilon}{3\gamma} + \frac{\varepsilon}{3} + \gamma \frac{\varepsilon}{3\gamma} = \varepsilon;$$

thus $(A_n y)$ is a Cauchy sequence in F and converges, since F is complete, to an element Ay in F. Thus (A_n) converges pointwise to the map $A: E \to F$. The linearity and continuity of A now follow from Proposition 33.1. ■

Exercises

1. We say that a sequence $(\xi_1, \xi_2, \ldots)$ has *finite support* if only finitely many of its terms are $\neq 0$. Let E be the linear subspace of l^∞ which consists of all sequences with finite support, and let $A \in \mathscr{S}(E)$ be defined by

$$A(\xi_1, \xi_2, \xi_3, \ldots) := (\xi_1, 2\xi_2, 3\xi_3, \ldots).$$

Show that A is not continuous but is the pointwise limit of a sequence of continuous linear maps $A_n: E \to E$.

2. Let (A_n) be a sequence of continuous linear maps from E into F, which converges pointwise to $A: E \to F$ (E, F normed spaces). Assume that given any closed ball K around 0 in F, the set $\bigcap_{n=1}^\infty A_n^{-1}(K)$ contains a ball around 0 in E. Then A is linear (trivial!) and continuous, and the sequence $(\|A_n\|)$ is bounded.

3. Let E be a complete, $F_\iota (\iota \in J)$ an arbitrary normed space, $A_\iota \in \mathscr{L}(E, F_\iota)$. If the family $(A_\iota: \iota \in J)$ is pointwise bounded (i.e., if for every $x \in E$ there exists an α_x such that $\|A_\iota x\| \leqq \alpha_x$ for all $\iota \in J$), then the family of norms $(\|A_\iota\|: \iota \in J)$ is bounded.

§34. Some applications of the principles of functional analysis to analysis

In this section we want to show by some examples how the principles of functional analysis can be applied to problems of classical analysis.

Example 34.1. If the coefficient functions f_0, f_1 are from $C[a, b]$, then the initial value problem

$$(34.1) \qquad x''(t) + f_1(t)x'(t) + f_0(t)x(t) = y(t), \qquad x(a) = \xi, x'(a) = \xi'$$

has exactly one solution x in $C^{(2)}[a, b]$ for every right-hand side y from $C[a, b]$ and every pair ξ, ξ' of initial values. As we shall show, x depends *continuously* on y and ξ, ξ', more precisely:

If $y_n \in C[a, b]$ and if

$$y_n(t) \to y(t) \text{ uniformly on } [a, b], \qquad \xi_n \to \xi \text{ and } \xi'_n \to \xi,$$

if furthermore for every subscript n

$$x''_n(t) + f_1(t)x'_n(t) + f_0(t)x_n(t) = y_n(t), \qquad x_n(a) = \xi_n,\ x'_n(a) = \xi'_n,$$

then uniformly on $[a, b]$

$$x_n(t) \to x(t), \qquad x'_n(t) \to x'(t) \quad \text{and} \quad x''_n(t) \to x''(t).$$

In the proof let E be the Banach space $C^{(2)}[a, b]$ with its canonical norm $\|x\| := \sum_{\nu=0}^{2} \max_{a \leqq t \leqq b} |x^{(\nu)}(t)|$, let F be the Banach space $C[a, b] \times \mathbf{K} \times \mathbf{K}$ with the product norm $\|(y, \xi, \xi')\| := \max_{a \leqq t \leqq b} |y(t)| + |\xi| + |\xi'|$, and $D: E \to C[a, b]$ the linear map defined by $(Dx)(t) := x''(t) + f_1(t)x'(t) + f_0(t)x(t)$. If we define the homomorphism $A: E \to F$ by $Ax := (Dx, x(a), x'(a))$, then the initial value problem (34.1) is equivalent to the operator equation $Ax = (y, \xi, \xi')$. The above remarks concerning problem (34.1) show that A is bijective. Since A is obviously continuous, it follows from Theorem 32.2 that A^{-1} must be continuous; from here we get immediately the above-asserted continuous dependence of the solution x on the right-hand side y and on the initial values ξ, ξ'.

Example 34.2. *Let $1 < p < \infty$ and $1/p + 1/q = 1$ (i.e., let q be the number* conjugate *to p). If the sequence (α_k) is such that $\sum_{k=1}^{\infty} \alpha_k \xi_k$ converges for* all *$x = (\xi_k) \in l^p$, then (α_k) lies in l^q.*

Proof. By

$$f_n(x) := \sum_{k=1}^{n} \alpha_k \xi_k \quad \text{and} \quad f(x) := \sum_{k=1}^{\infty} \alpha_k \xi_k \tag{34.2}$$

we define linear forms on l^p. According to the Hölder inequality

$$|f_n(x)| \leqq \left(\sum_{k=1}^{n} |\alpha_k|^q\right)^{1/q} \left(\sum_{k=1}^{n} |\xi_k|^p\right)^{1/p} \leqq \left(\sum_{k=1}^{n} |\alpha_k|^q\right)^{1/q} \|x\|$$

holds, hence f_n is continuous and

$$\|f_n\| \leqq \left(\sum_{k=1}^{n} |\alpha_k|^q\right)^{1/q}. \tag{34.3}$$

Since (f_n) converges pointwise to f, it follows from Proposition 33.1 that f is also continuous, hence for every sequence (ξ_k) in l^p we have

$$\left|\sum_{k=1}^{\infty} \alpha_k \xi_k\right| \leqq \|f\| \left(\sum_{k=1}^{\infty} |\xi_k|^p\right)^{1/p}. \tag{34.4}$$

If for a fixed $n \in \mathbf{N}$ we set

$$\xi_k := \begin{cases} \bar{\alpha}_k |\alpha_k|^{q-2} & \text{if} \quad 1 \leqq k \leqq n, \alpha_k \neq 0 \\ 0 & \text{otherwise,} \end{cases}$$

then $(\xi_k) \in l^p$, $|\xi_k|^p = |\alpha_k|^q = \alpha_k \xi_k$ for $k = 1, \ldots, n$ and thus—because of (34.4)—

$$\begin{aligned} \left(\sum_{k=1}^{n} |\alpha_k|^q \right)^{1/q} \left(\sum_{k=1}^{n} |\alpha_k|^q \right)^{1/p} &= \sum_{k=1}^{n} |\alpha_k|^q = \sum_{k=1}^{n} \alpha_k \xi_k \\ &\leqq \|f\| \left(\sum_{k=1}^{n} |\xi_k|^p \right)^{1/p} = \|f\| \left(\sum_{k=1}^{n} |\alpha_k|^q \right)^{1/p}, \end{aligned}$$

hence

$$\left(\sum_{k=1}^{n} |\alpha_k|^q \right)^{1/q} \leqq \|f\|. \tag{34.5}$$

Since this inequality holds for every natural number n, the sequence (α_k) lies in l^q. ■

Because of $\|f\| \leqq \liminf \|f_n\|$ (see Proposition 33.1) from (34.3) and (34.5) we obtain furthermore the equation

$$\|f\| = \left(\sum_{k=1}^{\infty} |\alpha_k|^q \right)^{1/q}. \tag{34.6}$$

Example 34.3. *If the sequence* (α_k) *is such that* $\sum_{k=1}^{\infty} \alpha_k \xi_k$ *converges for* every *sequence* (ξ_k) *which converges to zero, then* (α_k) *lies in* l^1.

For the proof we define again the linear forms f_n, f by (34.2) but this time on (c_0). Then f_n is continuous and

$$\|f_n\| \leqq \sum_{k=1}^{n} |\alpha_k|.$$

Since (f_n) converges pointwise to f, it follows from Proposition 33.1 that f is continuous, hence for every sequence (ξ_k) which converges to zero we have

$$\left| \sum_{k=1}^{\infty} \alpha_k \xi_k \right| \leqq \|f\| \sup_k |\xi_k|. \tag{34.7}$$

If for a fixed $n \in \mathbf{N}$ we set

$$\xi_k := \begin{cases} \dfrac{\bar{\alpha}_k}{|\alpha_k|} & \text{if} \quad 1 \leqq k \leqq n \quad \text{and} \quad \alpha_k \neq 0 \\ 0 & \text{otherwise,} \end{cases}$$

then it follows from (34.7) that

$$\sum_{k=1}^{n} |\alpha_k| \leqq \|f\|,$$

hence (α_k) lies in l^1 as asserted. ■

Just as in Example 34.2, we see furthermore that

$$\|f\| = \sum_{k=1}^{\infty} |\alpha_k|. \tag{34.8}$$

Example 34.4. Let $A := (\alpha_{ik})_{i,k=1,2,\ldots}$ be an infinite matrix. We say that a sequence of numbers (ξ_k) is *A-summable to the value* ξ if

(I) the series $\sum_{k=1}^{\infty} \alpha_{ik}\xi_k$ converge for $i = 1, 2, \ldots$ and
(II) the sequence of the numbers $\eta_i := \sum_{k=1}^{\infty} \alpha_{ik}\xi_k$ converges to ξ.

The matrix A (or the *summation method* A) is said to be *permanent* (or *regular*) if every convergent sequence (ξ_k) is A-summable to its limit $\lim \xi_k$. The *Toeplitz permanence theorem*, which we will now prove, asserts:

For the matrix A to be permanent it is necessary and sufficient that the following conditions be all satisfied:

(P1) $\sum_{k=1}^{\infty} |\alpha_{ik}| \leqq M$ *for* $i = 1, 2, \ldots$ *with a certain* $M > 0$,
(P2) $\lim_{i\to\infty} \alpha_{ik} = 0$ *for* $k = 1, 2, \ldots,$
(P3) $\lim_{i\to\infty} \sum_{k=1}^{\infty} \alpha_{ik} = 1.$

We first prove that these conditions are necessary. Assume that every sequence $x = (\xi_k)$ from (c) is A-summable to its limit $\lim \xi_k$. Because of (I) the equations $A_i(x) := \sum_{k=1}^{\infty} \alpha_{ik}\xi_k$ define linear forms $A_1, A_2, \ldots$ on (c). As in Example 34.3, we see that every A_i is continuous and

$$\sum_{k=1}^{\infty} |\alpha_{ik}| = \|A_i\|.$$

Since furthermore the sequence (A_i) converges pointwise on (c), there exists by Proposition 33.1 an $M > 0$ with $\|A_i\| \leqq M$ for $i = 1, 2, \ldots.$ Thus (P1) is proved. The sequence $e_k := (0, \ldots, 0, 1, 0, \ldots)$, which has 1 at the kth position and zeros everywhere else, lies in (c_0); hence the sequence $\eta_i = A_i(e_k) = \alpha_{ik}$ tends to 0 as $i \to \infty$ for $k = 1, 2, \ldots,$ which proves (P2). The sequence $e_0 := (1, 1, \ldots)$, whose terms are all equal to 1, has limit 1, hence $\eta_i = A_i(e_0) = \sum_{k=1}^{\infty} \alpha_{ik}$ tends to 1; thus also (P3) is proved. Now we show that conditions (P1) through (P3) are also sufficient. Because of (P1) the A_i are continuous linear forms on (c). If we take into account that every $x = (\xi_k)$ from (c) with $\lim \xi_k = \xi$ can be represented in the form

$$x = \xi e_0 + \sum_{k=1}^{\infty} (\xi_k - \xi)e_k, \tag{34.9}$$

where the vectors $e_0, e_1, \ldots$ were defined in the first part of the proof, then we see that

$$A_i(x) = \xi \cdot \sum_{k=1}^{\infty} \alpha_{ik} + \sum_{k=1}^{\infty} (\xi_k - \xi)\alpha_{ik}.$$

Because of (P3) the first term in the right-hand side sum tends to ξ as $i \to \infty$; thus we only need to prove that the second term tends to 0. This assertion is obtained from the estimate

$$\left| \sum_{k=1}^{\infty} (\xi_k - \xi)\alpha_{ik} \right| \leqq \sum_{k=1}^{r} |\xi_k - \xi| \cdot |\alpha_{ik}| + M \sup_{k>r} |\xi_k - \xi|,$$

which, due to (P1), holds for every natural number r, if we also apply (P2). ■

Example 34.5. Numerous *quadrature formulas* (e.g., the trapezoid rule, Simpson's rule, the formulas of Newton–Cotes) give the value of the definite integral $\int_a^b x(t)\mathrm{d}t$ approximately by an expression of the form $\sum_{k=0}^{n} \alpha_k x(t_k)$, where the points t_k satisfy $a \leqq t_0 < t_1 < \cdots < t_n \leqq b$, and are, together with the coefficients α_k, independent of the integrand x. To the question, under what hypotheses the sequence of approximate quadratures

$$(34.10) \qquad Q_n(x) := \sum_{k=0}^{n} \alpha_k^{(n)} x(t_k^{(n)}) \qquad (n = 1, 2, \ldots)$$

converges for continuous integrands x to $\int_a^b x(t)\mathrm{d}t$, the answer is given by *Szegö's convergence theorem*:

For the sequence of approximate quadratures $Q_n(x)$ in (34.10) *to converge for every $x \in C[a, b]$ to $\int_a^b x(t)\mathrm{d}t$ it is necessary and sufficient that the following conditions be simultaneously satisfied*:

$$\text{(Q1)} \qquad \sum_{k=0}^{n} |\alpha_k^{(n)}| \leqq M \qquad \textit{for} \quad n = 1, 2, \ldots \qquad \textit{with a certain} \quad M > 0,$$

$$\text{(Q2)} \qquad Q_n(p) \to \int_a^b p(t)\mathrm{d}t \qquad \textit{for every polynomial} \quad p.$$

Proof. If $\alpha_0, \ldots, \alpha_n$ are arbitrary scalars and if $a \leq t_0 < t_1 < \cdots < t_n \leqq b$, then by $f(x) := \sum_{k=0}^{n} \alpha_k x(t_k)$ we define a continuous linear form on $C[a, b]$ with $\|f\| \leqq \sum_{k=0}^{n} |\alpha_k|$. The reader can construct without difficulty a piecewise linear function $x \in C[a, b]$ for which $\|x\| \leqq 1$ and $|f(x)| = \sum_{k=0}^{n} |\alpha_k|$, so that we have $\|f\| = \sum_{k=0}^{n} |\alpha_k|$. With this observation Szegö's theorem follows immediately from Theorem 33.1, if we take into account that the set of all polynomials on $[a, b]$ is, by the Weierstrass approximation theorem, dense in $C[a, b]$. ■

Example 34.6. It is well known that the limit of a *uniformly* convergent sequence of continuous functions is itself continuous. If the convergence is *not* uniform, then the limit function can have discontinuities. We still have, however, as we shall show, the following theorem of Baire:

If the functions x_n are from $C[a, b]$ and if $x_n(t)$ converges to $x(t)$ for each $t \in [a, b]$, then the set of points of continuity of x is dense in $[a, b]$.

We first prove a *lemma*: *To every closed subinterval K of $[a, b]$ and every $\varepsilon > 0$ there exists a closed subinterval $\tilde{K}$ of K so that $|x(t_1) - x(t_2)| \leqq \varepsilon$ for every t_1, t_2 in $\tilde{K}$.* For the proof we define given $\eta := \varepsilon/3$ the sets

$$F_n := \{t \in K : |x_n(t) - x_m(t)| \leqq \eta \text{ for all } m \geqq n\}.$$

Obviously every F_n is closed and $K = \bigcup_{n=1}^{\infty} F_n$. It follows from the Baire category theorem (Principle 31.1) that at least one F_n, say F_p, contains a closed interval K'. For all $t \in K'$ and all $m \geqq p$ we have $|x_p(t) - x_m(t)| \leqq \eta$ and thus also $|x_p(t) - x(t)| \leqq \eta$. Since $t \mapsto x_p(t)$ is uniformly continuous on K', there exists furthermore a closed subinterval $\tilde{K}$ of K' so that $|x_p(t_1) - x_p(t_2)| \leqq \eta$ for all t_1, t_2 in $\tilde{K}$. For such points t_1, t_2 we have then

$$\begin{aligned} |x(t_1) - x(t_2)| &\leqq |x(t_1) - x_p(t_1)| + |x_p(t_1) - x_p(t_2)| + |x_p(t_2) - x(t_2)| \\ &\leqq 3\eta = \varepsilon. \end{aligned}$$ ■

From this lemma it follows immediately that for every closed subinterval K of $[a, b]$ we can construct a sequence of closed intervals $K_n := [a_n, b_n]$ with the following properties:

$$K_n \subset K, \qquad a_n < a_{n+1}, \qquad b_{n+1} < b_n, \qquad b_n - a_n < \frac{1}{n},$$

$$|x(t_1) - x(t_2)| \leqq \frac{1}{n} \qquad \text{for all} \quad t_1, t_2 \text{ in } \quad K_n.$$

At the common point t_0 of the intervals K_n (cf. Proposition 31.1), the function x is obviously continuous. Thus if t is an arbitrary point of $[a, b]$ and $\varepsilon > 0$, then $[t - \varepsilon, t + \varepsilon]$ contains a point where x is continuous, so the set of these points is indeed dense in $[a, b]$. ■

Example 34.7. We shall now show that *there exists a function which is continuous at every point of the interval* $[0, 1]$ *but is not differentiable at any point* (all functions which occur in what follows are assumed to be real-valued).

If $x \in C[0, 2]$ is differentiable at $t_0 \in [0, 1]$, then necessarily

$$\sup_{0<h<1} \frac{|x(t_0 + h) - x(t_0)|}{h} < \infty.$$

Now let F_n be the set of all $x \in C[0, 2]$ to which there exists a $t_0 \in [0, 1]$, depending on x, such that

$$\sup_{0<h<1} \frac{|x(t_0 + h) - x(t_0)|}{h} \leqq n.$$

We first show that F_n is closed in $C[0, 2]$. Let $x_k \in F_n$ and assume that $x_k \to x$ for the canonical norm of $C[0, 2]$, i.e., that $x_k(t) \to x(t)$ *uniformly* on $[0, 2]$. To every x_k there exists a $t_k \in [0, 1]$ such that

$$\sup_{0<h<1} \frac{|x_k(t_k + h) - x_k(t_k)|}{h} \leqq n.$$

The sequence (t_k) contains a subsequence which converges to a $t_0 \in [0, 1]$. Clearly we may assume that $t_k \to t_0$. For $h \in (0, 1)$ and $\varepsilon > 0$ we now determine

$$k_1 \quad \text{so that} \quad |x(t_0 + h) - x(t_k + h)| \leqq \frac{\varepsilon}{4} h \quad \text{for} \quad k \geqq k_1,$$

$$k_2 \geqq k_1 \quad \text{so that} \quad |x(t) - x_k(t)| \leqq \frac{\varepsilon}{4} h \quad \text{for} \quad k \geqq k_2 \quad \text{and all} \quad t \in [0, 2],$$

$$k_3 \geqq k_2 \quad \text{so that} \quad |x(t_k) - x(t_0)| \leqq \frac{\varepsilon}{4} h \quad \text{for} \quad k \geqq k_3.$$

From the inequality

$$\begin{aligned} |x(t_0 + h) - x(t_0)| \leqq {} & |x(t_0 + h) - x(t_k + h)| + |x(t_k + h) - x_k(t_k + h)| \\ & + |x_k(t_k + h) - x_k(t_k)| + |x_k(t_k) - x(t_k)| + |x(t_k) - x(t_0)| \end{aligned}$$

it follows immediately, choosing $k \geqq k_3$, that

$$\frac{|x(t_0 + h) - x(t_0)|}{h} \leqq \frac{\varepsilon}{4} + \frac{\varepsilon}{4} + n + \frac{\varepsilon}{4} + \frac{\varepsilon}{4} = n + \varepsilon.$$

Since h and ε were arbitrary, we obtain from this inequality that x lies in F_n. Thus F_n is closed, as has been asserted.

If now every $x \in C[0, 2]$ were differentiable at at least one point of $[0, 1]$ depending on x, then we would have $C[0, 2] = \bigcup_{n=1}^{\infty} F_n$ and thus, according to the Baire category theorem (Principle 31.1), at least one F_n, say F_m, would contain a closed ball. Because of the Weierstrass approximation theorem, F_m would also contain a closed ball $K_r[p]$ whose center is a polynomial p on $[0, 2]$, i.e., F_m would contain all $x \in C[0, 2]$ with

$$|x(t) - p(t)| \leqq r \qquad \text{for all} \quad t \in [0, 2].$$

In this r-strip around p there lies, however, a function $y \in C[0, 2] \setminus F_m$, e.g., a sawtooth function whose increasing segments have a slope $>m$ and whose decreasing segments have a slope $<-m$. This contradiction shows that there exists a function $x \in C[0, 2]$ which is not differentiable at any point of the interval $[0, 1]$. ■

Example 34.8. It is shown in [153] that the Hahn–Banach extension theorem implies the existence of the Green and Neumann functions in potential theory. We cannot give any further details concerning this interesting application of the extension theorem.

Exercises

*1. If $f(x) := \sum_{k=1}^{\infty} \alpha_k \xi_k$ exists for all (ξ_k) in l^1, then (α_k) lies in l^∞, and f is a continuous linear form on l^1 with $\|f\| = \sup_k |\alpha_k|$.

2. If E, F are sequence spaces, then the map $A: E \to F$ is called a *matrix transformation* if there exists an (infinite) matrix (α_{ik}) such that for every $x = (\xi_k)$ in E the numbers $\eta_i := \sum_{k=1}^{\infty} \alpha_{ik} \xi_k$ $(i = 1, 2, \ldots)$ exist, $y := (\eta_i)$ lies in F and $Ax = y$. Show that every matrix transformation from l^p into l^q $(1 \leqq p, q \leqq \infty)$ is continuous.

§35. Analytic representation of continuous linear forms

In order to test a continuous endomorphism of a normed space for normal solvability—for instance by means of Proposition 29.1 in connection with Lemma 29.2—it might be useful to dispose of an analytic representation of all continuous linear forms of our space. Also in other investigations, e.g., the approximation problem in §28, Exercise 3, such representations are advantageous. Therefore we will consider them in this section.

Example 35.1. If E is a finite-dimensional normed space over $\mathbf{K}$, then E' coincides with E^* because of Proposition 10.3. If we represent $x \in E$ by means of a basis $\{x_1, \ldots, x_n\}$ in the form $x = \sum_{k=1}^{n} \xi_k x_k$, then for every linear form f on E we have

$$(35.1) \qquad f(x) = \sum_{k=1}^{n} \alpha_k \xi_k \qquad \text{with} \quad \alpha_k := f(x_k).$$

Conversely, for arbitrary chosen scalars α_k a (continuous) linear form f is defined by (35.1) on E. With the aid of the correspondence $f \mapsto (\alpha_1, \ldots, \alpha_n)$ it is easy to see that E' is isomorphic to $\mathbf{K}^n$.

Example 35.2. We now determine the continuous linear forms on l^p, $1 \leqq p < \infty$. Let q be the number conjugate to p, i.e., let

$$\frac{1}{p} + \frac{1}{q} = 1, \qquad \text{if} \quad p > 1$$

$$q = \infty, \qquad \text{if} \quad p = 1.$$

Furthermore, let $e_k := (0, \ldots, 0, 1, 0, \ldots)$ be the sequence which has 1 at the kth position and zeros everywhere else. e_k lies in l^p, and every $x = (\xi_k) \in l^p$ can be represented in the form

$$x = \sum_{k=1}^{\infty} \xi_k e_k.$$

For a continuous linear form f on l^p we get therefore, if we set $\alpha_k := f(e_k)$,

$$f(x) = \sum_{k=1}^{\infty} \alpha_k \xi_k. \tag{35.2}$$

$a := (\alpha_1, \alpha_2, \ldots)$ lies in l^q according to Example 34.2 and Exercise 1 in §34, and we have

$$\|f\| = \|a\| = \begin{cases} \left(\sum_{k=1}^{\infty} |\alpha_k|^q \right)^{1/q} & \text{for} \quad p > 1 \\ \sup_k |\alpha_k| & \text{for} \quad p = 1. \end{cases} \tag{35.3}$$

Conversely, if $a := (\alpha_1, \alpha_2, \ldots)$ is an arbitrary vector in l^q, then (35.2) defines a continuous linear form f on l^p (in the case $p > 1$ use the Hölder inequality); the norm of f is given again by (35.3). With this we have proved the essential content of the following theorem (the reader will convince himself easily of the correctness of the assertions not yet proved):

Every continuous linear form f on l^p $(1 \leqq p < \infty)$ can be represented with the help of one and only one sequence $a := (\alpha_1, \alpha_2, \ldots)$ from l^q in the form (35.2); *the norm of f is given by* (35.3). *The correspondence $f \mapsto (\alpha_1, \alpha_2, \ldots)$ is an isomorphism of the normed spaces $(l^p)'$ and l^q; in the sense of this isomorphism of normed spaces we have therefore $(l^p)' = l^q$.*

Example 35.3. We now set ourselves the task to find an analytic representation of the continuous linear forms f on $C[a, b]$. The basic idea of our procedure is the following: We approximate the function $x \in C[a, b]$ by functions y for which $f(y)$ can be represented in a simple fashion and obtain then $f(x)$ by the passage to the limit $y \to x$. We will see that this simple idea requires a modification which will be made possible by the Hahn–Banach extension theorem.

For every partition $Z: a = t_0 < t_1 < \cdots < t_n = b$ of the interval $[a, b]$ we define the step function $y_Z \in B[a, b]$ by

$$y_Z(t) := \begin{cases} x(a) & \text{for} \quad t_0 \leqq t \leqq t_1 \\ x(t_{k-1}) & \text{for} \quad t_{k-1} < t \leqq t_k, \qquad k = 2, \ldots, n. \end{cases}$$

Since $x \in C[a, b]$ is even uniformly continuous on $[a, b]$, there exists for every $\varepsilon > 0$ a $\delta > 0$ such that

$$\sup_{a \leqq t \leqq b} |x(t) - y_Z(t)| \leqq \varepsilon \qquad \text{provided that } \mu(Z) := \max_{k=1}^{n} |t_k - t_{k-1}| \leqq \delta. \tag{35.4}$$

We now define a family of functions $u_s \in B[a, b]$ by the following equalities:

$$u_a(t) := 0 \qquad \text{for} \quad a \leqq t \leqq b$$

$$u_s(t) := \begin{cases} 1 & \text{for} \quad a \leqq t \leqq s \\ 0 & \text{for} \quad s < t \leqq b \end{cases}, \qquad \text{if} \quad a < s \leqq b.$$

With the help of these functions, y_Z can be represented in the form

(35.5) $$y_Z = \sum_{k=1}^{n} x(t_{k-1})[u_{t_k} - u_{t_{k-1}}].$$

Now the difficulty arises that the linear form f does not operate on the approximating function y_Z which is not continuous. This can be remedied by extending f according to the Hahn–Banach theorem (Theorem 28.1) to a continuous linear form F on $B[a, b]$ with $\|F\| = \|f\|$, since F does operate on y_Z and on the u_s. Setting

$$v(s) := F(u_s) \qquad \text{for} \quad a \leqq s \leqq b$$

we have, by (35.5),

$$F(y_Z) = \sum_{k=1}^{n} x(t_{k-1})[v(t_k) - v(t_{k-1})].$$

The sum on the right-hand side is a Riemann–Stieltjes sum. We first show that v is a function of bounded variation (which will ensure the existence of the Riemann–Stieltjes integral $\int_a^b x(t)\mathrm{d}v(t)$). To abbreviate, we define sgn α (the *sign* of α) by

$$\operatorname{sgn} \alpha := \begin{cases} 0 & \text{for} \quad \alpha = 0 \\ \dfrac{\alpha}{|\alpha|} & \text{for} \quad \alpha \neq 0; \end{cases}$$

thus $|\operatorname{sgn} \alpha| \leqq 1$ and $|\alpha| = \alpha \operatorname{sgn} \bar{\alpha}$.

With $\sigma_k := \operatorname{sgn}\overline{[v(t_k) - v(t_{k-1})]}$ we have now

$$\begin{aligned} &\sum_{k=1}^{n} |v(t_k) - v(t_{k-1})| \\ &\quad = \sum_{k=1}^{n} \sigma_k[v(t_k) - v(t_{k-1})] = \sum_{k=1}^{n} \sigma_k[F(u_{t_k}) - F(u_{t_{k-1}})] \\ &\quad = F\left(\sum_{k=1}^{n} \sigma_k[u_{t_k} - u_{t_{k-1}}]\right) \leqq \|F\| \left\| \sum_{k=1}^{n} \sigma_k[u_{t_k} - u_{t_{k-1}}] \right\| \\ &\quad \leqq \|F\| = \|f\|. \end{aligned}$$

Thus v is indced of bounded variation. For its total variation $V(v)$ on $[a, b]$ the estimate

(35.6) $$V(v) \leqq \|f\|$$

holds. If (Z_n) is a sequence of partitions such that $\mu(Z_n) \to 0$, then

$$F(y_{Z_n}) \to \int_a^b x(t)\mathrm{d}v(t);$$

but because of (35.4) also (y_{Z_k}) converges to x with respect to the norm of $B[a, b]$ and therefore $F(y_{Z_n}) \to F(x) = f(x)$. Consequently

(35.7) $$f(x) = \int_a^b x(t)\mathrm{d}v(t).$$

By a well-known theorem from the theory of Stieltjes integrals, we have

$$|f(x)| = \left|\int_a^b x(t)\mathrm{d}v(t)\right| \leqq V(v)\cdot\|x\|, \tag{35.8}$$

hence $\|f\| \leqq V(v)$. Together with (35.6) we obtain from here

$$\|f\| = V(v).$$

It also follows from (35.8) that for *every* function $v \in BV[a, b]$ a continuous linear form f with $\|f\| \leqq V(v)$ is defined on $C[a, b]$ by (35.7).

So we have proved the *Riesz representation theorem*:

For every continuous linear form f on $C[a, b]$ there exists a function $v \in BV[a, b]$ with which f can be represented in the form (35.7); *v can be chosen in such a fashion that $\|f\| = V(v)$. Conversely, for every $v \in BV[a, b]$ a continuous linear form f with $\|f\| \leqq V(v)$ is defined on $C[a, b]$ by* (35.7).

The function v which generates f with the aid of (35.7) is not determined uniquely; together with v, e.g., also $v + \alpha$, for an arbitrary scalar α, does what is required. It can be shown, however, that there exists exactly one 'normalized' function v in $BV[a, b]$ which generates f; a function $v \in BV[a, b]$ is said to be *normalized* if

$$v(a) = 0 \quad \text{and} \quad v(t + 0) = v(t) \qquad \text{for} \quad a < t < b.$$

For such a function v we have $V(v) = |v(a)| + V(v) = \|v\|$, where $\|v\|$ is the norm of v in $BV[a, b]$. The normalized functions form a linear subspace $N[a, b]$ of $BV[a, b]$; if we associate with every continuous linear form f on $C[a, b]$ its uniquely determined generating function $v \in N[a, b]$, then we obtain an isomorphism of normed spaces between the dual of $C[a, b]$ and the normed space $N[a, b]$. For the uniqueness problem see [13], p. 149.

Exercises

$^+$1. Represent the continuous linear forms on (c) and on (c_0) by infinite series and show that the dual of (c) and of (c_0) is isomorphic as a normed space to l^1. *Hint*: Use (34.9) and Example 34.3.

2. A continuous endomorphism A of $C[a, b]$ is normally solvable if and only if to every $x_0 \notin A(C[a, b])$ there exists a function $v \in BV[a, b]$ such that

$$\int_a^b (Ax)(t)\mathrm{d}v(t) = 0 \qquad \text{for all} \quad x \in C[a, b] \qquad \text{but} \qquad \int_a^b x_0(t)\mathrm{d}v(t) \neq 0.$$

3. *The moment problem* in $C[a, b]$: Let the numbers $\gamma > 0$ and $\alpha_1, \alpha_2, \ldots$ be given, as well as the functions $x_1, x_2, \ldots$ from $C[a, b]$. There exists a function $v \in BV[a, b]$ with $V(v) \leqq \gamma$ and $\int_a^b x_n(t)\mathrm{d}v(t) = \alpha_n$ for all n if and only if for arbitrary $n \in \mathbf{N}$ and $\beta_\nu \in \mathbf{K}$ the inequality $|\sum_{\nu=1}^n \beta_\nu\alpha_\nu| \leqq \gamma\|\sum_{\nu=1}^n \beta_\nu x_\nu\|$ holds. *Hint*: Exercise 4 of §28.

4. *The approximation problem* in $C[a, b]$: Let $x_0, x_1, x_2, \ldots$ be continuous

functions in $[a, b]$. There exists a sequence of linear combinations of the $x_1, x_2, \ldots$ which converges uniformly in $[a, b]$ to x_0 if and only if for every function $v \in BV[a, b]$ which satisfies $\int_a^b x_n(t)\mathrm{d}v(t) = 0$ $(n = 1, 2, \ldots)$ one also has $\int_a^b x_0(t)\mathrm{d}v(t) = 0$. *Hint*: Exercise 3 of §28.

§36. Operators with closed image space

According to Proposition 29.3 a continuous endomorphism of a normed space is normally solvable if and only if its image space is closed. The results of sections 31 and 32 make it possible to examine more closely the question when the image space of an endomorphism is closed. We start with a simple preliminary lemma.

Lemma 36.1. *If E, F are Banach spaces, then the operator $A \in \mathscr{L}(E, F)$ has a* continuous *inverse if and only if it is injective and its image space is closed.*

Indeed, if A is injective and $A(E)$ is closed, then the continuity of A^{-1} follows from Theorem 32.2 because $A(E)$ is complete as a closed subspace of the Banach space F. The converse has been proved already at the beginning of §31. ∎

The following consideration conduces us to associate with every $A \in \mathscr{L}(E, F)$ a number which gives information concerning the closedness of $A(E)$. Here E and F are assumed to be Banach spaces; let $\hat{E}$ be the Banach space $E/N(A)$ and $\hat{A}: \hat{E} \to F$ the continuous injection corresponding to A (Proposition 21.3). It follows from Lemma 36.1 that $A(E) = \hat{A}(\hat{E})$ is closed if and only if $(\hat{A})^{-1}$ is continuous; this again is, according to Proposition 7.5, the case if and only if with a certain constant $m > 0$ the inequality $m\|\hat{x}\| \leqq \|\hat{A}\hat{x}\|$ holds for all $\hat{x} \in \hat{E}$, i.e., if

$$\inf_{0 \neq \hat{x} \in \hat{E}} \frac{\|\hat{A}\hat{x}\|}{\|\hat{x}\|} > 0. \tag{36.1}$$

If we define the *distance* $d(x, N(A))$ of an element x from $N(A)$ by

$$d(x, N(A)) := \inf_{y \in N(A)} \|x - y\|,$$

then for every $x \in \hat{x}$ we have

$$\|\hat{x}\| = \inf_{z \in \hat{x}} \|z\| = \inf_{y \in N(A)} \|x - y\| = d(x, N(A)) \quad \text{and} \quad \|\hat{A}\hat{x}\| = \|Ax\|.$$

Consequently, the greatest lower bound in (36.1) coincides with the number

$$\gamma(A) := \inf_{\substack{x \in E \\ x \notin N(A)}} \frac{\|Ax\|}{d(x, N(A))}, \tag{36.2}$$

called the *minimal modulus* of A, and our considerations have proved the following proposition:

Proposition 36.1. *If E and F are Banach spaces, then the image space of the continuous linear operator $A: E \to F$ is closed if and only if its minimal modulus satisfies $\gamma(A) > 0$.*

From this proposition we obtain without effort the following *closedness criterion*:

Proposition 36.2. *Let A be a closed linear map from the Banach space E into the Banach space F. If there exists a closed subspace G of F such that $A(E) \cap G = \{0\}$ and $A(E) \oplus G$ is closed, then already $A(E)$ must be closed.*

Proof. G is complete as a closed subspace of the Banach space F, hence $E \times G$ is a Banach space (see the remarks following Theorem 32.2). We now define a continuous linear map $B: E \times G \to F$ by

$$B(x, y) := Ax + y \qquad (x \in E,\ y \in G).$$

The image space $B(E \times G) = A(E) \oplus G$ is closed by hypothesis, hence $\gamma(B) > 0$ (Proposition 36.1). We now show that $\gamma(A) \geqq \gamma(B)$ whereby, again because of Proposition 36.1, the proof will be concluded. Because of $A(E) \cap G = \{0\}$ we have $N(B) = N(A) \times \{0\}$; consequently for every $x \in E$ we have

$$d((x, 0), N(B)) = d(x, N(A)),$$

and so

$$\|Ax\| = \|B(x, 0)\| \geqq \gamma(B)d((x, 0), N(B)) = \gamma(B)d(x, N(A)),$$

from where $\gamma(A) \geqq \gamma(B)$ immediately follows. ■

Since by Proposition 10.5 a finite-dimensional subspace of a normed space is always closed, Proposition 36.2 yields directly

Proposition 36.3. *Let A be a continuous linear map from the Banach space E into the Banach space F. If the image space of A has finite codimension, then it is closed.*

For the sake of simplicity we formulate the next proposition for endomorphisms; see, however, Exercise 2 and also Theorem 97.1.

Proposition 36.4. *Let A be a continuous endomorphism of the Banach space E. If $A(E)$ is closed, then also $A'(E')$ is closed and*

$$A'(E') = N(A)^{\perp} := \{x' \in E' : \langle x, x' \rangle = 0 \textit{ for all } x \in N(A)\}. \tag{36.3}$$

Proof. $N(A)^{\perp}$ is obviously closed; hence the closedness of $A'(E')$ will have been shown as soon as we have proved (36.3).

The inclusion $A'(E') \subset N(A)^{\perp}$ can be seen immediately. We now show, conversely, that every $x' \in N(A)^{\perp}$ lies in $A'(E')$. For this we define with the help of such an x' a linear form f on $A(E)$ as follows: To $y \in A(E)$ we choose an $x \in E$ with $Ax = y$ and set

$$f(y) := \langle x, x' \rangle.$$

f is uniquely determined; indeed, if we also have $Ax_1 = y$, then $x_1 - x \in N(A)$, hence $\langle x_1, x'\rangle = \langle x, x'\rangle$. It is trivial that f is linear. f is also continuous: For every $u \in N(A)$ we have $f(y) = \langle x - u, x'\rangle$, hence $|f(y)| \leqq \|x'\|\,\|x - u\|$ and thus also $|f(y)| \leqq \|x'\| d(x, N(A))$. Since by Proposition 36.1 the minimal modulus satisfies $\gamma(A) > 0$, we get from here the estimate

$$|f(y)| \leqq \|x'\| \frac{\|Ax\|}{\gamma(A)} = \frac{\|x'\|}{\gamma(A)} \|y\|.$$

We now extend f according to the Hahn–Banach theorem (Theorem 28.1) to a continuous linear form y' on E. Then for every $x \in E$ we have

$$\langle x, x'\rangle = f(Ax) = y'(Ax) = \langle Ax, y'\rangle = \langle x, A'y'\rangle,$$

hence $x' = A'y'$: indeed, x' lies in $A'(E')$. ■

From Propositions 29.3 and 36.4 we obtain immediately the following assertions:

If the continuous endomorphism A of the Banach space E possesses a closed image space, *then we have*:

$$Ax = y \textit{ is solvable} \Leftrightarrow \langle y, x'\rangle = 0 \textit{ for all } x' \in N(A'),$$
$$A'x' = y' \textit{ is solvable} \Leftrightarrow \langle x, y'\rangle = 0 \textit{ for all } x \in N(A).$$

Thus operators with closed image spaces show the same solvability behavior as Fredholm operators (Proposition 27.3). We shall also see in Theorem 97.1 that Proposition 36.4 has a converse and that the above solvability criteria are also valid when A' has a closed image space.

For a surjective A we have $N(A') = \{0\}$; from this observation we obtain with the aid of Proposition 36.4 and Lemma 36.1 immediately the following proposition (cf. also Exercise 3):

Proposition 36.5. *If A is a surjective continuous endomorphism of the Banach space E, then A' has a continuous inverse on $A'(E')$.*

Also this proposition has a converse; see Proposition 97.1.

Exercises

$^+$1. The minimal modulus $\gamma(A)$ can be defined for every linear operator $A: E \to F$ with closed nullspace by (36.1) or (36.2) (we omit the uninteresting case $A = 0$). Show that if E and F are Banach spaces, if $\gamma(A) > 0$ and if $A(E)$ is closed, then A is continuous.

$^+$2. Let E be a Banach space, F a normed space, $A \in \mathscr{L}(E, F)$ and $A(E)$ complete. Then $A'(F') = N(A)^\perp$.

$^+$3. Let E be a normed space, F a Banach space and $A \in \mathscr{L}(E, F)$. If A is surjective, then A' has a continuous inverse on $A'(F')$. *Hint*: Extend A to the completion of E and use Exercise 2.

$^+$4. Let E, F be Banach spaces, D a linear subspace of E and $A: D \to F$ linear and closed. Show that if codim $A(D)$ is finite, then $A(D)$ is closed. *Hint*: Exercise 3 in §32 and Proposition 36.3.

$^+$5. Propositions 36.1 and 36.2 are valid also under the hypotheses of Exercise 4. Hint: Take into account Exercise 1 and the fact that the nullspace of a closed operator is closed.

§37. Fredholm operators on Banach spaces

We recall that a continuous endomorphism of a normed space E is called a Fredholm operator if it is relatively regular and has finite deficiency; according to Proposition 24.1 the relative regularity of an operator means that it is open and that its nullspace and image space are continuously projectable. With the help of the Hahn–Banach extension theorem we observed in Proposition 24.3 that $A \in \mathscr{L}(E)$ is a Fredholm operator exactly if it is open, has finite deficiency and a closed image space. If E is even a Banach space, then the fact that $A(E)$ has finite codimension already implies that $A(E)$ is closed (Proposition 36.3), and this in turn entails that A is open (Theorem 32.1). Consequently we have:

Proposition 37.1. *A continuous endomorphism of a* Banach space *is a Fredholm operator if and only if it has finite deficiency.*

From Propositions 25.3 and 23.3 we obtain a simple *perturbation theorem* for Fredholm operators on arbitrary normed spaces: *Together with A also $A + K$ is a Fredholm operator for every continuous finite-dimensional operator K, and, furthermore, the index is conserved*: $\operatorname{ind}(A + K) = \operatorname{ind}(A)$.

The following proposition is also a perturbation theorem: It shows that the Fredholm character (and the index) of an operator does not change if we add sufficiently *small* operators.

Proposition 37.2. *To every Fredholm operator A of the Banach space E there exists a number $\rho = \rho(A) > 0$ so that for all $S \in \mathscr{L}(E)$ with $\|S\| < \rho$ also $A + S$ is a Fredholm operator and* $\operatorname{ind}(A + S) = \operatorname{ind} A$. *Thus the set $\Phi(E)$ of the Fredholm operators on E is* open *in $\mathscr{L}(E)$.*

Proof. To A there exists by (24.5) a $B \in \mathscr{L}(E)$ and a $K \in \mathscr{F}(E)$ such that

$$BA = I - K. \tag{37.1}$$

Because of $\operatorname{ind}(I - K) = 0$ (Proposition 17.1), it follows with the aid of Lemma 23.1 that B has finite deficiency and

$$\operatorname{ind}(B) = -\operatorname{ind}(A). \tag{37.2}$$

If we set $\rho := 1/\|B\|$ and if $\|S\| < \rho$, then $\|BS\| \leqq \|B\|\,\|S\| < 1$; thus $I + BS$ is bijective by Theorem 8.1, and has therefore a vanishing index. Because of Proposition 23.3 we obtain from

$$B(A + S) = BA + BS = I - K + BS = (I + BS) - K$$

that also $\operatorname{ind}(B(A + S))$ vanishes. Since B has a finite deficiency, it follows with the aid of Lemma 23.1 that also $A + S$ has finite deficiency and that the equation

$$\operatorname{ind}(B) = -\operatorname{ind}(A + S) \tag{37.3}$$

must be valid. Thus $A + S$ is indeed a Fredholm operator (Proposition 37.1) whose index coincides with $\operatorname{ind}(A)$ because of (37.2) and (37.3). ■

Exercises

1. Let A be a Fredholm operator on the Banach space E and let B satisfy (37.1). Then $A + S$ is a Fredholm operator for every $S \in \mathscr{L}(E)$ with $\inf_{F \in \mathscr{F}(E)} \|S - F\| \cdot \inf_{F \in \mathscr{F}(E)} \|B - F\| < 1$, and $\operatorname{ind}(A + S) = \operatorname{ind}(A)$.

2. Under the hypotheses of Exercise 1 we have $\inf_{F \in \mathscr{F}(E)} \|B - F\| = \inf_{C \in \mathscr{L}(E)} \{\|C\|$: to C there exists an $F \in \mathscr{F}(E)$ with $CA = I - F\}$. In the case of an infinite-dimensional E this greatest lower bound is positive.

$^+$3. Let E be a normed space. An Atkinson operator in the (normal) algebra $\mathscr{L}(E)$ (see Exercise 8 and 9 in §25) is also called an Atkinson operator *on* E. Show that to every Atkinson operator A on the Banach space E there exists a number $\rho = \rho(A) > 0$ so that for all $S \in \mathscr{L}(E)$ with $\|S\| < \rho$ also $A + S$ is an Atkinson operator and $\operatorname{ind}(A + S) = \operatorname{ind}(A)$ (see also Exercise 4).

$^+$4. Let E be a Banach space and $A \in \mathscr{L}(E)$ a *semi-Fredholm operator*, i.e., let at least one of the deficiencies $\alpha(A)$, $\beta(A)$ be finite and let $A(E)$ be closed. Then there exists a number $\rho = \rho(A) > 0$ so that for all $S \in \mathscr{L}(E)$ with $\|S\| < \rho$ also $(A + S)(E)$ is closed, furthermore $\alpha(A + S) \leqq \alpha(A)$, $\beta(A + S) \leqq \beta(A)$ and $\operatorname{ind}(A + S) = \operatorname{ind}(A)$. For a proof see [54], p. 113.

VI

The Riesz–Schauder theory of compact operators

§38. Operators with finite chains

The nullspaces of the powers A^n of an endomorphism A on the vector space E form an increasing sequence $N(A^0) = \{0\} \subset N(A) \subset N(A^2) \subset \cdots$ which we call the *nullchain*. If for a certain $n \geqq 0$ we have $N(A^n) = N(A^{n+1})$, then also $N(A^{n+1}) = N(A^{n+2})$, and so $N(A^n) = N(A^{n+m})$ for $m = 1, 2, \ldots$; indeed from $x \in N(A^{n+2})$ it follows that $A^{n+1}Ax = 0$, hence $Ax \in N(A^{n+1}) = N(A^n)$ and thus $A^{n+1}x = 0$, i.e., $x \in N(A^{n+1})$. The smallest integer $n \geqq 0$ for which this occurs will be called the *length of the nullchain* of A and will be denoted by $p(A)$. If there is no such integer, i.e., if $N(A^n) \neq N(A^{n+1})$ for all n, then we set $p(A) = \infty$. The *image chain* of A is the decreasing sequence of image spaces $A^0(E) = E \supset A(E) \supset A^2(E) \supset \cdots$. If for a certain $n \geqq 0$ we have $A^n(E) = A^{n+1}(E)$, then $A^n(E) = A^{n+m}(E)$ for $m = 1, 2, \ldots$; the smallest integer $n \geqq 0$ for which this occurs is called the *length $q(A)$ of the image chain of A*. If always $A^n(E) \neq A^{n+1}(E)$, then we set $q(A) = \infty$.

Clearly $p(A) = 0$ means that A is *injective*, and $q(A) = 0$ that A is *surjective*.

For endomorphisms $A := I - K$ with finite-dimensional K the lengths of both chains are finite.

Indeed, the nullspace of

$$A^n = (I - K)^n = I - \left[nK - \binom{n}{2}K^2 + \cdots + (-1)^{n-1}K^n\right] =: I - K_n, \; n \geqq 1,$$

lies in the image space $K_n(E)$ and this space lies obviously in the finite-dimensional space $K(E)$, so that the nullchain of A must eventually break off. Since, furthermore, K_n is finite-dimensional, and thus by Proposition 17.1 one has $\alpha(I - K_n) = \beta(I - K_n)$, it follows that $\beta(A^n) = \beta(I - K_n)$ becomes a constant from the same exponent on as $\alpha(A^n) = \alpha(I - K_n)$. Thus we have $q(A) = p(A) < \infty$. ■

The following two propositions give precise conditions for the breaking off of the nullchain or the image chain of an endomorphism A of E.

Proposition 38.1. *We have $p(A) \leqq m < \infty$ if and only if $N(A^n) \cap A^m(E) = \{0\}$; here n is an arbitrary natural number.*

Indeed, if $p(A) \leqq m < \infty$, n is a natural number, and $y \in N(A^n) \cap A^m(E)$, then $y = A^m x$ and $A^n y = 0$, hence $A^{m+n}x = 0$. Thus x lies in $N(A^{m+n}) = N(A^m)$ and therefore $y = A^m x = 0$. Conversely, for a natural number n let $N(A^n) \cap A^m(E) = \{0\}$. Because of $N(A) \subset N(A^n)$ we have a fortiori $N(A) \cap A^m(E) = \{0\}$. From $x \in N(A^{m+1})$, i.e., $A(A^m x) = 0$, it follows therefore that $A^m x \in N(A) \cap A^m(E) = \{0\}$, thus x lies already in $N(A^m)$. So we have $N(A^m) = N(A^{m+1})$ and consequently $p(A) \leqq m$. ■

Proposition 38.2. *We have $q(A) \leqq m < \infty$ if and only if to $A^n(E)$ there exists a complementary subspace C_n in E which is contained in $N(A^m)$; here n is an arbitrary natural number.*

For the proof let $q := q(A) \leqq m < \infty$, let n be a natural number and C any complementary subspace to $A^n(E)$ in E (Proposition 4.1):

$$E = C \oplus A^n(E). \tag{38.1}$$

To every element x_ι of a basis $\{x_\iota : \iota \in J\}$ of C there exists because of $A^q(C) \subset A^q(E) = A^{q+n}(E)$ a $y_\iota \in E$ with $A^q x_\iota = A^{q+n} y_\iota$. If we set $z_\iota := x_\iota - A^n y_\iota$, then $A^q z_\iota = A^q x_\iota - A^{q+n} y_\iota = 0$. It follows that the linear hull C_n of the z_ι lies in $N(A^q)$ and a fortiori in $N(A^m)$. From (38.1) we obtain for every $x \in E$ a representation of the form

$$x = \sum \alpha_\iota x_\iota + A^n y = \sum \alpha_\iota (z_\iota + A^n y_\iota) + A^n y = \sum \alpha_\iota z_\iota + A^n z,$$

thus $E = C_n + A^n(E)$. This sum is even direct; indeed, for $x \in C_n \cap A^n(E)$ we have $x = \sum \beta_\iota z_\iota = A^n v$, hence

$$\sum \beta_\iota x_\iota = \sum \beta_\iota A^n y_\iota + A^n v \in A^n(E),$$

and thus, because of (38.1), $\beta_\iota = 0$ for all $\iota \in J$, and so $x = 0$. It follows that C_n is in fact a complementary subspace of $A^n(E)$ lying in $N(A^m)$. Now let n be a natural number and assume that to $A^n(E)$ there exists a complementary subspace C_n lying in $N(A^m)$, i.e., let $E = C_n \oplus A^n(E)$. Then $A^m(E) = A^m(C_n) + A^{m+n}(E) = A^{m+n}(E)$ and therefore $q(A) \leqq m$. ■

Proposition 38.3. *If both chain lengths of A are finite, then they are equal.*

Proof. We set $p := p(A)$, $q := q(A)$ and assume first that $p \leqq q$ so that $A^q(E) \subset A^p(E)$. Furthermore let $q > 0$ since otherwise there is nothing to prove.

From Proposition 38.2 follows the representation $E = N(A^q) + A^q(E)$; thus for every element $y := A^p x$ of $A^p(E)$ we have the decomposition $y = z + A^q w$

with $z \in N(A^q)$. The element $z = A^p x - A^q w$ lies in $A^p(E)$, hence $z \in N(A^q) \cap A^p(E)$. According to Proposition 38.1 this intersection contains only 0, so that $y = A^q w$ and y lies in $A^q(E)$. Thus we have shown the equality $A^p(E) = A^q(E)$, from where $p \geqq q$ follows. Altogether we have therefore $p = q$.

Let us now suppose that $q \leqq p$ and $p > 0$, so that $N(A^q) \subset N(A^p)$. We obtain from Proposition 38.2 the representation $E = N(A^q) + A^p(E)$, so that for an arbitrary element x of $N(A^p)$ we have the decomposition $x = u + A^p v$ with $u \in N(A^q)$. Because of $A^p x = A^p u = 0$ we obtain from here that $A^{2p} v = 0$. Thus $v \in N(A^{2p}) = N(A^p)$, so that already $A^p v = 0$ and therefore $x = u \in N(A^q)$. It follows that $N(A^q) = N(A^p)$, hence $q \geqq p$. Thus we have again $p = q$. ■

If the lengths of both chains of an endomorphism A are finite, then we say that A is *chain-finite* and the common length of the two chains (Proposition 38.3) is called the *chain length* of A.

Proposition 38.4. *If A has the chain length $p < \infty$, then the decomposition*

$$E = N(A^p) \oplus A^p(E) \tag{38.2}$$

is valid, and A maps the space $A^p(E)$ bijectively onto itself. Conversely, if for a natural number m we have

$$E = N(A^m) \oplus A^m(E), \tag{38.3}$$

then $p(A) = q(A) \leqq m$.

Proof. If $p(A) = q(A) = p < \infty$ (where we may assume $p > 0$), then decomposition (38.2) immediately follows from Propositions 38.1 and 38.2. If we denote by $\tilde{A}$ the restriction of A to $A^p(E)$, then $N(\tilde{A}) \subset N(A) \subset N(A^p)$ but also $N(\tilde{A}) \subset A^p(E)$, and it follows from (38.2) that $N(\tilde{A}) = \{0\}$, hence $\tilde{A}$ is indeed injective. Furthermore $\tilde{A}(A^p(E)) = A(A^p(E)) = A^{p+1}(E) = A^p(E)$, i.e., $\tilde{A}$ maps $A^p(E)$ onto itself. Conversely, if (38.3) is valid, then $p(A)$, $q(A) \leqq m$ (Propositions 38.1 and 38.2), and so $p(A) = q(A) \leqq m$ by Proposition 38.3. ■

There are certain relations between the chain lengths and the deficiencies of an endomorphism A on E, which we want to deduce now.

Proposition 38.5.

(a) *If $p(A) < \infty$, then $\alpha(A) \leqq \beta(A)$.*

(b) *If $q(A) < \infty$, then $\beta(A) \leqq \alpha(A)$.*

Proof.

(a) Let $p := p(A) < \infty$. If $\beta(A) = \infty$, then there is nothing to prove; therefore we may assume that $\beta(A) < \infty$. By Proposition 38.1 we have $N(A) \cap A^p(E) = \{0\}$; since together with $\beta(A)$ also $\beta(A^p)$ is finite (Exercise 1 in §23), it follows that $\alpha(A) < \infty$. Thus A is an endomorphism with finite deficiency, so that according to the index theorem (Theorem 23.1) we obtain for all $n \geqq p$ the following equality:

$$n \cdot \operatorname{ind}(A) = \operatorname{ind}(A^n) = \alpha(A^n) - \beta(A^n) = \alpha(A^p) - \beta(A^n).$$

If also $q := q(A) < \infty$, then we get from here for all $n \geqq \max(p, q)$ the relation $n \cdot \operatorname{ind}(A) = \alpha(A^p) - \beta(A^q) = \text{const.}$, so that $\operatorname{ind}(A) = 0$, i.e., $\alpha(A) = \beta(A)$. If, however, $q = \infty$, i.e., if $\beta(A^n) \to \infty$, then $n \cdot \operatorname{ind}(A)$ becomes eventually negative, hence $\operatorname{ind}(A) < 0$ and so $\alpha(A) < \beta(A)$.

(b) Now let $q := q(A) < \infty$. If $\alpha(A) = \infty$, then there is nothing to prove. We may therefore assume that $\alpha(A) < \infty$. Then also $\alpha(A^q)$ is finite (Exercise 1 in §23), and since by Proposition 38.2 we have $E = C \oplus A(E)$ with $C \subset N(A^q)$, it follows that $\beta(A) = \dim C \leqq \alpha(A^q) < \infty$. Thus A is in this case again an endomorphism with finite deficiency. If we apply with appropriate changes the index argument used in (a), then we obtain $\beta(A) = \alpha(A)$ if $p(A) < \infty$ and $\beta(A) < \alpha(A)$ if $p(A) = \infty$. ■

Proposition 38.6.
(a) *If both chain lengths of A are finite, then $\alpha(A) = \beta(A)$.*
(b) *If $\alpha(A) = \beta(A) < \infty$, and if one chain length is finite, then $p(A) = q(A)$.*

(a) follows immediately from Proposition 38.5, while (b) is an easy consequence of the equality $\alpha(A^n) - \beta(A^n) = \operatorname{ind}(A^n) = n \cdot \operatorname{ind}(A) = 0$ valid for $n = 0, 1, 2, \ldots$. ■

In the case $\alpha(A) < \infty$ we can deduce a useful criterion for the finiteness of the nullchain (Proposition 38.7). For this we need the following:

Lemma 38.1. *The endomorphism A of E maps the linear space $\bigcap_{n=1}^{\infty} A^n(E)$ into itself, and in the case of $\alpha(A) < \infty$ even onto itself.*

It is trivial that $U := \bigcap_{n=1}^{\infty} A^n(E)$ is mapped into itself by A. We now assume that $\alpha(A) < \infty$ and show that every element of U is the image of an element of U under A. From $N(A) \cap A^n(E) \supset N(A) \cap A^{n+1}(E)$ for $n = 0, 1, 2, \ldots$ it follows, because of $\alpha(A) < \infty$, that there exists a natural number m with

$$(38.4)\quad D := N(A) \cap A^m(E) = N(A) \cap A^{m+k}(E) \qquad \text{for} \quad k = 0, 1, 2, \ldots.$$

Obviously also $D = N(A) \cap U$. Let now y be an arbitrary element of U. Then for every $k = 0, 1, 2, \ldots$ there exists an $x_k \in E$ so that $y = A^{m+k}x_k$. If we set

$$(38.5)\qquad z_k := A^m x_1 - A^{m+k-1}x_k \qquad \text{for} \quad k = 1, 2, \ldots,$$

then z_k lies in $A^m(E)$ and, because of $Az_k = A^{m+1}x_1 - A^{m+k}x_k = y - y = 0$, also in $N(A)$, hence $z_k \in N(A) \cap A^m(E) = D$. From (38.4) it follows that z_k lies also in $A^{m+k-1}(E)$ and with the aid of (38.5) this implies

$$A^m x_1 = z_k + A^{m+k-1}x_k \in A^{m+k-1}(E) \qquad \text{for} \quad k = 1, 2, \ldots,$$

hence $A^m x_1 \in U$. Because of $A(A^m x_1) = A^{m+1} x_1 = y$, we see that y is indeed the image of an element of U under A. ■

Proposition 38.7. *For an endomorphism A on E with $\alpha(A) < \infty$ the following assertions are equivalent:*

(a) *The length of the nullchain is finite.*
(b) *On every subspace F of E, which is mapped by A onto itself, A is injective.*
(c) *A is injective on the subspace $U := \bigcap_{n=1}^{\infty} A^n(E)$.*

Proof. (a) ⇒ (b): If $A(F) = F$ and $\tilde{A}$ is the restriction of A to F, then $q(\tilde{A}) = 0$. From $N(\tilde{A}^n) = N(A^n) \cap F$ it follows because of (a) that $p(\tilde{A}) < \infty$. Because of Proposition 38.3 we thus have $p(\tilde{A}) = q(\tilde{A}) = 0$, and so $\tilde{A}$ is injective.

(b) ⇒ (c): This implication is trivial because of Lemma 38.1.

(c) ⇒ (a): From (c) it follows in the first place that $D = N(A) \cap U = \{0\}$. Because of (38.4) we have thus also $N(A) \cap A^m(E) = \{0\}$ for some natural number m. Assertion (a) is now a consequence of Proposition 38.1. ■

A generalization of the results from [105], discussed in this section, to operators, which are not defined on the whole space E, can be found in [111].

Exercises

1. Let $\tilde{I}$ and $\tilde{K}$ be the restrictions of I and $K \in \mathscr{S}(E)$ to $F := K(E)$. Show that the length of the nullchain or of the image chain of $I - K$ is finite whenever the length of the nullchain or of the image chain of $\tilde{I} - \tilde{K}$, respectively, is finite. *Hint*: Proposition 23.4.

$^+$2. If A, B commute, then AB is chain-finite with $\text{ind}(AB) = 0$ if and only if A and B are chain-finite with $\text{ind}(A) = \text{ind}(B) = 0$.

§39. Chain-finite Fredholm operators

In this section we give a representation of those Fredholm operators in saturated operator algebras which possess a finite nullchain and image chain. The investigations are based on a *bilinear system* (E, E^+).

Proposition 39.1. *Let A be an operator from the E^+-saturated algebra of endomorphisms $\mathscr{A}(E)$. If for A there exists a representation of the form*

$$A = R + K, \tag{39.1}$$

where $R \in \mathscr{A}(E)$ has an inverse in $\mathscr{A}(E)$, $K \in \mathscr{F}(\mathscr{A}(E))$ and $RK = KR$, then A is a Fredholm operator with vanishing index, whose chain lengths $p(A)$ and $q(A)$ are finite and equal. If, conversely, A is a Fredholm operator with finite length of the nullchain and of the image chain, then A has a representation of the form (39.1), *where R and K have the above-indicated properties.*

Proof. If the representation (39.1) is valid, then it follows from Proposition 26.2 that A is a Fredholm operator with vanishing index. To prove the assertion $p(A) = q(A) < \infty$, it is sufficient, because of Proposition 38.6, to show the finiteness of $p(A)$. Since R commutes with K, we get from $A = R(I + R^{-1}K)$ the representation $A^n = R^n(I + R^{-1}K)^n$. Thus $N(A^n) = N((I + R^{-1}K)^n)$ for $n = 1, 2, \ldots$ and since $R^{-1}K$ is finite-dimensional, these nullspaces become eventually constant (see the remark before Proposition 38.1).

Now let, conversely, A be a Fredholm operator with finite $p(A)$ and $q(A)$. According to Proposition 38.4 we have the decomposition

$$E = N(A^p) \oplus A^p(E) \quad \text{with} \quad p := p(A) = q(A). \tag{39.2}$$

A^p is a Fredholm operator by Proposition 25.3, hence $A^p(E)$ has finite codimension and is $\mathscr{A}(E)$-projectable (Proposition 25.1). From here and from (39.2) it follows, with the aid of Lemma 25.3, that there exists in $\mathscr{A}(E)$ a projector P which projects E parallel to $A^p(E)$ onto $N(A^p)$; furthermore P is finite-dimensional. With the projector $Q := I - P$, complementary to P, we set

$$R := AQ - P \quad \text{and} \quad K := AP + P.$$

R lies in $\mathscr{A}(E)$, K even in $\mathscr{F}(\mathscr{A}(E))$, and obviously

$$A = R + K.$$

Since $A^p(E)$ and $N(A^p)$ are mapped by A into themselves, P and Q commute with A (Proposition 5.4), hence R and K also commute. It remains to show that R has an inverse in $\mathscr{A}(E)$. If $Rx = 0$, then $AQx = Px$, hence also $QAx = Px$ and thus

$$QAx = 0 \quad \text{and} \quad Px = 0. \tag{39.3}$$

Because of the last one of these equations, x lies in $A^p(E)$, hence $Qx = x$; it follows with the help of the first equation (39.3) that $Ax = AQx = QAx = 0$, hence x lies also in $N(A)$. A glance at (39.2) now tells us that $x = 0$ and so $\alpha(R) = 0$. Since by Proposition 23.3 we have $\text{ind}(R) = \text{ind}(A - K) = \text{ind}(A)$, and since $\text{ind}(A)$ vanishes because of Proposition 38.6, we have also $\beta(R) = \alpha(R) - \text{ind}(R) = 0$, thus R is bijective. Furthermore, $R = A - K$ is by Proposition 25.3 a Fredholm operator, in particular it is relatively regular. Consequently we obtain from Proposition 26.1 that R has indeed an inverse in $\mathscr{A}(E)$. ■

Proposition 39.2. *If (E, E^+) is a* dual system *and $\mathscr{A}(E)$ an E^+-saturated algebra, then for every Fredholm operator A in $\mathscr{A}(E)$ we have*

$$p(A) = q(A^+) \quad \textit{and} \quad q(A) = p(A^+). \tag{39.4}$$

If furthermore A is chain-finite, then also A^+ is chain-finite and we have the following equalities:

$$\alpha(A) = \beta(A) = \alpha(A^+) = \beta(A^+), \qquad p(A) = q(A) = p(A^+) = q(A^+). \tag{39.5}$$

Since by Proposition 25.3 together with A also A^n is a Fredholm operator, (39.4) follows immediately from equations (27.8) if one replaces A by A^n. For the proof of the remaining assertions we only have to invoke Propositions 38.3 and 38.6. ■

§40. The Riesz theory of compact operators

In this section (with the exception of Proposition 40.2) let E be a normed space and K a compact self-map of E. We shall explain the essential traits of the theory which F. Riesz [133] has developed for operators of the form $I - K$, starting from the Fredholm integral equation of the second kind.

To abbreviate, we set

$$I - K =: A, \qquad N(A^n) =: N_n, \qquad A^n(E) =: B_n$$

and observe that because the compact operators form an ideal, the operator K_n in the expansion

$$A^n = (I - K)^n = I - \left[nK - \binom{n}{2}K^2 + \cdots + (-1)^{n+1}K^n\right] = I - K_n \tag{40.1}$$

is compact. From here it already follows with the aid of Propositions 13.3 and 30.1 that for $n = 1, 2, \ldots$ the nullspaces N_n are finite-dimensional and the image spaces B_n closed. We now show that both chain lengths of A are finite. If we had $p(A) = \infty$, then N_{n-1} would be a proper closed subspace of N_n for $n = 1, 2, \ldots$. According to the Lemma of F. Riesz (Lemma 10.2) there would then exist in each N_n a vector x_n such that $\|x_n\| = 1$ and $\|x_n - x\| \geqq \frac{1}{2}$ for all $x \in N_{n-1}$. Since

$$Kx_n - Kx_m = x_n - (x_m - Ax_m + Ax_n) \tag{40.2}$$

and for $m = 1, \ldots, n-1$ the element in parentheses lies in N_{n-1}, we would have for these subscripts m that $\|Kx_n - Kx_m\| \geqq \frac{1}{2}$, thus (Kx_n) could not have a convergent subsequence, in contradiction to the compactness of K: therefore $p(A)$ must be finite.

If we had $q(A) = \infty$, then because of the closedness of B_n for every n there would exist an $x_n \in B_n$ with $\|x_n\| = 1$ and $\|x_n - x\| \leqq \frac{1}{2}$ for all $x \in B_{n+1}$. Since for $m > n$ the element in parentheses in (40.2) belongs to B_{n+1}, we would have $\|Kx_n - Kx_m\| \geqq \frac{1}{2}$ for all $m > n$ and (Kx_n) could not contain a convergent subsequence. This contradiction to the definition of compactness proves that $q(A)$ is finite.

With the help of Propositions 38.3 and 38.6 we now see that

$$p(A) = q(A) \quad \text{and} \quad \alpha(A) = \beta(A), \tag{40.3}$$

and from Proposition 38.4 follows the decomposition

$$E = N_p \oplus B_p \quad \text{with} \quad p := p(A). \tag{40.4}$$

Since B_p has finite codimension and is closed, there exists a continuous (finite-dimensional) projector P which projects E along B_p onto N_p (Proposition 24.2); then $Q := I - P$ projects E along N_p onto B_p. Setting

$$S := PK \in \mathscr{F}(E), \qquad V := QK \in \mathscr{K}(E) \quad \text{and} \quad R := I - V$$

we obtain, because of $K = (P + Q)K = S + V$, the decomposition

$$A = I - K = R - S. \tag{40.5}$$

From $Rx = (I - V)x = 0$ it follows on the one hand that $x = Vx = QKx \in B_p$ and on the other that $Ax = (R - S)x = -Sx = -PKx \in N(A^p)$, hence $x \in N(A^{p+1}) = N(A^p) = N_p$, so that by (40.4), finally, $x = 0$; thus $R = I - V$ is injective. Since V is compact, it follows by what has been proved above—see the second equation in (40.3)—that R is even bijective. Furthermore, Proposition 11.6 shows that R^{-1} lies in $\mathscr{L}(E)$. And now it follows from (40.5) with the aid of Proposition 26.2 that $I - K$ is a Fredholm operator (the E^+-saturated algebra $\mathscr{A}(E)$ of that proposition is $\mathscr{L}(E)$ in the present case; E^+ is E'). We observe that $I - K$ is open by Proposition 24.3, and summarize our results:

Theorem 40.1. *If K is a compact endomorphism of the normed space E, then $I - K$ is a chain-finite Fredholm operator in $\mathscr{L}(E)$ with vanishing index, in particular, $I - K$ is open and $(I - K)(E)$ closed.*

From this result we obtain immediately, if we invoke Theorem 27.1 and Proposition 39.2:

Proposition 40.1. *If K is a compact endomorphism of the normed space E, then $I' - K'$ is a Fredholm operator in $\mathscr{L}(E')$ and the equations*

$$\alpha(I - K) = \beta(I - K) = \alpha(I' - K') = \beta(I' - K'), \tag{40.6}$$

$$p(I - K) = q(I - K) = p(I' - K') = q(I' - K') \tag{40.7}$$

are valid; in particular, the operators $I - K$ and $I' - K'$ are bijective whenever one of them is either injective or surjective. Furthermore, we have:

$$(I - K)x = y \text{ is solvable} \Leftrightarrow \langle y, x' \rangle = 0 \text{ for all } x' \in N(I' - K'),$$

$$(I' - K')x' = y' \text{ is solvable} \Leftrightarrow \langle x, y' \rangle = 0 \text{ for all } x \in N(I - K).$$

The above solvability criteria and (40.6) can, because of (40.5), also be obtained from Proposition 18.1.

Although the integral operator K in the Fredholm integral equation (19.1) is compact (Example 13.1), we cannot obtain Theorem 19.1, which describes the solvability behavior of the Fredholm integral equation, from Proposition 40.1, because this theorem is based on the dual system $(C[a, b], C[a, b])$, not on the topological dual system $(C[a, b], (C[a, b])')$. We shall see, however, that we can obtain Theorem 19.1, in fact even a much stronger assertion, solely by applying the fundamental index relation $\operatorname{ind}(I - K) = 0$. The next proposition is a preparation for this.

Proposition 40.2. *If (E, E^+) is a dual system and A an E^+-conjugable endomorphism of E with*

$$\operatorname{ind}(A) = -\operatorname{ind}(A^+), \tag{40.8}$$

then

$$\alpha(A) = \beta(A^+), \qquad \beta(A) = \alpha(A^+), \tag{40.9}$$

and the operators A, A^+ are normally solvable, i.e.,

$$Ax = y \text{ is solvable } \Leftrightarrow \langle y, x^+ \rangle = 0 \text{ for all } x^+ \in N(A^+), \tag{40.10}$$

$$A^+x^+ = y^+ \text{ is solvable} \Leftrightarrow \langle x, y^+ \rangle = 0 \text{ for all } x \in N(A). \tag{40.11}$$

Proof. We first show the inequality

$$\alpha(A) \leqq \beta(A^+). \tag{40.12}$$

We may assume that $n := \alpha(A) > 0$. Let $\{x_1, \ldots, x_n\}$ be a basis of $N(A)$. By Proposition 15.1 there exist in E^+ linearly independent elements $x_1^+, \ldots, x_n^+$ with

$$\langle x_i, x_k^+ \rangle = \delta_{ik} \qquad \text{for} \quad i, k = 1, \ldots, n. \tag{40.13}$$

We have

$$[x_1^+, \ldots, x_n^+] \cap A^+(E^+) = \{0\} \tag{40.14}$$

since from $\alpha_1 x_1^+ + \cdots + \alpha_n x_n^+ = A^+ y^+$ it follows, because of (40.13), that

$$\alpha_i = \langle x_i, \alpha_1 x_1^+ + \cdots + \alpha_n x_n^+ \rangle = \langle x_i, A^+ y^+ \rangle = \langle Ax_i, y^+ \rangle = \langle 0, y^+ \rangle = 0$$

for $i = 1, \ldots, n$. Since $\dim[x_1^+, \ldots, x_n^+] = n$, we obtain from (40.14) immediately the inequality (40.12).

If we let A and A^+ change roles, then we obtain the inequality

$$\alpha(A^+) \leqq \beta(A). \tag{40.15}$$

From (40.8), i.e., from the equality $\alpha(A) - \beta(A) = \beta(A^+) - \alpha(A^+)$, and from the inequalities (40.12) and (40.15) we obtain directly (40.9).

Now we see easily that also the solvability assertion (40.11) is valid, which we can write in the form $A^+(E^+) = N(A)^\perp$. In the case $\alpha(A) = 0$ it is trivial because of the first equation in (40.9). If, however, $n := \alpha(A) > 0$, then from the said equation, in connection with (40.14), the representation

$$E^+ = [x_1^+, \ldots, x_n^+] \oplus A^+(E^+)$$

follows. Hence, every $y^+ \in N(A)^\perp$ can be written in the form

$$y^+ = \alpha_1 x_1^+ + \cdots + \alpha_n x_n^+ + A^+ z^+.$$

Because of (40.13) we have for every x_i from the above basis $\{x_1, \ldots, x_n\}$ of $N(A)$ the equation $0 = \langle x_i, y^+ \rangle = \alpha_i + \langle x_i, A^+z^+ \rangle = \alpha_i + \langle Ax_i, z^+ \rangle = \alpha_i$, and consequently $y^+ = A^+z^+$. Thus we have proved the inclusion $N(A)^\perp \subset A^+(E^+)$. Since its converse is trivial, we have indeed $A^+(E^+) = N(A)^\perp$, i.e., the solvability assertion (40.11).

Letting A and A^+ interchange their roles, one sees that also statement (40.10) holds. Thus our proposition is completely proved. ■

From Theorem 40.1 and Proposition 40.2 we obtain immediately the following proposition, from which Theorem 19.1 can be deduced again (this time through a compactness argument in contrast to the approximation procedure used earlier).

Proposition 40.3. *Let E, E^+ be normed spaces which form a dual system (E, E^+). Let K be a compact, E^+-conjugable endomorphism of E and let also K^+ be compact. If we set $A := I - K$, then*

$$\alpha(A) = \beta(A) = \alpha(A^+) = \beta(A^+),$$

and the solvability assertions (40.10) *and* (40.11) *are valid.*

Proposition 40.3 leads to the question: under what conditions is K^+ automatically compact? We shall see in §42 that at any rate the dual transformation of a compact endomorphism of a normed space E is again compact, and that the converse is valid if E is complete. This converse is proved in the shortest way if one uses the concept of the bidual space; we introduce it in the next section.

Exercises

In the following exercises E is a normed space.

+1. $A \in \mathscr{L}(E)$ is a Fredholm operator if and only if its residue class $\hat{A}$ in $\mathscr{L}(E)/\mathscr{K}(E)$ is invertible, i.e., if and only if there exist $B, C \in \mathscr{L}(E)$ and $K_1, K_2 \in \mathscr{K}(E)$ so that $BA = I - K_1$ and $AC = I - K_2$ (cf. Propositions 25.2 and 51.6).

+2. If A is a Fredholm operator and K is compact, then $A + K$ is also a Fredholm operator and $\operatorname{ind}(A + K) = \operatorname{ind}(A)$. *Hint*: Exercise 1 and the proof of Proposition 23.3; see also Proposition 53.2.

+3. In Propositions 26.2 and 39.1 the finite-dimensional K can be replaced by a compact one in the characterizing representations $A = R + K$ of Fredholm operators A, provided that $\mathscr{A}(E) = \mathscr{L}(E)$. Use Exercise 2.

§41. The bidual of a normed space. Reflexivity

The dual E' of a normed space E is itself a normed space, consequently it has a dual which we denote by E'' and call the *bidual* of E. According to §15, we can imbed the vector space E into the algebraic dual $(E')^*$ of E'; the imbedding isomorphism J is given by

(41.1) $$Jx := F_x \quad \text{with} \quad F_x(x') := \langle x, x' \rangle \qquad \text{for} \quad x' \in E'.$$

Because of $|F_x(x')| \leqq \|x\| \, \|x'\|$, the linear form F_x is continuous on E', so that E is imbedded even into the bidual E''. Every $x \in E$ is, to put it briefly, a continuous linear form on E'. The question arises now whether x as an element of E has the same norm as x as a linear form on E', i.e., whether $\|x\| = \|F_x\|$. We shall be able to answer this question with the help of the next proposition.

Proposition 41.1. *For every vector x of the normed space E we have*

(41.2) $$\|x\| = \sup\{|\langle x, x' \rangle| : x' \in E', \|x'\| = 1\}.$$

For the proof we may assume that $x \neq 0$. For $\|x'\| = 1$ we have $|\langle x, x' \rangle| \leqq \|x\|$; equation (41.2) now follows because by Theorem 28.3 there exists an x' with $\|x'\| = 1$ and $|\langle x, x' \rangle| = \|x\|$. ∎

Observe the symmetry between (41.2) and the equation

$$\|x'\| = \sup\{|\langle x, x' \rangle| : x \in E, \|x\| = 1\}.$$

From Proposition 41.1 the relation

(41.3) $$\|F_x\| = \sup|F_x(x')| = \sup|\langle x, x' \rangle| = \|x\|$$

follows, where the least upper bounds are to be taken with respect to all $x' \in E'$ with $\|x'\| = 1$. Thus we have proved the following proposition:

Proposition 41.2. *The normed space E is imbedded into its bidual E'' by the map J defined in* (41.1) *in an isometrically isomorphic way. Thus E can be considered as a subspace of the normed space E''.*

If $J(E) = E''$, i.e., if in the sense of the canonical imbedding $E = E''$, then E is said to be *reflexive*. Since E'' is complete as a dual space (see the end of §13), *only Banach spaces can be reflexive.* In [57] Banach spaces are exhibited which are isomorphic to their biduals as normed spaces but which nevertheless are not reflexive.

(E, E') forms a dual system with respect to the (canonical) bilinear form $\langle x, x' \rangle := x'(x)$; the same is true for the system (E', E'') with the bilinear form $\langle x', x'' \rangle := x''(x')$ (in both cases we denote, without fear of confusion, the bilinear form by the usual bracket symbol $\langle \cdot, \cdot \rangle$). The bilinear forms of the two dual systems are connected through the canonical imbedding J: because of $(Jx)(x') = F_x(x') = \langle x, x' \rangle$, we have

$$\langle x, x' \rangle = \langle x', Jx \rangle,$$

or if we consider E as a subspace of E'' (i.e., if we identify Jx with x)

(41.4) $$\langle x, x' \rangle = \langle x', x \rangle,$$

where the first bilinear form is associated with the system (E, E') and the second with the system (E', E''). We recall now that we pass from the bilinear system (E, E') in a canonical way to the bilinear system (E', E) by equipping $E' \times E$ with the bilinear form $\langle x', x \rangle := \langle x, x' \rangle$ (see (14.10)). A glance at (41.4) now shows that we can obtain (E', E) by restricting the bilinear form $\langle x', x'' \rangle$ to the subspace $E' \times E$ of $E' \times E''$, i.e., that *the bilinear system (E', E) does not change if we consider E as a subspace of E''.*

For the operator dual to $A \in \mathscr{L}(E)$ we have

$$\langle Ax, x' \rangle = \langle x, A'x' \rangle \qquad \text{for all} \quad x \in E \quad \text{and} \quad x' \in E'.$$

A' can be conjugated in two ways: on one hand with respect to the dual system (E', E) with the bilinear form $\langle x', x \rangle$, on the other hand with respect to the dual system (E', E'') with the bilinear form $\langle x', x'' \rangle$. The E-conjugation of A' yields A, the E''-conjugation the operator $(A')'$ dual to A', which we denote briefly by A'' and which we call the *operator bidual* to A; it satisfies the equation

$$\langle A'x', x'' \rangle = \langle x', A''x'' \rangle \qquad \text{for all} \quad x' \in E' \quad \text{and} \quad x'' \in E''.$$

With the aid of (41.4) it follows for $x'' = x \in E$ and every $x' \in E'$ that

$$\langle x', A''x \rangle = \langle A'x', x \rangle = \langle x, A'x' \rangle = \langle Ax, x' \rangle = \langle x', Ax \rangle,$$

and so

$$A''x = Ax \qquad \text{for} \quad x \in E. \tag{41.5}$$

The proofs of the following two propositions will manifest the technical usefulness of our imbedding procedure.

Proposition 41.3. *Let E be a normed space and $A \in \mathscr{L}(E)$. A' is surjective if and only if A has a continuous inverse on $A(E)$.*

Proof. If A' is surjective, then because of Proposition 36.5 A'' has a continuous inverse on $A''(E'')$. With the aid of (41.5) it follows that A^{-1} exists on $A(E)$ and is continuous. Now let, conversely, A have a continuous inverse. Then for every $x' \in E'$ we define by $f(y) := \langle A^{-1}y, x' \rangle$ a continuous linear form f on $A(E)$. If we extend f according to Theorem 28.1 to a continuous linear form y' on E, then we have, for all $x \in E$,

$$\langle x, A'y' \rangle = \langle Ax, y' \rangle = \langle A^{-1}Ax, x' \rangle = \langle x, x' \rangle,$$

hence $x' = A'y'$ and thus A' is surjective. ■

The surjectivity of A will be discussed in Proposition 97.1.

Proposition 41.4. *The subset M of the normed space E is bounded if and only if every $x' \in E'$ is bounded on M.*

Proof. If every $x' \in E'$ is bounded on M, i.e., if for every x' there exists an $\alpha_{x'} > 0$ such that $|\langle x, x' \rangle| \leqq \alpha_{x'}$ for all $x \in M$, then the family of continuous linear forms $M \subset E''$ is pointwise bounded on the Banach space E', hence by Principle 33.2 there exists a $\gamma > 0$ such that for $x \in M$ we have $\|x\| \leqq \gamma$. The converse is trivial. ■

With the help of the proposition just proved, we show now that only the *continuous* homomorphisms $A: E \to F$ have dual transformations.

Proposition 41.5. *Suppose that the linear map $A: E \to F$ (E, F normed spaces) has a dual transformation, i.e., that there exists a linear map A' from F' into E' with $\langle Ax, y' \rangle = \langle x, A'y' \rangle$ for all $x \in E$, $y' \in F'$. Then A is continuous.*

Proof. For every $y' \in F'$ and all $x \in E$ with $\|x\| \leqq 1$ we have $|\langle Ax, y' \rangle| = |\langle x, A'y' \rangle| \leqq \|A'y'\|$. Thus every y' is bounded on the set $\{Ax: \|x\| \leqq 1\}$, so by Proposition 41.4 this set itself is bounded. This means that A is bounded and therefore continuous. ■

For subspaces of E' there are in a natural way *two* concepts of orthogonal closedness: first with respect to (E', E) and second with respect to (E', E''). Correspondingly there exist for endomorphisms of E' (in particular for the operator A' dual to $A \in \mathscr{L}(E)$) *two* concepts of normal solvability. Every E-orthogonally closed subspace of E' is, because of Lemma 29.2, also E''-orthogonally closed. The converse of this is trivially true if E is reflexive. Now let E be non-reflexive and let $x_0'' \in E'' \setminus E$. Then $N(x_0'')$ is E''-orthogonally closed but not E-orthogonally closed; otherwise there would exist by the lemma just quoted a non-trivial linear form $x \in E$ which vanishes on $N(x_0'')$; because of Lemma 15.1 we would have then $x = \beta x_0''$ with $\beta \neq 0$, though x_0'' does not lie in E. If we take into account Proposition 29.2, we can assert the following:

Proposition 41.6. *For a normed space E the following assertions are equivalent:*

(a) *E is reflexive.*
(b) *Every E''-orthogonally closed subspace of E' is also E-orthogonally closed.*
(c) *Every closed subspace of E' is E-orthogonally closed.*

Reflexivity is inherited by the dual (Exercise 3) and, as the next proposition shows, by closed subspaces.

Proposition 41.7. *Every closed subspace F of a reflexive Banach space E is reflexive.*

For the proof we need only to show, by the last proposition, that every F''-orthogonally closed subspace M of F' is also F-orthogonally closed, i.e., that—see Lemma 29.2—to every $y_0' \in F' \setminus M$ there exists an $x_0 \in F$ such that

$$(41.6) \qquad \langle x_0, y' \rangle = 0 \quad \text{for all} \quad y' \in M \quad \text{but} \quad \langle x_0, y_0' \rangle \neq 0.$$

Now by the lemma just quoted there exists to y'_0 a $y''_0 \in F''$ so that

$$(41.7) \qquad \langle y', y''_0 \rangle = 0 \quad \text{for all} \quad y' \in M \quad \text{but} \quad \langle y'_0, y''_0 \rangle \neq 0.$$

With the help of this y''_0 and the map $A \in \mathscr{L}(E', F')$, which associates with each $x' \in E'$ its restriction to the subspace F, we define a linear form $x''_0 \in E''$ by

$$\langle x', x''_0 \rangle := \langle Ax', y''_0 \rangle \qquad \text{for} \quad x' \in E'.$$

Since E is reflexive, x''_0 coincides with an $x_0 \in E$; thus the last equation can be written in the form

$$(41.8) \qquad \langle x_0, x' \rangle = \langle Ax', y''_0 \rangle \qquad \text{for} \quad x' \in E'.$$

x_0 even lies in F; otherwise there would exist by Theorem 28.2, since F is closed, an $x' \in E'$ with $\langle x_0, x' \rangle \neq 0$ and $Ax' = 0$; from (41.8) we get immediately a contradiction. If we extend $y' \in F'$ by Theorem 28.1 to an $x' \in E'$, then $Ax' = y'$, hence by (41.8)

$$\langle x_0, y' \rangle = \langle x_0, x' \rangle = \langle Ax', y''_0 \rangle = \langle y', y''_0 \rangle.$$

Because of (41.7) it follows that for every $y' \in M$:

$$\langle x_0, y' \rangle = \langle y', y''_0 \rangle = 0, \qquad \text{while} \quad \langle x_0, y'_0 \rangle = \langle y'_0, y''_0 \rangle \neq 0.$$

Thus we have indeed (41.6). ∎

We shall further pursue the topic of reflexivity in §62 and §105.

Exercises

*1. If E, F are normed spaces, then the *bidual* operator A'' to $A \in \mathscr{L}(E, F)$ is defined as the dual operator to A'; it lies in $\mathscr{L}(E'', F'')$ and satisfies the equation $\langle A'y', x'' \rangle = \langle y', A''x'' \rangle$ for all $y' \in F'$ and $x'' \in E''$. The restriction of A'' to E is A.

2. Let E be a normed space and $A \in \mathscr{L}(E)$. Show the following:

(a) If the dual operator A' is E-normally solvable, then it is also E''-normally solvable.

(b) If E is complete and A is E'-normally solvable, then A' is E-normally solvable (see Proposition 36.4).

$^+$3. A reflexive Banach space has a reflexive dual.

4. Every finite-dimensional Banach space is reflexive.

$^+$5. The spaces l^p are reflexive for $1 < p < \infty$.

6. (c_0), (c), l^1 and l^∞ are not reflexive. *Hint*: Show first the non-reflexivity of (c_0) with the help of Exercise 1 of §35, then use Proposition 41.7 and Exercise 3.

§42. The dual transformation of a compact operator

The following considerations prepare the answer to the question raised in §40 whether the dual transformation of a compact operator is again compact.

A set (or a sequence) M in a metric space E is said to be *totally bounded* if for every $\varepsilon > 0$ there exist finitely many open balls with radii ε and centers in M, whose union covers M.

A totally bounded set is bounded (in **K** the converse of this is also true). *A Cauchy sequence* (x_n) *is totally bounded* since for sufficiently large m every x_n lies in one of the balls $K_\varepsilon(x_\mu)$, $\mu = 1, \ldots, m$.

Lemma 42.1. *A set (or a sequence) M of a metric space is totally bounded if and only if every sequence* (x_n) *from M has a Cauchy subsequence.*

Proof. Let M be totally bounded. If we cover M with finitely many balls with radius 1, then in at least one of them there are infinitely many terms of the sequence (x_n); these form a subsequence $(x_{11}, x_{12}, \ldots)$ of (x_n) with $d(x_{1n}, x_{1m}) \leqq 2$. If we now cover M with finitely many balls with radius $\frac{1}{2}$, then one of them contains a subsequence $(x_{21}, x_{22}, \ldots)$ of (x_{1n}) with $d(x_{2n}, x_{2m}) \leqq 1$. One continues in this way. The diagonal sequence (x_{nn}) is a Cauchy subsequence of (x_n). If M is not totally bounded, then there exists an $\varepsilon_0 > 0$ so that M lies in no finite union of balls $K_{\varepsilon_0}(x)$, $x \in M$, Starting from an arbitrary $x_1 \in M$ we can therefore find successively points $x_n \in M$ such that $d(x_n, x_m) \geqq \varepsilon_0$ for $n \neq m$. Obviously (x_n) contains no Cauchy subsequence. ■

It follows from the Lemma that *every relatively compact set is totally bounded.* A metric space E is said to be *separable* if there exists an at most countable set, which is dense in E.

Lemma 42.2. *A totally bounded metric space E is separable.*

Indeed, if we cover for $k = 1, 2, \ldots$ the space E with finitely many balls with radius $1/k$, then the countable set of the centers of these balls is dense in E. ■

In the following investigations let E and F be normed spaces.

Proposition 42.1. *If* $K: E \to F$ *is compact, then the image of any bounded set is relatively compact and therefore totally bounded. The image space* $K(E)$ *is separable.*

The first assertion is immediately clear. Let $B_n := \{x \in E: \|x\| \leqq n\}$. Then $K(B_n)$ is totally bounded, hence separable (Lemma 42.2). It follows that $K(E) = \bigcup_{n=1}^{\infty} K(B_n)$ is also separable. ■

The following *theorem of Schauder* answers the question raised at the beginning of this section (see [137]).

Proposition 42.2. *The dual transformation* K' *of a compact operator* $K: E \to F$ *is compact.*

Proof. Let (y'_k) be a bounded sequence from F', i.e., let $\|y'_k\| \leqq \gamma$ for all k, and set $B_1 := \{x \in E: \|x\| \leqq 1\}$. For a given $\varepsilon > 0$ there exist, because of Proposition 42.1, elements $x_1, \ldots, x_n$ in B_1 with the following property: to every $x \in B_1$ there exists an x_i such that

$$\|Kx - Kx_i\| \leqq \varepsilon. \tag{42.1}$$

It follows from the estimate $|\langle Kx_i, y'_k\rangle| \leqq \|Kx_i\|\gamma$ that (y'_k) contains a subsequence (z'_k) such that $\lim_{k\to\infty} \langle Kx_i, z'_k\rangle$ exists for $i = 1, \ldots, n$. Thus the sequence $u_k := (\langle Kx_1, z'_k\rangle, \ldots, \langle Kx_n, z'_k\rangle)$ converges in $l^1(n)$ and is therefore totally bounded. Consequently there exist finitely many u_k, say $u_{k_1}, \ldots, u_{k_m}$, with the following property: to every u_k there exists an u_{k_μ} such that

$$\|u_k - u_{k_\mu}\| = \sum_{i=1}^{n} |\langle Kx_i, z'_k\rangle - \langle Kx_i, z'_{k_\mu}\rangle| \leqq \varepsilon.$$

It follows that for every z'_k there exists a z'_{k_μ} with

$$|\langle Kx_i, z'_k\rangle - \langle Kx_i, z'_{k_\mu}\rangle| \leqq \varepsilon \qquad \text{for} \quad i = 1, \ldots, n. \tag{42.2}$$

From (42.1) and (42.2) we obtain the assertion: There exist linear forms $z'_{k_1}, \ldots, z'_{k_m}$ with the following property: for every z'_k there exists a z'_{k_μ} so that for all $x \in B_1$ we have the following estimate:

$$\begin{aligned}
|\langle x, K'z'_k - K'z'_{k_\mu}\rangle| &= |\langle Kx, z'_k\rangle - \langle Kx, z'_{k_\mu}\rangle| \\
&\leqq |\langle Kx, z'_k\rangle - \langle Kx_i, z'_k\rangle| + |\langle Kx_i, z'_k\rangle - \langle Kx_i, z'_{k_\mu}\rangle| \\
&\quad + |\langle Kx_i, z'_{k_\mu}\rangle - \langle Kx, z'_{k_\mu}\rangle| \\
&\leqq \|Kx - Kx_i\|\,\|z'_k\| + |\langle Kx_i, z'_k\rangle - \langle Kx_i, z'_{k_\mu}\rangle| \\
&\quad + \|Kx_i - Kx\|\,\|z'_{k_\mu}\| \\
&\leqq \varepsilon\gamma + \varepsilon + \varepsilon\gamma = (2\gamma + 1)\varepsilon;
\end{aligned}$$

and this implies

$$\|K'z'_k - K'z'_{k_\mu}\| = \sup_{x\in B_1} |\langle x, K'z'_k - K'z'_{k_\mu}\rangle| \leqq (2\gamma + 1)\varepsilon.$$

Thus the sequence $(K'z'_k)$ is totally bounded, hence it contains, by Lemma 42.1, a Cauchy subsequence which even converges because of the completeness of E'. Thus the compactness of K' is proved. ■

Proposition 42.2 has a converse if F is complete:

Proposition 42.3. *Let E be a normed space, F a Banach space and $K \in \mathscr{L}(E, F)$. If K' is compact, then so is K.*

Proof. By Proposition 42.2 the bidual operator $K'': E'' \to F''$ is compact. Its restriction to the subspace E of E'' (see Proposition 41.2 and §41 Exercise 1) is then compact as a map from E into F''. Since, however, F is complete and thus a closed subspace of F'', we obtain the compactness of K as a map from E into F. ■

Exercises

$^+$1. A linear map $K: E \to F$ (E, F normed spaces) is said to be *precompact* if the image of every bounded sequence contains a Cauchy subsequence. Show the following:

(a) K is precompact if and only if the image of the unit ball (or of any bounded set) is totally bounded.

(b) A precompact operator is continuous.

(c) A compact operator is precompact; if $K: E \to F$ is precompact and F complete, then K is compact.

(d) If the sequence of precompact operators $K_n \in \mathscr{L}(E, F)$ converges uniformly to K, then also K is precompact.

(e) Sums and scalar multiples of precompact operators are precompact; the product of a precompact operator with a continuous operator is precompact.

(f) The precompact endomorphisms of a normed space E form a closed ideal in $\mathscr{L}(E)$.

(g) A continuous operator K is precompact if and only if K' is compact.

2. The following assertions concerning a normed space E are equivalent:

(a) E is finite-dimensional.

(b) Every bounded sequence in E has a Cauchy subsequence.

(c) Every bounded set is totally bounded.

(d) The unit ball is totally bounded.

3. A set M in the normed space E is totally bounded if and only if for every open ball U around 0 (for every neighborhood U of 0) there exist finitely many points $x_1, \ldots, x_n$ in M so that $M \subset \bigcup_{i=1}^n (x_i + U)$.

4. $A \in \mathscr{L}(E, F)$ maps every totally bounded set $M \subset E$ onto a totally bounded set.

5. Every subspace of a separable metric space is again separable.

6. The following spaces are separable: Finite-dimensional normed spaces, l^p for $1 \leqq p < \infty$, (c), (c_0), $C[a, b]$.

7. The following spaces are not separable: l^∞, and more generally $B(T)$ if T is an infinite set, $BV[a, b]$.

*8. A normed space E is separable if its dual E' is separable. The converse is false. *Hint*: Let $\{x'_1, x'_2, \ldots\}$ be dense on $\{x' \in E': \|x'\| = 1\}$. Choose $x_n \in E$ with $\|x_n\| = 1$ and $|\langle x_n, x'_n \rangle| \geqq \frac{3}{4}$. Show with the help of Theorem 28.2 that $[x_1, x_2, \ldots] = E$.

§43. Singular values and eigenvalues of a compact operator

When boundary value problems of mathematical physics lead to a Fredholm integral equation, then frequently a parameter λ occurs in these in the following way (see (3.28)):

$$x(s) - \lambda \int_a^b k(s, t)x(t)dt = y(s).$$

Therefore we consider operator equations of the form

$$(I - \lambda K)x = y, \tag{43.1}$$

where K is an endomorphism of a vector space E. The solvability behavior of equation (43.1) will be different for different values of λ. We say that λ is a *regular* value of K when (43.1) is uniquely solvable for all y; a non-regular value is said to be *singular*. For $\lambda \neq 0$ (this is the only interesting case) and $\mu := 1/\lambda$ we have

$$I - \lambda K = \lambda(\mu I - K) \quad \text{and} \quad \alpha(I - \lambda K) = \alpha(\mu I - K). \tag{43.2}$$

Every number μ (even $\mu = 0$) with $\alpha(\mu I - K) \neq 0$ is called an *eigenvalue* (or *proper value*) of K. *A necessary and sufficient condition for μ to be an eigenvalue of K is that there exist in E an $x \neq 0$ with $(\mu I - K)x = 0$, i.e., $Kx = \mu x$*; every such x is called an *eigenvector* (*proper vector*) or an *eigensolution* (*proper solution*) of K corresponding to the eigenvalue μ. The space $N(\mu I - K)$ is called the *eigenspace* of K corresponding to the eigenvalue μ, and dim $N(\mu I - K)$ is the *multiplicity* of μ. Clearly $N(\mu I - K)\backslash\{0\}$ is the set of all eigensolutions of K corresponding to the eigenvalue μ. Observe that 'eigenvalue of a boundary value problem' (§3) and the here defined 'eigenvalue of an operator' are two different concepts.

For a compact endomorphism K every multiple λK is compact, therefore the singular values of such a K are exactly the λ with $\alpha(I - \lambda K) \neq 0$ (Theorem 40.1), and $\lambda \neq 0$ is singular exactly if $1/\lambda$ is an eigenvalue. Thus instead of studying the singular values of a compact operator, we can just as well investigate the eigenvalues. We start with some algebraic considerations.

For a polynomial $f(\lambda) := \alpha_0 + \alpha_1\lambda + \cdots + \alpha_n\lambda^n$ in the variable $\lambda \in \mathbf{K}$ with coefficients $\alpha_\nu \in \mathbf{K}$ and for an endomorphism A of the vector space E over $\mathbf{K}$ we set

$$f(A) := \alpha_0 I + \alpha_1 A + \cdots + \alpha_n A^n.$$

The correspondence $f \mapsto f(A)$ is a homomorphism of the algebra of all polynomials f over $\mathbf{K}$ onto the (commutative) algebra of all operators $f(A)$.

Proposition 43.1. *For an endomorphism A of the vector space E the following assertions are valid*:

(a) *If $f_1, \ldots, f_n$ are pairwise relatively prime polynomials and if $f := f_1 \cdots f_n$, then*

$$N(f(A)) = N(f_1(A)) \oplus \cdots \oplus N(f_n(A)).$$

(b) *Eigensolutions of A corresponding to different eigenvalues are linearly independent.*

(c) *For relatively prime polynomials f_1, f_2 we have*

$$N(f_1(A)) \subset f_2(A)(E).$$

Proof.

(a) Let first $n = 2$, $N := N(f(A))$ and $N_i := N(f_i(A))$. Because of $N_i \subset N$ we have $N_1 + N_2 \subset N$. The converse inclusion is obtained as follows. Since f_1, f_2 are relatively prime, there exist polynomials g_1, g_2 so that $g_1(\lambda)f_1(\lambda) + g_2(\lambda)f_2(\lambda) = 1$, hence $g_1(A)f_1(A) + g_2(A)f_2(A) = I$. Every $x \in E$ admits thus a decomposition

$$x = g_1(A)f_1(A)x + g_2(A)f_2(A)x. \tag{43.3}$$

For $x \in N$ we have $f_1(A)f_2(A)x = f_2(A)f_1(A)x = f(A)x = 0$, hence

$$f_2(A)x \in N_1 \quad \text{and} \quad f_1(A)x \in N_2,$$

and since every polynomial in A maps the spaces N_1, N_2 into themselves, we obtain

$$x_1 := g_2(A)f_2(A)x \in N_1 \quad \text{and} \quad x_2 := g_1(A)f_1(A)x \in N_2.$$

With the aid of (43.3) it follows that $x = x_1 + x_2$, thus indeed $N \subset N_1 + N_2$ and therefore $N = N_1 + N_2$. Equation $N_1 \cap N_2 = \{0\}$ follows immediately from (43.3). For $n > 2$ one proves (a) by mathematical induction; we only have to observe that if $f_1, \ldots, f_n$ are pairwise relatively prime, then the polynomials $f_1 \cdots f_{n-1}$ and f_n are relatively prime.

(b) Let $\lambda_1, \ldots, \lambda_n$ be (pairwise) different eigenvalues of A and $x_1, \ldots, x_n$ the corresponding eigensolutions. Since the polynomials $f_1(\lambda) := \lambda - \lambda_1, \ldots, f_n(\lambda) := \lambda - \lambda_n$ are pairwise relatively prime, by (a) the sum $N(\lambda_1 I - A) + \cdots + N(\lambda_n I - A)$ is direct. From $\alpha_1 x_1 + \cdots + \alpha_n x_n = 0$ it follows therefore that $\alpha_i x_i = 0$ and thus also $\alpha_i = 0$ for $i = 1, \ldots, n$.

(c) Since for an arbitrary $x \in E$ we have the decomposition (43.3), we have in particular for every $x \in N(f_1(A))$ the relation

$$x = g_2(A)f_2(A)x = f_2(A)g_2(A)x \in f_2(A)(E).$$

Thus indeed $N(f_1(A)) \subset f_2(A)(E)$. ∎

The next proposition shows that the eigenvalues (hence also the singular values) of a compact operator have a very simple distribution.

Proposition 43.2. *For every eigenvalue μ of the* continuous *operator K on the normed space E we have*

$$|\mu| \leqq \|K\|; \tag{43.4}$$

if K is even compact, *then its eigenvalues form either a finite (possibly empty) set or a sequence which converges to zero.*

Proof. From $Kx = \mu x$, $x \neq 0$, it follows that $|\mu| \leqq \|Kx\|/\|x\| \leqq \|K\|$ which already proves (43.4). The second assertion of the proposition is equivalent to saying that the eigenvalues of K have no limit point $\neq 0$. If ξ were such a limit point, then there would exist a sequence (μ_n) of eigenvalues with $\mu_n \neq \xi$ and $\mu_n \to \xi$, furthermore a sequence (x_n) with $Kx_n = \mu_n x_n$ and $\|x_n\| = 1$. Then we would have

$$(43.5) \qquad (\xi I - K)x_n = (\mu_n I - K)x_n + (\xi - \mu_n)x_n = (\xi - \mu_n)x_n \to 0.$$

If p is the chain length of $\xi(I - (1/\xi)K) = \xi I - K$, which according to Theorem 40.1 is finite, then by Proposition 43.1c we have $x_n \in F := (\xi I - K)^p(E)$. The subspace F is mapped into itself by K. If we denote by $\tilde{I}$, $\tilde{K}$ the restrictions of I, K to F, then it follows from the closedness of F (§40) that $\tilde{K}$ is compact as an endomorphism of F and Proposition 38.4 implies the existence of the inverse $(\xi\tilde{I} - \tilde{K})^{-1}$ on F. By Proposition 11.6 the inverse is continuous. From (43.5) it would follow that

$$x_n = (\xi\tilde{I} - \tilde{K})^{-1}(\xi I - K)x_n \to 0$$

in contradiction to $\|x_n\| = 1$. ∎

Exercises

1. If A is a chain-finite continuous endomorphism with finite deficiency of a Banach space, then $\alpha(\lambda I - A) = \beta(\lambda I - A) = 0$ for all $\lambda \neq 0$ in a certain neighborhood of 0.

2. For every eigenvalue μ of $K \in \mathscr{L}(E)$ we have $|\mu| \leqq \lim \|K^n\|^{1/n}$ (this improves (43.4)).

3. Let A be an endomorphism of the finite-dimensional complex vector space E (thus A has at least one eigenvalue!). Then there exists exactly one polynomial $m(\lambda)$—the *minimal polynomial* of A—with the following properties:
   (a) the highest coefficient of m is 1,
   (b) $m(A) = 0$,
   (c) $m(\lambda)$ divides every polynomial $f(\lambda)$ such that $f(A) = 0$.
If $m(\lambda) = (\lambda - \lambda_1)^{p_1} \cdots (\lambda - \lambda_k)^{p_k}$ is the canonical decomposition into factors, then $\{\lambda_1, \ldots, \lambda_k\}$ is the set of eigenvalues of A and p_i the chain length of $\lambda_i I - A$.

$^+$4. Assume that some power K^n $(n \geqq 1)$ of the continuous endomorphism K of the complex normed space E is compact. Then $\mu I - K$ is a chain-finite Fredholm operator with vanishing index for every $\mu \neq 0$, and the eigenvalues of K form, if there are infinitely many of them, a sequence converging to zero. *Hint*: $\lambda^n - \mu^n = (\lambda - \mu)\prod_{\nu=2}^{n}(\lambda - \lambda_\nu)$, hence

$$K^n - \mu^n I = (K - \mu I)\prod_{\nu=2}^{n}(K - \lambda_\nu I).$$

Use Exercise 1 in §40 to establish the Fredholm property of $K - \mu I$. The finiteness of the chain lengths is obtained with the help of Exercise 2 of §38 or of Proposition 43.1.

5. If $f_1, \ldots, f_n$ are pairwise relatively prime polynomials, if $f := f_1 \cdots f_n$ and $A \in \mathscr{S}(E)$, then $f(A)(E) = \bigcap_{\nu=1}^{n} f_\nu(A)(E)$, $f_\nu(A)(E) + f_\mu(A)(E) = E$ for $1 \leqq \nu < \mu \leqq n$ and

$$\operatorname{codim} f(A)(E) = \sum_{\nu=1}^{n} \operatorname{codim} f_\nu(A)(E).$$

6. The eigenvalues of the boundary value problem (3.25) can accumulate only at infinity if 0 is not an eigenvalue.

VII

Spectral theory in Banach spaces and Banach algebras

§44. The resolvent

In §43 we investigated the invertibility of the operator $\lambda I - K$ where λ varies; there K was compact. In this section we study the same question for an arbitrary *continuous* K on a Banach space. It will turn out to be not essential that K is a *map*; what is decisive is that K lies in a Banach algebra. We shall therefore situate our investigation very soon in a Banach algebra. First we define some basic concepts:

The *resolvent set* $\rho(K)$ of a continuous endomorphism K on the Banach space E consists of all scalars λ for which $\lambda I - K$ has an inverse $R_\lambda \in \mathscr{L}(E)$. We call R_λ the *resolvent operator* of K; the map $\lambda \mapsto R_\lambda$ defined on $\rho(K)$ is called the *resolvent* of K. The set $\sigma(K) := \mathbf{K} \backslash \rho(K)$ is the *spectrum* of K.

Because of Theorem 32.2, *λ lies in $\rho(K)$ if and only if $\lambda I - K$ is bijective.* With the help of the deficiencies and chain lengths $\rho(K)$ can be described as follows:

(44.1)
$$\begin{aligned}\rho(K) &= \{\lambda \in \mathbf{K} : \alpha(\lambda I - K) = \beta(\lambda I - K) = 0\} \\ &= \{\lambda \in \mathbf{K} : p(\lambda I - K) = q(\lambda I - K) = 0\}.\end{aligned}$$

The *completeness* of E enters in an essential way into this characterization of the resolvent set.

The eigenvalues of K all lie in $\sigma(K)$; they form the *point spectrum* $\sigma_p(K)$ of K.

Theorem 40.1 implies directly:

Proposition 44.1. *For the* compact *endomorphism K of a Banach space we have $\sigma(K) \backslash \{0\} = \sigma_p(K) \backslash \{0\}$.*

We get very easily also

Proposition 44.2. *For the continuous endomorphism A of the Banach space E we have $\sigma(A) = \sigma(A')$.*

Proof. From the conjugation rules of Proposition 16.3 we get immediately $\rho(A) \subset \rho(A')$. Now let $\lambda \in \rho(A')$. Then $\lambda I - A$ has a continuous inverse according to Proposition 41.3, thus $(\lambda I - A)(E)$ is closed by Lemma 36.1, and from here it follows with the aid of Proposition 29.3 that $\lambda I - A$ is normally solvable, hence that $(\lambda I - A)(E) = N(\lambda I' - A')^\perp = \{0\}^\perp = E$. Thus λ lies also in $\rho(A)$, hence taken together we have $\rho(A) = \rho(A')$ and thus $\sigma(A) = \sigma(A')$. ■

Resolvent set, resolvent and spectrum are concepts which make sense also for elements of a Banach algebra E (the unit element of E will always be denoted by e):

For $x \in E$ the *resolvent set* is $\rho(x) := \{\lambda \in \mathbf{K}: \lambda e - x \text{ is invertible in } E\}$, the *resolvent* is the map $\lambda \mapsto r_\lambda := (\lambda e - x)^{-1}$ defined on $\rho(x)$, the *spectrum* $\sigma(x)$ is the complement of $\rho(x)$ in $\mathbf{K}$.

From the propositions and formulas of §9 we obtain immediately assertions (a) and (b) of the following theorem; we only have to take into consideration the trivial transformations

$$(\lambda e - x) - (\lambda_0 e - x) = (\lambda - \lambda_0)e, \qquad \lambda e - x = \lambda\left(e - \frac{1}{\lambda}x\right).$$

(c) follows immediately from (a) and (b); (d) will be proved following the statement of the theorem.

Theorem 44.1. *For an element x of a* Banach algebra *E the following assertions are valid:*

(a) *If $\lambda_0 \in \rho(x)$ and $|\lambda - \lambda_0| < 1/\|r_{\lambda_0}\|$, then $\lambda \in \rho(x)$,*

$$r_\lambda = \sum_{n=0}^{\infty} (-1)^n (\lambda - \lambda_0)^n r_{\lambda_0}^{n+1} \tag{44.2}$$

and

$$\|r_\lambda - r_{\lambda_0}\| \leqq \frac{|\lambda - \lambda_0|}{1 - |\lambda - \lambda_0|\,\|r_{\lambda_0}\|} \|r_{\lambda_0}\|^2; \tag{44.3}$$

in particular, $\rho(x)$ is open, *and r_λ depends* continuously *on λ.*

(b) *For $|\lambda| > \lim \|x^n\|^{1/n}$ we have $\lambda \in \rho(x)$ and*

$$r_\lambda = \sum_{n=0}^{\infty} \frac{x^n}{\lambda^{n+1}}. \tag{44.4}$$

For $|\lambda| > \|x\|$ we have therefore

$$\|r_\lambda\| \leqq \frac{1}{|\lambda|}\left(1 - \frac{\|x\|}{|\lambda|}\right)^{-1} \tag{44.5}$$

and, in particular, r_λ tends to 0 for $|\lambda| \to \infty$.

(c) *$\sigma(x)$ is closed and because of*

$$|\lambda| \leqq \lim \|x^n\|^{1/n} \quad \textit{for all} \quad \lambda \in \sigma(x) \tag{44.6}$$

also bounded, thus compact.

(d) *If λ and μ lie in $\rho(x)$, then we have the* resolvent equation

$$r_\lambda - r_\mu = -(\lambda - \mu) r_\lambda r_\mu, \tag{44.7}$$

furthermore

$$r_\lambda r_\mu = r_\mu r_\lambda. \tag{44.8}$$

We still need to prove (d). We have

$$r_\lambda = r_\lambda(\mu e - x) r_\mu = r_\lambda[(\mu - \lambda)e + \lambda e - x] r_\mu = (\mu - \lambda) r_\lambda r_\mu + r_\mu,$$

hence the resolvent equation (44.7) holds. From it follows the commutation relation (44.8). ∎

If F is a non-trivial Banach space, then $E := \mathscr{L}(F)$ is a Banach algebra; thus Theorem 44.1 can be applied to $\mathscr{L}(F)$.

Exercises

1. The spectrum of a compact operator on an infinite-dimensional Banach space contains at least 0.

$^+$2. It is occasionally useful to denote the resolvent of $T \in \mathscr{L}(E)$ more precisely by $R_\lambda(T)$. Show that on $\rho(A) = \rho(A')$ we have $R_\lambda(A') = (R_\lambda(A))'$.

3. (44.8) also follows from the fact that $e - \lambda x$ and $e - \mu x$ commute.

$^+$4. If $\delta(\mu) := \inf_{\xi \in \sigma(x)} |\mu - \xi|$ is the distance of the point $\mu \in \rho(x)$ from $\sigma(x)$, then $\|r_\mu\| \geqq 1/\delta(\mu)$; in particular, $\|r_\mu\| \to \infty$ when μ approaches $\sigma(x)$. *Hint*: Theorem 44.1a.

$^+$5. Besides the point spectrum $\sigma_p(A)$ of an operator A, we have as another interesting subset of $\sigma(A)$ the *continuous spectrum* $\sigma_c(A) := \{\lambda \in \sigma(A) \colon \lambda I - A$ is injective and $(\lambda I - A)(E)$ is dense in E but $\neq E\}$ and the *residual spectrum* $\sigma_r(A) := \{\lambda \in \sigma(A) \colon \lambda I - A$ is injective and $(\lambda I - A)(E)$ is not dense in $E\}$. $\sigma_p(A)$, $\sigma_c(A)$ and $\sigma_r(A)$ are pairwise disjoint and their union is $\sigma(A)$. To any $\lambda \in \sigma_c(A)$ there exists a sequence (x_n) with $\|x_n\| = 1$ such that $(\lambda I - A)x_n \to 0$. *Hint*: Proposition 36.1.

$^+$6. If A is normally solvable then 0 does not lie in $\sigma_c(A)$.

§45. The spectrum

An endomorphism of a *finite*-dimensional space is already bijective when it is injective, thus its spectrum consists only of eigenvalues and may therefore be empty if the space is real. In the case of a complex space, however, the fundamental theorem of algebra (which is a theorem of complex function theory) ensures the existence of eigenvalues. If we want to prove that $\sigma(x)$ is not empty, we will have to assume that the Banach algebra E containing x is *complex*; this assumption will open the possibility to put to work far-reaching theorems from complex function theory.

If $\sigma(x)$ is not empty, we define the *spectral radius* of x by

$$r(x) := \sup_{\lambda \in \sigma(x)} |\lambda|.$$

Points of the spectrum are also called *spectral points.*

Theorem 45.1. *For every element x of a* complex *Banach algebra E we have*

$$\sigma(x) \neq \varnothing \quad \textit{and} \quad r(x) = \lim \|x^n\|^{1/n}. \tag{45.1}$$

Proof. With the continuous linear form x' on E we set $f(\lambda) := x'(r_\lambda)$. From the resolvent equation (44.7) and the continuous dependence of r_λ on λ (Theorem 44.1a) it follows that

$$\frac{f(\lambda) - f(\mu)}{\lambda - \mu} = -x'(r_\lambda r_\mu) \to -x'(r_\mu^2) \qquad \text{as} \quad \lambda \to \mu.$$

Thus f is a holomorphic function on $\rho(x)$ which vanishes at infinity because of (44.5) and the inequality $|f(\lambda)| \leqq \|x'\| \, \|r_\lambda\|$. If $\sigma(x)$ were empty, i.e., $\rho(x) = \mathbf{C}$, then Liouville's theorem would imply that $f(\lambda) = x'(r_\lambda) = 0$ for all λ—and this for *all* $x' \in E'$. Thus r_λ would vanish on $\mathbf{C}$ (Theorem 28.3). However, in an algebra with $e \neq 0$ every inverse is $\neq 0$. This contradiction shows that $\sigma(x)$ is not empty. Because of (44.6) we have $r(x) \leqq r := \lim \|x^n\|^{1/n}$. Assume now that $r(x) < \mu < r$. By Theorem 44.1b, for every $x' \in E'$ the series

$$f(\lambda) = x'(r_\lambda) = \sum_{n=0}^{\infty} \frac{x'(x^n)}{\lambda^{n+1}} \qquad \text{converges for} \quad |\lambda| > r; \tag{45.2}$$

it is the Laurent expansion of the function $f(\lambda)$ which, according to the above, is holomorphic in $\rho(x)$, hence certainly for $|\lambda| > r(x)$. The theorem concerning the domain of convergence of a Laurent expansion tells us that (45.2) must converge even for $|\lambda| > r(x)$, hence for $\lambda = \mu$. But then $x'(x^n/\mu^{n+1}) \to 0$ and so by Proposition 41.4 we have $\|x^n/\mu^{n+1}\| \leqq \gamma < \infty$ for all n, hence $r = \lim \|x^n\|^{1/n} \leqq \lim (\gamma \mu^{n+1})^{1/n} = \mu < r$. The false assertion '$r < r$' shows that we must have $r(x) = r$. ∎

The closed disc $\{\lambda \in \mathbf{C} : |\lambda| \leqq r(x)\}$ is called the *spectral disk* of x (even if $r(x) = 0$). Since $\sigma(x)$ is not empty and closed (Theorems 45.1, 44,1), *on the boundary of the spectral disk there lies at least one spectral point.*

Proposition 45.1. *For two* commuting *elements x, y of a complex Banach algebra E we have*

$$r(x + y) \leqq r(x) + r(y) \quad \textit{and} \quad r(xy) \leqq r(x) r(y). \tag{45.3}$$

Proof. From Theorem 45.1 it follows immediately that

$$r(xy) = \lim\|(xy)^n\|^{1/n} = \lim\|x^n y^n\|^{1/n} \leqq \lim(\|x^n\|^{1/n}\|y^n\|^{1/n}) = r(x)r(y).$$

For the proof of the first inequality in (45.3) let the numbers $\alpha > r(x)$ and $\beta > r(y)$ be given arbitrarily and let $m \in \mathbf{N}$ be such that for $n \geqq m$ we have

$$\|x^n\| \leqq \alpha^n, \qquad \|y^n\| \leqq \beta^n.$$

Setting $\xi := \|x\|$, $\eta := \|y\|$ we have

$$\|x^k\| \leqq \xi^k, \qquad \|y^k\| \leqq \eta^k \qquad \text{for} \quad k = 0, 1, \ldots .$$

For $n \geqq 2m$ we obtain

$$\begin{aligned}
\|(x + y)^n\| &= \left\| \sum_{\nu=0}^{n} \binom{n}{\nu} x^\nu y^{n-\nu} \right\| \leqq \sum_{\nu=0}^{n} \binom{n}{\nu} \|x^\nu\| \, \|y^{n-\nu}\| \\
&\leqq \sum_{\nu=0}^{m-1} \binom{n}{\nu} \xi^\nu \beta^{n-\nu} + \sum_{\nu=m}^{n-m} \binom{n}{\nu} \alpha^\nu \beta^{n-\nu} + \sum_{\nu=n-m+1}^{n} \binom{n}{\nu} \alpha^\nu \eta^{n-\nu} \\
&= \sum_{\nu=0}^{m-1} \binom{n}{\nu} \alpha^\nu \beta^{n-\nu} \left(\frac{\xi}{\alpha}\right)^\nu + \sum_{\nu=m}^{n-m} \binom{n}{\nu} \alpha^\nu \beta^{n-\nu} \\
&\quad + \sum_{\nu=n-m+1}^{n} \binom{n}{\nu} \alpha^\nu \beta^{n-\nu} \left(\frac{\eta}{\beta}\right)^{n-\nu} \\
&\leqq \gamma \sum_{\nu=0}^{n} \binom{n}{\nu} \alpha^\nu \beta^{n-\nu} = \gamma(\alpha + \beta)^n,
\end{aligned}$$

where

$$\gamma := \max_{0 \leqq \nu \leqq m-1} \left(\frac{\xi}{\alpha}\right)^\nu + 1 + \max_{0 \leqq \nu \leqq m-1} \left(\frac{\eta}{\beta}\right)^\nu .$$

From here it follows that $r(x + y) \leqq (\alpha + \beta)\lim\sqrt[n]{\gamma} = \alpha + \beta$; thus also $r(x + y) \leqq r(x) + r(y)$. ∎

For every natural m we have $\lim_{n\to\infty} \|(x^m)^n\|^{1/n} = (\lim_{n\to\infty} \|x^{mn}\|^{1/mn})^m$, hence

(45.4) $$r(x^m) = [r(x)]^m.$$

x is said to be *nilpotent* if $x^n = 0$ for a natural number n, it is *quasi-nilpotent* if $r(x) = 0$. Nilpotent elements and Volterra integral operators (see (8.9)) are quasi-nilpotent. *The spectrum of an element x in a complex Banach algebra is reduced to* 0 *if and only if x is quasi-nilpotent.* Quasi-nilpotent endomorphisms of a Banach space are also called *Volterra operators.*

The structure of those complex Banach algebras which are at the same time *division algebras*, i.e., in which every element $x \neq 0$ is invertible, is surprisingly simple:

Proposition 45.2. *A complex Banach algebra E, which is at the same time a division algebra, consists of the multiples of the unit element*: $E = \{\lambda e : \lambda \in \mathbf{C}\}$.

Proof. If a certain $x \in E$ were not a multiple of e, then $\lambda e - x$ would be different from zero for all $\lambda \in \mathbf{C}$ and thus invertible. In contradiction to Theorem 45.1, the element would have no spectral points. ■

Exercises

$^+$1. For commuting x, y we have $|r(x) - r(y)| \leqq r(x - y)$.

2. The nilpotent elements of an arbitrary commutative algebra R form an ideal in R.

3. The quasi-nilpotent elements in a commutative Banach algebra E form a closed ideal in E. Use Exercise 1.

4. If x is a nilpotent element in an arbitrary algebra R with unit element e, then $\lambda e - x$ has an inverse for all $\lambda \neq 0$ which can be represented by a terminating Neumann series.

§46. Vector-valued holomorphic functions. Weak convergence

For the proof of Theorem 45.1 we have put into action twice essential theorems of the theory of holomorphic functions. The passage from the vector-valued function $\lambda \mapsto r_\lambda$ to a complex-valued one was accomplished with the help of continuous linear forms. In this section we set ourselves the task to develop the fundamental concepts (derivative, integral) and essential assertions (Cauchy's theorem and integral formula, Taylor and Laurent expansions, Liouville's theorem) of complex function theory for the case of functions which depend on a complex variable and whose values lie in a complex Banach space.

In principle it would be sufficient to assert that the above-mentioned concepts and theorems (together with their proofs) can be obtained from classical function theory by a purely formal transfer; we only need to replace absolute values by norms. But we can also use linear forms to obtain from the classical theorems—without repeating their proofs—the corresponding assertions concerning vector-valued functions. We shall illustrate this method of transfer by some examples. For the needed facts from complex function theory we refer to [177].

Let E be a normed space over $\mathbf{K}$, Δ a (non-empty) open subset of $\mathbf{K}$. The vector-valued function $f: \Delta \to E$ is said to be *differentiable* at the point $\lambda_0 \in \Delta$ if there exists an $f'(\lambda_0) \in E$ such that

$$\left\| \frac{f(\lambda) - f(\lambda_0)}{\lambda - \lambda_0} - f'(\lambda_0) \right\| \to 0 \quad \text{as} \quad \lambda \to \lambda_0. \tag{46.1}$$

$f'(\lambda_0)$ is called the *first derivative* of f at the point λ_0.

It is evident how the *higher derivatives* are defined.

It follows from (46.1) that for every $x' \in E'$ the limit

$$\lim_{\lambda \to \lambda_0} x'\left[\frac{f(\lambda) - f(\lambda_0)}{\lambda - \lambda_0}\right] = \lim_{\lambda \to \lambda_0} \frac{x'[f(\lambda)] - x'[f(\lambda_0)]}{\lambda - \lambda_0} \tag{46.2}$$

exists. This fact is described by the locution that f is *weakly differentiable* at λ_0. We shall show that conversely weak differentiability implies differentiability if E is complex and complete. First we must define and explain the concept of weak convergence.

Let (E, E^+) be a bilinear system. We say that a sequence (x_n) from E *converges* E^+*-weakly* to $x \in E$ (in symbol: $x_n \rightharpoonup x$) if

$$\langle x_n, x^+\rangle \to \langle x, x^+\rangle \qquad \text{for all} \quad x^+ \in E^+. \tag{46.3}$$

If (E, E^+) *is a* right *dual system, then the* E^+*-weak limit is uniquely determined.* In this case the E^+-weak convergence is nothing but the pointwise convergence of the *sequence of linear forms* (x_n) to the linear form x, because by §15 we can consider E as a subspace of $(E^+)^*$.

We call (x_n) an E^+*-weak Cauchy sequence* if $(\langle x_n, x^+\rangle)$ is a Cauchy sequence for every x^+, or equivalently, if $\lim\langle x_n, x^+\rangle$ exists for every x^+. Observe that an E^+-weak Cauchy sequence does not need to have an E^+-weak limit. The E-*weak convergence* of a sequence (x_n^+) from E^+ to $x^+ \in E^+$ is defined analogously and we have analogous statements.

If E is a *normed* space and we fix (E, E') as our basic dual system, then we call E'-weakly convergent sequences from E more shortly *weakly convergent*, and E'-weak Cauchy sequences more shortly *weak Cauchy sequences.* Then $x_n \rightharpoonup x$ means that the sequence of continuous linear forms x_n on E' tends pointwise to the linear form x (see Proposition 41.2). From here and from Proposition 41.4 and 33.1 we obtain:

Proposition 46.1. *A weak Cauchy sequence* (x_n) *in the normed space* E *is bounded. From* $x_n \rightharpoonup x$ *it follows that* $\|x\| \leqq \liminf \|x_n\|$.

If we want to prove a corresponding theorem about E-weak convergent sequences of continuous linear forms $x_n' \in E'$, then we must assume the completeness of E; see Exercise 6.

Convergence in the sense of the norm implies weak convergence; the converse, however, is not true in general. It is the more remarkable therefore that weakly convergent *power series* in Banach spaces are even convergent in the sense of the norm. We prepare this assertion (Proposition 46.3) by the following proposition (where we agree, in order to stay close to the classical notation, that $x\alpha := \alpha x$ for $x \in E$, $\alpha \in \mathbf{K}$).

Proposition 46.2. *Let* E *be a Banach space, and assume that the power series*

$$\sum_{k=0}^{\infty} a_k(\lambda - \lambda_0)^k \qquad \text{with coefficients} \quad a_k \in E \tag{46.4}$$

is for $\lambda_1 \neq \lambda_0$ *a* weak Cauchy series, *i.e., that*

$$\sum_{k=0}^{\infty} x'(a_k)(\lambda - \lambda_0)^k \qquad \text{exists for all} \quad x' \in E'. \tag{46.5}$$

Then $\sum_{k=0}^{\infty} a_k(\lambda - \lambda_0)^k$ *converges* in the sense of the norm *for* $|\lambda - \lambda_0| < |\lambda_1 - \lambda_0|$.

Proof. From (46.5) we get $\lim x'[a_k(\lambda_1 - \lambda_0)^k] = 0$ for all $x' \in E'$. By Proposition 46.1 there exists therefore a $\gamma > 0$ such that $\|a_k(\lambda_1 - \lambda_0)^k\| \leqq \gamma$ for $k = 1, 2, \ldots$. If $q := |\lambda - \lambda_0|/|\lambda_1 - \lambda_0| < 1$, then we have for all k

(46.6) $$\|a_k(\lambda - \lambda_0)^k\| = \|a_k(\lambda_1 - \lambda_0)^k\| q^k \leqq \gamma q^k,$$

from where the assertion follows with the aid of Lemma 8.1. ∎

Proposition 46.3. *If the power series* $\sum_{k=0}^{\infty} a_k(\lambda - \lambda_0)^k$ *with coefficients from a Banach space* E *is a weak Cauchy series in an open* ε*-neighborhood*

$$U := \{\lambda \in \mathbf{K} : |\lambda - \lambda_0| < \varepsilon\}$$

of λ_0, *then it converges there in the sense of the norm; furthermore even*

$$\sum_{k=0}^{\infty} \|a_k\| \, |\lambda - \lambda_0|^k$$

converges for $\lambda \in U$ (absolute *convergence*). *It defines a vector-valued function*

(46.7) $$f(\lambda) := \sum_{k=0}^{\infty} a_k(\lambda - \lambda_0)^k$$

which is differentiable arbitrarily often *on* U; *the derivatives can be obtained by* termwise differentiation:

(46.8) $$f^{(n)}(\lambda) = \sum_{k=n}^{n} (k - n + 1) \cdots k a_k(\lambda - \lambda_0)^{k-n}.$$

In particular, $f^{(n)}(\lambda_0) = n! a_n$, *hence*

(46.9) $$f(\lambda) = \sum_{k=0}^{\infty} \frac{f^{(k)}(\lambda_0)}{k!} (\lambda - \lambda_0)^k.$$

Proof. The first two assertions follow very easily from Proposition 46.2 and the estimate (46.6). We prove the differentiability assertion only for the case $\lambda_0 = 0$ and $n = 1$. For an arbitrary $x' \in E'$ we define $F: U \to \mathbf{K}$ by

$$F(\lambda) := x'[f(\lambda)] = \sum_{k=0}^{\infty} x'(a_k)\lambda^k.$$

It is well known that for $\lambda \in U$

$$F'(\lambda) = \sum_{k=1}^{\infty} x'[k a_k]\lambda^{k-1} \quad \text{and} \quad F''(\lambda) = \sum_{k=2}^{\infty} x'[(k-1)k a_k]\lambda^{k-2},$$

hence, by what we have just proved, the series

$$\sum_{k=1}^{\infty} k a_k \lambda^{k-1} \quad \text{and} \quad \sum_{k=2}^{\infty} (k-1)k a_k \lambda^{k-2}$$

converge for all $\lambda \in U$; one even has

$$(46.10) \qquad \sum_{k=2}^{\infty} (k-1)k\|a_k\| \, |\lambda|^{k-2} < \infty \qquad \text{for} \quad |\lambda| < \varepsilon.$$

For two different points $\lambda, \lambda_1 \in U$ we have therefore the expansion

$$(46.11) \qquad \frac{f(\lambda) - f(\lambda_1)}{\lambda - \lambda_1} - \sum_{k=1}^{\infty} k a_k \lambda_1^{k-1} = \sum_{k=2}^{\infty} a_k \left[\frac{\lambda^k - \lambda_1^k}{\lambda - \lambda_1} - k\lambda_1^{k-1} \right].$$

Now let $|\lambda_1| < \mu < \varepsilon$. Then for $|\lambda - \lambda_1| < \mu - |\lambda_1|$ we also have $|\lambda| < \mu$ and from

$$\frac{\lambda^m - \lambda_1^m}{\lambda - \lambda_1} = \lambda^{m-1} + \lambda^{m-2}\lambda_1 + \cdots + \lambda\lambda_1^{m-2} + \lambda_1^{m-1} \qquad (m = 1, 2, \ldots)$$

it follows for these λ that

$$|\lambda^m - \lambda_1^m| \leqq |\lambda - \lambda_1| m\mu^{m-1}$$

and thus

$$\begin{aligned} \left| \frac{\lambda^k - \lambda_1^k}{\lambda - \lambda_1} - k\lambda_1^{k-1} \right| &= |(\lambda^{k-1} - \lambda_1^{k-1}) + \lambda_1(\lambda^{k-2} - \lambda_1^{k-2}) + \cdots + \lambda_1^{k-2}(\lambda - \lambda_1)| \\ &\leqq |\lambda - \lambda_1| [(k-1)\mu^{k-2} + \mu(k-2)\mu^{k-3} + \cdots + \mu^{k-2}1] \\ &= |\lambda - \lambda_1| \frac{(k-1)k}{2} \mu^{k-2}. \end{aligned}$$

From here we obtain with the help of (46.10) that for $0 < |\lambda - \lambda_1| < \mu - |\lambda_1|$ the series on the right-hand side of the equation (46.11) converges absolutely and that

$$\left\| \frac{f(\lambda) - f(\lambda_1)}{\lambda - \lambda_1} - \sum_{k=1}^{\infty} k a_k \lambda_1^{k-1} \right\| \leqq \tfrac{1}{2} |\lambda - \lambda_1| \sum_{k=2}^{\infty} (k-1)k\|a_k\|\mu^{k-2}.$$

With this the assertion concerning the differentiability of f and the value of the first derivative are proved. ■

If Γ is an oriented, closed, rectifiable curve in **C** (such curves will be called henceforth *integration paths*) and E a complex Banach space, then we define for continuous functions $f : \Gamma \to E$ the *path integral*

$$\int_{\Gamma} f(\lambda) \mathrm{d}\lambda,$$

as in the classical case, as the limit of the Riemann sums

$$\sum f(\xi_k)(\lambda_k - \lambda_{k-1})$$

(the passage to the limit is to be performed in the usual sense; the existence of the integral is proved as in complex function theory). From this definition we obtain immediately the following properties:

$$\int_\Gamma \alpha f(\lambda)\mathrm{d}\lambda = \alpha \int_\Gamma f(\lambda)\mathrm{d}\lambda, \int_\Gamma (f(\lambda) + g(\lambda))\mathrm{d}\lambda = \int_\Gamma f(\lambda)\mathrm{d}\lambda + \int_\Gamma g(\lambda)\mathrm{d}\lambda, \tag{46.12}$$

$$\left\| \int_\Gamma f(\lambda)\mathrm{d}\lambda \right\| \leqq \max_{\lambda\in\Gamma} \|f(\lambda)\| \cdot (\text{length of } \Gamma), \tag{46.13}$$

$$x'\left[\int_\Gamma f(\lambda)\mathrm{d}\lambda\right] = \int_\Gamma x'(f(\lambda))\mathrm{d}\lambda \qquad \text{for all} \quad x' \in E' \tag{46.14}$$

and more generally

$$A \int_\Gamma f(\lambda)\mathrm{d}\lambda = \int_\Gamma Af(\lambda)\mathrm{d}\lambda \qquad \text{for all} \quad A \in \mathscr{L}(E, F), \tag{46.15}$$

where F is also a complex Banach space. Equation (46.14) will be our most important tool in what follows. Finally another basic definition:

If Δ is a (non-empty) open subset of $\mathbf{C}$ and E a complex Banach space, then the function $f : \Delta \to E$ is *locally holomorphic* in Δ if it is differentiable at every point of Δ; if Δ is furthermore connected, i.e., a *region*, then we say that f is holomorphic in Δ. *Weak local holomorphy* is defined by means of weak differentiability.

By Proposition 46.3 power series $\sum a_k(\lambda - \lambda_0)^k$ define holomorphic functions in the interior of their domain of convergence.

Proposition 46.4. (Cauchy's integral theorem). *If $f : \Delta \to E$ is holomorphic in the region Δ and if Γ_1, Γ_2 are two integration paths with the same initial and final points, which can be deformed into each other continuously in Δ, then*

$$\int_{\Gamma_1} f(\lambda)\mathrm{d}\lambda = \int_{\Gamma_2} f(\lambda)\mathrm{d}\lambda;$$

in particular,

$$\int_\Gamma f(\lambda)\mathrm{d}\lambda = 0$$

if Γ is a closed curve whose interior contains only points of Δ.

The proposition is obtained simply from the fact that for every continuous linear form x' on E we have

$$x'\left[\int_{\Gamma_1} f(\lambda)\mathrm{d}\lambda\right] = \int_{\Gamma_1} x'[f(\lambda)]\mathrm{d}\lambda = \int_{\Gamma_2} x'[f(\lambda)]\mathrm{d}\lambda = x'\left[\int_{\Gamma_2} f(\lambda)\mathrm{d}\lambda\right],$$

where the middle equality follows from the Cauchy integral theorem for the complex-valued holomorphic function $x'[f(\lambda)]$. ■

Proposition 46.5. *If $f:\Delta \to E$ is holomorphic, or only* weakly *holomorphic, in the simply connected region Δ, then f is arbitrarily often differentiable and for the derivatives the* Cauchy integral formulas

$$(46.16) \qquad f^{(n)}(\lambda) = \frac{n!}{2\pi i}\int_{\Gamma}\frac{f(\zeta)}{(\zeta-\lambda)^{n+1}}\,d\zeta \qquad (n = 0, 1, \ldots)$$

are valid; here Γ is a simple, closed, positively oriented integration path in Δ which encloses λ. Furthermore f can be expanded into a power series

$$f(\lambda) = \sum_{k=0}^{\infty} a_k(\lambda - \lambda_0)^k \qquad \textit{with} \quad a_k \in E$$

around every point λ_0 of Δ; the series converges at least in the largest open disk around λ_0 which contains only points of Δ.

Proof. For an arbitrary $x' \in E'$ we set again $F(\lambda) := x'[f(\lambda)]$. By hypothesis F is holomorphic in Δ, hence the Cauchy integral formulas hold:

$$(46.17) \qquad F^{(n)}(\lambda) = \frac{n!}{2\pi i}\int_{\Gamma}\frac{F(\zeta)}{(\zeta-\lambda)^{n+1}}\,d\zeta \qquad \text{for} \quad n = 0, 1, \ldots;$$

furthermore $F(\lambda) = x'[f(\lambda)]$ can be expanded around λ_0 into the power series

$$(46.18) \qquad x'[f(\lambda)] = \sum_{k=0}^{\infty}\left[\frac{1}{2\pi i}\int_{C}\frac{x'[f(\zeta)]}{(\zeta-\lambda_0)^{k+1}}\,d\zeta\right](\lambda-\lambda_0)^k;$$

here C is the positively oriented boundary of an arbitrary disk around λ_0 lying entirely in Δ. The series (46.18) converges in the interior U of every such disk. If we move x' in front of the integral, we see that the power series

$$\sum_{k=0}^{\infty}\left[\frac{1}{2\pi i}\int_{C}\frac{f(\zeta)}{(\zeta-\lambda_0)^{k+1}}\,d\zeta\right](\lambda-\lambda_0)^k$$

converges in U weakly, therefore by Proposition 46.3 also in the sense of the norm. Because of (46.18) its sum is necessarily $f(\lambda)$. Again by Proposition 46.3 the function f is infinitely differentiable in U. The reader will easily prove the remaining assertions. ■

Proposition 46.6 (Liouville's theorem). *A function which is holomorphic in all* **C**, *and is bounded, is constant.*

Proof. If f is such a function and $F(\lambda) := x'[f(\lambda)]$ with $x' \in E'$, then also F is holomorphic in **C** and is bounded, hence by the classical Liouville theorem it is constant: $x'[f(\lambda)] = x'[f(0)]$ for all λ. Since this equation holds for arbitrary $x' \in E'$, we have $f(\lambda) = f(0)$ for all $\lambda \in \mathbf{C}$. ■

After these samples of proofs, the reader will easily obtain the following expansion theorem:

Proposition 46.7 (Laurent expansion). *A function f holomorphic in $0 < |\lambda - \lambda_0| < r$ with values in E can be represented in the form*

$$f(\lambda) = \sum_{k=0}^{\infty} a_k(\lambda - \lambda_0)^k + \sum_{k=1}^{\infty} \frac{b_k}{(\lambda - \lambda_0)^k} \quad \text{with} \quad a_k, b_k \in E. \tag{46.19}$$

This expansion is valid for $0 < |\lambda - \lambda_0| < r$. The coefficients are given by the formulas

$$a_k = \frac{1}{2\pi i} \int_C \frac{f(\lambda)}{(\lambda - \lambda_0)^{k+1}} \, d\lambda, \qquad b_k = \frac{1}{2\pi i} \int_C f(\lambda)(\lambda - \lambda_0)^{k-1} \, d\lambda, \tag{46.20}$$

where C is a positively oriented circle $|\lambda - \lambda_0| = \rho$ with $0 < \rho < r$.

Under the hypotheses of Proposition 46.7 we call λ_0 a *removable singularity* (of f) if all b_k vanish, a *pole of order p* if $b_p \neq 0$ and $b_n = 0$ for $n > p$, an *essential singularity* if infinitely many b_n are $\neq 0$. A pole of order 1 is also called a *simple pole.*

The resolvent r_λ of an element x in a complex Banach algebra is holomorphic in $\rho(x)$ because of Theorem 44.1. *An isolated point λ_0 of the spectrum of x is,* as we shall show now, *a pole or an essential singularity of r_λ*. Otherwise it would follow from the Laurent expansion $r_\lambda = \sum_{k=0}^{\infty} a_k(\lambda - \lambda_0)^k$ for every sequence $(\lambda_n) \subset \rho(x)$ with $\lambda_n \neq \lambda_0$ and $\lambda_n \to \lambda_0$ that $r_{\lambda_n} \to a_0$, hence that

$$r_{\lambda_n}(\lambda_0 e - x) \to a_0(\lambda_0 e - x).$$

Since

$$r_{\lambda_n}(\lambda_0 e - x) = r_{\lambda_n}[(\lambda_n e - x) + (\lambda_0 - \lambda_n)e] = e + (\lambda_0 - \lambda_n)r_{\lambda_n}$$

converges to e, we would have $a_0(\lambda_0 e - x) = e$ and similarly $(\lambda_0 e - x)a_0 = e$; thus λ_0 would lie in $\rho(x_0)$ which contradicts our original assumption. ∎

Exercises

1. The n-th derivative of the resolvent is $r_\lambda^{(n)} = (-1)^n n! r_\lambda^{n+1}$.

+2. The resolvent cannot be extended beyond $\rho(x)$ as a holomorphic function. *Hint*: Exercise 4 in §44.

+3. A power series $\sum_{k=0}^{\infty} a_k(\lambda - \lambda_0)^k$ is said to be *nowhere convergent* if it converges only for $\lambda = \lambda_0$; it is said to be *everywhere convergent* if it converges for all λ (the coefficients are from an arbitrary Banach space). If it is neither nowhere convergent nor everywhere convergent, then there exists a (finite) number $r > 0$ so that it converges for $|\lambda - \lambda_0| < r$ and diverges for $|\lambda - \lambda_0| > r$. The number r is called the *radius of convergence*, $\{\lambda \in \mathbf{K}: |\lambda - \lambda_0| < r\}$ the *interval of convergence* or the *disk of convergence* of the series (according as $\mathbf{K} = \mathbf{R}$ or $= \mathbf{C}$). r can be calculated by the *Cauchy–Hadamard formula*

$$r = \frac{1}{\limsup \sqrt[k]{\|a_k\|}}.$$

For nowhere convergent power series we have $\limsup \sqrt[k]{\|a_k\|} = \infty$, for everywhere convergent ones $\limsup \sqrt[k]{\|a_k\|} = 0$; it is reasonable to set in these cases $r = 0$ or $= \infty$, respectively.

*4. Let E be a complex Banach space. If α, β are two points in the disk of convergence of the power series $\sum a_k(\lambda - \lambda_0)^k$ with $a_k \in E$ (see Exercise 3) and Γ is an integration path joining the two, then the power series may be integrated termwise, more precisely: We have

$$\begin{aligned}\int_\Gamma \sum_{k=0}^{\infty} a_k(\lambda - \lambda_0)^k \,\mathrm{d}\lambda &= \sum_{k=0}^{\infty} \int_\Gamma a_k(\lambda - \lambda_0)^k \,\mathrm{d}\lambda \\ &= \sum_{k=0}^{\infty} \frac{a_k}{k+1}\left[(\beta - \lambda_0)^{k+1} - (\alpha - \lambda_0)^{k+1}\right].\end{aligned}$$

More generally: If the functions $f_k\colon \Gamma \to E$ are continuous on the integration path Γ and if $\sum_{k=0}^{\infty} f_k(\lambda)$ converges uniformly on Γ, then

$$\int_\Gamma \sum_{k=0}^{\infty} f_k(\lambda)\mathrm{d}\lambda = \sum_{k=0}^{\infty} \int_\Gamma f_k(\lambda)\mathrm{d}\lambda.$$

+5. If f is holomorphic in the domain $\Delta \subset \mathbf{C}$ and $\|f(\lambda)\|$ is not constant there, then $\|f(\lambda)\|$ has no absolute maximum in Δ (*maximum principle*).

*6. Let the sequence of continuous linear forms x_n' on the Banach space E converge E-weakly to $x' \in E'$ (i.e., let $x_n'(x) \to x'(x)$ for all $x \in E$). Then

$$\sup_n \|x_n'\| < \infty \quad \text{and} \quad \|x'\| \leqq \liminf \|x_n'\|.$$

7. Let E, F be normed spaces, $A_n \in \mathscr{S}(E, F)$ and let $A_n x \rightharpoonup Ax$ for every $x \in E$. Then $A\colon E \to F$ is linear and we say that the sequence (A_n) *converges weakly* to A, in symbols $A_n \rightharpoonup A$. If E is complete and $A_n \in \mathscr{L}(E, F)$, then it follows from $A_n \rightharpoonup A$ that $(\|A_n\|)$ is bounded and A is continuous. *Hint*: Proposition 46.1 and Principle 33.2.

8. If E, F are normed spaces and $x_n \rightharpoonup x$, then for every $A \in \mathscr{L}(E, F)$ also $Ax_n \rightharpoonup Ax$. If A is compact, then we even have $Ax_n \to Ax$.

9. (a) In a finite-dimensional Banach space weak convergence is equivalent to convergence in the sense of the norm.

(b) In l^p, $1 < p < \infty$, the sequence (x_n) converges weakly to x if and only if it is bounded and converges componentwise to x.

(c) In l^1 the sequence (x_n) converges weakly to x if and only if it converges to x in the sense of the norm (see [1], p. 137; [47]).

10. (a) Let $A_n \in \mathscr{L}(E)$ and let $A_n \Rightarrow A$. Then also $A'_n \Rightarrow A'$.

(b) Let E be a complex Banach space and $A(\lambda) \in \mathscr{L}(E)$ for all λ of an integration path Γ; let the function $\lambda \mapsto A(\lambda)$ be continuous on Γ. Show with the help of (a) that $(\int_\Gamma A(\lambda)\mathrm{d}\lambda)' = \int_\Gamma A'(\lambda)\mathrm{d}\lambda$. In a corresponding way we may conjugate termwise series which converge in the sense of the norm.

§47. Power series in Banach algebras

In a Banach algebra E we can consider two types of power series, namely series of the form

$$\sum_{n=0}^{\infty} a_n(\lambda - \lambda_0)^n \quad \text{with} \quad a_n \in E \quad \text{and} \quad \lambda, \lambda_0 \in \mathbf{K}$$

as well as series of the form

$$\sum_{n=0}^{\infty} \alpha_n(x - x_0)^n \quad \text{with} \quad \alpha_n \in \mathbf{K} \quad \text{and} \quad x, x_0 \in E.$$

The first type was studied in the preceding section; we now turn to the second. For the sake of simplicity we assume that $x_0 = 0$. The general case can be obtained from this only seemingly more special one by a trivial substitution.

One can think of $\sum_{n=0}^{\infty} \alpha_n x^n$ as obtained from the *numerical* power series $\sum_{n=0}^{\infty} \alpha_n \lambda^n$ by substituting x for the variable λ. Since we want to make use of the analytic properties of this power series, we shall assume that it converges not only for $\lambda = 0$, i.e., that $\limsup \sqrt[n]{|\alpha_n|} < \infty$. Incidentally, for nowhere convergent power series the corresponding series $\sum \alpha_n x^n$ can very well exist; this is for instance the case when x is nilpotent. A less trivial example is furnished by the Volterra integral operator (see Exercise 6).

With the help of the root test (before Proposition 8.1) the reader will prove very easily the following convergence theorem:

Proposition 47.1. *Let the complex power series*

$$\sum_{n=0}^{\infty} \alpha_n \lambda^n \tag{47.1}$$

have a non-vanishing radius of convergence

$$r = \frac{1}{\limsup \sqrt[n]{|\alpha_n|}} \tag{47.2}$$

and let x be an element in the complex Banach algebra E with spectral radius $r(x)$. Then the series

$$\sum_{n=0}^{\infty} \alpha_n x^n \tag{47.3}$$

converges or diverges according as $r(x) < r$ *or* $r(x) > r$, *respectively, i.e., according as the spectrum* $\sigma(x)$ *lies entirely in the (open) disk of convergence of the power series* (47.1) *or parts of* $\sigma(x)$ *protrude outside its perimeter.*

Observe that for $\limsup \sqrt[n]{|\alpha_n|} = 0$ (everywhere convergent power series) we must set $r = \infty$; then (47.3) converges for every $x \in E$.

In the case of the Banach algebra **C**, Proposition 47.1 reduces to the well-known convergence theorem for complex power series since then $r(x) = |x|$. Since already for such series no general assertion concerning their convergence can be made for the $x \in \mathbf{C}$ with $|x| = r$, a fortiori the question, whether (47.3) converges in the case $r(x) = r$, will not have a final answer. If one considers, however, the geometric series $\sum_{n=0}^{\infty} \lambda^n$ with radius of convergence $r = 1$, then we have

Proposition 47.2. *For the* compact *endomorphism* K *of the complex Banach space* E *the* Neumann series $\sum_{n=0}^{\infty} K^n = I + K + K^2 + \cdots$ *converges if and only if* $r(K) < 1$, *i.e., if and only if all the eigenvalues of* K *lie in the interior of the unit disk.*

Convergence is meant in the sense of the norm of operators (uniform convergence). We recall that the sum of the Neumann series is $(I - K)^{-1}$ (Proposition 8.2).

Now the *proof* of the proposition: In the sense of Proposition 47.1 the Neumann series originates from the geometric series which has radius of convergence $r = 1$. Thus we have only to prove that $r(K) = 1$ implies divergence. From $r(K) = 1$ it follows that there exists an $\alpha \in \sigma(K)$ with $|\alpha| = 1$; according to Proposition 44.1 this α is an eigenvalue of K. Let $Kx = \alpha x$, $x \neq 0$. Then we have $K^n x = \alpha^n x$ for $n = 0, 1, 2, \ldots$, hence $\sum_{n=0}^{\infty} K^n x = \sum_{n=0}^{\infty} \alpha^n x$. This series is divergent because of $\|\alpha^n x\| = \|x\| \neq 0$, hence the Neumann series is also divergent. To see the last assertion, apply again Proposition 44.1. ■

Proposition 47.2 can be applied in particular to every endomorphism K of the vector space $\mathbf{C}^n$ equipped with any norm, since such a K is always compact (Propositions 10.3 and 13.2). We want to examine this case more closely.

It is well-known that to every $K \in \mathscr{S}(\mathbf{C}^n)$ there exists exactly one matrix $\mathfrak{K} := (\alpha_{ik})$, $i, k = 1, \ldots, n$, so that the components of the image vector

$$(\eta_1, \ldots, \eta_n) := K(\xi_1, \ldots, \xi_n)$$

are given by the formulas

$$\eta_i = \sum_{k=1}^{n} \alpha_{ik} \xi_k \qquad (i = 1, \ldots, n); \tag{47.4}$$

conversely, every (n, n)-matrix (α_{ik}) defines by means of (47.4) an endomorphism of $\mathbf{C}^n$. The correspondence $K \mapsto \mathfrak{K}$ is an isomorphism between the algebra

$\mathscr{S}(\mathbf{C}^n)$ and the n^2-dimensional algebra of all (n, n)-matrices. For the sake of simplicity we identify these two algebras; thus we do not differentiate between the matrix $\mathfrak{K}$ and the map K it generates. If we equip $\mathbf{C}^n$ with a norm, then we denote by $\|\mathfrak{K}\|$ the norm of the linear map $\mathfrak{K}$ and call it the *map-norm* of $\mathfrak{K}$ to distinguish it from the matrix-norms soon to be introduced.

The vector space $\mathscr{S}(\mathbf{C}^n)$ for all (n, n)-matrices (α_{ik}) is identical with the n^2-dimensional linear space $\mathbf{C}^{n^2}$ of all vectors

$$(\alpha_{11}, \ldots, \alpha_{1n}, \alpha_{21}, \ldots, \alpha_{2n}, \ldots, \alpha_{n1}, \ldots, \alpha_{nn}),$$

and the map-norm $\|\mathfrak{K}\|$ turns $\mathbf{C}^{n^2}$ into a normed space. Since in finite-dimensional normed spaces convergence is equivalent to componentwise convergence, we obtain from Proposition 47.2 the following assertion (where $\mathfrak{E}$ denotes the unit matrix (δ_{ik})).

Proposition 47.3. *If $\mathfrak{K}$ is an (n, n)-matrix over $\mathbf{C}$, then the* Neumann series

$$\sum_{n=0}^{\infty} \mathfrak{K}^n \quad \textit{and} \quad \sum_{n=0}^{\infty} \mathfrak{K}^n y \tag{47.5}$$

converge componentwise to $(\mathfrak{E} - \mathfrak{K})^{-1}$ and to $(\mathfrak{E} - \mathfrak{K})^{-1}y$ for all $y \in \mathbf{C}^n$, respectively, if and only if every eigenvalue of $\mathfrak{K}$ lies in the interior of the unit circle.

Observe that no norms on $\mathbf{C}^n$ or $\mathscr{S}(\mathbf{C}^n)$ figure in the formulation of this proposition. Compare this result with the assertion after (3.11) in connection with Exercise 2 in §7 and Proposition 43.2. For practical applications the quoted assertion has the advantage that to ensure the convergence of (47.5) we do not need to know the eigenvalues of $\mathfrak{K}$, only the quantities

$$q_1 := \max_{k=1}^{n} \sum_{i=1}^{n} |\alpha_{ik}|, \qquad q_2 := \sqrt{\sum_{i,k=1}^{n} |\alpha_{ik}|^2}, \qquad q_\infty := \max_{i=1}^{n} \sum_{k=1}^{n} |\alpha_{ik}|. \tag{47.6}$$

These quantities behave, as we can see immediately, like norms, more precisely: If for $\mathfrak{K} = (\alpha_{ik})$ we set

$$\|\mathfrak{K}\|_j := q_j \qquad (j = 1, 2, \infty),$$

then $\|\mathfrak{K}\|_j$ is a norm on the matrix space $\mathscr{S}(\mathbf{C}^n)$. In particular, $\|\mathfrak{K}\|_2$ is nothing but the l^2-norm on $\mathscr{S}(\mathbf{C}^n)$. From (3.4) through (3.6) we obtain the following inequalities, where $\|x\|_j$ denotes the l^j-norm on $\mathbf{C}^n$:

$$\|\mathfrak{K}x\|_j \leqq \|\mathfrak{K}\|_j \|x\|_j \qquad \text{for all} \quad x \in \mathbf{C}^n \quad \text{and} \quad j = 1, 2, \infty.$$

If we call a norm $\|\mathfrak{K}\|_{\mathscr{S}}$ on $\mathscr{S}(\mathbf{C}^n)$ a *matrix-norm* provided that for some norm on $\mathbf{C}^n$ it satisfies for all x the estimate

$$\|\mathfrak{K}x\| \leqq \|\mathfrak{K}\|_{\mathscr{S}} \|x\|, \tag{47.7}$$

then the above norms $\|\mathfrak{K}\|_j$ are matrix-norms. Between a matrix-norm $\|\mathfrak{K}\|_{\mathscr{S}}$

and a map-norm $\|\mathfrak{K}\|$, which corresponds to a given norm on $\mathbf{C}^n$, we have obviously the inequality

$$(47.8) \qquad \|\mathfrak{K}\| \leqq \|\mathfrak{K}\|_{\mathscr{S}}$$

(see also Exercise 2 in §7). Since all norms on a finite-dimensional space are equivalent (Proposition 10.2), there exists by Proposition 7.6 a $\gamma > 0$ with

$$\|\mathfrak{K}\|_{\mathscr{S}} \leqq \gamma \|\mathfrak{K}\| \qquad \text{for all} \quad \mathfrak{K} \in \mathscr{S}(\mathbf{C}^n).$$

Consequently, for every natural number m

$$r(\mathfrak{K}) \leqq \|\mathfrak{K}^m\|^{1/m} \leqq \|\mathfrak{K}^m\|_{\mathscr{S}}^{1/m} \leqq \gamma^{1/m} \|\mathfrak{K}^m\|^{1/m}$$

(Theorem 45.1 and Proposition 8.1) and thus

$$r(\mathfrak{K}) = \lim \|\mathfrak{K}^m\|_{\mathscr{S}}^{1/m}.$$

Thus we have $r(\mathfrak{K}) < 1$, i.e., all eigenvalues of $\mathfrak{K}$ lie in the interior of the unit disk, if and only if one of the numbers $\|\mathfrak{K}^m\|_{\mathscr{S}}$ is <1. Proposition 47.3 now turns into:

Proposition 47.4. *If $\mathfrak{K}$ is an (n, n)-matrix over $\mathbf{C}$ and $\|\cdot\|_{\mathscr{S}}$ is a matrix-norm on $\mathscr{S}(\mathbf{C}^n)$, then the convergence assertions of Proposition* 47.3 *are valid if and only if one of the numbers $\|\mathfrak{K}^m\|_{\mathscr{S}}$ is <1.*

We return to our general considerations and define now functions which depend on elements of a Banach algebra.

If the complex-valued function f is *holomorphic* for $|\lambda| < r$, i.e., if for these λ

$$(47.9) \qquad f(\lambda) = \sum_{n=0}^{\infty} \alpha_n \lambda^n,$$

if, furthermore, x is an element of a complex Banach algebra E with $r(x) < r$, then we define $f(x)$ by

$$(47.10) \qquad f(x) := \sum_{n=0}^{\infty} \alpha_n x^n;$$

the convergence of the series is ensured by Proposition 47.1.

In §43 we had set $f(A) := \alpha_0 I + \alpha_1 A + \cdots + \alpha_n A^n$ for polynomials $f(\lambda) := \alpha_0 + \alpha_1 \lambda + \cdots + \alpha_n \lambda^n$ and endomorphisms A of a vector space; for an A from the Banach algebra of all continuous endomorphisms of a Banach space this definition coincides with the one given above since a polynomial is a power series with an infinite radius of convergence.

One of the most important functions of analysis is the exponential function

$$\exp(\lambda) := \sum_{n=0}^{\infty} \frac{\lambda^n}{n!}.$$

It is holomorphic in all of **C**, consequently for every x of a Banach algebra we have, by definition,

$$\exp(x) := \sum_{n=0}^{\infty} \frac{x^n}{n!}. \tag{47.11}$$

Obviously $\exp(0) = e$. In the exercises we shall prove the *functional equation of the exponential function*:

(47.12) $\exp(x + y) = \exp(x)\exp(y)$ for *commuting* elements x, y.

For an element A of the special Banach algebra $\mathscr{L}(E)$ ($E \neq \{0\}$ a Banach space) one writes usually e^A instead of $\exp(A)$. On the other hand it is not advisable to use the symbol e^A in an unspecified Banach algebra because of the possibility of confusion with the unit element e.

The function f in (47.9), which is holomorphic in $|\lambda| < r$, can be represented for all λ inside a positively oriented circle Γ around 0 with radius $<r$ also by the Cauchy integral formula

$$f(\lambda) = \frac{1}{2\pi i} \int_\Gamma f(\zeta)(\zeta - \lambda)^{-1}\, d\zeta. \tag{47.13}$$

We ask whether the equation

$$f(x) = \frac{1}{2\pi i} \int_\Gamma f(\zeta)(\zeta e - x)^{-1}\, d\zeta = \frac{1}{2\pi i} \int_\Gamma f(\zeta) r_\zeta\, d\zeta,$$

obtained from (47.13) by substituting x for λ, is correct; here we suppose of course that $r(x) < r$ and the radius of Γ is not only $<r$ but $>r(x)$, so that Γ lies in the disk of holomorphy of f and contains the whole spectrum of x in its interior. The answer is positive: In the first place we have for $n = 0, 1, 2, \ldots$

$$\frac{1}{2\pi i} \int_\Gamma \zeta^n r_\zeta\, d\zeta = \frac{1}{2\pi i} \int_\Gamma \zeta^n \left(\frac{e}{\zeta} + \frac{x}{\zeta^2} + \frac{x^2}{\zeta^3} + \cdots \right) d\zeta = x^n, \tag{47.14}$$

as can be seen immediately by termwise integration (cf. Exercise 4 in §46) taking into consideration the integral formula

$$\int_\Gamma \frac{1}{\zeta^k}\, d\zeta = \begin{cases} 2\pi i & \text{for } k = 1 \\ 0 & \text{for every other integer } k; \end{cases} \tag{47.15}$$

thus we get, again by termwise integration,

$$\frac{1}{2\pi i} \int_\Gamma f(\zeta) r_\zeta\, d\zeta = \frac{1}{2\pi i} \int_\Gamma \left(\sum_{n=0}^{\infty} \alpha_n \zeta^n r_\zeta \right) d\zeta = \frac{1}{2\pi i} \sum_{n=0}^{\infty} \alpha_n \int_\Gamma \zeta^n r_\zeta\, d\zeta = \sum_{n=0}^{\infty} \alpha_n x^n = f(x).$$

We state this result:

Proposition 47.5. *If the complex-valued function f is holomorphic for $|\lambda| < r$ and if x is an element of the complex Banach algebra E with $r(x) < r$, then the equation*

$$f(x) = \frac{1}{2\pi i} \int_\Gamma f(\lambda) r_\lambda\, d\lambda \tag{47.16}$$

is valid; here r_λ is the resolvent of x and Γ is a positively oriented circle around 0 whose radius lies strictly between $r(x)$ and r.

In the subsequent investigations we shall direct our attention to the dependence of the expression $f(x)$ on f when x is *fixed*. It will be useful to define $f(x)$ not only for functions f which are holomorphic in sufficiently large disks around 0. This will be done in the next section based on equation (47.16).

Exercises

In what follows E will always be a complex Banach algebra; x, y, a_n, b_n etc., will be elements from E.

1. If the series $\sum_{n=0}^{\infty} \|a_n\|$, $\sum_{n=0}^{\infty} \|b_n\|$ converge, then $\sum_{n=0}^{\infty} \|a_0 b_n + a_1 b_{n-1} + \cdots + a_n b_0\|$ also converges, and we have

$$\left(\sum_{n=0}^{\infty} a_n\right)\left(\sum_{n=0}^{\infty} b_n\right) = \sum_{n=0}^{\infty} (a_0 b_n + a_1 b_{n-1} + \cdots + a_n b_0) \text{ (\textit{Cauchy multiplication})}.$$

Hint: Because of $\|a_0 b_n + \cdots + a_n b_0\| \leqq \|a_0\|\,\|b_n\| + \cdots + \|a_n\|\,\|b_0\|$ one can use the known theorem concerning multiplication of absolutely convergent series.

2. If $\sum_{n=0}^{\infty} \alpha_n \lambda^n$ converges for $|\lambda| < r$ and if $r(x) < r$, then also $\sum_{n=0}^{\infty} |\alpha_n|\,\|x^n\|$ converges.

3. Use Exercises 1 and 2 to prove the functional equation (47.12) of the exponential function.

4. If $f(\lambda) := \sum_{n=0}^{\infty} \alpha_n \lambda^n$ and $g(\lambda) := \sum_{n=0}^{\infty} \beta_n \lambda^n$ are holomorphic for $|\lambda| < r$ and if $r(x) < r$, then we have

$$(\alpha f)(x) = \alpha f(x), \qquad (f + g)(x) = f(x) + g(x) \quad \text{and} \quad (fg)(x) = f(x)g(x),$$

in particular, $f(x)g(x) = g(x)f(x)$. The correspondence $f \mapsto f(x)$ is thus a homomorphism from the algebra H_r of all functions holomorphic in

$$\{\lambda: |\lambda| < r\}$$

onto the commutative algebra $\{f(x): f \in H_r;$ fixed $x \in E$ with $r(x) < r\}$. This homomorphism is continuous in the following sense: if $f_n, f \in H_r$ and if

$$f_n(\lambda) \to f(\lambda)$$

uniformly in a disk $\{\lambda: |\lambda| \leqq r_0\}$ with $r(x) < r_0 < r$, then $f_n(x) \to f(x)$. *Hint*: Exercise 1 and 2 for the conservation of products, Proposition 47.5 for continuity.

5. If under the assumptions of Exercise 4 $f(\lambda) \neq 0$ for $|\lambda| < r$ or only for $\lambda \in \sigma(x)$, then the inverse $f(x)^{-1}$ exists and is given by

$$f(x)^{-1} = \frac{1}{2\pi i} \int_\Gamma \frac{1}{f(\lambda)} r_\lambda \, d\lambda = \sum_{n=0}^{\infty} \gamma_n x^n,$$

where $\sum_{n=0}^{\infty} \gamma_n \lambda^n$ is the expansion of $1/f(\lambda)$. Briefly (but perhaps ambiguously): $f(x)^{-1} = f^{-1}(x)$ (f^{-1} is here the *reciprocal* not the inverse, function). *Hint*: Exercise 4 and a Heine–Borel covering argument; cf. also Theorem 48.1e.

6. The power series $\sum_{n=0}^{\infty} n!\lambda^n$ is nowhere convergent. If K is the Volterra integral operator defined by (8.9), then $\sum_{n=0}^{\infty} n!K^n$ converges, however, if $\mu < 1/(b-a)$ (the meaning of μ, a, b is defined in connection with (8.9); there the reader will also find the further tools to solve the exercise).

$^+$7. Let F be a Banach space, $u_0 \in F$ and $A \in \mathscr{L}(F)$. Then $u(t) := e^{tA}u_0$ is the unique solution of the initial value problem $u'(t) = Au(t)$ $(t \geqq 0)$, $u(0) = u_0$. Specialize this assertion to linear systems of differential equations with constant coefficients ($F = \mathbf{C}^n$, A a matrix operator).

§48. The functional calculus

Taking our cue from the last remark of the last section, we now describe a class of functions f for which $f(x)$ is to be defined. We denote by $\Delta(f) \subset \mathbf{C}$ the set of definition of f.

If x is an element of a complex Banach algebra, then let $\mathscr{H}(x)$ be the set of all complex-valued functions f which are locally holomorphic on an (open) set $\Delta(f) \supset \sigma(x)$. To $\mathscr{H}(x)$ belong in particular all functions which are holomorphic on an open disk containing $\sigma(x)$. Observe that the set of definition $\Delta(f)$ may vary with f.

For functions f from $\mathscr{H}(x)$ we shall define $f(x)$ by the integral formula (47.16). For this purpose we must, however, make some preliminary remarks concerning the technique of integration.

The region $B \subset \mathbf{C}$ is said to be *admissible* (with respect to x) if the following hold:

(a) $\sigma(x) \subset B$;

(b) B is open and bounded;

(c) the boundary ∂B of B consists of finitely many closed, rectifiable Jordan curves $C_1, \ldots, C_n$ which are pairwise disjoint;

(d) the (positive) orientation of ∂B is given by the orientation of each C_i: we will describe C_i counterclockwise if the points of B which are adjacent to C_i lie in the interior of C_i, otherwise C_i will be oriented clockwise.

It is easy to see that the points of C_i cannot come arbitrarily close to the spectrum of x; thus C_i lies completely in the resolvent set $\rho(x)$.

Figures 1 and 2 give examples of admissible domains B, their boundary curves C_i and their orientation. B is crosshatched simply, $\sigma(x)$ doubly.

For every $f \in \mathscr{H}(x)$ there exists an admissible region B with $\sigma(x) \subset B \subset \bar{B} \subset \Delta(f)$. If B' is a second region of this kind, then according to the Cauchy integral theorem (Proposition 46.4)

$$\int_{\partial B} f(\lambda) r_\lambda \, d\lambda = \int_{\partial B'} f(\lambda) r_\lambda \, d\lambda.$$

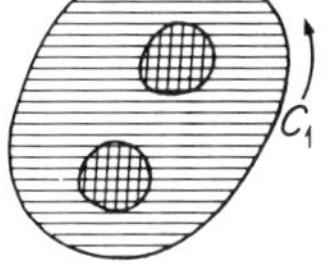

Figure 1

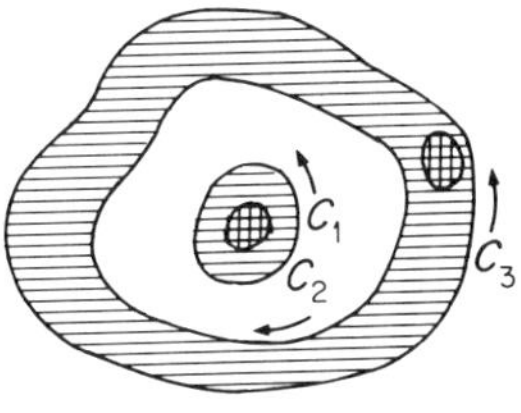

Figure 2

These facts make the following definition possible:

If $f \in \mathscr{H}(x)$ and B is an admissible region with $\sigma(x) \subset B \subset \bar{B} \subset \Delta(f)$, then let

$$f(x) := \frac{1}{2\pi i} \int_{\partial B} f(\lambda) r_\lambda \, d\lambda. \tag{48.1}$$

Because of Proposition 47.5 this definition agrees with the definition of $f(x)$ when f is holomorphic in a disk around 0 containing $\sigma(x)$. We will, however, not be satisfied with functions of this kind, because particularly valuable, namely idempotent, elements $f(x)$ are obtained only from such functions f which are equal to 1 on certain parts of $\sigma(x)$ and equal to 0 on others; such an f can be locally holomorphic but not holomorphic. If two functions f and g from $\mathscr{H}(x)$ coincide at every point of an open set containing $\sigma(x)$—in which case we say that they are *equivalent* (with respect to x)—then obviously $f(x) = g(x)$.

The usual algebraic operations are at first not defined in $\mathscr{H}(x)$ because its elements do not have a common domain of definition. However, the observation we just made concerning equivalent functions suggests the following definitions of αf, $f + g$ and fg ($f, g \in \mathscr{H}(x)$):

$$(\alpha f)(\lambda) := \alpha f(\lambda) \qquad \text{for} \quad \lambda \in \Delta(f)$$

$$(f + g)(\lambda) := f(\lambda) + g(\lambda) \quad \text{and} \quad (fg)(\lambda) := f(\lambda)g(\lambda) \qquad \text{for} \quad \lambda \in \Delta(f) \cap \Delta(g).$$

After these definitions we can formulate the central theorem of the functional calculus:

Theorem 48.1. *The correspondence $f \mapsto f(x)$ defined in* (48.1) *has the following properties:*

(a) $(\alpha f)(x) = \alpha f(x)$.
(b) $(f + g)(x) = f(x) + g(x)$.
(c) $(fg)(x) = f(x)g(x)$, *hence $f(x)$ commutes with $g(x)$.*
(d) *For $f(\lambda) = \lambda^n$ we have $f(x) = x^n$ ($n = 0, 1, 2, \ldots$).*
(e) *If $f(\lambda) \neq 0$ for all $\lambda \in \sigma(x)$, then $f(x)$ has an inverse $f(x)^{-1} = (1/f)(x)$.*

Proof. (a) and (b) are trivial, (d) was shown in (47.14). For the proof of (c) we choose admissible regions B_f, B_g with

$$\bar{B}_f \subset B_g \subset \bar{B}_g \subset \Delta(f) \cap \Delta(g). \tag{48.2}$$

Then

$$f(x)g(x) = \left[\frac{1}{2\pi i}\int_{\partial B_f} f(\lambda)r_\lambda\,\mathrm{d}\lambda\right]\left[\frac{1}{2\pi i}\int_{\partial B_g} g(\mu)r_\mu\,\mathrm{d}\mu\right]$$
$$= \frac{1}{2\pi i}\int_{\partial B_f} f(\lambda)\left[\frac{1}{2\pi i}\int_{\partial B_g} g(\mu)r_\lambda r_\mu\,\mathrm{d}\mu\right]\mathrm{d}\lambda.$$

From the resolvent equation $r_\lambda - r_\mu = (\mu - \lambda)r_\lambda r_\mu$ (Theorem 44.1) we obtain for the product $r_\lambda r_\mu$ in the second integral the representation

$$r_\lambda r_\mu = \frac{r_\lambda}{\mu - \lambda} + \frac{r_\mu}{\lambda - \mu}$$

and thus the following continuation of our computation:

$$f(x)g(x) = \frac{1}{2\pi i}\int_{\partial B_f} f(\lambda)r_\lambda\left[\frac{1}{2\pi i}\int_{\partial B_g}\frac{g(\mu)}{\mu - \lambda}\,\mathrm{d}\mu\right]\mathrm{d}\lambda$$
$$+ \frac{1}{(2\pi i)^2}\int_{\partial B_g} g(\mu)r_\mu\left[\int_{\partial B_f}\frac{f(\lambda)}{\lambda - \mu}\,\mathrm{d}\lambda\right]\mathrm{d}\mu$$

(why is the interchange of the order of integration in the second term permitted?). Because of $\lambda \in \partial B_f$ and $\mu \in \partial B_g$ we obtain from (48.2), with the aid of Cauchy's formulas,

$$\frac{1}{2\pi i}\int_{\partial B_g}\frac{g(\mu)}{\mu - \lambda}\,\mathrm{d}\mu = g(\lambda) \quad \text{and} \quad \int_{\partial B_f}\frac{f(\lambda)}{\lambda - \mu}\,\mathrm{d}\lambda = 0.$$

Thus we have, finally,

$$f(x)g(x) = \frac{1}{2\pi i}\int_{\partial B_f} f(\lambda)g(\lambda)r_\lambda\,\mathrm{d}\lambda,$$

which proves (c). For the proof of (e) we observe that $f(\lambda)$ does not vanish on an open set $\Delta \supset \sigma(x)$, hence $1/f$ is locally holomorphic there and belongs thus to $\mathscr{H}(x)$; (e) now follows with the help of (c) and (d) from the equation

$$f(\lambda)(1/f)(\lambda) = 1 \qquad \text{for} \quad \lambda \in \Delta. \qquad \blacksquare$$

The first important consequence of Theorem 48.1 is the *spectral mapping theorem*:

Theorem 48.2. *For every $f \in \mathscr{H}(x)$ we have $\sigma(f(x)) = f(\sigma(x))$.*

Proof. Let $\mu \in \sigma(f(x))$. If we had $\mu \notin f(\sigma(x))$, i.e., $\mu - f(\lambda) \neq 0$ for all $\lambda \in \sigma(x)$, then $\mu e - f(x)$ would have an inverse (Theorem 48.1e) in contradiction to the hypothesis. Thus $\sigma(f(x)) \subset f(\sigma(x))$. Now let, conversely, $\mu \in f(\sigma(x))$, i.e., $\mu = f(\zeta)$ with $\zeta \in \sigma(x)$. The function g defined on $\Delta(f)$ by

$$g(\lambda) := \frac{f(\lambda) - f(\zeta)}{\lambda - \zeta} \qquad \text{for } \lambda \neq \zeta, \qquad g(\zeta) := f'(\zeta)$$

lies in $\mathscr{H}(x)$; from $g(\lambda)(\zeta - \lambda) = f(\zeta) - f(\lambda)$ it follows therefore that $g(x)(\zeta e - x) = f(\zeta)e - f(x) = \mu e - f(x)$ (Theorem 48.1c). If μ were in $\rho(f(x))$, then we would have

$$[(\mu e - f(x))^{-1}g(x)](\zeta e - x) = (\zeta e - x)[(\mu e - f(x))^{-1}g(x)] = e,$$

thus $\zeta e - x$ would have an inverse, which is impossible because by hypothesis ζ lies in $\sigma(x)$. Therefore we have also $f(\sigma(x)) \subset \sigma(f(x))$. ∎

Now we are in the position to investigate *composite functions*.

Proposition 48.1. *If f lies in $\mathscr{H}(x)$, g in $\mathscr{H}(f(x))$ and if we define h by*

$$h(\lambda) := g[f(\lambda)]$$

for all λ with $f(\lambda) \in \Delta(g)$, then h belongs to $\mathscr{H}(x)$ and we have

$$h(x) = g[f(x)].$$

Proof. $\Delta(h) := \{\lambda \in \Delta(f): f(\lambda) \in \Delta(g)\}$ is open (Exercise 10 in §7). From $\lambda \in \sigma(x)$ it follows by Theorem 48.2 that $f(\lambda) \in f(\sigma(x)) = \sigma(f(x))$, thus $f(\lambda)$ lies a fortiori in $\Delta(g)$, hence $\sigma(x) \subset \Delta(h)$ and h belongs to $\mathscr{H}(x)$. Let now B_x and B_y be regions which are admissible with respect to x and $y := f(x)$, respectively, and which satisfy the following conditions:

$$\sigma(y) \subset B_y \subset \bar{B}_y \subset \Delta(g), \qquad \sigma(x) \subset B_x \subset \bar{B}_x \subset \Delta(f) \text{ and } f(\bar{B}_x) \subset B_y.$$

Then

$$\begin{aligned}
h(x) &= \frac{1}{2\pi i}\int_{\partial B_x} g(f(\zeta))(\zeta e - x)^{-1}\,\mathrm{d}\zeta \\
&= \frac{1}{2\pi i}\int_{\partial B_x}\left[\frac{1}{2\pi i}\int_{\partial B_y}\frac{g(\lambda)}{\lambda - f(\zeta)}\,\mathrm{d}\lambda\right](\zeta e - x)^{-1}\,\mathrm{d}\zeta \\
&= \frac{1}{2\pi i}\int_{\partial B_y} g(\lambda)\left[\frac{1}{2\pi i}\int_{\partial B_x}\frac{(\zeta e - x)^{-1}}{\lambda - f(\zeta)}\,\mathrm{d}\zeta\right]\mathrm{d}\lambda \\
&= \frac{1}{2\pi i}\int_{\partial B_y} g(\lambda)[\lambda e - f(x)]^{-1}\,\mathrm{d}\lambda \\
&= g[f(x)].
\end{aligned}$$

∎

A subset σ of $\sigma(x)$ is called a *spectral set* of x if σ and $\sigma(x)\setminus\sigma$ are closed. It is clearly equivalent to say that σ has a positive distance from $\sigma(x)\setminus\sigma$, or that *there exist open sets $\Delta \supset \sigma$ and $\Omega \supset \sigma(x)\setminus\sigma$ which do not intersect*. With σ also $\sigma(x)\setminus\sigma$ is a spectral set of x. It is called the spectral set *complementary* to σ.

If σ_1, σ_2 are complementary spectral sets of x and Δ_1, Δ_2 open sets which cover σ_1, σ_2, respectively, and do not intersect, then we define on $\Delta := \Delta_1 \cup \Delta_2$ the functions f_1, f_2 by

$$(48.3) \qquad f_1(\lambda) := \begin{cases} 1 & \text{for } \lambda \in \Delta_1 \\ 0 & \text{for } \lambda \in \Delta_2 \end{cases}, \qquad f_2(\lambda) := \begin{cases} 0 & \text{for } \lambda \in \Delta_1 \\ 1 & \text{for } \lambda \in \Delta_2 \end{cases}.$$

Both functions belong to $\mathscr{H}(x)$, consequently the elements

$$p_1 := f_1(x) \quad \text{and} \quad p_2 := f_2(x) \tag{48.4}$$

are defined. By Theorem 48.1 we have $p_k^2 = p_k$ (thus p_k is idempotent), $p_1 p_2 = p_2 p_1 = 0$ and $p_1 + p_2 = e$. We state this result with a modification which is only formal:

Proposition 48.2. *Let σ_1, σ_2 be complementary spectral sets of the element x of a complex Banach algebra E, and let Γ_1, Γ_2 be simple, closed integration paths, oriented counterclockwise, which lie in $\rho(x)$ and contain in their interior σ_1 and σ_2, respectively, but no further parts of $\sigma(x)$. Then the elements*

$$p_k := \frac{1}{2\pi i} \int_{\Gamma_k} r_\lambda \, d\lambda, \qquad k = 1, 2 \tag{48.5}$$

lie in E and satisfy the following equations:

$$p_k^2 = p_k, \qquad p_1 p_2 = p_2 p_1 = 0, \qquad p_1 + p_2 = e. \tag{48.6}$$

It is permitted that one of the spectral sets be empty.

Exercises

+1. Formulate and prove a continuity property of the correspondence $f \mapsto f(x)$ defined by (48.1) (cf. Exercise 4 in §47).

2. The equivalence of two functions from $\mathscr{H}(x)$ defined in §48 establishes an equivalence relation in $\mathscr{H}(x)$. If we define the algebraic operations in the corresponding set of equivalence classes $\hat{\mathscr{H}}(x)$ in the usual fashion (namely by means of representatives), then $\hat{\mathscr{H}}(x)$ becomes an algebra over **C**, which is homomorphic to a commutative subalgebra of E (cf. Theorem 48.1).

3. If $f \in \mathscr{H}(x)$ and $f(x) = 0$, then $f(\lambda)$ vanishes for all $\lambda \in \sigma(x)$.

4. For $f \in \mathscr{H}(x)$ we have $r(f(x)) = \max_{\lambda \in \sigma(x)} |f(\lambda)| \leqq \max_{|\lambda| \leqq r(x)} |f(\lambda)|$.

5. Let x be an element of the complex Banach algebra $C[a, b]$ (see Exercise 1 in §9). Then $\sigma(x) = \{x(t) : t \in [a, b]\}$, $r(x) = \|x\|$, and for $f \in \mathscr{H}(x)$ we have $[f(x)](t) = f[x(t)]$. For instance $[e^x](t) = e^{x(t)}$.

6. Define the continuous endomorphism A of the complex Banach algebra $C[a, b]$ by $(Ax)(t) := g(t)x(t)$, where g is a fixed element of $C[a, b]$. Then $\sigma(A) = \{g(t) : t \in [a, b]\}$, $r(A) = \|g\|$, and for $f \in \mathscr{H}(A)$ we have $[f(A)x](t) = f(g(t))x(t)$ (cf. Exercise 5).

§49. Spectral projectors

Let E be a non-trivial complex Banach space. We now apply the functional calculus to the Banach algebra $\mathscr{L}(E)$.

If σ is a (possibly empty) spectral set, Γ_σ a curve enclosing σ of the kind described in Proposition 48.2 and R_λ is the resolvent of $A \in \mathscr{L}(E)$, then

$$P_\sigma := \frac{1}{2\pi i} \int_{\Gamma_\sigma} R_\lambda \, d\lambda \tag{49.1}$$

is a continuous projector of E, the *spectral projector* or *Riesz projector* associated with σ; occasionally we shall denote it more precisely by $P_\sigma(A)$. The most important properties of projectors can be found in §5.

The projector complementary to P_σ is

$$I - P_\sigma = P_\tau = \frac{1}{2\pi i}\int_{\Gamma_\tau} R_\lambda \, \mathrm{d}\lambda \quad \text{with} \quad \tau := \sigma(A)\backslash\sigma.$$

For an empty σ we have $P_\sigma = 0$ and $P_\tau = I$. Let

$$M_\sigma := P_\sigma(E) = \{y \in E: P_\sigma y = y\} \quad \text{and} \quad N_\sigma := N(P_\sigma);$$

M_τ and N_τ are defined correspondingly. These spaces are all closed and we have

$$M_\tau = N_\sigma \quad \text{and} \quad N_\tau = M_\sigma.$$

E can be represented as the direct sum $E = M_\sigma \oplus N_\sigma$; if $x = y + z$ is the corresponding decomposition of x, then—since P_σ and P_τ commute with A, just like any $f(A)$—

$$Ay = AP_\sigma x = P_\sigma Ax \in M_\sigma \qquad \text{and correspondingly} \qquad Az \in M_\tau = N_\sigma.$$

Thus the spaces M_σ and N_σ are invariant under A. The restrictions A_σ, A_τ of A to M_σ and $N_\sigma = M_\tau$, are consequently continuous endomorphisms of M_σ and N_σ, respectively. The same proof shows, by the way, that M_σ and N_σ are invariant even under every $f(A)$, a fact we shall soon need. To a given $\mu \notin \sigma$ there exist open, mutually disjoint sets Δ_σ, Δ_τ covering σ and τ, respectively, and such that $\mu \notin \Delta_\sigma$. We now define f_σ and f by

$$f_\sigma(\lambda) = \begin{cases} 1 & \text{for} \quad \lambda \in \Delta_\sigma \\ 0 & \text{for} \quad \lambda \in \Delta_\tau \end{cases}, \qquad f(\lambda) = \begin{cases} \dfrac{1}{\mu - \lambda} & \text{for} \quad \lambda \in \Delta_\sigma \\ 0 & \text{for} \quad \lambda \in \Delta_\tau \end{cases}.$$

f_σ and f lie in $\mathscr{H}(A)$ and we have $(\mu - \lambda)f(\lambda) = f_\sigma(\lambda)$ for all $\lambda \in \Delta_\sigma \cup \Delta_\tau$, and consequently by Theorem 48.1

$$(\mu I - A)f(A) = f(A)(\mu I - A) = f_\sigma(A) = P_\sigma.$$

Since M_σ is invariant under all the operators of this equation and since P_σ is the identity on M_σ, we obtain by restriction to M_σ that μ lies in $\rho(A_\sigma)$. Thus we have $\sigma(A_\sigma) \subset \sigma$ and analogously $\sigma(A_\tau) \subset \tau$. Since, however, as it is very easy to see,

(49.2) $$\rho(A) = \rho(A_\sigma) \cap \rho(A_\tau), \quad \text{hence} \quad \sigma(A) = \sigma(A_\sigma) \cup \sigma(A_\tau),$$

we must have $\sigma(A_\sigma) = \sigma$ and $\sigma(A_\tau) = \tau$. Thus we have proved the following *decomposition theorem*:

Theorem 49.1. *Let E be a complex Banach space and σ a (possibly empty) spectral set of the continuous endomorphism A. Then the projector P_σ in* (49.1) *generates a decomposition*

(49.3) $$E = M_\sigma \oplus N_\sigma \quad \textit{where} \quad M_\sigma := P_\sigma(E) \quad \textit{and} \quad N_\sigma := N(P_\sigma);$$

the spaces M_σ and N_σ are invariant under A, and even under every $f(A)$ with $f \in \mathscr{H}(A)$; the spectrum of the restriction of A to M_σ is σ and the spectrum of the restriction of A to N_σ is $\sigma(A)\backslash\sigma$.

If the spectral set σ lies in a particularly simple way in $\sigma(A)$, then M_σ can be characterized analytically. We say that a subset σ of $\sigma(A)$ is *circularly isolated* if there exists a circle

$$\Gamma := \{\lambda : |\lambda - \alpha| = r\} \tag{49.4}$$

with the following properties:

(a) Γ lies in $\rho(A)$;
(b) the interior of Γ contains σ but no further points of $\sigma(A)$.

Obviously such a σ is a spectral set of A and

$$(\alpha I - A)^n P_\sigma = \frac{1}{2\pi i} \int_\Gamma (\alpha - \lambda)^n R_\lambda \, d\lambda \qquad \text{for} \quad n = 0, 1, 2, \ldots \tag{49.5}$$

(cf. Exercise 6), hence also

$$(\alpha I - A)^n P_\sigma x = \frac{1}{2\pi i} \int_\Gamma (\alpha - \lambda)^n R_\lambda x \, d\lambda \tag{49.6}$$

(see Exercise 1). For $x \in M_\sigma$ we have $P_\sigma x = x$ and thus by (46.13)

$$\|(\alpha I - A)^n x\| \leqq \frac{1}{2\pi} 2\pi r \cdot r^n \max_{\lambda \in \Gamma} \|R_\lambda\| \, \|x\|.$$

Since this estimate also holds for some $r' < r$ (Γ lies in $\rho(A)$), we have

$$\limsup \|(\alpha I - A)^n x\|^{1/n} < r. \tag{49.7}$$

Conversely, if this inequality is valid, then the series $\sum_{n=0}^{\infty} [(\alpha I - A)/(\alpha - \lambda)]^n x$ converges for all $\lambda \in \Gamma$ and for its sum $y(\lambda)$ we have, according to Proposition 11.1,

$$\left(I - \frac{\alpha I - A}{\alpha - \lambda}\right) y(\lambda) = x,$$

so that, as a simple calculation shows, $y(\lambda) = (\lambda - \alpha) R_\lambda x$ and therefore

$$R_\lambda x = -\sum_{n=0}^{\infty} \frac{(\alpha I - A)^n x}{(\alpha - \lambda)^{n+1}} \qquad \text{for all} \quad \lambda \in \Gamma.$$

A licit termwise integration now yields

$$P_\sigma x = \frac{1}{2\pi i} \int_\Gamma R_\lambda x \, d\lambda = -\frac{1}{2\pi i} \int_\Gamma \frac{1}{\alpha - \lambda} x \, d\lambda = x$$

(use an integration formula corresponding to (47.15)). Thus x lies in M_σ. We summarize:

Proposition 49.1. *If under the hypotheses of Theorem* 49.1 *the set $\sigma \subset \sigma(A)$ is circularly isolated through $|\lambda - \alpha| = r$, then the vector x lies in M_σ if and only if* (49.7) *is valid. In particular, if α is an* isolated point *of the spectrum of A, then*

$$M_{\{\alpha\}} = \{x : \lim \|(\alpha I - A)^n x\|^{1/n} = 0\}. \tag{49.8}$$

Exercises

*1. Formulate and prove a theorem which justifies the passage from (49.5) to (49.6); the theorem has the form $(\int_\Gamma A(\lambda)\mathrm{d}\lambda)x = \int_\Gamma A(\lambda)x\,\mathrm{d}\lambda$, $A(\lambda) \in \mathscr{L}(E)$.

*2. Let the Banach space E be the direct sum of the closed subspaces F_1, F_2 invariant under $A \in \mathscr{L}(E)$; let A_k be the restriction of A to F_k. Then $\rho(A) = \rho(A_1) \cap \rho(A_2)$, hence $\sigma(A) = \sigma(A_1) \cup \sigma(A_2)$ (generalization of (49.2)).

3. Under the hypotheses of Theorem 49.1 the eigenvalues of A_σ are precisely the eigenvalues of A lying in σ (A_σ is the restriction of A to M_σ).

4. A spectral set σ of A is also a spectral set of A', and we have $P_\sigma(A') = (P_\sigma(A))'$. *Hint*: Proposition 44.2, Exercise 2 in §44 and Exercise 10 in §46.

*5. If $\sigma_1, \ldots, \sigma_n$ are pairwise disjoint spectral sets of an operator T, if $P_1, \ldots, P_n$ are the corresponding spectral projectors and if

$$\sigma(T) = \sigma_1 \cup \cdots \cup \sigma_n,$$

then we have:

(a) $P_\nu P_\mu = \delta_{\nu\mu} P_\nu$ and $I = \sum_{\nu=1}^{n} P_\nu$,

(b) $E = P_1(E) \oplus \cdots \oplus P_n(E)$,

(c) $P_\nu(E)$ is invariant under T $(\nu = 1, \ldots, n)$.

*6. We assume the hypotheses necessary for the definition of P_σ in (49.1) and we use the notation employed there. Show that for every $f \in \mathscr{H}(A)$ we have

$$\frac{1}{2\pi i}\int_{\Gamma_\sigma} f(\lambda)R_\lambda\,\mathrm{d}\lambda = f(A)P_\sigma.$$

§50. Isolated points of the spectrum

We begin this section with a representation theorem. In it we speak of 'functions of A'. By this we mean operators of the form $f(A)$ with $f \in \mathscr{H}(A)$. For the notation employed (e.g., P_σ, M_σ) we refer to Theorem 49.1. *In the whole section let E be a complex Banach space and $A \in \mathscr{L}(E)$.*

Proposition 50.1. *Assume that A is not quasi-nilpotent and that it has the spectral set $\sigma := \{\lambda_0\}$ consisting of one point. Then there is a representation*

(50.1) $$A = U + S$$

of A with the following properties:

(a) *U and S are functions of A, hence they commute with each other and with A;*

(b) *λ_0 lies in $\rho(U)$ and in $\sigma(S)$, so that, in particular, $R := \lambda_0 I - U$ has an inverse in $\mathscr{L}(E)$ and $\lambda_0 I - A$ can be represented in the form*

(50.2) $$\lambda_0 I - A = R - S \quad \textit{with} \quad R^{-1} \in \mathscr{L}(E);$$

(c) *S maps E, and already the subspace M_σ, onto M_σ;*

(d) *we have*

(50.3) $$\lim \|(\lambda_0 I - A)^n S x\|^{1/n} = 0 \qquad \textit{for all} \quad x \in E.$$

For quasi-nilpotent operators a representation of the described kind is not possible.

Proof. Let Δ_σ, Δ_τ be open, mutually disjoint sets which cover σ and $\tau := \sigma(A)\backslash\sigma$, respectively, and let the functions f_σ, f_τ from $\mathscr{H}(A)$ be defined as follows:

$$f_\sigma(\lambda) := \begin{cases} 1 & \text{for } \lambda \in \Delta_\sigma \\ 0 & \text{for } \lambda \in \Delta_\tau \end{cases}, \qquad f_\tau(\lambda) := \begin{cases} 0 & \text{for } \lambda \in \Delta_\sigma \\ 1 & \text{for } \lambda \in \Delta_\tau \end{cases};$$

finally, let $P := P_\sigma$ and $Q := P_\tau = I - P$. For the proof of the representation (50.1) and of assertion (a) we now set

$$(50.4) \qquad U := AQ \quad \text{and} \quad S := AP \qquad \text{if} \quad \lambda_0 \neq 0;$$

$$(50.5) \qquad U := AQ + P \quad \text{and} \quad S := AP - P \qquad \text{if} \quad \lambda_0 = 0.$$

With the operators so defined (50.1) holds, furthermore U and S are functions of A (e.g., in the case $\lambda_0 = 0$ obviously $U = f(A)$ with $f(\lambda) := \lambda f_\tau(\lambda) + f_\sigma(\lambda)$).

(b) For $\lambda_0 \neq 0$ we have $g(\lambda) := \lambda_0 - \lambda f_\tau(\lambda) \neq 0$ for $\lambda \in \sigma(A)$, thus $g(A) = \lambda_0 I - AQ = \lambda_0 I - U$ has an inverse in $\mathscr{L}(E)$ by Theorem 48.1e, i.e., λ_0 lies in $\rho(U)$; and since $S = h(A)$ with $h(\lambda) := \lambda f_\sigma(\lambda)$, the set

$$\sigma(S) = \{\lambda f_\sigma(\lambda) : \lambda \in \sigma(A)\}$$

contains, because of Theorem 48.2, the point $\lambda_0 = \lambda_0 f_\sigma(\lambda_0)$. In the case $\lambda_0 = 0$ let $g(\lambda) := \lambda f_\tau(\lambda) + f_\sigma(\lambda)$ and $h(\lambda) := \lambda f_\sigma(\lambda) - f_\sigma(\lambda)$. Just like above, one sees that $g(A) = AQ + P = U$ has an inverse in $\mathscr{L}(E)$, and $\sigma(S) = \sigma(h(A)) = h(\sigma(A))$ contains the point $0 = \lambda_0$ (observe that $\tau \neq \varnothing$ because A is not quasi-nilpotent!).

(c) The inclusion $S(E) \subset M_\sigma = P(E)$ is trivial; the equation $S(M_\sigma) = M_\sigma$ becomes obvious if we take into account that $\sigma(A_\sigma) = \{\lambda_0\}$ (Theorem 49.1) and that S coincides on M_σ with A_σ or $A_\sigma - I$, respectively.

(d) now follows from Proposition 49.1.

Let us suppose, finally, that for a quasi-nilpotent A we have the representation (50.1) with the described properties, where of course $\{\lambda_0\} = \sigma(A) = \{0\}$. Then in particular S is a function of A, say $S = f(A)$, consequently $\sigma(S)$ is the set $\{f(0)\}$ consisting of one point. Since 0 lies in $\sigma(S)$ by hypothesis, we have $\sigma(S) = \{0\}$ and thus S is quasi-nilpotent. From Proposition 45.1 it follows now that $r(U) = r(A - S) \leqq r(A) + r(S) = 0$, hence also U is quasi-nilpotent in contradiction to the assumption that 0 should lie in $\rho(U)$. Thus Proposition 50.1 is completely proved. ■

An isolated point λ_0 of the spectrum of A is a non-removable singularity of the resolvent R_λ (end of §46). According to Proposition 46.7, there exists a

Laurent expansion of R_λ in a neighborhood of λ_0:

(50.6) $$R_\lambda = \sum_{n=1}^{\infty} \frac{P_n}{(\lambda-\lambda_0)^n} + \sum_{n=0}^{\infty} Q_n(\lambda-\lambda_0)^n \quad \text{for} \quad 0 < |\lambda-\lambda_0| < r;$$

the coefficients are calculated according to the formulas

(50.7) $$P_n = \frac{1}{2\pi i}\int_\Gamma (\lambda-\lambda_0)^{n-1} R_\lambda \, d\lambda,$$

(50.8) $$Q_n = \frac{1}{2\pi i}\int_\Gamma \frac{R_\lambda}{(\lambda-\lambda_0)^{n+1}} \, d\lambda,$$

where Γ is a sufficiently small, positively oriented circle around λ_0.

We turn our attention to the principal part of the Laurent expansion. For $\sigma := \{\lambda_0\}$ it follows, with the help of the functional calculus, from (50.7) immediately that

(50.9) $$P_1 = P_\sigma \quad \text{and} \quad P_n = (A - \lambda_0 I)^{n-1} P_\sigma \qquad (n = 1, 2, \ldots).$$

These equations show that either all $P_n \neq 0$, or that there exists a natural number p such that

$$P_n \neq 0 \quad \text{for} \quad n = 1, \ldots, p, \quad \text{but} \quad P_n = 0 \quad \text{for} \quad n = p+1, p+2, \ldots.$$

In the first case λ_0 is an essential singularity of R_λ, in the second case a pole of order p. The next proposition shows that in the discussion of the poles of R_λ the *chain conditions* play a decisive role. We recall again that an endomorphism T is injective or surjective if the length $p(T)$ of its nullchain or the length $q(T)$ of its image chain vanish, respectively (§38).

Proposition 50.2. *λ_0 is a pole of the resolvent of A if and only if $A - \lambda_0 I$ has* positive finite *chain lengths; the common length p of the chains is the order of the pole. In this case λ_0 is an eigenvalue of A; for the spectral projector P corresponding to $\{\lambda_0\}$ we have*

(50.10) $$P(E) = N[(A - \lambda_0 I)^p], \qquad N(P) = (A - \lambda_0 I)^p(E).$$

Proof. Let λ_0 be a pole of order p of R_λ; setting $T := A - \lambda_0 I$ we have thus

(50.11) $$T^{p-1}P \neq 0, \qquad T^p P = 0.$$

It follows that $P(E) \subset N(T^p)$, $P(E) \neq N(T^{p-1})$. Since on the other hand $N(T^p)$ lies in $P(E)$ by Proposition 49.1, we have altogether $N(T^{p-1}) \neq N(T^p) = P(E)$, thus the nullchain of T has length $p > 0$ (consequently λ_0 is an eigenvalue of A), and the first one of the equations (50.10) holds. If A is not quasi-nilpotent, we can, according to Proposition 50.1, represent T in the form

(50.12) $$T = R + S \quad \text{with bijective} \quad R \quad \text{and} \quad S = PB,$$

where $B \in \mathscr{L}(E)$ is determined by (50.4) or (50.5); but such a representation also holds trivially for a quasi-nilpotent A: In this case we have, namely, $P = I$, hence $A = (A - I) + S$ with a bijective $A - I$ and $S := P = I$. From (50.11) and (50.12) it follows that $T^{p+1} = T^pR + T^pPB = T^pR$; because of $R(E) = E$ we obtain $T^{p+1}(E) = T^p(E)$, hence the image chain of T has also finite length. The two chain lengths of T coincide by Proposition 38.3. Because of Theorem 49.1 and Proposition 38.4 the decompositions

$$E = P(E) \oplus N(P) \tag{50.13}$$

$$E = N(T^p) \oplus T^p(E) \tag{50.14}$$

hold, where T is bijective on $N(P)$ and therefore $N(P) = T^p[N(P)] \subset T^p(E)$. The decomposition $x = y + z$ of an element x from $T^p(E)$ according to (50.13) is because of $P(E) = N(T^p)$ and $N(P) \subset T^p(E)$ at the same time a decomposition according to (50.14); but then $y = 0$ and $x = z \in N(P)$, hence $T^p(E) \subset N(P)$, so that also the second equation in (50.10) is valid.

Let us now suppose conversely that T has the finite chain length $p > 0$. Then λ_0 is an eigenvalue of A and we have the decomposition (50.14) with the closed nullspace $N(T^p)$. With the aid of Proposition 36.2 it follows that also $T^p(E)$ is closed. Let T_1, T_2 be the restrictions of T to the Banach spaces $N(T^p)$, $T^p(E)$, respectively. Since T_1 is nilpotent and T_2 bijective (Proposition 38.4), $\rho(T_1)$ contains $\mathbf{C}\setminus\{0\}$ and $\rho(T_2)$ contains a disk around 0, and it follows because of $\rho(T) = \rho(T_1) \cap \rho(T_2)$ (Exercise 2 in §49) that 0 is an isolated point of $\sigma(T)$, hence λ_0 is an isolated spectral point of A. As above, let P be the spectral projector corresponding to $\{\lambda_0\}$. Because of (50.9), we only have to prove that $P(E) \subset N(T^m)$ for some natural number m. If $x \in P(E)$ and $x = u + v$ is the decomposition of x according to (50.14) with $u \in N(T^p)$ and $v \in T^p(E)$, then $T^nx = T^nv$ for $n \geqq p$; with the aid of Proposition 49.1 it follows that v lies in $P(E)$, thus taken together we have $v \in T^p(E) \cap P(E)$. If we can prove that this intersection is equal to $\{0\}$, then $x = u \in N(T^p)$ and our proof is ended. We set $D := T^p(E) \cap P(E)$ and prove the equation $D = \{0\}$ by showing that the restriction A_0 of A to the Banach space D, invariant under A, has an empty spectrum (see Theorem 45.1). To every $x \in D$ there exists exactly one $y \in T^p(E)$ with $x = Ty$ (Proposition 38.4); with the aid of Proposition 49.1 it follows that y lies in D, i.e., that T is bijective on D and so λ_0 is in $\rho(A_0)$. Now let $\lambda \neq \lambda_0$. Then Theorem 49.1 implies that λ lies in the resolvent set of the restriction of A to $P(E)$, consequently to every $x \in D$ there exists exactly one $y \in P(E)$ such that $x = (A - \lambda I)y$. If $y = u + v$ with $u \in N(T^p)$, $v \in T^p(E)$ is the decomposition of y according to (50.14), then $x - (A - \lambda I)(u + v) = 0$, hence $x - (A - \lambda I)v = (A - \lambda I)u$, and since the element on the left-hand side belongs to $T^p(E)$, the one on the right-hand side to $N(T^p)$, we have $(A - \lambda I)u = 0$, hence $u \in N[(A - \lambda_0 I)^p] \cap N(A - \lambda I)$. With the aid of Proposition 43.1a it follows that $u = 0$, thus $y = v \in T^p(E)$, and so y lies in D. This means that also $A_0 - \lambda I$ is bijective, i.e., that λ lies in $\rho(A_0)$. Therefore we have indeed $\rho(A_0) = \mathbf{C}$ and thus $\sigma(A_0) = \varnothing$. Our proposition is now completely proved. ∎

Proposition 50.3. *$A - \lambda_0 I$ is a Fredholm operator with* positive finite *chain length if and only if λ_0 is an isolated spectral point of A and the corresponding spectral projector P is finite-dimensional. In this case λ_0 is a pole of R_λ.*

Proof. If λ_0 is isolated and P finite-dimensional, then because of

$$N[(A - \lambda_0 I)^n] \subset P(E)$$

—see Proposition 49.1—$\alpha(A - \lambda_0 I)$ and $p(A - \lambda_0 I)$ are finite. Let A_1 be the restriction of A to $N(P)$. According to Theorem 49.1 we have $\lambda_0 \in \rho(A_1)$, hence for $n = 1, 2, \ldots$

$$N(P) = (A_1 - \lambda_0 I)^n(N(P)) \subset (A - \lambda_0 I)^n(E). \tag{50.15}$$

If we take into account that in the decomposition $E = P(E) \oplus N(P)$ the first summand has a finite dimension, i.e., that codim $N(P)$ is finite, then it follows from (50.15) that also $\beta(A - \lambda_0 I)$ and $q(A - \lambda_0 I)$ must be finite. With the aid of Proposition 37.1 we obtain the assertion in one direction. The other direction follows immediately from Proposition 50.2. ■

From the Riesz theory of compact operators we obtain very easily:

Proposition 50.4. *Every spectral point λ_0 different from zero of a compact operator K on E is a pole of the resolvent, and the corresponding spectral projector is finite-dimensional.*

For the proof we observe that $\lambda_0 I - K$ is a chain-finite Fredholm operator (Theorem 40.1) whose chain length must be positive because λ_0 is a spectral point. The assertion now follows from Proposition 50.3. ■

Exercises

$^+$1. λ_0 is a pole of the resolvent of A if and only if λ_0 is a pole of the resolvent of A'; in this case the orders of the poles are the same. *Hint*: Proposition 44.2, Exercise 2 in §44 and Exercise 10 in §46.

2. Investigate the case when 0 is a pole of the resolvent of a compact operator. In §52, Exercise 2 a similar problem will be discussed.

§51. The Fredholm region

In this section let E be a Banach space over **K** *and let A be from $\mathscr{L}(E)$ (unless something else is explicitly said).* We shall investigate the so-called *Fredholm region* of A, i.e., the set

$$\Phi_A := \{\lambda \in \mathbf{K} : \lambda I - A \in \Phi(E)\}.$$

Because of Proposition 37.1 we have $\Phi_A = \{\lambda \in \mathbf{K} : \lambda I - A \text{ has finite deficiency}\}$. Obviously $\rho(A) \subset \Phi_A$. For a compact A we have $\Phi_A \supset \mathbf{K}\backslash\{0\}$. The points of the Fredholm region Φ_A are called the *Fredholm points* of A. Before we formulate the next proposition, we recall that an open subset $M \neq \varnothing$ of $\mathbf{K}$ can be decomposed into maximal open, connected, pairwise disjoint non-empty sets, the *components* of M.

Proposition 51.1. *The set Φ_A is* open, *and on every component of Φ_A the function* $\operatorname{ind}(\lambda I - A)$ *is* constant.

The openness of Φ_A follows immediately from Proposition 37.2. The assertion concerning the index follows from the same proposition with the help of the method of chains of disks, familiar from the theory of analytic continuation: Join a fixed point λ_0 of the component C with an arbitrary point $\lambda_1 \in C$ by a polygonal line $P \subset C$ and associate with each $\mu \in P$ a disk in which $\operatorname{ind}(\lambda I - A)$ is equal to $\operatorname{ind}(\mu I - A)$. Since by the Heine–Borel theorem already finitely many of these disks cover P, we have $\operatorname{ind}(\lambda_1 I - A) = \operatorname{ind}(\lambda_0 I - A)$. ∎

For further investigation and classification of the components of Φ_A we draw on the length $p(\lambda I - A)$ of the nullchain of $\lambda I - A$. We shall make ample use of the propositions of §38. The reader should remember that λ is an eigenvalue of A if one—and thus each—of the assertions

$$\text{'}\alpha(\lambda I - A) \neq 0\text{'}, \qquad \text{'}p(\lambda I - A) \neq 0\text{'}$$

holds.

Proposition 51.2. *For $\lambda_0 \in \Phi_A$ we have the following alternative:*

Either $p(\lambda_0 I - A)$ is finite*—then λ_0 is not a limit point of eigenvalues of A—or $p(\lambda_0 I - A)$ is* infinite*—then there exists a neighborhood U on λ_0 whose points are all eigenvalues, and in such a fashion that for $\lambda \in U \backslash \{\lambda_0\}$ one has*

$$\alpha(\lambda I - A) = \text{const} \leqq \alpha(\lambda_0 I - A).$$

Proof. We set $T := \lambda_0 I - A$. Obviously λ is an eigenvalue of A if and only if $\mu := \lambda_0 - \lambda$ is an eigenvalue of T. In the case $p(T) < \infty$ we must therefore prove that the eigenvalues of T cannot accumulate at the point 0, in the case $p(T) = \infty$, however, that there exists a neighborhood V of 0 which consists entirely of eigenvalues, and that

$$\alpha(\mu I - T) = \text{const} \leqq \alpha(T) \qquad \text{for all} \quad \mu \in V \backslash \{0\}.$$

Let $F := \bigcap_{n=1}^{\infty} T^n(E)$ and $\tilde{T}$ the restriction of T to F. F is a Banach space because the powers T^n lie all in $\Phi(E)$ and so their image spaces are closed (Propositions 25.3 and 24.3). In the first case of the alternative ($p(T) < \infty$) we have by Proposition 38.7 and Lemma 38.1 that $\alpha(\tilde{T}) = \beta(\tilde{T}) = 0$, so that 0 and with it also a neighborhood $\{\mu \in \mathbf{K} : |\mu| < r\}$ of 0 lie in $\rho(\tilde{T})$. None of these values $\mu \neq 0$ is an eigenvalue of T; otherwise we would have $Tx = \mu x$ for some $x \neq 0$, hence $T^n x = \mu^n x$ $(n = 1, 2, \ldots)$. Thus x would lie in F and we would have $\tilde{T}x = \mu x$ which is impossible. Thus 0 is indeed not a limit point of eigenvalues of T. In the second case of the alternative ($p(T) = \infty$) we have, by Proposition 38.7, that $0 < \alpha(\tilde{T})$ and trivially $\alpha(\tilde{T}) \leqq \alpha(T)$, furthermore by Lemma 38.1 that $\beta(\tilde{T}) = 0$. Thus $\tilde{T} \in \Phi(F)$ and

$$(51.1) \qquad 0 < \operatorname{ind}(\tilde{T}) = \alpha(\tilde{T}) - \beta(\tilde{T}) = \alpha(\tilde{T}) \leqq \alpha(T).$$

Because of Proposition 37.2 there exists an $r_1 > 0$ with

$$(51.2) \qquad \operatorname{ind}(\mu\tilde{I} - \tilde{T}) = \operatorname{ind}(\tilde{T}) > 0 \qquad \text{for} \quad |\mu| < r_1.$$

Since Proposition 26.1 ensures the existence of a right inverse $R \in \mathscr{L}(F)$ of $\tilde{T}$, there exists an $r_2 > 0$ with

$$(51.3) \qquad \beta(\mu\tilde{I} - \tilde{T}) = 0 \qquad \text{for} \quad |\mu| < r_2$$

(Exercise 4 in §9). For $|\mu| < \min(r_1, r_2)$ we have then because of (51.1) through (51.3)

$$0 < \alpha(\mu\tilde{I} - \tilde{T}) = \operatorname{ind}(\mu\tilde{I} - \tilde{T}) = \operatorname{ind}(\tilde{T}) \leqq \alpha(T).$$

From this estimate the assertion follows, since for $\mu \neq 0$ we have

$$\alpha(\mu I - T) = \alpha(\mu\tilde{I} - \tilde{T});$$

indeed, for these μ we have $N(\mu I - T) \subset F$, as we have already seen above. ■

Proposition 51.3. *$p(\lambda I - A)$ is finite either for every point or for no point of a component C of Φ_A.*

Proof. Let $M := \{\lambda \in C: p(\lambda I - A) < \infty\} \neq \varnothing$. A point $\lambda_0 \in M$ is according to Proposition 51.2 not a limit point of eigenvalues. In a certain disk around λ_0, which can be chosen so small that it lies in C (C is open!), we have therefore $p(\lambda I - A) = 0$. Consequently M is open in C. Now let μ be a limit point of M lying in C, and let (λ_n) be a sequence from M with $\lambda_n \neq \mu$, $\lambda_n \to \mu$. If we had $\mu \notin M$, then according to case 2 of the alternative of Proposition 51.2 a neighborhood of μ would consist of eigenvalues, hence for sufficiently large n also a neighborhood of λ_n would consist of eigenvalues, in contradiction to case 1 of the alternative of the proposition quoted. Thus μ lies in M, hence M is not only open but also closed in C. Since C is connected, we must have $M = C$ (see [179]).

Proposition 51.4. *In every component C of Φ_A there exists a (possibly empty) subset M with the following properties:*

(a) *M has no limit point in C;*
(b) *there exists a constant $\gamma \geqq 0$ such that*

$$(51.4) \quad \alpha(\lambda I - A) = \gamma \text{ for } \lambda \in C \setminus M \quad \text{and} \quad \alpha(\lambda I - A) > \gamma \quad \text{for} \quad \lambda \in M.$$

The eigenvalues of A have no limit point in C if and only if $p(\lambda_0 I - A)$ is finite for some $\lambda_0 \in C$.

According to Proposition 51.3 we distinguish two cases in the proof:

1. $p(\lambda I - A) < \infty$ on C: let M be the set of eigenvalues in C. Then (a) holds because of Proposition 51.2, (b) holds trivially with $\gamma = 0$.

2. $p(\lambda I - A) = \infty$ on C: now let M be the set of eigenvalues $\lambda_0 \in C$ with the following property (cf. Proposition 51.2): There exists a neighborhood $U \subset C$ of λ_0 so that with a $\gamma > 0$ the following assertion holds:

$$\alpha(\lambda I - A) = \gamma < \alpha(\lambda_0 I - A) \quad \text{for} \quad \lambda \in U \backslash \{\lambda_0\}. \tag{51.5}$$

If μ were a limit point of M in C, then by Proposition 51.2 there would exist a neighborhood $V \subset C$ of μ such that

$$\alpha(\lambda I - A) = \text{const} \quad \text{for} \quad \lambda \in V \backslash \{\mu\}; \tag{51.6}$$

but V also contains a point $\lambda_0 \neq \mu$ from M, hence (51.6) contradicts (51.5). Thus (a) is again valid. If λ_1, λ_2 are from $C \backslash M$, then the equation $\alpha(\lambda_1 I - A) = \alpha(\lambda_2 I - A)$ is obtained with the method of chains of disks (see the proof of Proposition 51.1). ■

Together with T also T' is a Fredholm operator and we have $\beta(T) = \alpha(T')$ and $q(T) = p(T')$ (Propositions 27.3 and 39.2). With the help of this fact we can immediately obtain from the last three propositions assertions concerning the quantities $\beta(\lambda I - A)$ and $q(\lambda I - A)$. If we call λ a *deficiency value* of A when $\beta(\lambda I - A) \neq 0$, then we can summarize these assertions as follows:

Proposition 51.5. *Propositions* 51.2 *through* 51.4 *remain true if we replace* p *by* q, α *by* β *and eigenvalue by deficiency value.*

From the last five propositions and from Propositions 38.3, 38.5, 38.6 we now obtain in a very simple way, if we classify the components of Φ_A according to the sign of the index and the finiteness or infiniteness of the lengths of the chains of $\lambda I - A$ the, following:

Theorem 51.1. *For the behavior of the index, the chain lengths, the eigenvalues and the deficiency values in a component C of Φ_A there are exactly the following six possibilities, where the equations and inequalities listed hold for all $\lambda \in C$:*

(a_1) $\text{ind}(\lambda I - A) = 0$ *and* $p(\lambda I - A) = q(\lambda I - A) < \infty$; *eigenvalues and deficiency values do not have a limit point in C. This case occurs exactly when $C \cap \rho(A)$ is non-empty.*

(a_2) $\text{ind}(\lambda I - A) = 0$ *and* $p(\lambda I - A) = q(\lambda I - A) = \infty$; *every point of C is an eigenvalue and a deficiency value.*

(b_1) $\text{ind}(\lambda I - A) < 0$ *and* $p(\lambda I - A) < \infty$, $q(\lambda I - A) = \infty$; *the eigenvalues do not have a limit point in C, every point of C is a deficiency value;*

(b_2) $\text{ind}(\lambda I - A) < 0$ *and* $p(\lambda I - A) = q(\lambda I - A) = \infty$; *every point of C is an eigenvalue and a deficiency value.*

(c$_1$) $\operatorname{ind}(\lambda I - A) > 0$ *and* $p(\lambda I - A) = \infty$, $q(\lambda I - A) < \infty$; *every point of* C *is an eigenvalue, the deficiency values do not have a limit point in* C.

(c$_2$) $\operatorname{ind}(\lambda I - A) > 0$ *and* $p(\lambda I - A) = q(\lambda I - A) = \infty$; *every point of* C *is an eigenvalue and a deficiency value.*

The details of the proof can be left to the reader.

Every $\lambda \in \mathbf{K}$ with $|\lambda| > \lim \|A^n\|^{1/n}$ lies in $\rho(A)$ and thus also in Φ_A. The (possibly zero) radius $r_\Phi(A)$ of the smallest circle about 0 outside which there lie only Fredholm points is called the *Fredholm radius* of A; thus

$$r_\Phi(A) := \inf\{r > 0 : \lambda I - A \in \Phi(E) \text{ for all } |\lambda| > r\}.$$

In order to present a formula for the Fredholm radius, we quote the definition of a Φ-ideal from Exercise 7 in §25 and prove the result formulated there.

A (two-sided) ideal $\mathscr{J}$ in an operator algebra $\mathscr{A}(E)$ on the vector space E is said to be Φ-*ideal* if $\mathscr{J} \supset \mathscr{F}(\mathscr{A}(E))$ and $I - K \in \Phi(\mathscr{A}(E))$ for all $K \in \mathscr{J}$.

Proposition 51.6. *If* $\mathscr{J}$ *is a* Φ-*ideal in the saturated operator algebra* $\mathscr{A}(E)$, *then* $T \in \mathscr{A}(E)$ *is a Fredholm operator in* $\mathscr{A}(E)$ *if and only if the class* $\tilde{T}$ *of* T *is invertible in* $\mathscr{A}(E)/\mathscr{J}$.

Indeed, if $T \in \Phi(\mathscr{A}(E))$, then the assertion follows immediately from Theorem 25.1 because the finite-dimensional operators K_1, K_2 occurring in (25.6) lie also in $\mathscr{J}$. Conversely, let $\tilde{T}$ be invertible. Then there exist $S_1, S_2 \in \mathscr{A}(E)$ and $K_1, K_2 \in \mathscr{J}$ so that $S_1 T = I - K_1$ and $TS_2 = I - K_2$. Since by hypothesis $I - K_1, I - K_2$ are Fredholm operators, i.e., have invertible residue classes in $\hat{\mathscr{A}} := \mathscr{A}(E)/\mathscr{F}(\mathscr{A}(E))$, these equations imply that the residue class of T in $\hat{\mathscr{A}}$ is left- as well as right-invertible. Consequently T lies in $\Phi(\mathscr{A}(E))$. ■

Proposition 51.7. *If* E *is a Banach space, then the closure of a* Φ-*ideal* $\mathscr{J}$ *in* $\mathscr{L}(E)$ *is again a* Φ-*ideal.*

Proof. Certainly $\bar{\mathscr{J}}$ is an ideal in $\mathscr{L}(E)$. To $R \in \bar{\mathscr{J}}$ there exists a $K \in \mathscr{J}$ such that $\|R - K\| < 1$. The operator $S := I - (R - K)$ is invertible in $\mathscr{L}(E)$, thus it is a Fredholm operator and therefore $I - R = I - (R - K) - K = S - K = S(I - S^{-1}K)$ lies in $\Phi(E)$ (Proposition 25.3a). ■

If E is a normed space, then $\mathscr{F}(E)$ and $\mathscr{K}(E)$ are Φ-ideals; if E is complete, then, according to the last proposition, also $\overline{\mathscr{F}(E)}$ is a Φ-ideal. We now show that the Fredholm radius can be computed with the help of Φ-ideals.

Proposition 51.8. *If* E *is a complex Banach space and* $\mathscr{J}$ *a* Φ-*ideal in* $\mathscr{L}(E)$, *then*

$$r_\Phi(A) = \lim \left[\inf_{K \in \mathscr{J}} \|A^n - K\|\right]^{1/n} \tag{51.7}$$

and

$$(51.8) \qquad r_\Phi(A) \leqq \|A - K\| \quad \textit{for every} \quad K \in \mathscr{J}.$$

Proof. The assertions of the proposition are trivial for finite-dimensional spaces (in this case $\Phi_A = \mathbf{C}$ and $r_\Phi(A) = 0$). Let now $\dim E = \infty$. Then I does not lie in $\mathscr{J}$, the quotient algebra $\tilde{\mathscr{L}} := \mathscr{L}(E)/\mathscr{J}$ has thus a unit element $\tilde{I}$ different from $\tilde{0}$, and if we first assume that $\mathscr{J}$ is *closed*, then $\tilde{\mathscr{L}}$ is a Banach algebra with respect to the usual quotient norm. The residue class of T in $\tilde{\mathscr{L}}$ will be denoted by $\tilde{T}$. Because of Proposition 51.6 we have

$$(51.9) \qquad \Phi_A = \rho(\tilde{A}), \qquad \text{hence} \quad r_\Phi(A) = r(\tilde{A}),$$

thus with the help of Theorem 45.1 and the definition of the quotient norm we obtain

$$r_\Phi(A) = \lim \|\tilde{A}^n\|^{1/n} = \lim \left[\inf_{K \in \mathscr{J}} \|A^n - K\| \right]^{1/n}.$$

(51.8) follows from the estimate $r(\tilde{A}) \leqq \|\tilde{A}\|$. (51.7) is still true if $\mathscr{J}$ is *not* closed: compute first $r_\Phi(A)$ with the help of the Φ-ideal $\bar{\mathscr{J}}$ (Proposition 51.7) and take into account that $\inf_{R \in \mathscr{J}} \|T - R\| = \inf_{R \in \bar{\mathscr{J}}} \|T - R\|$. One sees analogously that also (51.8) is valid in the case of a non-closed $\mathscr{J}$. ■

From (51.9) and Theorem 45.1 we also obtain immediately:

Proposition 51.9. *In the case of an infinite-dimensional complex Banach space we have* $\Phi_A \neq \mathbf{C}$.

Exercises

1. If E is an infinite-dimensional complex Banach space, then there exists on $|\lambda| = r_\Phi(A)$ at least one point λ_0 for which either $\alpha(\lambda_0 I - A)$ or $\beta(\lambda_0 I - A)$ is infinite. With Exercise 4 in §37 one obtains the stronger assertion: On $|\lambda| = r_\Phi(A)$ there exists a λ_0 with $\beta(\lambda_0 I - A) = \infty$.

2. In the case of a complex Banach space the poles of the resolvent belonging to Φ_A lie precisely in those components of Φ_A which intersect $\rho(A)$.

3. For all λ on the boundary of a component of Φ_A either $\alpha(\lambda I - A)$ or $\beta(\lambda I - A)$ is infinite. What can be said about the chain lengths $p(\lambda I - A)$ and $q(\lambda I - A)$?

$^+$4. $A \in \mathscr{L}(E)$ is said to be *quasicompact* if there exists a natural number m and a $K \in \mathscr{K}(E)$ such that $\|A^m - K\| < 1$. Show that in the case of a complex Banach space E the following properties are equivalent:

(a) A is quasicompact;

(b) $r(\tilde{A}) < 1$, where $\tilde{A}$ is the residue class of A in $\mathscr{L}(E)/\mathscr{K}(E)$;

(c) there exists a positive $\eta < 1$ such that for $|\lambda| > \eta$ we have $\lambda I - A \in \Phi(E)$.

Replace in the above definition $\mathscr{K}(E)$ by a Φ-ideal $\mathscr{J}$, and investigate the situation so obtained.

5. Let E be a Banach space over $\mathbf{K}$ and $A \in \mathscr{L}(E)$. Investigate the *Atkinson region* $\{\lambda \in \mathbf{K}: \lambda I - A$ is an Atkinson operator$\}$. *Hint*: Exercise 3 in §37.
6. If a projector belongs to a Φ-ideal, then it is finite-dimensional.

§52. Riesz operators

In this section let A be a continuous endomorphism of the normed space E over $\mathbf{K}$.

If A is even compact, then $\lambda I - A$ possesses for all $\lambda \neq 0$ the following properties:

(a) $\lambda I - A$ is open,
(b) $(\lambda I - A)(E)$ is closed,
(c) $\operatorname{ind}(\lambda I - A) = 0$,
(d) $p(\lambda I - A) = q(\lambda I - A) < \infty$;

furthermore we also have

(e) the eigenvalues of A form, in case there are infinitely many of them, a sequence which converges to zero (Theorem 40.1 and Proposition 43.2).

λ is called a *Riesz point* of the (arbitrary) continuous operator A, if for $\lambda I - A$ the assertions (a) through (d) hold. The set of all Riesz points of A forms the *Riesz region* P_A, the number

$$r_{\mathrm{P}}(A) := \inf\{r > 0: \lambda \text{ is a Riesz point of } A \text{ for all } |\lambda| > r\}$$

is the *Riesz radius*. A is called a *Riesz operator* if $r_{\mathrm{P}}(A) = 0$, i.e., if every $\lambda \neq 0$ is a Riesz point of A, and if, furthermore, the eigenvalues of A are distributed according to (e). Let $\mathrm{P}(E)$ be the set of all Riesz operators on E. We have $\mathscr{K}(E) \subset \mathrm{P}(E)$. Every endomorphism on a finite-dimensional space is a Riesz operator.

Proposition 52.1. *Let E be* complete. *Then $r_{\mathrm{P}}(A) = r_{\Phi}(A)$, and P_A consists precisely of those components of Φ_A which intersect $\rho(A)$. Furthermore A is a Riesz operator if and only if one of the following assertions holds*:

(I) *$r_{\mathrm{P}}(A)$ vanishes.*
(II) *For all $\lambda \neq 0$ the deficiencies $\alpha(\lambda I - A)$, $\beta(\lambda I - A)$ are finite.*

Proof. The assertions concerning $r_{\mathrm{P}}(A)$ and P_A follow easily from Theorem 51.1; observe only that because of the completeness of E the Riesz points are already characterized by properties (c) and (d) (Proposition 36.3 and Theorem 32.1). If A is a Riesz operator, then (I) and (II) are trivially satisfied. Conversely, if one of these assertions holds, then Φ_A has only one component $C \supset \mathbf{K}\setminus\{0\}$; by Theorem 51.1 all $\lambda \in C$ are Riesz points and the eigenvalues can have a limit point only at 0. ■

From the last proposition we obtain with the aid of Proposition 50.3, Proposition 51.8 and (51.9) immediately the following characterization of Riesz operators on complex Banach spaces:

Proposition 52.2. *Let E be a* complex *Banach space, $\mathscr{J}$ a closed Φ-ideal and $\tilde{A}$ the residue class of A in $\mathscr{L}(E)/\mathscr{J}$. Then the following assertions are equivalent:*

(a) *A is a Riesz operator,*
(b) $\lim[\inf_{K \in \mathscr{J}} \|A^n - K\|]^{1/n} = 0$,
(c) $r(\tilde{A}) = 0$, *or equivalently: $\tilde{A}$ is quasi-nilpotent,*
(d) *every spectral point $\lambda \neq 0$ of A is isolated, and the corresponding spectral projector is finite-dimensional.*

In (b) one may use also non-closed Φ-ideals. In (c) at first $r(\tilde{A})$ is not defined when E has finite dimension (Banach algebras should have unit elements different from zero!). Of course, $\tilde{A}$ exists in the algebra $\mathscr{L}(E)/\mathscr{J} = \{\tilde{0}\}$. If we set $r(\tilde{A}) := 0$, then the proposition trivially also holds in the finite-dimensional case.

We can now invoke Proposition 45.1 in order to make assertions concerning sums and products of Riesz operators. Here the following generalization of the concept of commutativity will be useful: If $\mathscr{J}$ is an ideal in $\mathscr{L}(E)$, then we say that A *$\mathscr{J}$-commutes* with B if $AB - BA \in \mathscr{J}$, i.e., if the residue classes of A and B commute in $\mathscr{L}(E)/\mathscr{J}$. We can leave the proof of the following proposition to the reader; observe only that operators which $\mathscr{J}$-commute also $\bar{\mathscr{J}}$-commute, and that the closure of a Φ-ideal is again a Φ-ideal (Proposition 51.7).

Proposition 52.3. *Let E be a complex Banach space and $\mathscr{J}$ a Φ-ideal in $\mathscr{L}(E)$. Then the following invariance assertions hold:*

(a) *Scalar multiples, sums and products of $\mathscr{J}$-*commuting *Riesz operators are again Riesz operators; in the case of products it is even sufficient that only one of the factors lies in* P(E).

(b) *The sum of a Riesz operator and of an arbitrary operator from $\mathscr{J}$ is a Riesz operator.*

The Riesz property is to a large extent stable with respect to the formation of functions. More precisely we have

Proposition 52.4. *Let E be a complex Banach space, $f \in \mathscr{H}(A)$ and $f(0) = 0$. Then together with A also $f(A)$ is a Riesz operator. The converse is true whenever f vanishes* only *at* 0.

Proof. Let A be a Riesz operator. Because of $f(0) = 0$ there exists a $g \in \mathscr{H}(A)$ such that $f(\lambda) = \lambda g(\lambda)$, i.e., $f(A) = Ag(A)$. Since A commutes with $g(A)$, it follows from here with the aid of Proposition 52.3a that $f(A)$ is a Riesz operator.

Now let $f(A)$ be a Riesz operator and assume that f vanishes only at 0. Then there exists an $h \in \mathscr{H}(A)$ and a natural number m such that for all λ of the set of definition of f we have:

$$f(\lambda) = \lambda^m h(\lambda) \quad \text{and} \quad h(\lambda) \neq 0.$$

Consequently $f(A) = A^m h(A)$, and $h(A)$ has an inverse in $\mathscr{L}(E)$. Since $f(A)$ and $h(A)^{-1}$ commute, we obtain from here with the help of Proposition 52.3a that $A^m = f(A)h(A)^{-1}$ is a Riesz operator. It now follows from Proposition 52.2c in combination with (45.4) that A itself is a Riesz operator. ■

If K is from a Φ-ideal, then $I - \lambda K$ is a Fredholm operator for all λ. With the aid of Proposition 52.1 we obtain therefore immediately:

Proposition 52.5. *If E is* complete, *then every operator from a Φ-ideal is a Riesz operator.*

Because of this proposition Φ-ideals are also often called *Riesz ideals* or *ideals of Riesz operators.* In general, $\mathrm{P}(E)$ is not a Riesz ideal.

The requirement of $\mathscr{J}$-commutation is the weaker—and thus the efficiency of Proposition 52.3 the greater—the larger $\mathscr{J}$ is. It is therefore reasonable to ask whether there exists a Riesz ideal containing all Φ-ideals. The answer is positive and we are led to it as follows.

In the first place we may disregard finite-dimensional spaces since for them $\mathrm{P}(E) = \mathscr{L}(E)$. Thus let $\dim E = \infty$ so that $\hat{\mathscr{L}} := \mathscr{L}(E)/\mathscr{F}(E)$ is an algebra with a unit element $\neq$ the zero element. If $\mathscr{J}$ is a Φ-ideal and $K \in \mathscr{J}$, then $I - AK$ is a Fredholm operator for every $A \in \mathscr{L}(E)$, consequently $\hat{I} - \hat{A}\hat{K}$ is invertible for every $\hat{A} \in \hat{\mathscr{L}}$ (Proposition 25.2) (we denote by $\hat{T}$ the residue class of $T \in \mathscr{L}(E)$ in $\hat{\mathscr{L}}$). The set

$$\mathfrak{r} := \{\hat{R} \in \hat{\mathscr{L}} : \hat{I} - \hat{A}\hat{R} \text{ is invertible for every } \hat{A} \in \hat{\mathscr{L}}\}$$

is the 'radical' of $\hat{\mathscr{L}}$; namely, if $\mathscr{A}$ is an algebra with unit element $e \neq 0$, then we call

$$\{r \in \mathscr{A} : e - ar \text{ is invertible for every } a \in \mathscr{A}\}$$

the *radical* of $\mathscr{A}$. We are interested here in the fact that the radical is a proper (two-sided) ideal of $\mathscr{A}$ (see e.g. [14]). If h is the canonical homomorphism from $\mathscr{L}(E)$ onto $\hat{\mathscr{L}}$, then the image $h(\mathscr{J})$ of every Φ-ideal $\mathscr{J}$ lies in the radical $\mathfrak{r}$ of $\hat{\mathscr{L}}$; conversely $h^{-1}(\mathfrak{r})$ is obviously a (proper) Φ-ideal. Thus $h^{-1}(\mathfrak{r})$ is a maximal Φ-ideal, containing all Φ-ideals, and is so uniquely determined and closed (otherwise the closure $\overline{h^{-1}(\mathfrak{r})}$ would be, according to Proposition 51.7, a Φ-ideal which contains $h^{-1}(\mathfrak{r})$ properly). We summarize:

Proposition 52.6. *If E is an infinite-dimensional Banach space over* **K**, *then there exists in $\mathscr{L}(E)$ exactly one maximal Riesz ideal containing all Riesz ideals. This ideal is closed and equal to $h^{-1}(\mathfrak{r})$, where h is the canonical homomorphism from $\mathscr{L}(E)$ onto $\mathscr{L}(E)/\mathscr{F}(E)$ and $\mathfrak{r}$ is the radical of the latter algebra.*

The next proposition deals with restrictions of Riesz operators.

Proposition 52.7. *Let A be a Riesz operator on the complex Banach space E, and F a closed subspace of E invariant under A. Then the restriction $\tilde{A}$ of A to F is a Riesz operator and $\sigma(\tilde{A}) \subset \sigma(A)$.*

Proof. For $|\lambda| > r(A)$ the resolvent R_λ of A possesses the representation $R_\lambda = \sum_{n=0}^{\infty} A^n/\lambda^{n+1}$ (Theorem 44.1) from where $R_\lambda(F) \subset F$ immediately follows. For every $x \in F$ and $x' \in F^\perp = \{y' \in E' : \langle y, y' \rangle = 0 \text{ for all } y \in F\}$ the function $\lambda \mapsto \langle R_\lambda x, x' \rangle$, which is holomorphic on $\rho(A)$, vanishes outside the spectral disk of A. Since $\rho(A)$ is connected, it follows from here, by the identity theorem for holomorphic functions, that $\langle R_\lambda x, x' \rangle = 0$ for all $\lambda \in \rho(A)$. Therefore $R_\lambda x$ lies in $F^{\perp\perp} = F$ (Proposition 29.2). We conclude from here that every element $x \in F$ is the image of an element of F under $\lambda I - A$, namely of $R_\lambda x$. Since on the other hand F is invariant under $\lambda I - A$, we obtain the equation $(\lambda I - A)(F) = F$ for all $\lambda \in \rho(A)$. It follows immediately that $\rho(A) \subset \rho(\tilde{A})$ and therefore $\sigma(\tilde{A}) \subset \sigma(A)$, and furthermore we obtain that for $\lambda \in \rho(A)$ the resolvent $\tilde{R}_\lambda$ of $\tilde{A}$ is the restriction of R_λ to F. Now let $\lambda_0 \neq 0$ be a spectral point of $\tilde{A}$. Because of $\sigma(\tilde{A}) \subset \sigma(A)$, the point λ_0 is isolated in $\sigma(\tilde{A})$ and in $\sigma(A)$. The spectral projector $\tilde{P} \in \mathscr{L}(F)$ corresponding to $\{\lambda_0\}$ and to $\tilde{A}$ is given by

$$\tilde{P} = \frac{1}{2\pi i} \int_\Gamma \tilde{R}_\lambda \, d\lambda,$$

where Γ is a circle around λ_0 which isolates λ_0 from $\sigma(A)\backslash\{\lambda_0\}$. For every $x \in F$ we thus have, according to Exercise 1 in §49,

$$\tilde{P}x = \frac{1}{2\pi i} \int_\Gamma \tilde{R}_\lambda x \, d\lambda = \frac{1}{2\pi i} \int_\Gamma R_\lambda x \, d\lambda = Px,$$

where P is the spectral projector corresponding to $\{\lambda_0\}$ and A. Thus $\tilde{P}$ is the restriction of P to F. It now follows immediately from Proposition 52.2 that $\tilde{A}$ is a Riesz operator. ■

We can now clarify without effort the behavior of Riesz operators with respect to duality:

Proposition 52.8. *A continuous endomorphism A of the complex Banach space is a Riesz operator if and only if the dual transformation A' is a Riesz operator.*

Proof. Let A be a Riesz operator. Then $\lambda I - A$ is for every $\lambda \neq 0$ a Fredholm operator in $\mathscr{L}(E)$ (Proposition 37.1), consequently $\lambda I' - A'$ is for all these λ a Fredholm operator in $\mathscr{L}(E')$ (Proposition 27.3). Since E' is complete, A' must be a Riesz operator by Proposition 52.1.

Now let conversely A' be a Riesz operator. By what we have just proved, the bidual transformation A'' is then also a Riesz operator. Since its restriction to the closed subspace E of E'' is exactly A because of (41.5), it follows from Proposition 52.7 that A itself must be a Riesz operator. ■

Exercises

$^{+}$1. Prove with the help of only Proposition 52.2 the following assertion, which follows of course also from Proposition 52.4: If a power A^n is compact, then A is a Riesz operator (we are in a complex Banach space).

2. Cf. §50 Exercise 2. The spectrum of a Riesz operator A on a complex, infinite-dimensional Banach space contains at least the point 0. The point 0 is a pole of the resolvent of A if and only if a power A^n ($n \geqq 1$) is finite-dimensional.

$^{+}$3. The limit of a uniformly convergent sequence of commuting Riesz operators on a complex Banach space is a Riesz operator. Can the assumption of commutation be weakened? *Hint*: Exercise 1 in §45.

4. In a saturated operator algebra $\mathscr{A}(E)$ there exists exactly one Φ-ideal which contains all Φ-ideals.

$^{+}$5. A continuous endomorphism A of a Banach space is a Riesz operator if and only if $\lambda I - A$ is an Atkinson operator for every $\lambda \neq 0$. *Hint*: Exercise 3 in §37. If one invokes Exercise 4 in §37 then one can even say: A is a Riesz operator if and only if $\lambda I - A$ is a semi-Fredholm operator for all $\lambda \neq 0$. As a consequence, because of Proposition 36.3, the following proposition is valid: A is a Riesz operator if and only if $\beta(\lambda I - A)$ is finite for all $\lambda \neq 0$.

§53. Essential spectra

In this section let E be a Banach space over $\mathbf{C}$ *and* $A \in \mathscr{L}(E)$. The points of the resolvent set of A are in a certain sense the 'good' points of A: For $\lambda \in \rho(A)$, and only such a λ, the equation $(\lambda I - A)x = y$ is uniquely solvable for an arbitrary y (furthermore the solution depends continuously on the right-hand side). Now one could think to consider, besides the elements of $\rho(A)$, also such points λ as 'good' for which $\lambda I - A$ (or the equation $(\lambda I - A)x = y$) has a behavior which can be easily described; the points which are not 'good' would then form a subset of the spectrum, an 'essential spectrum'. Examples of sets of 'good' points are for instance the following:

$$\rho_1(A) := \{\lambda: \operatorname{ind}(\lambda I - A) = 0 \text{ and } p(\lambda I - A) = q(\lambda I - A) < \infty\} = \mathrm{P}_A,$$

$$\rho_2(A) := \{\lambda: \operatorname{ind}(\lambda I - A) = 0\},$$

$$\rho_3(A) := \{\lambda: \operatorname{ind}(\lambda I - A) \text{ exists}\} = \Phi_A,$$

$$\rho_4(A) := \{\lambda: \lambda I - A \text{ is normally solvable}\}.$$

Setting $\sigma_k(A) := \mathbf{C} \backslash \rho_k(A)$ $(k = 1, \ldots, 4)$, we then obtain four 'essential spectra'. We have

$$\rho(A) \subset \rho_1(A) \subset \rho_2(A) \subset \rho_3(A) \subset \rho_4(A),$$

hence

$$\sigma(A) \supset \sigma_1(A) \supset \sigma_2(A) \supset \sigma_3(A) \supset \sigma_4(A).$$

Each of these 'essential spectra' is obtained by removing from $\sigma(A)$ points which according to a certain criterion are still considered to be 'good'. Obviously

$$\sigma_3(A) = \sigma(\hat{A}), \qquad \hat{A} \text{ the residue class of } A \text{ in } \mathscr{L}(E)/\mathscr{K}(E).$$

If E is finite-dimensional, then all 'essential spectra' are empty; in the case of infinite dimension

$$\sigma_k(A) = \{0\} \qquad (k = 1, 2, 3) \text{ for every Riesz operator } A.$$

In this section we want to consider $\sigma_2(A)$ more closely. We call $\sigma_2(A)$ *the essential spectrum* of A and denote it henceforth by $\sigma_e(A)$. Correspondingly, let $\rho_e(A) := \rho_2(A)$, i.e.,

$$\rho_e(A) := \{\lambda\colon \operatorname{ind}(\lambda I - A) = 0\}, \qquad \sigma_e(A) = \mathbf{K} \backslash \rho_e(A).$$

$\sigma_e(A)$ is also called the *Weyl spectrum* of A.

Proposition 53.1. *$\sigma_e(A)$ is closed and lies entirely in the closed disk around* 0 *with radius $r_\Phi(A)$* (see (51.7)). *If E is infinite-dimensional, then at least one point of $\sigma_e(A)$ lies on the boundary of this disk.*

Proof. $\rho_e(A)$ consists of all the components of the Fredholm region Φ_A on which $\operatorname{ind}(\lambda I - A)$ vanishes (Theorem 51.1), it is therefore open; consequently $\sigma_e(A)$ is closed and lies in the disk indicated. Now let $\dim E = \infty$. Then, with the notation of the proof of Proposition 51.8, we have $r_\Phi(A) = r(\tilde{A})$ (see (51.9)), on the circle $|\lambda| = r(\tilde{A})$ there lies a point λ_0 from $\sigma(\tilde{A})$ (see §45), λ_0 does not belong to Φ_A (Proposition 51.6), and a fortiori not to $\rho_e(A)$. ∎

The next proposition will show that the points of the essential spectrum of A are firmly rooted in $\sigma(A)$: They cannot be removed from $\sigma(A)$ by adding to A operators from a Φ-ideal (e.g., compact operators). For the proof we need a simple proposition concerning the preservation of the index:

Proposition 53.2. *For every $T \in \Phi(E)$ and for every K from a Φ-ideal in $\mathscr{L}(E)$ we have*

$$\operatorname{ind}(T + K) = \operatorname{ind} T.$$

This proposition is proved exactly like Proposition 23.3; observe only that all operators which figure there can in the present case be chosen to be continuous and that $\operatorname{ind}(I - K) = 0$ (Proposition 52.5). ∎

Proposition 53.3. *For any Φ-ideal $\mathscr{J}$ in $\mathscr{L}(E)$ we have*

$$\sigma_e(A) = \bigcap_{K \in \mathscr{J}} \sigma(A + K).$$

We prove the proposition in the following formulation: $\lambda \in \rho_e(A) \Leftrightarrow$ there exists a $K \in \mathcal{J}$ with $\lambda \in \rho(A + K)$. If $\lambda \in \rho_e(A)$, then by Proposition 26.2 there exists a bijective $R \in \mathcal{L}(E)$ and a $K \in \mathcal{F}(E) \subset \mathcal{J}$ with $\lambda I - A = R + K$. Consequently $\lambda I - (A + K) = R$ is bijective, hence $\lambda \in \rho(A + K)$. Now let conversely $\lambda \in \rho(A + K)$ for some $K \in \mathcal{J}$. Then $\operatorname{ind}(\lambda I - A - K) = 0$, hence by Proposition 53.2 we also have $\operatorname{ind}(\lambda I - A) = 0$ and thus λ lies in $\rho_e(A)$. ■

Exercises

$^+$1. A Φ-ideal $\mathcal{J}$ in the operator algebra $\mathcal{A}(E)$ (E an arbitrary vector space) is said to be a Φ_0-ideal if $\operatorname{ind}(I - K)$ vanishes for all $K \in \mathcal{J}$. If $\mathcal{A}(E)$ is saturated, then for every Fredholm operator A in $\mathcal{A}(E)$ and for every K from a Φ-ideal $\mathcal{J}$ also $A + K \in \Phi(A(E))$; if $\mathcal{J}$ is even a Φ_0-ideal, then we have $\operatorname{ind}(A + K) = \operatorname{ind}(A)$. *Hint*: Proposition 51.6 and proof of Proposition 23.3. Examples of Φ_0-ideals are $\mathcal{F}(E)$ and $\mathcal{K}(E)$; cf. also Proposition 52.5.

2. Let $\mathcal{A}(E)$ be a saturated operator algebra on E and

$$\Phi_0 := \{A \in \Phi(\mathcal{A}(E)) : \operatorname{ind}(A) = 0\}.$$

Show the following: (a) An ideal $\mathcal{J}$ in $\mathcal{A}(E)$ is a Φ_0-ideal $\Leftrightarrow \Phi_0 + \mathcal{J} \subset \Phi_0$. (b) There exists exactly one maximal Φ_0-ideal $\mathcal{J}_m$ in $\mathcal{A}(E)$; every Φ_0-ideal is contained in $\mathcal{J}_m$. *Hint* for (b): Apply Zorn's lemma to the non-empty set of Φ_0-ideals ordered by inclusion and use (a) for the proof of uniqueness.

3. The representation theorems 26.2 and 39.1 for Fredholm operators in saturated operator algebras are valid without change if we replace the Φ_0-ideal $\mathcal{F}(\mathcal{A}(E))$ by an arbitrary Φ_0-ideal. Use Exercise 1.

$^+$4. Let $\mathcal{A}(E)$ be a saturated operator algebra on the vector space E over **K** and $A \in \mathcal{A}(E)$. Let the *spectrum* $\sigma(A)$ of A be the complement of

$$\rho(A) := \{\lambda : (\lambda I - A)^{-1} \in \mathcal{A}(E)\}$$

in **K**, the *essential spectrum* $\sigma_e(A)$ the complement of

$$\rho_e(A) := \{\lambda : \lambda I - A \in \Phi(\mathcal{A}(E)), \operatorname{ind}(\lambda I - A) = 0\}.$$

Show that for every Φ-ideal $\mathcal{J}$ in $\mathcal{A}(E)$ one has $\sigma_e(A) = \bigcap_{K \in \mathcal{J}} \sigma(A + K)$. *Hint*: Exercises 1 and 3.

§54. Normaloid operators

Let A be throughout a continuous endomorphism of the complex Banach space E. The inequality

$$r(A) \leqq \|A^n\|^{1/n} \leqq \|A\| \tag{54.1}$$

suggests to consider operators A for which

$$r(A) = \|A\| \qquad \text{or equivalently} \qquad \|A^n\| = \|A\|^n \text{ for } n = 1, 2, \ldots; \tag{54.2}$$

such A are called *normaloid*. The explanation of the name is that so-called normal operators on complex Hilbert spaces satisfy equation (54.2), as we shall see later. At present we will content ourselves with proving that *hermitian matrices (or operators) A on $E = l^2(n)$ are normaloid* (we identify again (n, n)-matrices with the operators they generate; see the remarks following the proof of Proposition 47.2). As known, $A := (\alpha_{jk})$ is said to be *hermitian* if one of the following equivalent conditions is satisfied:

$$\alpha_{jk} = \bar{\alpha}_{kj} \quad \text{for} \quad j, k = 1, \ldots, n$$

(54.3)

$$\text{or} \quad (Ax \mid y) = (x \mid Ay) \quad \text{for all} \quad x, y \in l^2(n);$$

here $(x \mid y) := \sum_{\nu=1}^{n} \xi_\nu \bar{\eta}_\nu$ is the canonical inner product of the vectors

$$x = (\xi_1, \ldots, \xi_n), \qquad y = (\eta_1, \ldots, \eta_n).$$

The inner product is related to the norm on $l^2(n)$ by the equation $\|x\|^2 = \sum_{\nu=1}^{n} |\xi_\nu|^2 = \sum_{\nu=1}^{n} \xi_\nu \bar{\xi}_\nu = (x \mid x)$ and by the Cauchy–Schwarz inequality

$$|(x \mid y)| \leqq \sum_{\nu=1}^{n} |\xi_\nu| |\bar{\eta}_\nu| \leqq \left(\sum_{\nu=1}^{n} |\xi_\nu|^2\right)^{1/2} \left(\sum_{\nu=1}^{n} |\eta_\nu|^2\right)^{1/2} = \|x\| \, \|y\|.$$

For every continuous endomorphism A we have $\|A^2\| \leqq \|A\|^2$; if A is furthermore hermitian on $l^2(n)$ then $\|Ax\|^2 = (Ax \mid Ax) = (A^2 x \mid x) \leqq \|A^2 x\| \, \|x\| \leqq \|A^2\| \, \|x\|^2$ and therefore $\|A\|^2 \leqq \|A^2\|$, thus $\|A^2\| = \|A\|^2$. Since together with A obviously every power A^n is hermitian, it follows that $\|A^{2n}\| = \|A\|^{2n}$, hence $r(A) = \lim \|A^{2^n}\|^{1/2^n} = \|A\|$. Thus a hermitian A is indeed normaloid.

Hermitian operators on $l^2(n)$ have the valuable property that an expansion theorem holds for them, more precisely: There exist n eigensolutions $y_1, \ldots, y_n$ to the eigenvalues $\alpha_1, \ldots, \alpha_n$ so that $(y_\nu \mid y_\mu) = \delta_{\nu\mu}$ and that for every $x \in l^2(n)$

$$\text{(54.4)} \qquad x = \sum_{\nu=1}^{n} (x \mid y_\nu) y_\nu \quad \text{hence} \quad Ax = \sum_{\nu=1}^{n} \alpha_\nu (x \mid y_\nu) y_\nu;$$

the second equation is called the *expansion theorem.* In the finite sequence $\alpha_1, \ldots, \alpha_n$ each eigenvalue of A occurs as often as its multiplicity indicates. If $\lambda_1, \ldots, \lambda_m$ are the different ones among the $\alpha_1, \ldots, \alpha_n$ and if in (54.4) the order is fixed in such a way that $\lambda_1 = \alpha_1 = \cdots = \alpha_{k_1}$, $\lambda_2 = \alpha_{k_1+1} = \cdots = \alpha_{k_2}, \ldots$ then we define by

$$P_1 x := \sum_{\nu=1}^{k_1} (x \mid y_\nu) y_\nu, \qquad P_2 x := \sum_{\nu=k_1+1}^{k_2} (x \mid y_\nu) y_\nu, \ldots$$

projectors $P_1, \ldots, P_m$; P_μ projects $l^2(n)$ onto the eigenspace of λ_μ. Thus we have $AP_\mu x = \lambda_\mu P_\mu x$ and $Ax = \sum_{\mu=1}^{m} \lambda_\mu P_\mu x$ or shorter:

$$\text{(54.5)} \qquad A = \sum_{\mu=1}^{m} \lambda_\mu P_\mu \quad \text{and} \quad AP_\mu = \lambda_\mu P_\mu \quad \text{for} \quad \mu = 1, \ldots, m.$$

The question arises whether for normaloid operators on an arbitrary complex Banach space E there also exist projectors P_μ so that equations of the form

(54.5) are valid (where, of course, the finite sums are replaced by infinite series). We shall show that under certain conditions this is indeed the case. First we present a new characterization of normaloid operators.

Proposition 54.1. *$A \neq 0$ is normaloid if and only if the sequence of numbers $\|A^n x\|/r^n(A)$ $(n = 0, 1, \ldots)$ is* monotone *decreasing.*

Namely, if A is normaloid, then

$$\frac{\|A^{n+1}x\|}{r^{n+1}(A)} \leqq \frac{\|A\|\,\|A^n x\|}{r(A)r^n(A)} = \frac{\|A^n x\|}{r^n(A)} \qquad \text{for} \quad n = 0, 1, \ldots.$$

Conversely, assume that the monotonicity condition is satisfied. Then in particular $\|Ax\|/r(A) \leqq \|x\|$ for all x, hence $\|A\| \leqq r(A)$ and thus $\|A\| = r(A)$. ■

The set $\sigma_\pi(A) = \sigma(A) \cap \{\lambda \in \mathbf{C} : |\lambda| = r(A)\}$ is called the *peripheral spectrum* of A. Interesting assertions can be made concerning the peripheral spectrum of a normaloid operator.

Proposition 54.2. *If $A \neq 0$ is normaloid, then for $\lambda \in \sigma_\pi(A)$ we have*

$$p(\lambda I - A) \leqq 1 \qquad \textit{and} \qquad \beta(\lambda I - A) > 0.$$

Proof. If $(\lambda I - A)^2 x = 0$ but $(\lambda I - A)x \neq 0$ for a certain $x \in E$, then $y := (1/\lambda)(\lambda I - A)x$ is different from zero and

$$A^n x = [\lambda I - (\lambda I - A)]^n x = \lambda^n x - n(\lambda I - A)\lambda^{n-1} x = n\lambda^n\left(\frac{x}{n} - y\right),$$

hence

$$\frac{\|A^n x\|}{r^n(A)} = \frac{\|A^n x\|}{|\lambda|^n} = n\left\|\frac{x}{n} - y\right\| \geqq n\left|\frac{\|x\|}{n} - \|y\|\right|,$$

from where it follows because of $y \neq 0$ that $\|A^n x\|/r^n(A) \to \infty$. Since this contradicts Proposition 54.1, there cannot exist an x of the above kind, we have rather $N[(\lambda I - A)^2] \subset N(\lambda I - A)$ and therefore $p(\lambda I - A) \leqq 1$. If we had $\beta(\lambda I - A) = 0$, then it would follow with the help of Proposition 38.5 that also $\alpha(\lambda I - A) = 0$, and therefore the spectral point λ would lie in $\rho(A)$. This contradiction shows that we must have $\beta(\lambda I - A) > 0$. ■

From Proposition 54.2 and 50.2 we get immediately:

Proposition 54.3. *If $A \neq 0$ is normaloid, then $\lambda \in \sigma_\pi(A)$ is a pole of the resolvent R_λ if and only if $q(\lambda I - A)$ is finite; in this case the pole λ has order* 1.

The operator A is said to be *meromorphic* if its non-zero spectral points are poles of the resolvent. *Compact and, more generally, Riesz operators are meromorphic.* A meromorphic operator possesses at most countably many spectral points $\lambda_1, \lambda_2, \ldots$ which we always consider arranged according to decreasing absolute values:

(54.6) $$|\lambda_1| \geqq |\lambda_2| \geqq \cdots .$$

With this ordering we have obviously $\sigma_\pi(A) = \{\lambda_1, \ldots, \lambda_n\}$ for an appropriate index n. Each λ_ν is an eigenvalue of A (Proposition 50.2). Let P_ν be the spectral projector associated with $\{\lambda_\nu\}$ in what follows. In the case $\lambda_\nu \in \sigma_\pi(A)$ we can make an important assertion concerning its norm:

Proposition 54.4. *If $A \neq 0$ is normaloid and meromorphic, then the spectral projectors P_ν corresponding to the poles $\lambda_\nu \in \sigma_\pi(A)$ have norm* 1, *furthermore*

(54.7) $$\|x\| \leqq \|x + y\| \qquad \text{for} \quad x \in N(\lambda_\nu I - A), \qquad y \in (\lambda_\nu I - A)(E).$$

This inequality is valid in particular if y is a linear combination of eigenvectors belonging to eigenvalues $\neq \lambda_\nu$.

Proof. Let $\sigma_\pi(A) = \{\lambda_1, \ldots, \lambda_n\}$ and C_ν a circle around λ_ν with such a small radius r that C_ν lies entirely in $\rho(A)$ and λ_ν is the only spectral point in its interior. Since the pole λ_ν is simple according to Proposition 54.3, it follows from (50.10) for $\lambda_0 = \lambda_\nu$ and $p = 1$ that $AP_\nu x = \lambda_\nu P_\nu x$ for all $x \in E$; consequently, $AP_\nu = \lambda_\nu P_\nu$ and more generally

(54.8) $$A^k P_\nu = \lambda_\nu^k P_\nu \quad \text{hence} \quad \frac{1}{2\pi i} \int_{C_\nu} \lambda^k R_\lambda \, d\lambda = \lambda_\nu^k P_\nu \qquad (k = 0, 1, \ldots)$$

(see Exercise 6 in §49). We can represent A^k in the form

$$A^k = \sum_{\nu=1}^{n} \frac{1}{2\pi i} \int_{C_\nu} \lambda^k R_\lambda \, d\lambda + \frac{1}{2\pi i} \int_C \lambda^k R_\lambda \, d\lambda = \sum_{\nu=1}^{n} \lambda_\nu^k P_\nu + \frac{1}{2\pi i} \int_C \lambda^k R_\lambda \, d\lambda;$$

here C is an appropriate circle around 0 with radius $\rho < |\lambda_1| = r(A)$. If as an abbreviation we set $A_k := (1/2\pi i) \int_C \lambda^k R_\lambda \, d\lambda$, then

(54.9) $$A^k = \sum_{\nu=1}^{n} \lambda_\nu^k P_\nu + A_k, \qquad \|A_k\| \leqq \rho\rho^k \max_{\lambda \in C} \|R_\lambda\| = \gamma\rho^k$$

and therefore

$$\left\| \sum_{k=1}^{m} \frac{A_k}{\lambda_1^k} \right\| \leqq \gamma \sum_{k=1}^{m} \left(\frac{\rho}{|\lambda_1|} \right)^k \leqq \frac{\gamma}{1 - \dfrac{\rho}{|\lambda_1|}} \qquad \text{for all natural} \quad m.$$

Furthermore, setting $\alpha := \min_{\nu=2}^{n} |1 - (\lambda_\nu/\lambda_1)|$ and $\mu := \max_{\nu=2}^{n} \|P_\nu\|$, we have for $m = 1, 2, \ldots$

$$\left\| \sum_{k=1}^{m} \sum_{\nu=2}^{n} \left(\frac{\lambda_\nu}{\lambda_1} \right)^k P_\nu \right\| = \left\| \sum_{\nu=2}^{n} \sum_{k=1}^{m} \left(\frac{\lambda_\nu}{\lambda_1} \right)^k P_\nu \right\| = \left\| \sum_{\nu=2}^{n} \frac{\lambda_\nu}{\lambda_1} \frac{1 - \left(\dfrac{\lambda_\nu}{\lambda_1} \right)^m}{1 - \dfrac{\lambda_\nu}{\lambda_1}} P_\nu \right\| \leqq \frac{2n\mu}{\alpha}.$$

From (54.9) it now follows with the aid of the last two estimates that

$$(54.10)\qquad \frac{1}{m}\sum_{k=1}^{m}\left(\frac{A}{\lambda_1}\right)^k = P_1 + \frac{1}{m}\sum_{k=1}^{m}\sum_{\nu=2}^{n}\left(\frac{\lambda_\nu}{\lambda_1}\right)^k P_\nu + \frac{1}{m}\sum_{k=1}^{m}\frac{A_k}{\lambda_1^k} \Rightarrow P_1$$

as $m \to \infty$. Since

$$\left\|\frac{1}{m}\sum_{k=1}^{m}\left(\frac{A}{\lambda_1}\right)^k\right\| \leqq \frac{1}{m}\sum_{k=1}^{m}\left(\frac{\|A\|}{|\lambda_1|}\right)^k = \frac{1}{m}m = 1,$$

we obtain from (54.10) that also $\|P_1\| \leqq 1$. But then we have even $\|P_1\| = 1$ since because of $0 < \|P_1\| = \|P_1^2\| \leqq \|P_1\|^2$ we have $\|P_1\| \geqq 1$. Thus the first assertion of the proposition is proved. From it follows (54.7) with the aid of Propositions 50.2 and 54.3, and, if we further invoke Proposition 43.1c, also the last assertion of the proposition. ∎

In order to be able to make assertions also concerning spectral points which do not lie on the boundary of the spectral disk, we must restrict the class of normaloid operators. We say that A is *spectrally normaloid* if for every spectral set $\sigma \subset \sigma(A)$ the restriction of A to the invariant subspace $M_\sigma := P_\sigma(E)$ (see Theorem 49.1) is again normaloid. *Hermitian operators on $l^2(n)$ are spectrally normaloid*: indeed, because of the second equation in (54.3) they are hermitian even on every invariant subspace. A second example will be given in Proposition 54.6.

Proposition 54.5. *If A is spectrally normaloid and meromorphic, then* every *pole of the resolvent R_λ is simple. Furthermore*

$$(54.11)\qquad \|P_n x\| \leqq \|x\| \quad \text{for} \quad x \in \bigcap_{\nu=1}^{n-1} N(P_\nu), \qquad n > 1, \quad \text{and}$$

$$(54.12)\qquad \|P_n\| \leqq 2^{n-1} \quad \text{for} \quad n = 1, 2, \ldots.$$

Proof. Obviously we may assume that $A \neq 0$. By Proposition 54.3 the pole λ_1 has order 1. We first prove the analogous fact for the poles λ_n $(n > 1)$ which are $\neq 0$. Let $\sigma := \sigma(A)\setminus\{\lambda_1, \ldots, \lambda_{n-1}\}$. According to Theorem 49.1, λ_n is an isolated, peripheral spectral point of the restriction A_σ of A to M_σ. By restricting the Laurent expansion of R_λ around λ_n to M_σ we see that λ_n is a pole of the resolvent of A_σ and that λ_n as a pole of the resolvent of A has the same order as it has as a pole of the resolvent of A_σ. But the latter order is 1 since $A \neq 0$ is normaloid (Proposition 54.3). Now we consider the case that 0 is a pole of R_λ. Let p be its order and set $\sigma := \{0\}$. Then by Proposition 50.2 we have $M_\sigma = N(A^p)$, the normaloid restriction A_σ of A to M_σ is nilpotent and thus $\|A_\sigma\| = r(A_\sigma) = 0$. It follows that $N(A^p) = N(A)$; hence $p = 1$. By a similar splitting-off procedure, whose details can be left to the reader, we obtain with the aid of Proposition 54.4 and Exercise 5 in §49 the estimate (54.11). We now pass to the proof of (54.12). For $n = 1$ we obtain the assertion from Proposition

54.4. Now let $n > 1$. Because of $P_\mu P_\nu = \delta_{\mu\nu} P_\mu$ the vector $(I - \sum_{\nu=1}^{n-1} P_\nu)x$ lies in $\bigcap_{\mu=1}^{n-1} N(P_\mu)$; from §49 Exercise 5 and (54.11) it follows therefore that

$$\|P_n x\| = \left\| P_n\left(I - \sum_{\nu=1}^{n-1} P_\nu\right)x \right\| \leqq \left\| \left(I - \sum_{\nu=1}^{n-1} P_\nu\right)x \right\| \qquad \text{for all} \quad x \in E,$$

and thus

$$\|P_n\| \leqq \left\| I - \sum_{\nu=1}^{n-1} P_\nu \right\| \leqq 1 + \sum_{\nu=1}^{n-1} \|P_\nu\|.$$

With the help of this estimate we obtain (54.12) by mathematical induction on n. ■

We now state the announced expansion theorem which represents the analogue to (54.4):

Theorem 54.1. *If $A \neq 0$ is spectrally normaloid and meromorphic, and if P_n is the spectral projector corresponding to the eigenvalue $\lambda_n \neq 0$ onto the eigenspace $N(\lambda_n I - A)$, then the* uniformly *convergent expansion*

$$A = \sum_{n=1}^{\infty} \lambda_n P_n \tag{54.13}$$

is valid whenever one of the following conditions is satisfied:

(a) *The sequence of norms $\|\sum_{\nu=1}^{n} P_\nu\|$ $(n = 1, 2, \ldots)$ is bounded;*
(b) *$2^n \lambda_n \to 0$ as $n \to \infty$;*
(c) *$\|P_n\| = 1$ for $n = 1, 2, \ldots$, and $n\lambda_n \to 0$.*

The last two conditions are meaningful only if A has infinitely many eigenvalues. If the set of eigenvalues is finite, then the series (54.13) is to be interpreted as a finite sum.

Proof. We assume that A has infinitely many eigenvalues. If

$$\sigma := \sigma(A) \setminus \{\lambda_1, \ldots, \lambda_{n-1}\}$$

and A_σ is the normaloid restriction of A to M_σ, then $\|A_\sigma\| = r(A_\sigma) = |\lambda_n|$ (see Theorem 49.1). It follows that

$$\left\| A\left(I - \sum_{\nu=1}^{n-1} P_\nu\right)x \right\| \leqq |\lambda_n| \left\| \left(I - \sum_{\nu=1}^{n-1} P_\nu\right)x \right\| = \left\| \lambda_n\left(I - \sum_{\nu=1}^{n-1} P_\nu\right)x \right\|,$$

consequently

$$\left\| Ax - \sum_{\nu=1}^{n-1} \lambda_\nu P_\nu x \right\| \leqq |\lambda_n| \left(1 + \sum_{\nu=1}^{n-1} \|P_\nu\|\right) \|x\|.$$

From here we obtain directly or with the help of the estimate (54.12) the uniform convergence of the expansion (54.13) in case (a), (c) or (b) are verified, respectively. The reader will now easily settle the case of a finite set of eigenvalues. ■

A is said to be *paranormal* if for every $x \in E$ the inequality

$$\|Ax\|^2 \leq \|A^2x\| \, \|x\| \tag{54.14}$$

holds. At the beginning of this section we have seen that hermitian operators on $l^2(n)$ are paranormal. We shall now show that paranormal operators are normaloid on every closed invariant subspace; a fortiori they are spectrally normaloid. We first prove a lemma.

Lemma 54.1. *If A is paranormal, then so is every power A^n $(n = 1, 2, \ldots)$.*

Proof. From (54.14) we get for $k = 0, 1, \ldots$ the estimate

$$\frac{\|A^{k+1}x\|}{\|A^kx\|} \leqq \frac{\|A^{k+2}x\|}{\|A^{k+1}x\|}.$$

From here follows that

$$\frac{\|A^nx\|}{\|x\|} = \frac{\|Ax\|}{\|x\|}\frac{\|A^2x\|}{\|Ax\|}\cdots\frac{\|A^nx\|}{\|A^{n-1}x\|}$$

$$\leqq \frac{\|A^{n+1}x\|}{\|A^nx\|}\frac{\|A^{n+2}x\|}{\|A^{n+1}x\|}\cdots\frac{\|A^{2n}x\|}{\|A^{2n-1}x\|} = \frac{\|A^{2n}x\|}{\|A^nx\|},$$

and with it the asserted inequality $\|A^nx\|^2 \leqq \|(A^n)^2x\| \, \|x\|$. ■

With this lemma we obtain exactly like in the discussion of hermitian operators at the beginning of this section:

Proposition 54.6. *A paranormal operator A is normaloid, and since A is obviously paranormal on every closed invariant subspace, A must be normaloid also on all these subspaces. In particular, A is spectrally normaloid.*

In this section we presented essentially unpublished investigations of F. V. Atkinson and H. Heuser.

Exercises

1. A spectrally normaloid Riesz operator, which satisfies one of the conditions (a) through (c) of Theorem 54.1, is compact.
2. The inverse of a paranormal operator is also paranormal if it exists on all of E.

VIII

Approximation problems in normed spaces

§55. An approximation problem

Our functional analytic studies have been motivated so far mainly by the problem of solving equations. We want to approach now a completely different group of questions and begin by formulating two important problems of analysis and by extracting their functional analytic core.

The Weierstrass approximation theorem states that a continuous function x on the interval $[a, b]$ can be approximated arbitrarily well by polynomials, uniformly on $[a, b]$. For computational reasons, in applied mathematics it is not so much the *arbitrarily close* approximability by *all* polynomials on what the interest is centered, but rather on the problem for a given natural number n to approximate x uniformly *as well as possible* by polynomials $p(t) := \alpha_0 + \alpha_1 t + \cdots + \alpha_n t^n$ *of degree* $\leqq n$; we ask therefore whether in the set P_n of all polynomials of degree $\leqq n$ there exists a q such that

$$\max_{a \leqq t \leqq b} |x(t) - q(t)| \leqq \max_{a \leqq t \leqq b} |x(t) - p(t)| \qquad \text{for all} \quad p \in P_n. \tag{55.1}$$

If we use the canonical maximum-norm on $C[a, b]$, then the question can be restated: Does there exist in the finite-dimensional subspace P_n of $C[a, b]$ a q such that

$$\|x - q\| \leqq \|x - p\| \qquad \text{for all} \quad p \in P_n, \tag{55.2}$$

or, expressed differently, can the *variational problem*

$$\|x - p\| = \min, \qquad p \in P_n \tag{55.3}$$

be solved by an element of P_n?

The next problem is one of the starting points for the theory of Fourier series. Let $C := C[-\pi, \pi]$ be the real vector space of the real-valued continuous functions on $[-\pi, \pi]$. The question is: Given an $x \in C$, does there exist in the set T_n of trigonometric polynomials

$$p(t) := \frac{\alpha_0}{2} + \sum_{\nu=1}^{n} (\alpha_\nu \cos \nu t + \beta_\nu \sin \nu t)$$

of degree $\leqq n$ a q, such that

$$(55.4) \qquad \int_{-\pi}^{\pi} [x(t) - q(t)]^2 \, dt \leqq \int_{-\pi}^{\pi} [x(t) - p(t)]^2 \, dt \qquad \text{for all} \quad p \in T_n?$$

If, contrary to our convention, we introduce on C the norm

$$\|x\| := \sqrt{\int_{-\pi}^{\pi} [x(t)]^2 \, dt}$$

(Exercise 4 in §1), then we can formulate this problem of *approximation in the square mean* as follows: Does there exist a q in the finite-dimensional subspace T_n of C such that

$$(55.5) \qquad \|x - q\| \leqq \|x - p\| \qquad \text{for all} \quad p \in T_n,$$

or, expressed differently, can the variational problem

$$(55.6) \qquad \|x - p\| = \min, \qquad p \in T_n$$

be solved by an element of T_n? It is very simple to deal with this problem computationally. If we set

$$(55.7) \qquad \begin{aligned} a_\nu &:= \frac{1}{\pi} \int_{-\pi}^{\pi} x(t) \cos \nu t \, dt \qquad (\nu = 0, 1, \ldots) \\ b_\nu &:= \frac{1}{\pi} \int_{-\pi}^{\pi} x(t) \sin \nu t \, dt \qquad (\nu = 1, 2, \ldots) \end{aligned}$$

(these numbers are called the *Fourier coefficients* of x) and if we take into account the so-called orthogonality relations

$$(55.8) \qquad \begin{aligned} &\int_{-\pi}^{\pi} \cos \nu t \sin \mu t \, dt = 0 \qquad \text{for} \quad \nu, \mu = 0, 1, \ldots, \\ &\int_{-\pi}^{\pi} \cos \nu t \cos \mu t \, dt = \int_{-\pi}^{\pi} \sin \nu t \sin \mu t \, dt = \begin{cases} 0 & \text{for} \quad \nu \neq \mu \\ \pi & \text{for} \quad \nu = \mu \geqq 1, \end{cases} \end{aligned}$$

then we obtain (omitting the variable t)

$$\begin{aligned} \int_{-\pi}^{\pi} (x - p)^2 \, dt &= \int_{-\pi}^{\pi} x^2 \, dt - 2 \int_{-\pi}^{\pi} xp \, dt + \int_{-\pi}^{\pi} p^2 \, dt = \int_{-\pi}^{\pi} x^2 \, dt \\ &\quad + \pi[\tfrac{1}{2}(\alpha_0 - a_0)^2 + \cdots + (\alpha_n - a_n)^2 + (\beta_1 - b_1)^2 + \cdots \\ &\quad + (\beta_n - b_n)^2] \\ &\quad - \pi[\tfrac{1}{2}a_0^2 + a_1^2 + \cdots + a_n^2 + b_1^2 + \cdots + b_n^2]. \end{aligned}$$

This expression assumes its smallest value exactly if $\alpha_\nu = a_\nu$ for $\nu = 0, 1, \ldots, n$ and $\beta_\nu = b_\nu$ for $\nu = 1, \ldots, n$, i.e., exactly if

$$p(t) = \frac{a_0}{2} + \sum_{\nu=1}^{n} (a_\nu \cos \nu t + b_\nu \sin \nu t)$$

is the nth partial sum of the Fourier series of x (as it is well-known, the *Fourier series* of x is the trigonometric series $a_0/2 + \sum_{\nu=1}^{\infty} (a_\nu \cos \nu t + b_\nu \sin \nu t)$ formed with the Fourier coefficients a_ν, b_ν, which, however, does not need to converge for all $t \in [-\pi, \pi]$.) Thus problem (55.5) or (55.6) has a unique solution.

Both approximation problems have the following structure: In a normed space E over $\mathbf{K}$ elements $x, x_1, \ldots, x_n$ are given and it is asked whether in the set of all linear combinations $\alpha_1 x_1 + \cdots + \alpha_n x_n$, i.e., in the finite-dimensional subspace $F := [x_1, \ldots, x_n]$, there exists an element $\beta_1 x_1 + \cdots + \beta_n x_n$ such that

$$(55.9) \qquad \left\| x - \sum_{\nu=1}^{n} \beta_\nu x_\nu \right\| \leqq \left\| x - \sum_{\nu=1}^{n} \alpha_\nu x_\nu \right\| \qquad \text{for all} \quad (\alpha_1, \ldots, \alpha_n) \in \mathbf{K}^n,$$

or, equivalently, whether the variational problem

$$(55.10) \qquad \left\| x - \sum_{\nu=1}^{n} \alpha_\nu x_\nu \right\| = \min$$

can be solved by an n-tuple $(\alpha_1, \ldots, \alpha_n)$. Such a problem is called a *Čebišov approximation problem.* Geometrically this problem can be formulated as follows: Given a vector $x \in E$, does there exist in a given finite-dimensional subspace F of E an element which has the shortest distance from x? The answer is positive:

Proposition 55.1. *In a normed space E the Čebišov approximation problem* (55.9), *or equivalently* (55.10), *is* always *solvable. Thus a finite-dimensional subspace F of E always contains an element which has the shortest distance from a given element $x \in E$ (every such element is called a* best approximation *of the point x in F).*

Proof. To $\gamma := \inf_{(\alpha_1, \ldots, \alpha_n)} \|x - \sum_{\nu=1}^{n} \alpha_\nu x_\nu\|$ there exists a 'minimizing sequence'

$$(y_k) := \left(\sum_{\nu=1}^{n} \alpha_\nu^{(k)} x_\nu \right) \qquad \text{such that} \quad \|x - y_k\| \to \gamma.$$

Because of $\|y_k\| - \|x\| \leqq \|y_k - x\| \leqq \gamma + 1$ for $k \geqq k_0$ the sequence (y_k) is bounded, thus by Proposition 10.4 it has a convergent subsequence (y_{k_l}), which according to Proposition 10.5 converges to a $y := \sum_{\nu=1}^{n} \beta_\nu x_\nu$. Since $\|x - y_{k_l}\|$ converges to γ as well as to $\|x - y\|$, we have $\|x - \sum_{\nu=1}^{n} \beta_\nu x_\nu\| = \gamma$; therefore $y = \sum_{\nu=1}^{n} \beta_\nu x_\nu$ is a best approximation of the desired kind. ∎

We want to consider three more very simple but informative approximation problems. We base our considerations on the space $\mathbf{R}^2$ and provide it with different norms:

(a) $\mathbf{R}^2$ with the l^2-norm: Setting $x := (1, 1)$, $x_1 := (1, 0)$, we have:

$$\|x - \alpha x_1\| = \|(1 - \alpha, 1)\| = \sqrt{(1 - \alpha)^2 + 1} = \min \Leftrightarrow \alpha = 1.$$

Thus x_1 is the unique best approximation in $[x_1]$.

(b) $\mathbf{R}^2$ with the l^∞-norm: With the above elements x and x_1 we have:

$$\|x - \alpha x_1\| = \max(|1 - \alpha|, 1) = \min \Leftrightarrow |1 - \alpha| \leqq 1 \Leftrightarrow 0 \leqq \alpha \leqq 2.$$

There exist infinitely many best approximations in $[x_1]$; they fill the segment $\{(\alpha, 0) \in \mathbf{R}^2 : 0 \leqq \alpha \leqq 2\}$.

(c) $\mathbf{R}^2$ with the l^1-norm: Set $x := (0, 1)$, $x_1 := (1, 1)$. The reader should convince himself that the best approximations in $[x_1]$ form again a segment in $\mathbf{R}^2$.

The behavior of the best approximations in the last two examples corresponds essentially to the general situation: If y_1 and y_2 are best approximations, then every point of the 'segment joining them' is again a best approximation; here the *segment joining the points* y_1, y_2 is the set $\{\lambda_1 y_1 + \lambda_2 y_2 : \lambda_1, \lambda_2 \geqq 0, \lambda_1 + \lambda_2 = 1\}$, a concept which is meaningful in every vector space.

We prove the assertion made above. For the minimal distance $\gamma := \|x - y_1\| = \|x - y_2\|$ we have, if $\lambda_1, \lambda_2 \geqq 0$, $\lambda_1 + \lambda_2 = 1$, the estimate

$$\begin{aligned} \gamma &\leqq \|x - (\lambda_1 y_1 + \lambda_2 y_2)\| = \|(\lambda_1 + \lambda_2)x - (\lambda_1 y_1 + \lambda_2 y_2)\| \\ &= \|\lambda_1(x - y_1) + \lambda_2(x - y_2)\| \leqq \lambda_1\|x - y_1\| + \lambda_2\|x - y_2\| \\ &= (\lambda_1 + \lambda_2)\gamma = \gamma, \end{aligned}$$

hence $\|x - (\lambda_1 y_1 + \lambda_2 y_2)\| = \gamma$ which finishes the proof. ■

If we call a set K in an arbitrary vector space *convex*, provided that it contains with any two points also the segment joining them, then we can state our result as follows:

Proposition 55.2. *In a normed space the set of best approximations* (*solutions of problem* (55.10)) *is* convex.

§56. Strictly convex spaces

If y_1, y_2 are two different best approximations (solutions of problem (55.10)), then according to Proposition 55.2

$$\|x - (\lambda_1 y_1 + \lambda_2 y_2)\| = \gamma \quad \text{for} \quad \lambda_1, \lambda_2 \geqq 0, \quad \lambda_1 + \lambda_2 = 1,$$

thus the surface of the sphere $K_\gamma[x]$ contains the segment joining the points y_1, y_2. It follows that

$$\left\| \lambda_1 \frac{x - y_1}{\gamma} + \lambda_2 \frac{x - y_2}{\gamma} \right\| = 1 \quad \text{for} \quad \lambda_1, \lambda_2 \geqq 0, \qquad \lambda_1 + \lambda_2 = 1,$$

i.e., that the surface of the unit ball $K_1[0]$ contains the segment joining the points $(x - y_1)/\gamma$, $(x - y_2)/\gamma$ (which themselves belong to the surface. Why is $\gamma > 0$?). In particular, it contains the midpoint

$$\frac{1}{2}\left(\frac{x - y_1}{\gamma} + \frac{x - y_2}{\gamma}\right)$$

Conversely, it follows that if for any two different points of the surface of $K_1[0]$ the midpoint of the segment joining them is always in the *interior* of $K_1[0]$, then the approximation problem (55.10) has a unique solution. If the unit ball $K_1[0]$ of the normed space E has the property just described, i.e., if

(56.1) 'from $\|x\| = \|y\| = 1$ and $x \neq y$ it follows that $\|\frac{1}{2}(x + y)\| < 1$
(or $\|x + y\| < 2$)',

then the unit ball or the space E is said to be *strictly convex*; for a better understanding of the terminology, observe that the unit ball, like any (closed or open) ball, is convex. We state our results:

Proposition 56.1. *In a strictly convex space the Čebišov approximation problem* (55.10) *has always* exactly one *solution.*

The uniqueness considerations in Examples (a) through (c) in §55 (after the proof of Proposition 55.1) become now geometrically transparent; we only have to draw the unit balls of the spaces $l^p(2)$ over $\mathbf{R}$ for $p = 1, 2, \infty$ (Figs. 3–5) (and to translate them parallelly).

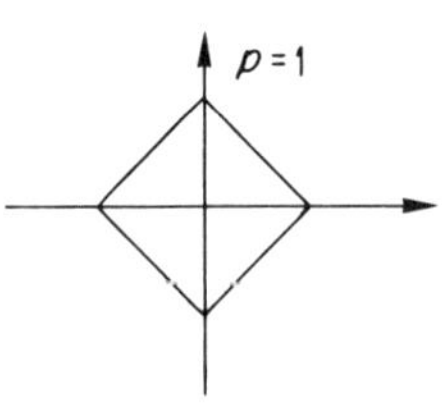

Figure 3

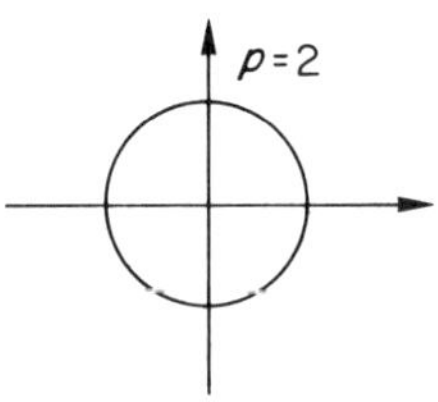

Figure 4

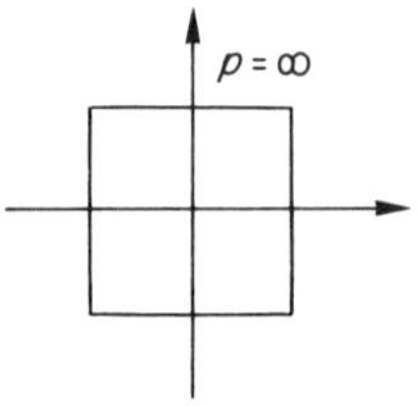

Figure 5

Strictly convex spaces can be characterized by the validity of the *strict triangle inequality*:

Proposition 56.2. *A normed space E is strictly convex if and only if the following holds*:

(56.2) '*from* $\|x + y\| = \|x\| + \|y\|$ *it follows that* $x = \alpha y$ *or* $y = \alpha x$ *with* $\alpha > 0$'.

Proof. Assume that E is strictly convex and let $\|x + y\| = \|x\| + \|y\|$. Without restricting the generality we may assume that $\|y\| - \|x\| \geqq 0$ and $x \neq 0$. Then

$$2 \geqq \left\| \frac{x}{\|x\|} + \frac{y}{\|y\|} \right\| \geqq \left\| \frac{x}{\|x\|} + \frac{y}{\|x\|} \right\| - \left\| \frac{y}{\|x\|} - \frac{y}{\|y\|} \right\|$$

$$= \frac{\|x\| + \|y\|}{\|x\|} - \left\| \frac{(\|y\| - \|x\|)y}{\|x\| \, \|y\|} \right\| = 2,$$

hence $\|(x/\|x\|) + (y/\|y\|)\| = 2$. Because of (56.1) we have $x/\|x\| = y\|y\|$ and therefore (56.2) holds. Now assume conversely that (56.2) is valid and let $\|x\| = \|y\| = \|(x + y)/2\| = 1$. Then $\|x + y\| = 2 = \|x\| + \|y\|$ from where we get, by hypothesis, that $x = \alpha y$ with $\alpha \geqq 0$. It follows that $\alpha = \|x\|/\|y\| = 1$, hence $x = y$. Thus E is strictly convex. ∎

If we take into account the remark concerning the validity of the equal sign in the Minkowski inequality, then we obtain from Proposition 56.2 immediately:

Proposition 56.3. *The spaces l^p are strictly convex for* $1 < p < \infty$.

The reader can convince himself very easily that the following spaces are not strictly convex: $l^1(n)$ for $n \geqq 2$, l^1, $l^\infty(n)$ for $n \geqq 2$, l^∞, (c), (c_0), $B(T)$ if T has at least 2 elements, $C[a, b]$.

Exercises

1. A normed space is strictly convex if and only if the surface of the unit ball does not contain any segments.

$^+$2. A normed space is strictly convex whenever its dual is strictly convex. For reflexive spaces the converse is also valid.

$^+$3. Linear subspaces of strictly convex spaces are strictly convex.

$^+$4. Let F be a finite-dimensional subspace of the strictly convex space E. Show that the *perpendicular projector* L, which associates with each $x \in E$ its best approximation Lx in F, is continuous (L is not necessarily linear).

§57. Inner product spaces

The strict convexity of l^2 can also be proved as follows. For $x = (\xi_n)$ and $y = (\eta_n)$ from l^2 we have

$$\begin{aligned} \|x + y\|^2 &= \sum |\xi_n + \eta_n|^2 = \sum (\xi_n + \eta_n)(\bar{\xi}_n + \bar{\eta}_n) \\ &= \sum |\xi_n|^2 + \sum \xi_n \bar{\eta}_n + \sum \eta_n \bar{\xi}_n + \sum |\eta_n|^2, \\ \|x - y\|^2 &= \sum |\xi_n - \eta_n|^2 = \sum (\xi_n - \eta_n)(\bar{\xi}_n - \bar{\eta}_n) \\ &= \sum |\xi_n|^2 - \sum \xi_n \bar{\eta}_n - \sum \eta_n \bar{\xi}_n + \sum |\eta_n|^2, \end{aligned}$$

from where the so-called *parallelogram law*

$$\|x + y\|^2 + \|x - y\|^2 = 2\|x\|^2 + 2\|y\|^2 \tag{57.1}$$

follows. For $\|x\| = \|y\| = 1$ and $x \neq y$ (that is, $\|x - y\| > 0$) we obtain from it that $\|x + y\|^2 < 4$, which proves strict convexity.

If we equip $C[a, b]$ with the norm $\|x\| := (\int_a^b |x(t)|^2 \, dt)^{1/2}$, and compute as above $\|x + y\|^2$ and $\|x - y\|^2$, then we see that also in this case (57.1) holds, so that the space $C[a, b]$ with the *euclidean* norm is again strictly convex.

In both examples the decisive parallelogram law follows from the formal properties of the expression

$$(x|y) := \begin{cases} \sum_{n=1}^{\infty} \xi_n \bar{\eta}_n & \text{for} \quad x, y \in l^2 \\ \int_a^b x(t)\overline{y(t)}\mathrm{d}t & \text{for} \quad x, y \in C[a, b] \end{cases}$$

and its relation to the norm: For all vectors x, y, z and all scalars α we have namely:

(IP1) $(x + y|z) = (x|z) + (y|z)$
(IP2) $(\alpha x|y) = \alpha(x|y)$
(IP3) $(x|y) = \overline{(y|x)}$
(IP4) $(x|x) \geqq 0$, where $(x|x) = 0$ if and only if $x = 0$,

furthermore

$$\|x\| = \sqrt{(x|x)}. \tag{57.2}$$

From (IP1) and (IP2) it follows with the help of (IP3) that

$$(x|y + z) = (x|y) + (x|z) \quad \text{and} \quad (x|\alpha y) = \bar{\alpha}(x|y). \tag{57.3}$$

From these rules we obtain

$$\|x + y\|^2 = (x + y|x + y) = (x|x) + (x|y) + (y|x) + (y|y)$$
$$\|x - y\|^2 = (x - y|x - y) = (x|x) - (x|y) - (y|x) + (y|y)$$

from where indeed the parallelogram law follows.

A vector space E over $\mathbf{K}$ is called an *inner product space* or a *prehilbert space* if with every pair (x, y) of elements of E a number $(x|y)$ from $\mathbf{K}$ is associated so that (IP1) through (IP4) are valid. $(x|y)$ is called the *inner product* of the vectors x, y. The space E equipped with the inner product $(\cdot|\cdot)$ is sometimes denoted more carefully by $(E, (\cdot|\cdot))$.

Our considerations can now be summarized as follows: If an inner product space E is at the same time a normed space, and if the inner product and the norm are related by (57.2), then in E the parallelogram law (57.1) is valid and E is strictly convex.

It is now a fundamental fact that in every inner product space E a norm can be *defined* by (57.2); equipped with this norm E is thus strictly convex.

That by (57.2) a real number is associated with every $x \in E$ and that this correspondence satisfies the norm axioms (N1), (N2) in §6, follows immediately from (IP4), (IP2) and (57.3). Before proving the triangle inequality (N3), we prove the capital *Schwarz inequality*: *For all vectors x, y of an inner product space we have*

$$|(x|y)| \leqq \sqrt{(x|x)}\sqrt{(y|y)}. \tag{57.4}$$

This assertion is trivial for $y = 0$ (observe that $(x|0) = (x|0 \cdot 0) = 0 \cdot (x|0) = 0$); therefore we assume $y \neq 0$. For all α we have $0 \leqq (x + \alpha y|x + \alpha y) = (x|x) + \alpha(y|x) + \bar{\alpha}(x|y) + \alpha\bar{\alpha}(y|y)$; from here we obtain (57.4) by setting $\alpha = -(x|y)/(y|y)$. ∎

From the definition (57.2) we now obtain, with the help of the Schwarz inequality,

$$\begin{aligned}\|x + y\|^2 &= (x + y|x + y) = \|x\|^2 + (x|y) + (y|x) + \|y\|^2\\ &= \|x\|^2 + 2\mathrm{Re}(x|y) + \|y\|^2 \leqq \|x\|^2 + 2\|x\|\,\|y\| + \|y\|^2\\ &= (\|x\| + \|y\|)^2,\end{aligned}$$

hence also (N3) holds. We can therefore summarize our results in the following.

Proposition 57.1. *In every inner product space E, a norm, the* canonical *norm of E, is defined by* $\|x\| := \sqrt{(x|x)}$. *For this norm the parallelogram law* (57.1) *holds; thus E is not only a normed space but even* strictly convex.

One can show that a normed space E over $\mathbf{K}$, in which (57.1) holds, is an inner product space, i.e., that one can introduce an inner product $(\cdot|\cdot)$ in E which generates the existing norm $\|\cdot\|$ according to $\|x\| = \sqrt{(x|x)}$. For this purpose one writes

$$(x|y) := \begin{cases} \left\|\dfrac{x+y}{2}\right\|^2 - \left\|\dfrac{x-y}{2}\right\|^2 & \text{if } \mathbf{K} = \mathbf{R}\\[2ex] \left\|\dfrac{x+y}{2}\right\|^2 - \left\|\dfrac{x-y}{2}\right\|^2 + i\left\|\dfrac{x+iy}{2}\right\|^2 - i\left\|\dfrac{x-iy}{2}\right\|^2 & \text{if } \mathbf{K} = \mathbf{C}.\end{cases}$$

We cannot go into the proof which employs elementary but long winded computations (see [80]). Summarized briefly, we can say that *exactly those normed spaces are inner product spaces in which the parallelogram law is valid.*

For euclidean $\mathbf{R}^2$ the parallelogram law expresses the theorem from elementary geometry that in a parallelogram the sum of the squares of the sides is equal to the sum of the squares of the diagonals.

An inner product space will be equipped from now on always with its canonical norm.

The Schwarz inequality (57.4) can be written more briefly in the form

(57.5) $$|(x \mid y)| \leqq \|x\| \, \|y\|.$$

From it we obtain the *continuity of the scalar product*:

Proposition 57.2. *In an inner product space $x_n \to x$, $y_n \to y$ implies $(x_n \mid y_n) \to (x \mid y)$. In particular, we have $(x \mid y) = \sum_{n=1}^{\infty} (x_n \mid y)$ whenever $x = \sum_{n=1}^{\infty} x_n$.*

The proof follows from the estimate

$$\begin{aligned} |(x_n \mid y_n) - (x \mid y)| &= |(x_n - x \mid y) + (x_n \mid y_n - y)| \\ &\leqq \|x_n - x\| \, \|y\| + \|x_n\| \, \|y_n - y\|. \end{aligned}$$ ■

An inner product space, which as a normed space is complete, is called a *Hilbert space*. $l^2(n)$, l^2 and $L^2(a, b)$ with the inner products $(x \mid y) := \sum_{\nu=1}^{n} \xi_\nu \bar{\eta}_\nu$, $(x \mid y) := \sum_{\nu=1}^{\infty} \xi_\nu \bar{\eta}_\nu$ and $(x \mid y) := \int_a^b x(t)\overline{y(t)}dt$ are Hilbert spaces; $C[a, b]$ with $(x \mid y) := \int_a^b x(t)\overline{y(t)}dt$ is a noncomplete inner product space.

A linear subspace F of the inner product space E becomes in a natural way—by the restriction of the inner product to F—an inner product space, whose canonical norm coincides with the one induced by E.

The product $E_1 \times \cdots \times E_n$ of finitely many inner product spaces $(E_\nu, (\cdot \mid \cdot)_\nu)$ becomes with the inner product

$$(x \mid y) := \sum_{\nu=1}^{n} (x_\nu \mid y_\nu)_\nu, \qquad \text{where} \quad x = (x_1, \ldots, x_n), \qquad y = (y_1, \ldots, y_n),$$

an inner product space. Its norm is $\|x\| = \left(\sum_{\nu=1}^{n} \|x_\nu\|_\nu^2\right)^{1/2}$; it is a Hilbert space if and only if each E_ν is complete.

Exercises

1. If $\alpha_1, \ldots, \alpha_n$ are given positive numbers, then an inner product will be defined on $\mathbf{K}^n$ by $(x \mid y) := \sum_{\nu=1}^{n} \alpha_\nu \xi_\nu \bar{\eta}_\nu$; in this connection $(\alpha_1, \ldots, \alpha_n)$ is called a weight vector. Define a new inner product on l^2 by means of a 'weight sequence'.

2. If $p(t) > 0$ is continuous on $[a, b]$ ('weight function'), then $(x \mid y) := \int_a^b p(t)x(t)\overline{y(t)}dt$ is an inner product on $C[a, b]$.

$^+$3. Let T be an arbitrary non-empty set and $l^2(T)$ the set of all functions $x\colon T \to \mathbf{K}$ with the following properties:

(a) $x(t) \neq 0$ for at most countably many $t \in T$;

(b) $\sum_{t \in T} |x(t)|^2$ converges (the sum is taken with respect to only those t for which $x(t) \neq 0$).

Then $l^2(T)$ is a linear function space over $\mathbf{K}$ and becomes a Hilbert space through the introduction of an inner product $(x \mid y) := \sum_{t \in T} x(t)\overline{y(t)}$. For every $s \in T$ let $y_s \in l^2(T)$ be defined by $y_s(t) = \delta_{st}$. We have $(y_r \mid y_s) = \delta_{rs}$, and $(x \mid y_s) = 0$ for all $s \in T$ implies $x = 0$.

4. In the Schwarz inequality the equality sign holds if and only if x, y are linearly dependent.

5. A normed space, which is isometrically isomorphic to an inner product space, is itself an inner product space.
6. An algebraic isomorphism A between two inner product spaces E, F is an isomorphism of normed spaces if and only if $(Ax|Ay) = (x|y)$ for all $x, y \in E$.
7. The vectors $x_1, \ldots, x_n$ of the inner product space E are linearly dependent if and only if their *Gram determinant* $|(x_i|x_k)|$ vanishes.

§58. Orthogonality

As is well known, two vectors $x = (\xi_1, \xi_2, \xi_3)$, $y = (\eta_1, \eta_2, \eta_3)$ of $\mathbf{R}^3$ are perpendicular if their inner product $(x|y) = \sum_{\nu=1}^{3} \xi_\nu \eta_\nu$ vanishes. This suggests the following definition of a new structural element which distinguishes the inner product spaces among the normed spaces: Two vectors x and y of an inner product space are said to be *orthogonal* or *perpendicular* to each other if $(x|y) = 0$. The symmetry in the expression is justified by (IP3).

Because of the importance of the matter, we will not refrain from documenting with a few examples that the concept of orthogonality penetrates many mathematical considerations.

Example 58.1. If we want to represent the vector x from $\mathbf{K}^n$ as a linear combination of n basis vectors $y_1, \ldots, y_n$, then we must calculate in general the coefficients α_ν in $x = \alpha_1 y_1 + \cdots + \alpha_n y_n$ from a system of linear equations. If, however, $\{y_1, \ldots, y_n\}$ is a so-called *orthonormal basis*, i.e., if $(y_\nu | y_\mu) = \delta_{\nu\mu}$ for all ν, μ, then $(x|y_\mu) = (\sum_{\nu=1}^{n} \alpha_\nu y_\nu | y_\mu) = \sum_{\nu=1}^{n} \alpha_\nu (y_\nu | y_\mu) = \alpha_\mu$ and x has the representation which can be written down immediately:

$$x = \sum_{\nu=1}^{n} (x|y_\nu) y_\nu. \tag{58.1}$$

Example 58.2. D. Bernoulli and L. Euler were led through their investigations of the vibrating string to the problem of *harmonic analysis*: Given a real-valued function $x \in C[-\pi, \pi]$ with $x(-\pi) = x(\pi)$, determine real numbers a_ν, b_ν so that for all t in $[-\pi, \pi]$ the representation

$$x(t) = \frac{a_0}{2} + \sum_{\nu=1}^{\infty} (a_\nu \cos \nu t + b_\nu \sin \nu t) \tag{58.2}$$

is valid. If we assume for a moment that the expansion (58.2) is possible and even that it converges uniformly in $[-\pi, \pi]$, then we can determine immediately the coefficients a_ν, b_ν multiplying the series by $\cos \nu t$ or $\sin \nu t$, respectively, integrating termwise and taking into account the orthogonality relations (55.8) of the trigonometric functions; we obtain

$$a_\nu = \frac{1}{\pi} \int_{-\pi}^{\pi} x(t) \cos \nu t \, dt \qquad (\nu = 0, 1, \ldots),$$

$$b_\nu = \frac{1}{\pi} \int_{-\pi}^{\pi} x(t) \sin \nu t \, dt \qquad (\nu = 1, 2, \ldots),$$

i.e., the Fourier coefficients we defined already in (55.7). Decisive for this procedure was that the functions

$$y_1(t) := \frac{1}{\sqrt{2\pi}},\ y_2(t) := \frac{\cos t}{\sqrt{\pi}},\ y_3(t) := \frac{\sin t}{\sqrt{\pi}},\ y_4(t) := \frac{\cos 2t}{\sqrt{\pi}},\ y_5(t) := \frac{\sin 2t}{\sqrt{\pi}}, \ldots$$

form an *orthonormal sequence* in the (real) function space $C[-\pi, \pi]$, equipped with the inner product $(x|y) := \int_{-\pi}^{\pi} x(t)y(t)dt$, i.e., that $(y_\nu|y_\mu) = \delta_{\nu\mu}$ for all ν, μ. The result of our considerations can now be formulated as follows: If G is the linear subspace of all functions x from $C[-\pi, \pi]$ which can be expanded on $[-\pi, \pi]$ into a uniformly convergent series $\sum_{\nu=1}^{\infty} \alpha_\nu y_\nu$, then we have

$$x = \sum_{\nu=1}^{\infty} (x|y_\nu)y_\nu. \tag{58.3}$$

This expansion of x is so similar to (58.1) that we may call the sequence $(y_1, y_2, \ldots)$ an 'orthonormal basis' of the space G. By the way, also the numbers $(x|y_\mu)$ are called Fourier coefficients of x.

In the theory of Fourier series one inverts these considerations. One starts with a function x, integrable on $[-\pi, \pi]$, forms with the Fourier coefficients $(x|y_\nu)$ the so-called Fourier series $\sum_{\nu=1}^{\infty} (x|y_\nu)y_\nu$ and asks whether it converges in the interval $[-\pi, \pi]$ to x. As long as one has *pointwise* convergence in mind, this question is extraordinarily difficult. It is, however, reasonable to replace pointwise convergence by convergence in the sense of the norm $\|x\| := (\int_{-\pi}^{\pi} x^2(t)dt)^{1/2}$, which is generated by the inner product (*convergence in quadratic mean*), i.e., to ask whether for $n \to \infty$ one has

$$\left\| x - \sum_{\nu=1}^{n} (x|y_\nu)y_\nu \right\|^2 = \int_{-\pi}^{\pi} \left[x(t) - \sum_{\nu=1}^{n} (x|y_\nu)y_\nu(t) \right]^2 dt \to 0.$$

We shall discuss this problem later in a very general form. Compare with these considerations also the approximation of continuous functions by trigonometric polynomials, which we treated in §55, and where the orthogonality relations (55.8) of the trigonometric functions ensured the explicit solvability of the approximation problem and the uniqueness of the solution.

Example 58.3. We now investigate the distribution of temperature $u(x, t)$ of a finite thin rod going from $x = 0$ to $x = L$, whose left end-point is kept at the constant temperature 0, while the right end-point is subject to an exchange of heat with a surrounding medium of temperature $\neq 0$; at time $t = 0$ let the rod have the given temperature $f(x)$ at the point x (observe that in this example x is a space variable, not a function). The distribution of temperature is then that solution of the heat equation

$$u_t = a^2 u_{xx} \tag{58.4}$$

which satisfies the boundary conditions

$$u(0, t) = 0, \qquad u_x(L, t) + \sigma u(L, t) = 0 \qquad \text{for all} \quad t \geqq 0 \tag{58.5}$$

and the initial condition

$$u(x, 0) = f(x) \quad \text{for} \quad 0 \leqq x \leqq L; \tag{58.6}$$

a and σ are positive constants depending on the material of the rod. If we substitute into (58.4) a solution which we assume tentatively to be of the form $u(x, t) = v(x)w(t)$, then we obtain the equation

$$\frac{v''(x)}{v(x)} = \frac{1}{a^2}\frac{\dot{w}(t)}{w(t)}$$

which can hold only if each of its sides is equal to the same constant $-\lambda$. Taking into consideration the boundary conditions (58.5), we obtain that v is a solution of the boundary value problem

$$v'' + \lambda v = 0, \qquad v(0) = 0, \qquad v'(L) + \sigma v(L) = 0, \tag{58.7}$$

while w has only to satisfy the differential equation

$$\dot{w} + a^2\lambda w = 0 \tag{58.8}$$

without further conditions. Conversely, we can check immediately that a solution v of (58.7) and a solution w of (58.8) always yields a solution $u := vw$ of (58.4) which satisfies the boundary conditions (58.5), and which is continuous together with its partial derivatives u_t and u_{xx}.

The boundary value problem (58.7) does, however, not have a non-trivial solution for all λ; if, namely, v is a non-trivial solution, thus because of $v(0) = 0$ also a non-constant one, then it follows from $\lambda v = -v''$ that

$$\lambda \int_0^L v^2\,\mathrm{d}x = -\int_0^L vv''\,\mathrm{d}x = -[vv']_0^L + \int_0^L (v')^2\,\mathrm{d}x = \sigma v^2(L) + \int_0^L (v')^2\,\mathrm{d}x,$$

hence that λ must be >0. For positive λ the general solution of the differential equation (58.7) is given by $v(x) := A\sin(\sqrt{\lambda}x) + B\cos(\sqrt{\lambda}x)$ with arbitrary constants A, B. Because of $v(0) = 0$ we have $B = 0$ and so $v(x) = A\sin(\sqrt{\lambda}x)$. With the second boundary condition $v'(L) + \sigma v(L) = 0$ we obtain $\tan(\sqrt{\lambda}L) = -\sqrt{\lambda}/\sigma$. This equation for λ has countably many solutions $\lambda_n > 0$ which increase strictly to $+\infty$. Only for these eigenvalues of the boundary value problem (58.7) do there exist non-trivial solutions; the eigenfunction v_n corresponding to the eigenvalue λ_n is given up to a multiplicative constant by $v_n(x) := \sin\sqrt{\lambda_n}x$. These eigensolutions have, like the trigonometric functions, the basic orthogonality property

$$\int_0^L v_n(x)v_m(x)\mathrm{d}x = \begin{cases} 0 & \text{if} \quad n \neq m \\ \beta_n \neq 0 & \text{if} \quad n = m. \end{cases} \tag{58.9}$$

From $v_m(v_n'' + \lambda_n v_n) = 0$, $v_n(v_m'' + \lambda_m v_m) = 0$ we obtain first $(\lambda_n - \lambda_m)v_n v_m + (v_m v_n'' - v_n v_m'') = 0$ and hence, using the boundary conditions,

$$\begin{aligned}0 &= (\lambda_n - \lambda_m)\int_0^L v_n v_m \,\mathrm{d}x + \int_0^L (v_m v_n'' - v_n v_m'')\mathrm{d}x \\ &= (\lambda_n - \lambda_m)\int_0^L v_n v_m \,\mathrm{d}x + \int_0^L \frac{d}{dx}(v_m v_n' - v_n v_m')\mathrm{d}x \\ &= (\lambda_n - \lambda_m)\int_0^L v_n v_m \,\mathrm{d}x + v_m(L)v_n'(L) - v_n(L)v_m'(L) - v_m(0)v_n'(0) \\ &\quad + v_n(0)v_m'(0) \\ &= (\lambda_n - \lambda_m)\int_0^L v_n v_m \,\mathrm{d}x,\end{aligned}$$

thus $\int_0^L v_n v_m \,\mathrm{d}x = 0$ when $n \neq m$ and so $\lambda_n \neq \lambda_m$. ■

The general solution corresponding to λ_n of the differential equation (58.8) is given by $C_n \exp(-\lambda_n a^2 t)$. For every choice of the constant C_n the function u_n defined by

(58.10) $$u_n(x, t) := C_n \sin(\sqrt{\lambda_n}\, x)\exp(-\lambda_n a^2 t)$$

is a solution of the heat equation which satisfies the boundary conditions (58.5). It will, however, in general not verify the initial condition (58.6). In order to satisfy (58.6) we will reason as follows: Equation (58.4) and the boundary conditions (58.5) are linear and homogeneous, thus every (finite) sum $\sum u_n$ of solutions of the form (58.10) satisfies again (58.4) and (58.5) for every choice of the C_n. The same is true for

(58.11) $$u(x, t) := \sum_{n=1}^{\infty} u_n(x, t) = \sum_{n=1}^{\infty} C_n \sin(\sqrt{\lambda_n}\, x)\cdot\exp(-\lambda_n a^2 t)$$

provided that the coefficients C_n are chosen in such a way that (58.11) converges and the derivatives u_t, u_{xx} can be obtained by termwise differentiation. And the decisive question, whether the solution u can be fitted to the initial condition $u(x, 0) = f(x)$, amounts now to thc problem of determining the C_n so that

$$f(x) = \sum_{n=1}^{\infty} C_n \sin(\sqrt{\lambda_n}\, x).$$

Thus we face again the problem *to expand a given function according to pairwise orthogonal functions*, where orthogonality in $C[0, L]$ is defined by means of the inner product $(g\,|\,h) := \int_0^L g(x)h(x)\mathrm{d}x$.

After these examples we return to general considerations. Let E be an inner product space throughout. We express orthogonality by writing $x \perp y$. The symbol $N \perp M$ means that every vector of N is orthogonal to every vector in M. If N consists of only one element x, then we write more briefly $x \perp M$. The set

$$M^{\perp} := \{x \in E : x \perp M\} \tag{58.12}$$

is a closed linear subspace of E, the *space orthogonal* to M; there is no danger of confusing it with the orthogonal space defined in §29: In prehilbert spaces we shall only use the orthogonal space (58.12). If M is itself a linear subspace of E, then we have obviously $M \cap M^{\perp} = \{0\}$; we shall often make use of this simple fact.

The zero vector is the only vector orthogonal to all elements of E. If x is perpendicular to the vectors $x_1, \ldots, x_n$, then $x \perp [x_1, \ldots, x_n]$. From the continuity of the inner product it follows that a vector which is orthogonal to all terms of a convergent sequence is also orthogonal to the limit. From these remarks it follows that *$x \perp M$ implies $x \perp \overline{[M]}$.*

We conclude this section with the *Theorem of Pythagoras*. First the simplest version:

From $u_j \perp u_k$ for $j \neq k$ it follows that

$$\|u_1 + \cdots + u_n\|^2 = \|u_1\|^2 + \cdots + \|u_n\|^2. \tag{58.13}$$

We have indeed

$$\left\|\sum_{\nu=1}^{n} u_\nu\right\|^2 = \left(\sum_{\nu=1}^{n} u_\nu \,\middle|\, \sum_{\mu=1}^{n} u_\mu\right) = \sum_{\nu=1}^{n}\sum_{\mu=1}^{n} (u_\nu | u_\mu) = \sum_{\nu=1}^{n} (u_\nu | u_\nu) = \sum_{\nu=1}^{n} \|u_\nu\|^2. \quad \blacksquare$$

A non-empty subset S of E is called an *orthogonal system* if two different elements of S are always orthogonal to each other. If furthermore the vectors of S are *normalized*, i.e., if $\|u\| = 1$ for every $u \in S$, then S is called an *orthonormal system.* A countable orthogonal system is also called an *orthogonal sequence*; it is now clear what is meant by an *orthonormal sequence.* An orthogonal system S does not have to be linearly independent, since it can contain the zero vector; if 0 does not lie in S (if e.g., S is an orthonormal system), then the linear independence is ensured: From $\alpha_1 u_1 + \cdots + \alpha_n u_n = 0$ $(u_k \in S)$ it then follows that

$$0 = \left(\sum_{k=1}^{n} \alpha_k u_k \,\middle|\, u_m\right) = \sum_{k=1}^{n} \alpha_k (u_k | u_m) = \alpha_m (u_m | u_m),$$

hence $\alpha_m = 0$ for $m = 1, \ldots, n$.

We can now generalize considerably the *Theorem of Pythagoras*:

Proposition 58.1. *If $(u_1, u_2, \ldots)$ is an orthogonal sequence and if $\sum_{k=1}^{\infty} u_k$ converges, then also $\sum_{k=1}^{\infty} \|u_k\|^2$ converges, and we have*

$$\left\|\sum_{k=1}^{\infty} u_k\right\|^2 = \sum_{k=1}^{\infty} \|u_k\|^2.$$

For the proof let $s_n := u_1 + \cdots + u_n$ and $u := \sum_{k=1}^{\infty} u_k = \lim s_n$. Because of the continuity of the inner product (Proposition 57.2), we have $(s_n | s_n) = \sum_{k=1}^{n} \|u_k\|^2 \to (u|u) \|u\|^2$, which proves the assertions. ■

Exercises

1. The unit vectors $e_k := (\delta_{jk})$ form an orthonormal sequence in l^2. From $x \perp e_k$ for all k it follows that $x = 0$.

2. The functions y_s $(s \in T)$ of Exercise 3 of §57 form an orthonormal system in $l^2(T)$. From $x \perp y_s$ for all s it follows that $x = 0$.

3. The trigonometric functions $y_1, y_2, \ldots$ in Example 58.2 form, as we already observed there, an orthonormal sequence in the real (and also in the complex) space $C[-\pi, \pi]$. The functions $(1/\sqrt{2\pi})e^{int}$ $(n = 0, \pm 1, \pm 2, \ldots)$ form an orthonormal sequence in the complex space $C[-\pi, \pi]$. Every x, which is orthonormal to this sequence, vanishes; the reader can find a proof for this theorem, which is valid also in $L^2(-\pi, \pi)$, and which is basic in the theory of Fourier series, e.g., in [181].

$^+$4. For $(M^\perp)^\perp$ we write more briefly $M^{\perp\perp}$. Show:
(a) $M_1 \subset M_2 \Rightarrow M_2^\perp \subset M_1^\perp$.
(b) $M \subset M^{\perp\perp}$, even $\overline{[M]} \subset M^{\perp\perp}$.
(c) $M^\perp = M^{\perp\perp\perp}$.

5. We consider the Fredholm integral equation (19.1) under the hypotheses stated there; let K be the corresponding integral operator. If $k(s, t) = k(t, s)$ or $k(s, t) = \overline{k(t, s)}$ for all s, t, according as $C[a, b]$ is real or complex, then $(Kx|y) = (x|Ky)$ for all x, y. It follows from here that every eigenvalue of K is real (trivial if $C[a, b]$ is real) and that eigensolutions corresponding to different eigenvalues are perpendicular to each other (see Proposition 68.1).

$^+$6. Orthogonality in an inner product space E over $\mathbf{K}$ can be characterized alone with the help of the norm:

$$x \perp y \Leftrightarrow \|x\| \leqq \|x - \alpha y\| \qquad \text{for all} \quad \alpha \in \mathbf{K}$$

(geometric meaning?). By the inequality on the right we can define the orthogonality of x to y, in symbol: $x \perp y$, in an arbitrary normed space. Introduce in $\mathbf{R}^2$ the maximum norm, determine to which $x \in \mathbf{R}^2$ with $\|x\| = 1$ the vector $(0, 1)$ is orthogonal to, and show that $x \perp y$ does not imply $y \perp x$ and that $x \perp y_1, y_2$ does not imply $x \perp (y_1 + y_2)$.

§59. The Gauss approximation

We saw in §55 that the problem to approximate in quadratic mean as closely as possible a continuous function by trigonometric polynomials of degree $\leqq n$ could be solved easily and explicitly because the trigonometric polynomials form an orthogonal system. Also the uniqueness of the best approximation was obtained from the orthogonality relations (by now we know that every Čebišov approximation problem (55.10) in an inner product space E has only one solution because E is strictly convex). It is therefore reasonable to consider *orthonormal* vectors in connection with the Čebišov approximation in an inner product space. In this case we speak of a *Gauss approximation problem.* The following theorem is essentially nothing but an abstract repetition of the trigonometric approximation in §55.

Proposition 59.1. *Let $S := \{u_1, \ldots, u_n\}$ be an orthonormal system in the inner product space E. Then the Gauss approximation problem to determine, for a given $x \in E$, numbers $\alpha_1, \ldots, \alpha_n$ so that $\|x - \sum_{\nu=1}^n \alpha_\nu u_\nu\|$ is minimal, has $\alpha_\nu := (x|u_\nu)$ as its unique solution. The vector $x - \sum_{\nu=1}^n (x|u_\nu)u_\nu$ is perpendicular to S and thus also to $[u_1, \ldots, u_n]$. Furthermore we have the* Bessel identity

$$\left\|x - \sum_{\nu=1}^{n} (x|u_\nu)_\nu\right\|^2 = \|x\|^2 - \sum_{\nu=1}^{n} |(x|u_\nu)|^2 \tag{59.1}$$

and the Bessel inequality

$$\sum_{\nu=1}^{n} |(x|u_\nu)|^2 \leqq \|x\|^2. \tag{59.2}$$

Proof. For arbitrary α_ν we have

$$\begin{aligned}
0 \leqq \left\|x - \sum_{\nu=1}^{n} \alpha_\nu u_\nu\right\|^2 &= \left(x - \sum_{\nu=1}^{n} \alpha_\nu u_\nu \,\middle|\, x - \sum_{\mu=1}^{n} \alpha_\mu u_\mu\right) \\
&= (x|x) - \sum_{\nu=1}^{n} \alpha_\nu(u_\nu|x) - \sum_{\mu=1}^{n} \bar{\alpha}_\mu(x|u_\mu) + \sum_{\nu,\mu=1}^{n} \alpha_\nu\bar{\alpha}_\mu(u_\nu|u_\mu) \\
&= \|x\|^2 - \sum_{\nu=1}^{n} \alpha_\nu\overline{(x|u_\nu)} - \sum_{\nu=1}^{n} \bar{\alpha}_\nu(x|u_\nu) + \sum_{\nu=1}^{n} \alpha_\nu\bar{\alpha}_\nu \\
&= \|x\|^2 - \sum_{\nu=1}^{n} |(x|u_\nu)|^2 + \sum_{\nu=1}^{n} [(x|u_\nu) - \alpha_\nu][\overline{(x|u_\nu)} - \bar{\alpha}_\nu] \\
&= \|x\|^2 - \sum_{\nu=1}^{n} |(x|u_\nu)|^2 + \sum_{\nu=1}^{n} |(x|u_\nu) - \alpha_\nu|^2.
\end{aligned}$$

The distance $\|x - \sum_{\nu=1}^n \alpha_\nu u_\nu\|$ is therefore minimal if we choose $\alpha_\nu = (x|u_\nu)$ for every ν. The Bessel identity follows immediately from our computation, the Bessel inequality follows from (59.1) because the left-hand side is non-negative. $y := x - \sum_{\nu=1}^n (x|u_\nu)u_\nu$ is perpendicular to S because $(y|u_\mu) = (x|u_\mu) - \sum_{\nu=1}^n (x|u_\nu)(u_\nu|u_\mu) = (x|u_\mu) - (x|u_\mu) = 0$ for $\mu = 1, \ldots, n$. ■

The converse of the last assertion also holds: If $x - \sum_{\nu=1}^n \alpha_\nu u_\nu$ is perpendicular to S, then $\alpha_\nu = (x|u_\nu)$; indeed,

$$0 = \left(x - \sum_{\mu=1}^{n} \alpha_\mu u_\mu \,\middle|\, u_\nu\right) = (x|u_\nu) - \sum_{\mu=1}^{n} \alpha_\mu(u_\mu|u_\nu) = (x|u_\nu) - \alpha_\nu.$$

The best approximation $y \in [u_1, \ldots, u_n]$ *of the point* x *is thus characterized by the fact that* $x - y$ *is orthogonal to* $[u_1, \ldots, u_n]$ (y is the 'foot of the perpendicular' dropped from x onto $[u_1, \ldots, u_n]$).

In applications the case occurs frequently that though the vectors $x_1, \ldots, x_n$ of problem (55.10) are linearly independent, they do not form an orthonormal system. This situation can be remedied by determining an orthonormal basis $\{u_1, \ldots, u_n\}$ in $[x_1, \ldots, x_n]$, obtaining the best approximation y of the point x in the form $y = \sum_{\nu=1}^{n} (x|u_\nu)u_\nu$ and—if desired—writing y as a linear combination of the original vectors $x_1, \ldots, x_n$ (for which it is necessary to know how the vectors u_ν can be expressed as linear combinations of the $x_1, \ldots, x_n$). That, starting from the elements $x_1, \ldots, x_n$, one can easily determine such an orthonormal basis, and expand each u_ν simply in terms of the basis $\{x_1, \ldots, x_n\}$, is ensured by the *Gram–Schmidt orthogonalization process*, which is described in the following proposition.

Proposition 59.2. *From an at most countable and linearly independent subset* $\{x_1, x_2, \ldots\}$ *of the inner product space* E *we can obtain an orthonormal system* $\{u_1, u_2, \ldots\}$ *whose construction will be given in the proof, and which has the following properties: For every* $n = 1, 2, \ldots$ *we have*

$$u_n = \alpha_{n1}x_1 + \cdots + \alpha_{nn}x_n, \qquad x_n = \beta_{n1}u_1 + \cdots + \beta_{nn}u_n. \tag{59.3}$$

Thus, in particular, $\{u_1, \ldots, u_n\}$ *or* $\{u_1, u_2, \ldots\}$ *generates the same subspace as* $\{x_1, \ldots, x_n\}$ *or* $\{x_1, x_2, \ldots\}$, *respectively.*

Proof. We define the sequence $\{u_1, u_2, \ldots\}$ inductively:

(a) Set $u_1 := x_1/\|x_1\|$. This is possible because of $x_1 \neq 0$. Now (59.3) is trivially true for $n = 1$.

(b) If an orthogonal system $\{u_1, \ldots, u_{r-1}\}$ is already defined so that (59.3) holds for $n = 1, 2, \ldots, r-1$, then

$$z_r := x_r - \sum_{\rho=1}^{r-1} (x_r|u_\rho)u_\rho \perp \{u_1, \ldots, u_{r-1}\}. \tag{59.4}$$

z_r does not vanish since otherwise we would have by our construction

$$x_r = \sum_{\rho=1}^{r-1} (x_r|u_\rho)u_\rho = \sum_{\rho=1}^{r-1} (x_r|u_\rho) \sum_{\sigma=1}^{\rho} \alpha_{\rho\sigma}x_\sigma = \sum_{\rho=1}^{r-1} \gamma_\rho x_\rho,$$

which contradicts the linear independence of the set $\{x_1, x_2, \ldots\}$. If we set $u_r := z_r/\|z_r\|$, then $\{u_1, \ldots, u_r\}$ is an orthonormal system according to (59.4), and (59.3) is valid for $n = 1 \ldots, r$. ∎

In [178] Proposition 59.2 is applied to obtain from simple functions the polynomials of *Legendre*, *Čebišov*, *Jacobi*, *Hermite* and *Laguerre* (see also Exercise 2).

Exercises

1. Let $x_1, \dots, x_n$ be linearly independent vectors in the inner product space E. We have $\|x - \sum_{\nu=1}^{n} \alpha_\nu x_\nu\| = \min$ for a fixed given $x \in E$ if and only if all α_ν satisfy the system of the so-called *normal equations* $\sum_{\mu=1}^{n} \alpha_\mu (x_\mu | x_\nu) = (x | x_\nu)$ $(\nu = 1, \dots, n)$. The solvability of the system is ensured by Exercise 7 of §57. Discuss the case of linearly dependent vectors $x_1, \dots, x_n$ applying the normal equations.

2. Define for real-valued continuous functions on $[-1, 1]$ an inner product by $(x|y) := \int_{-1}^{1} x(t)y(t)\mathrm{d}t$ and obtain by orthogonalization of the elementary polynomials $1, t, t^2, \dots$ the polynomials $\varphi_n(t) := \sqrt{(2n+1)/2}\,P_n(t)$ $(n = 0, 1, \dots, 4)$, where $P_0(t) := 1$, $P_1(t) := t$, $P_2(t) := \frac{3}{2}t^2 - \frac{1}{2}$, $P_3(t) := \frac{5}{2}t^3 - \frac{3}{2}t$, $P_4(t) := \frac{35}{8}t^4 - \frac{15}{4}t^2 + \frac{3}{8}$. The polynomials $P_n(t)$ are called Legendre polynomials.

§60. The general approximation problem

The Čebišov approximation problem seeks to a given point $x \in E$ a point y lying closest in a *finite-dimensional subspace* $F \subset E$. In the general approximation problem we may take F to be an *arbitrary* non-empty subset of E. In this generality the problem is, however, not solvable; we must subject F to certain restrictions. In prehilbert space it is enough if e.g., F is convex and complete (completeness is to be understood in the sense of the theory of metric spaces: every Cauchy sequence in F converges to an element of F):

Theorem 60.1. *If $K \neq \varnothing$ is a* convex *and* complete *subset of the inner product space E (e.g., a complete subspace of E), then for every $x \in E$ the problem*

$$\|x - y\| = \min, \qquad y \in K \tag{60.1}$$

is uniquely solvable in K, i.e., there exists exactly one $y_0 \in K$ such that

$$\|x - y_0\| \leqq \|x - y\| \qquad \textit{for all} \quad y \in K. \tag{60.2}$$

Proof. Let $\gamma := \inf_{y \in K} \|x - y\|$ and let (y_n) be a minimizing sequence in K, i.e., $\lim \|x - y_n\| = \gamma$. In the parallelogram law $\|u + v\|^2 + \|u - v\|^2 = 2\|u\|^2 + 2\|v\|^2$ we set $u := x - y_m$ and $v := x - y_n$. Since

$$u + v = (x - y_m) + (x - y_n) = 2\left[x - \frac{y_m + y_n}{2}\right],$$

$$u - v = (x - y_m) - (x - y_n) = y_n - y_m,$$

and since $(y_m + y_n)/2$ lies in K because of the convexity of K, we have

$$\|y_n - y_m\|^2 = 2\|x - y_m\|^2 + 2\|x - y_n\|^2 - 4\left\|x - \frac{y_m + y_n}{2}\right\|^2$$

$$\leqq 2\|x - y_m\|^2 + 2\|x - y_n\|^2 - 4\gamma^2,$$

hence if $n, m \to \infty$ then $\|y_n - y_m\| \to 0$. Thus (y_n) is a Cauchy sequence in K and has therefore a limit $y_0 \in K$. Since $\|x - y_n\|$ converges to γ and to $\|x - y_0\|$, we must have $\|x - y_0\| = \gamma$ and y_0 is a solution of (60.1).

We have obtained in particular that every minimizing sequence in K (with respect to x) is a Cauchy sequence. If now $u, v \in K$ are two solutions of (60.1), i.e., if $\|x - u\| = \|x - v\| = \gamma$, then obviously the sequence $u, v, u, v, \ldots$ is a minimizing sequence, thus a Cauchy sequence, and therefore $u = v$. ∎

The hypotheses of the proposition are satisfied in particular when *E is complete and K a convex, closed, non-empty subset of E.*

In the case of a Gauss approximation (K finite-dimensional subspace) y_0 is the 'foot of the perpendicular' from x to K ($x - y_0 \perp K$). This holds also for arbitrary subspaces, more precisely:

Proposition 60.1. *If F is an arbitrary linear subspace of the inner product space E and if for an $x \in E$ the problem*

$$\|x - y\| = \min, \qquad y \in F \tag{60.3}$$

has a solution $y_0 \in F$, then y_0 is the unique solution in F and $x - y_0$ is perpendicular *to F.*

Proof. We first show that $x - y_0 \perp y$ for all $y \neq 0$ in F. For all $\alpha \in \mathbf{K}$, $y \in F$ we have

$$\begin{aligned}\|x - y_0\|^2 &\leqq \|x - (y_0 + \alpha y)\|^2 = ((x - y_0) - \alpha y \,|\, (x - y_0) - \alpha y)\\ &= \|x - y_0\|^2 - \bar{\alpha}(x - y_0 \,|\, y) - \alpha(y \,|\, x - y_0) + \alpha\bar{\alpha}\|y\|^2.\end{aligned}$$

If $y \neq 0$, we set

$$\alpha := \frac{(x - y_0 \,|\, y)}{\|y\|^2} \quad \text{thus} \quad \bar{\alpha} = \frac{(y \,|\, x - y_0)}{\|y\|^2},$$

and obtain from the above estimate that

$$\|x - y_0\|^2 \leqq \|x - y_0\|^2 - 2\frac{|(x - y_0 \,|\, y)|^2}{\|y\|^2} + \frac{|(x - y_0 \,|\, y)|^2}{\|y\|^2}$$

hence

$$\frac{|(x - y_0 \,|\, y)|^2}{\|y\|^2} \leqq 0$$

and therefore $(x - y_0 \,|\, y) = 0$, as asserted. For any further solution $y_1 \in F$ of (60.3) we now have $y_0 - y_1 = (x - y_1) - (x - y_0) \in F \cap F^{\perp} = \{0\}$, which proves the uniqueness. ∎

Exercise

Proposition 60.1 has a converse: If $y_0 \in F$ and $x - y_0 \perp F$, then y_0 is the (unique) solution of (60.3). Taken together, and expressed briefly: the feet of the perpendiculars (provided they exist) are the best approximations.

§61. Approximation in uniformly convex spaces

The proof of the Approximation Theorem 60.1 was based on the fact that every minimizing sequence is a Cauchy sequence: this fact in turn followed from a property of the inner product norm, namely the parallelogram law. One would therefore think of looking for normed spaces whose norms have the property to make a Cauchy sequence out of every minimizing sequence of the problem (60.1).

From the parallelogram law in the form

$$\left\|\frac{x+y}{2}\right\|^2 + \left\|\frac{x-y}{2}\right\|^2 = \frac{1}{2}\|x\|^2 + \frac{1}{2}\|y\|^2$$

we obtain

$$\text{'from } \|x_n\| \leqq 1,\ \|y_n\| \leqq 1 \quad \text{and} \quad \left\|\frac{x_n + y_n}{2}\right\| \to 1$$

(61.1) $\qquad$ it follows that $\|x_n - y_n\| \to 0$'.

This property is, similarly to strict convexity, a *rotundity property* of the closed unit ball: If the midpoints of segments in $K_1[0]$ crowd to the surface, the end-points will be pushed together arbitrarily.

A normed space is said to be *uniformly convex* if (61.1) is valid in it. In uniformly convex spaces the problem (60.1) will turn out to be solvable.

The proof of the following proposition can be left to the reader:

Proposition 61.1. *Inner product spaces are uniformly convex. Uniformly convex spaces are strictly convex.*

Figure 6 illustrates the hierarchy of the types of spaces studied so far.

We arrive now at the decisive approximation assertion:

Proposition 61.2. *If $K \neq \varnothing$ is a convex and complete subset of the uniformly convex space E, then for every $x \in E$ the problem*

(61.2) $$\|x - y\| = \min, \qquad y \in K$$

is uniquely solvable in K.

Proof. Let $\gamma := \inf_{y \in K} \|x - y\|$. If $\gamma = 0$, then $x \in K$ and x itself is the unique solution of the problem. Let now $\gamma > 0$ and let (y_n) be a minimizing sequence from K, i.e., $\lim\|x - y_n\| = \gamma$. If we set

$$x_n := x - y_n \quad \text{and} \quad \sigma_n := \|x_n\| - \gamma,$$

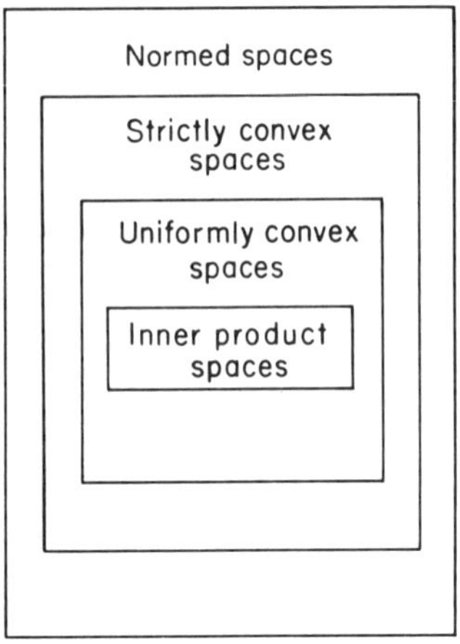

Figure 6

then $x_n \neq 0$, $\sigma_n \geqq 0$, furthermore

$$\|x_n\| \to \gamma \quad \text{and} \quad \sigma_n \to 0.$$

Because of the convexity of K the point $(y_n + y_m)/2$ lies in K, and consequently

$$\gamma \leqq \left\| x - \frac{y_n + y_m}{2} \right\| = \left\| \frac{x_n + x_m}{2} \right\| \leqq \frac{1}{2}\|x_n\| + \frac{1}{2}\|x_m\|,$$

from where $\|(x_n + x_m)/2\| \to \gamma$ follows, as $n, m \to \infty$. For $\tilde{x}_n := x_n/\|x_n\|$ we have thus

$$\|\tilde{x}_n\| = \|\tilde{x}_m\| = 1 \quad \text{and} \quad \left\| \frac{\tilde{x}_n + \tilde{x}_m}{2} \right\| \to 1 \qquad \text{for} \quad n, m \to \infty.$$

Since E is uniformly convex, we obtain $\|\tilde{x}_n - \tilde{x}_m\| \to 0$.

Because of

$$\begin{aligned} \|y_n - y_m\| &= \|x_n - x_m\| = \| \|x_n\|\tilde{x}_n - \|x_m\|\tilde{x}_m\| = \|(\gamma + \sigma_n)\tilde{x}_n - (\gamma + \sigma_m)\tilde{x}_m\| \\ &\leqq \gamma\|\tilde{x}_n - \tilde{x}_m\| + \sigma_n\|\tilde{x}_n\| + \sigma_m\|\tilde{x}_m\| \end{aligned}$$

it follows that the minimizing sequence (y_n) is indeed a Cauchy sequence. The remaining demonstration proceeds now exactly as in the proof of Theorem 60.1. ■

For the proof of the next proposition we need the inequality

(61.3) $\quad \|x + y\|^p + \|x - y\|^p \leqq 2^{p-1}(\|x\|^p + \|y\|^p)$ for $\quad x, y \in l^p$, $2 \leqq p < \infty$,

which we will verify now.

For $s, t > 0$, $\alpha := (s^2 + t^2)^{1/2}$ and $p \geqq 2$ we have

$$\left(\frac{s}{\alpha}\right)^2 + \left(\frac{t}{\alpha}\right)^2 = 1 \geqq \left(\frac{s}{\alpha}\right)^p + \left(\frac{t}{\alpha}\right)^p$$

hence also

$$\left[\left(\frac{s}{\alpha}\right)^2 + \left(\frac{t}{\alpha}\right)^2\right]^{1/2} = 1 \geqq \left[\left(\frac{s}{\alpha}\right)^p + \left(\frac{t}{\alpha}\right)^p\right]^{1/p}$$

and thus

$$(61.4) \qquad (s^2 + t^2)^{1/2} \geqq (s^p + t^p)^{1/p} \qquad \text{for} \quad s, t \geqq 0, \qquad p \geqq 2.$$

Consequently for all complex ξ, η

$$(61.5) \qquad (|\xi + \eta|^p + |\xi - \eta|^p)^{1/p} \leqq (|\xi + \eta|^2 + |\xi - \eta|^2)^{1/2} = \sqrt{2}(|\xi|^2 + |\eta|^2)^{1/2}.$$

Furthermore we have

$$(61.6) \qquad |\xi|^2 + |\eta|^2 \leqq (|\xi|^p + |\eta|^p)^{2/p} \cdot 2^{(p-2)/p},$$

an estimate which is trivial for $p = 2$ and for $p > 2$ follows from the Hölder inequality (in which p is to be replaced by $p/2$ and q by $p/(p - 2)$). From (61.5) and (61.6) we obtain

$$(|\xi + \eta|^p + |\xi - \eta|^p)^{1/p} \leqq 2^{(p-1)/p}(|\xi|^p + |\eta|^p)^{1/p}$$

and thus also the inequality

$$(61.7) \qquad |\xi + \eta|^p + |\xi - \eta|^p \leqq 2^{p-1}(|\xi|^p + |\eta|^p) \qquad \text{for} \quad p \geqq 2.$$

If we replace in it ξ, η by the components ξ_n, η_n of the vectors $x, y \in l^p$, then we obtain by addition of the inequalities the estimates (61.3). ■

Proposition 61.3. *The spaces l^p are uniformly convex for $1 < p < \infty$.*

For $p \geqq 2$ the proposition follows immediately from (61.3). For the case $1 < p < 2$ we refer the reader to [8], p. 358, where also the uniform convexity of the spaces $L^p(1 < p < \infty)$ is proved. ■

Exercises

1. The spaces l^1, l^∞, (c), (c_0), $C[a, b]$ and $B(T)$ (T should contain at least 2 elements) are not uniformly convex.

$^+$2. The normed space E is uniformly convex if and only if to every ε, $0 < \varepsilon \leqq 2$, there exists a $\delta(\varepsilon) > 0$ such that

$$\|x\| \leqq 1, \|y\| \leqq 1 \qquad \text{and} \quad \|x - y\| \geqq \varepsilon \qquad \text{imply} \quad \left\|\frac{x + y}{2}\right\| \leqq 1 - \delta(\varepsilon)$$

($\delta(\varepsilon)$ is called a *convexity modulus* of E).

$^+$3. If E is uniformly convex, then every subspace, every quotient space and the completion of E is uniformly convex.

4. If we introduce on $C[a, b]$ the norm $\|x\| := (\int_a^b |x(t)|^p \, dt)^{1/p}$, $1 \leqq p < \infty$, (see Exercise 8 in §6), then inequality (61.3) also holds in this case; thus $C[a, b]$ is uniformly convex with respect to the above 'L^p-norm' at least for $p \geqq 2$.

5. One proves the inequality $(s^p + t^p)^{1/p} \leqq (s^q + t^q)^{1/q}$ for $s, t \geqq 0, 1 \leqq q < p$ exactly like (61.4). With its help one can see that l^q is a proper linear subspace of l^p for $1 \leqq q < p < \infty$, and that $\|x\|_p \leqq \|x\|_q$ for all $x \in l^q$.

6. If $K \neq \emptyset$ is a convex subset of the strictly convex space E, then problem (61.2) has at most one solution in K.

§62. Approximation in reflexive spaces

The solution of the approximation problem (55.10) was based on the fact that every minimizing sequence is bounded and thus—since it lies in a *finite*-dimensional subspace—has a convergent subsequence. If we look for best approximations in an *infinite*-dimensional subspace, then this selection procedure cannot be applied any more because of Theorem 10.1. (Problems (60.1) and (61.2) were treated consequently by a different method: We used the fact that, under the hypotheses given there, a minimizing sequence is always a Cauchy sequence.) The spaces l^p $(1 < p < \infty)$ suggest, however, to use a selection procedure with respect to a notion of convergence which is different from convergence with respect to the norm. Namely, if (y_n) is a minimal sequence in a set $M \subset l^p$, hence a bounded sequence, then with the help of the diagonal process we see immediately that (y_n) contains a subsequence which at any rate converges componentwise to an element $y_0 \in l^p$; and about this y_0 we can prove that it is a best approximation provided it lies in M at all. By Exercise 9b of §46 a bounded and componentwise convergent sequence in l^p—and only such a sequence—is weakly convergent; we should therefore try to obtain the solution of the general approximation problem with the help of a selection procedure with respect to *weak convergence*. Indeed, this way leads to success in reflexive spaces. We first make some preparations.

Proposition 62.1. *Every bounded sequence (x_n') in the dual E' of a separable normed space E has an E-weakly convergent subsequence.*

Proof. Let $\{x_1, x_2, \ldots\}$ be a dense subset of E. The sequence of numbers $x_n'(x_1)$ is bounded and contains therefore a convergent subsequence $(x_{n1}'(x_1))$. For the same reason (x_{n1}') contains a subsequence (x_{n2}') which converges at x_2. One continues in this way. The diagonal sequence $(y_n') := (x_{nn}')$ converges then at every x_k. Since $\{x_1, x_2, \ldots\}$ is also dense in the completion $\tilde{E}$ of E, and since y_n' can be extended to $\tilde{E}$ with conservation of the norm (Proposition 12.3), it follows with the help of Theorem 33.1 that (y_n') converges pointwise on E to a certain $y' \in E'$. ■

We can now prove without great effort a 'weak' selection theorem for reflexive spaces.

Proposition 62.2. *Every bounded sequence (x_n) in a reflexive Banach space E has a weakly convergent subsequence.*

Proof. The closed subspace $F := \overline{[x_1, x_2, \ldots]}$ of E is obviously separable and by Proposition 41.7 it is reflexive, it can therefore be identified through the canonical imbedding with its bidual F''. Consequently F'' and thus also F' is separable (Exercise 8 in §42). Since (x_n) is a bounded sequence in F'', it follows from Proposition 62.1 that there exists a subsequence (x_{n_k}) and an $x \in F'' = F$ such that $\langle y', x_{n_k}\rangle \to \langle y', x\rangle$ for every $y' \in F'$. If x' is an arbitrary continuous linear form on E and if y' is its restriction to F, then

$$\langle x_{n_k}, x'\rangle = \langle x_{n_k}, y'\rangle = \langle y', x_{n_k}\rangle \to \langle y', x\rangle = \langle x, y'\rangle = \langle x, x'\rangle,$$

hence (x_{n_k}) converges weakly to x. ■

We now have the means at hand to prove at least part of the following approximation theorem.

Proposition 62.3. *If F is a closed linear subspace or only a closed convex subset $\neq \emptyset$ of the reflexive Banach space E, then for every $x \in E$ the problem*

$$\|x - y\| = \min, \qquad y \in F \tag{62.1}$$

is solvable in F.

Here we prove the proposition only under the hypothesis that F is linear; the case of a convex F will be considered at the end of §97. Let (y_n) be a minimizing sequence in F, i.e., $\|x - y_n\| \to \inf_{y \in F} \|x - y\| =: \gamma$. Since (y_n) is a bounded subset in the *reflexive* Banach space F (Proposition 41.7), there exists by Proposition 62.2 a subsequence (y_{n_k}) and an $y_0 \in F$ such that $y_{n_k} \rightharpoonup y_0$. Then $x - y_{n_k} \rightharpoonup x - y_0$, so that by Proposition 46.1 we have

$$\gamma \leqq \|x - y_0\| \leqq \liminf \|x - y_{n_k}\| = \gamma;$$

thus y_0 is indeed a best approximation for x in F. ■

Propositions 61.2 and 62.3 are, in spite of the totally different methods of proof, closely connected. One can prove, namely, that *every uniformly convex Banach space is also reflexive* (see e.g., [8], p. 354).

Exercises

1. A subset M of the normed space E is said to be *weakly sequentially complete* if every weak Cauchy sequence in M has a weak limit in M. Discuss the approximation problem (62.1) for the case that F is a weakly sequentially complete subset of the reflexive Banach space E.

2. Proposition 62.3 remains true if E is only a normed space but F is a reflexive subspace of E.

$^+$3. Every reflexive Banach space is weakly sequentially complete. So is l^1. But $C[a, b]$ is not weakly sequentially complete (for the definition see Exercise 1).

IX

Orthogonal decomposition in Hilbert spaces

§63. Orthogonal complements

From Theorem 60.1 and Proposition 60.1 we obtain immediately:

Theorem 63.1. *If F is a* complete *subspace of the inner product space E, then every vector $x \in E$ can be written uniquely in the form*

$$x = y + z \quad \text{with} \quad y \in F, z \in F^{\perp}. \tag{63.1}$$

In other words: we have the orthogonal decomposition

$$E = F \oplus F^{\perp}. \tag{63.2}$$

The hypotheses of the theorem are satisfied in particular if *E is a Hilbert space and F a closed subspace of E.*

If for a given subspace F the representation (63.2) is possible, we call the orthogonal space $F^{\perp}$ the *orthogonal complement* of F (in E) and we also say that F has an orthogonal complement. In this case one sees easily that $F^{\perp\perp} = F$, i.e., F is the orthogonal complement of $F^{\perp}$. In particular, F is then closed as an orthogonal space: *Only closed subspaces can have orthogonal complements* (and they do indeed, when E is complete).

If the orthogonal decomposition (63.2) is valid, then we call the projector P from E onto F along $F^{\perp}$ the *orthogonal projector* from E onto F. From (63.1) it follows with the aid of the theorem of Pythagoras that $\|Px\|^2 = \|y\|^2 \leqq \|y\|^2 + \|z\|^2 = \|x\|^2$, consequently P is continuous and $\|P\| \leqq 1$. Since in the case $P \neq 0$ we have also $\|P\| \geqq 1$, every orthogonal projector $P \neq 0$ has norm 1 (cf. Exercise 6 in §32). We decompose now an element $u \in E$ according to (63.1): $u = v + w\,(v \in F, w \in F^{\perp})$. It follows that $(Px\,|\,u) = (y\,|\,v + w) = (y\,|\,v) = (x\,|\,Pu)$. If we call an endomorphism A of E *symmetric* provided that for $x, y \in E$ it satisfies $(Ax\,|\,y) = (x\,|\,Ay)$, then P is symmetric. Conversely, if P is a symmetric projector and $E = P(E) \oplus N(P)$ is the corresponding decomposition, then one sees immediately that $P(E) \perp N(P)$, hence P is an orthogonal projector. We summarize:

Proposition 63.1. *Every orthogonal projector* $\neq 0$ *has norm* 1. *A projector is orthogonal if and only if it is symmetric.*

It is shown in [81] how the orthogonal decomposition of a Hilbert space leads to the solution of Dirichlet's boundary value problem (*Method of orthogonal projection*).

Exercises

$^+$1. Every finite-dimensional subspace of an inner product space has an orthogonal complement.

$^+$2. If the linear subspace F is not dense in the Hilbert space E, then there exists $x_0 \neq 0$ in E such that $x_0 \perp F$.

$^+$3. For an orthogonal projector P of E one has $P(E) = \{y \in E: \|Py\| = \|y\|\}$.

*4. If F, G are closed subspaces of the Hilbert space E and $F \perp G$, then the sum $F \oplus G$ is also closed.

$^+$5. If $E = F \oplus F^\perp$ and $A \in \mathscr{S}(E)$ is symmetric, then together with F also $F^\perp$ is invariant under A.

6. Every symmetric operator A on a Hilbert space is continuous (however, on a non-complete inner product space there may exist non-continuous symmetric operators. Examples?). *Hint*: Show that A is closed (cf. also Exercise 5 in §32).

$^+$7. The continuous endomorphism A of the Hilbert space E is relatively regular if and only if $A(E)$ is closed. *Hint*: Proposition 24.1.

8. A continuous endomorphism of a Hilbert space is a semi-Fredholm operator if and only if it is an Atkinson operator (see Exercise 7). For Hilbert spaces the characterization of Riesz operators in Exercise 5 of §52 (A is a Riesz operator $\Leftrightarrow \lambda I - A$ is a semi-Fredholm operator for all $\lambda \neq 0 \Leftrightarrow \beta(\lambda I - A) < \infty$ for all $\lambda \neq 0$) can be proved therefore much more simply than it is possible for Banach spaccs, namely with the help of the first assertion in Exercise 5 in §52.

§64. Orthogonal series

In Examples 58.1 through 58.3 we were led to the problem to expand a given element of an inner product space with respect to a (finite or infinite) sequence of pairwise orthogonal vectors. The problem will not change if we assume that these vectors are even normalized i.e., form an orthonormal sequence. We want to consider in a general way the questions connected with such expansions. *Let E be an inner product space throughout.*

Proposition 64.1. *Let $\{u_1, u_2, \dots\}$ be a countable orthonormal system in E. Then for the orthogonal series*

(64.1) $$\sum_{\nu=1}^{\infty} \alpha_\nu u_\nu$$

the following assertions hold:

(a) (64.1) *is a Cauchy series if and only if*

(64.2) $$\sum_{\nu=1}^{\infty} |\alpha_\nu|^2$$

converges. Thus in a complete *space E the series* (64.1) *and* (64.2) *converge at the same time.*

(b) *If* $\sum_{\nu=1}^{\infty} \alpha_\nu u_\nu = x$, *then* $\alpha_\nu = (x|u_\nu)$ *for all* ν *and the series converges to x for every rearrangement of its terms* (unconditional convergence).

(c) $\sum_{\nu=1}^{\infty} (x|u_\nu)u_\nu = x$ *holds if and only if*

(64.3) $$\sum_{\nu=1}^{\infty} |(x|u_\nu)|^2 = \|x\|^2$$

holds.

(d) *If E is complete, then for all* $x \in E$ *the series* $\sum_{\nu=1}^{\infty} (x|u_\nu)u_\nu$ *converges to a* $z \in E$ *and* $x - z$ *is orthogonal to* $\overline{[u_1, u_2, \dots]}$.

Proof. By (58.13) one has $\|\sum_{\nu=m}^{n} \alpha_\nu u_\nu\|^2 = \sum_{\nu=m}^{n} |\alpha_\nu|^2$, from where (a) already follows. From $\sum_{\nu=1}^{\infty} \alpha_\nu u_\nu = x$ it follows with the aid of Proposition 57.2 that $(x|u_\mu) = \sum_{\nu=1}^{\infty} \alpha_\nu(u_\nu|u_\mu) = \alpha_\mu$ for all μ, i.e., the first assertion in (b). (c) follows immediately from Bessel's identity (59.1). Let us now turn to the second assertion in (b): if $\sum_{\nu=1}^{\infty} \alpha_\nu u_\nu = x$, then by what we have already proved in (b), we have $\alpha_\nu = (x|u_\nu)$, hence (64.3) is valid because of (c). However, this equation is valid for any ordering of the u_ν; hence again by (c), for any rearrangement $\sum_{\nu=1}^{\infty} \alpha_\nu u_\nu$ converges to x.

Because of the Bessel inequality (59.2) the series $\sum_{\nu=1}^{\infty} |(x|u_\nu)|^2$ converges for every $x \in E$; if E is complete, then by (a) $\sum_{\nu=1}^{\infty} (x|u_\nu)u_\nu$ converges to an element $z \in E$. It follows from (b) that $(x|u_\nu) = (z|u_\nu)$, hence $(x - z|u_\nu) = 0$ for all ν. But then $x - z$ is also perpendicular to $\overline{[u_1, u_2, \dots]}$ (Remark before (58.13)). Thus (d) is proved. ■

We can free ourselves from the restriction to *countable* orthonormal systems with the help of the following Proposition.

Proposition 64.2. *If S is an arbitrary orthonormal system in E, then for any* $x \in E$ *at most countably many of its so-called* Fourier coefficients $(x|u)$, $u \in S$, *are different from zero, and the (generalized)* Bessel inequality

(64.4) $$\sum_{u \in S} |(x|u)|^2 \leqq \|x\|^2$$

holds, where only the Fourier coefficients different from zero are to figure in the series. In particular, the series in (64.4) *is always convergent.*

Proof. Let $\varepsilon > 0$ be given, and let $v_1, \ldots, v_n$ be elements of S with $|(x|v_\nu)| \geqq \varepsilon$. By (59.2) we have then $n\varepsilon^2 \leqq \sum_{\nu=1}^n |(x|v_\nu)|^2 \leqq \|x\|^2$, hence $n \leqq \varepsilon^{-2}\|x\|^2$. There can therefore exist only finitely many $u \in S$ with $|(u|x)| \geqq \varepsilon$. If we set ε successively equal to $1, \frac{1}{2}, \frac{1}{3}, \ldots$, we see that there are at most countably many $u \in S$, say $u_1, u_2, \ldots$, for which $(x|u) \neq 0$. For these u_ν we have, according to (59.2), for arbitrary n the inequality $\sum_{\nu=1}^n |(x|u_\nu)|^2 \leqq \|x\|^2$ from which we obtain for $n \to \infty$ the estimate (64.4). ■

If we agree once for all that in any series in which the Fourier coefficients $(x|u)$ with respect to an uncountable orthonormal system S enter (e.g., $\sum_{u\in S}(x|u)u$ or $\sum_{u\in S}|(x|u)|^2$), we only consider the coefficients $(x|u_1)$, $(x|u_2), \ldots$ which are *different from zero*, we obtain from Proposition 64.1 almost immediately

Proposition 64.3. *If S is an arbitrary orthonormal system in E, then the following assertions are valid*:

(a) *The orthogonal series $\sum_{u\in S}(x|u)u$ is either divergent or it converges at* every *reordering of the Fourier coefficients $(x|u) \neq 0$ to the same element z* (unconditional convergence). *It converges to x if and only if the equality*

$$\sum_{u\in S}|(x|u)|^2 = \|x\|^2 \tag{64.5}$$

holds.

(b) *If E is* complete, *then for every $x \in E$ the series $\sum_{u\in S}(x|u)u$ converges unconditionally to an element z of E, and $x - z$ is perpendicular to the closed linear hull of S; the latter is precisely the set $\{\sum_{u\in S}(x|u)u\colon x\in E\}$.*

We only need to prove the assertion concerning orthogonality and the characterization of the hull in (b). Let $x \in E$ be given and let $S_0 := \{u_1, u_2, \ldots\}$ be the set of those $u \in S$ which yield non-vanishing Fourier coefficients $(x|u)$ of x. For every $u \in S \backslash S_0$ we have then $(x|u) = 0$ and furthermore $(z|u) = \sum (x|u_\nu)(u_\nu|u) = 0$, hence $x - z \perp (S\backslash S_0)$. But since $x - z$ is also perpendicular to S_0 by Proposition 64.1d, we have $x - z \perp S$ and thus $x - z \perp \overline{[S]}$. Let $G := \{\sum_{u\in S}(x|u)u\colon x\in E\}$. Trivially $G \subset \overline{[S]}$. If now x is an arbitrary element from $\overline{[S]}$ and $z = \sum_{u\in S}(x|u)u$, then z and therefore $x - z$ lies in $\overline{[S]}$. On the other hand, by what we have just proved, $x - z$ is perpendicular to $\overline{[S]}$. It follows that $x - z = 0$, i.e., $\sum_{u\in S}(x|u)u = x$, and this means that x lies in G. Altogether we have $G = \overline{[S]}$. ■

The series $\sum_{u\in S}(x|u)u$ is called the *Fourier series* of x with respect to the orthonormal system S, even if it does not converge to x. If $x = \sum_{u\in S}(x|u)u$, then we say that x is expanded into its Fourier series with respect to S.

Exercises

1. If (u_n) is an orthonormal sequence, then for every x the Fourier coefficients $(x|u_n)$ tend to 0.

2. A family $(x_\iota\colon \iota \in I)$ of vectors in a normed space E is said to be *summable* to the sum $x \in E$, in symbols $\sum_{\iota \in I} x_\iota = x$, if for every $\varepsilon > 0$ there exists a finite set $J(\varepsilon)$ of indices such that for every finite set of indices $J \supset J(\varepsilon)$ one has $\|x - \sum_{\iota \in J} x_\iota\| \leqq \varepsilon$. Show the following:

(a) From $\sum_{\iota \in I} x_\iota = x, \sum_{\iota \in I} y_\iota = y$ it follows that $\sum_{\iota \in I} (x_\iota + y_\iota) = x + y$ and $\sum_{\iota \in I} \alpha x_\iota = \alpha x$ for all $\alpha \in \mathbf{K}$.

(b) At most countably many terms of a summable family are $\neq 0$.

(c) If the countable family $(x_n\colon n \in \mathbf{N})$ is summable to the sum x, then the series $\sum_{n=1}^{\infty} x_n$ converges unconditionally (i.e., for every rearrangement of its terms) to x.

(d) The family $(x_\iota : \iota \in I)$ in a Banach space E is summable if and only if for every $\varepsilon > 0$ there exists a finite index set $J(\varepsilon)$ so that for every finite index set J disjoint from $J(\varepsilon)$ one has $\|\sum_{\iota \in J} x_\iota\| \leqq \varepsilon$ (Cauchy criterion).

(e) A family $(\alpha_\iota\colon \iota \in I)$ of positive numbers is summable if and only if there exists a number γ so that for every finite index set J one has $\sum_{\iota \in J} \alpha_\iota \leqq \gamma$. In this case one has also $\sum_{\iota \in I} \alpha_\iota \leqq \gamma$.

(f) An orthogonal system $(u_\iota\colon \iota \in I)$ in a Hilbert space is summable if and only if the family $(\|u_\iota\|^2\colon \iota \in I)$ of numbers is summable.

§65. Orthonormal bases

An orthonormal system S in the inner product space E is called an *orthonormal basis* (of E) if for every $x \in E$ the Fourier expansion $x = \sum_{u \in S} (x|u)u$ holds.

From Proposition 64.3a we obtain immediately the following criterion:

Proposition 65.1. *The orthonormal system S in the inner product space E is an orthonormal basis if and only if for every $x \in E$ the* Parseval identity

$$\|x\|^2 = \sum_{u \in S} |(x|u)|^2$$

holds.

The question arises now whether every inner product space has an orthonormal basis. The answer is positive if E is separable or complete (Proposition 65.2 and Theorem 65.1).

Proposition 65.2. *Every* separable *inner product space $E \neq \{0\}$ has an at most countable orthonormal basis.*

Proof. Let $M := \{x_1, x_2, \ldots\}$ be dense in E, x_{n_1} the first non-zero element in M, x_{n_2} the first element in M which does not lie in $[x_{n_1}]$, in general x_{n_k} the first element in M which is not contained in $[x_{n_1}, \ldots, x_{n_{k-1}}]$. Obviously $\{x_{n_1}, x_{n_2}, \ldots\}$ is a linearly independent set whose linear hull is dense in E. The Gram–Schmidt orthogonalization process (Proposition 59.2) yields now an orthonormal system $\{u_1, u_2, \ldots\}$ whose linear hull is also dense in E. For every $x \in E$ there exists therefore a sequence of elements $y_n := \sum_{\nu=1}^{\infty} \alpha_{n\nu} u_\nu$ converging to x, where only finitely many of the elements $\alpha_{n1}, \alpha_{n2}, \ldots$ are different from zero. Consequently, we can determine natural numbers $m_1 < m_2 < \cdots$ so that

$$y_n = \sum_{\nu=1}^{m_n} \alpha_{n\nu} u_\nu \quad \text{and} \quad y_n \to x.$$

By Proposition 59.1 we have

$$0 \leqq \|x\|^2 - \sum_{\nu=1}^{m_n} |(x|u_\nu)|^2 = \left\| x - \sum_{\nu=1}^{m_n} (x|u_\nu)u_\nu \right\|^2 \leqq \left\| x - \sum_{\nu=1}^{m_n} \alpha_{n\nu} u_\nu \right\|^2$$
$$= \|x - y_n\|^2;$$

hence for $n \to \infty$ the sequence of partial sums $\sum_{\nu=1}^{m_n} |(x|u_\nu)|^2$ converges to $\|x\|^2$, and since by Proposition 64.2 the series $\sum_{\nu=1}^{\infty} |(x|u_\nu)|^2$ converges, it follows that $\sum_{\nu=1}^{\infty} |(x|u_\nu)|^2 = \|x\|^2$. Proposition 65.1 shows now that $\{u_1, u_2, \ldots\}$ is an orthonormal basis of E. ■

If S is an orthonormal basis of E and $x \perp S$, then $x = \sum_{u \in S} (x|u)u = \sum_{u \in S} 0 \cdot u = 0$, hence S is a *maximal* orthonormal system, i.e., S is not properly contained in any orthonormal system of E. If conversely S is maximal, then S is an orthonormal basis whenever E is complete (see Proposition 64.3b). We have therefore:

Proposition 65.3. *An orthonormal system S in a Hilbert* space *is an orthonormal basis if and only if it is* maximal.

The question, which suggests itself now, about the existence of a maximal orthonormal system is answered by

Theorem 65.1. *An inner product space $E \neq \{0\}$ has maximal orthonormal systems, hence, if it is complete, also orthonormal bases. Every orthonormal system can be extended to a maximal orthonormal system.*

We first prove the second assertion. Let S_0 be an orthonormal system in E and let $\mathfrak{M}$ be the set of all orthonormal systems $S \supset S_0$ in E. If we order $\mathfrak{M}$ by inclusion, then we obtain the assertion from Zorn's lemma. The existence of maximal orthonormal systems follows now if one extends an orthonormal system of the form $S_0 = \{x/\|x\|\}$, $x \neq 0$, to a maximal one. ■

A *maximal* orthonormal system S can be useful even when it is not an orthonormal basis, because *the Fourier coefficients* $(x|u)$ *of an element x with respect* *to S determine this element uniquely*. Indeed, from $(x|u) = (y|u)$, i.e., $(x - y|u) = 0$ for all $u \in S$, it follows namely that $x - y = 0$.

In [71] examples of non-complete inner product spaces are given which have no orthonormal basis.

Exercises

$^+$1. The following assertions concerning the orthonormal system S in the inner product space E are equivalent: (a) S is an orthonormal basis. (b) S is a *basic set* (i.e., $\overline{[S]} = E$). (c) For all x, y in E one has $(x|y) = \sum_{u \in S} (x|u)\overline{(y|u)}$ (also this equality is called the *Parseval identity*).

2. Let $S := \{u_1, u_2, \ldots\}$ be a maximal orthonormal system in the inner product space E. Show that E is complete if and only if $\sum_{\nu=1}^{\infty} \alpha_\nu u_\nu$ converges for every sequence of numbers (α_ν) for which $\sum_{\nu=1}^{\infty} |\alpha_\nu|^2 < \infty$. *Hint* for the proof of the sufficiency: Show that every $x \in E$ can be expanded with respect to S and construct an isomorphism of normed spaces from E onto l^2.

3. Let $\{u_t : t \in T\}$ be a maximal orthonormal system in the inner product space E. Show that E is complete if and only if $\sum_{t \in T} \alpha_t u_t$ converges for every family of numbers $(\alpha_t : t \in T)$ which has the following properties: (a) At most countably many terms of the family (α_t) are $\neq 0$ (and only these terms shall occur effectively in the above series.) (b) $\sum_{t \in T} |\alpha_t|^2 < \infty$. *Hint*: Proceed as in Exercise 2, replace, however, l^2 by the space $l^2(T)$ (see §57 Exercise 3).

$^+$4. A Hilbert space E with the orthonormal basis S is isomorphic as a normed space to $l^2(S)$ (see Exercise 3 in §57). Special cases: (a) If $\dim E = n < \infty$, then E is isomorphic as a normed space to $l^2(n)$. (b) If E is infinite-dimensional but separable, then E is isomorphic as a normed space to l^2.

$^+$5. Two maximal orthonormal systems S_1, S_2 in the inner product space E have the same cardinality (which is also called the *Hilbert dimension* of E). *Hint*: With every $u \in S_1$ associate the (at most countable) set $S_2(u) := \{v \in S_2 : (u|v) \neq 0\}$ and show that $S_2 = \bigcup_{u \in S_1} S_2(u)$.

§66. The dual of a Hilbert space

The dual of a Hilbert space is described by the following *representation theorem* of Fréchet and F. Riesz:

Proposition 66.1. *For each fixed element z of the Hilbert space E a continuous linear form is defined on E by $f(x) := (x|z)$. Conversely, for each continuous linear form f on E there exists exactly one vector $z \in E$ such that $f(x) = (x|z)$ holds for all $x \in E$. Furthermore $\|f\| = \|z\|$.*

Proof. Trivially, for a fixed $z \in E$ a linear form is defined on E by $f(x) := (x|z)$ which is continuous because of $|f(x)| = |(x|z)| \leqq \|x\| \|z\|$, and satisfies $\|f\| \leqq \|z\|$. Because of $|f(z)| = |(z|z)| = \|z\| \|z\|$, we even have $\|f\| = \|z\|$ (this part of the proof is valid for arbitrary inner product spaces). Now let, conversely, f be an element of E'. If $f = 0$, then $z = 0$ satisfies the requirements. If $f \neq 0$, then because of Theorem 63.1 there exists a vector $x_0 \neq 0$ which is orthogonal to the closed nullspace $N(f)$; in particular $f(x_0) \neq 0$. By (15.7) every $x \in E$ can be written in the form $x = \alpha x_0 + y$ with uniquely determined $\alpha \in \mathbf{K}$ and $y \in N(f)$. Thus we have

$$f(x) = \alpha f(x_0). \tag{66.1}$$

Setting $z := [\overline{f(x_0)}/(x_0|x_0)]x_0 \in N(f)^\perp$, we have

$$(x|z) = (\alpha x_0 + y|z) = \alpha(x_0|z) = \alpha\left(x_0 \,\Big|\, \overline{f(x_0)}\frac{x_0}{(x_0|x_0)}\right) = \alpha f(x_0) = f(x);$$

the last equality in this chain is nothing but (66.1). z thus generates f, and is obviously the only vector in E which does so. ■

For greater clarity we denote the vector generating $f \in E'$ by z_f. By Proposition 66.1 the correspondence $f \mapsto z_f$ defines a norm-preserving, bijective map A from E' onto E. Obviously $z_{f+g} = z_f + z_g$, hence A is also additive. But because of $(x|z_{\alpha f}) = (\alpha f)(x) = \alpha f(x) = (x|\bar{\alpha} z_f)$, we have $z_{\alpha f} = \bar{\alpha} z_f$, i.e., $A(\alpha f) = \bar{\alpha} A f$, hence the map A is only in the case of a real Hilbert space linear, and thus an isomorphism of normed spaces between E' and E. If E is complex, then we define by means of an orthonormal basis S of E (Theorem 65.1) a self-map K of E (a *conjugation*) by $Kx := \sum_{u \in S} \overline{(x|u)} u$. Obviously K has the following properties:

(a) $K(x + y) = Kx + Ky$,
(b) $K(\alpha x) = \bar{\alpha} Kx$,
(c) $K^2 = I$,
(d) $\|Kx\| = \|x\|$,
(e) K is bijective and $K^{-1} = K$.

It now follows that $f \mapsto K z_f$ is an isomorphism of normed spaces between E' and E. We state this result:

Proposition 66.2. *Every Hilbert space is isometrically isomorphic to its dual.*

On the dual E' of a Hilbert space E one defines by $(f|g) := (z_g|z_f)$ an inner product which generates the norm of E' by Proposition 66.1: thus E' is a Hilbert space. Consequently every continuous linear form F on E' is generated by a well-determined $g \in E$ (Proposition 66.1): $F(f) = (f|g)$ for all $f \in E'$. By the definition of the inner product $(f|g)$ and of the imbedding isomorphism $J: E \to E''$ (see (41.1)), we have thus $F(f) = (z_g|z_f) = f(z_g) = (Jz_g)(f)$; thus $E'' = J(E)$ and we obtain

Proposition 66.3. *Every Hilbert space is reflexive.*

Because of Proposition 66.1, weak convergence $x_n \rightharpoonup x$ in a Hilbert space E means that $(x_n|z) \to (x|z)$ for every $z \in E$.

From Propositions 62.2 and 66.3 we obtain immediately:

Proposition 66.4. *Every bounded sequence (x_n) in a Hilbert space E contains a weakly convergent subsequence, i.e., there exists a subsequence (x_{n_k}) and an $x \in E$ such that $(x_{n_k} | z)$ converges to $(x | z)$ for every $z \in E$.*

Some important propositions of the present and the preceding section could be proved only under completeness hypotheses. In the case of a non-complete space E we will often be able to remedy the situation by constructing the completion of E; it is a Hilbert space, more precisely:

Proposition 66.5. *For the non-complete inner product space E there exists a Hilbert space $\tilde{E}$, unique up to an isomorphism of normed spaces, so that E is a dense subspace of $\tilde{E}$ (in particular, the inner product of E is induced by the one in $\tilde{E}$).*

Proof. Let $\tilde{E}$ be the completion of the normed space E according to Proposition 12.2. If x, y are vectors in $\tilde{E}$ and (x_n), (y_n) are approximating sequences in E, then from

$$|(x_n | y_n) - (x_m | y_m)| \leqq |(x_n - x_m | y_n)| + |(x_m | y_n - y_m)|$$
$$\leqq \|x_n - x_m\| \, \|y_n\| + \|x_m\| \, \|y_n - y_m\|$$

it follows that $\lim(x_n | y_n)$ exists. A similar estimate shows that this limit is independent of the particular choice of the approximating sequences. Consequently, we have the right to define $(x | y) := \lim(x_n | y_n)$. It is easy to see that hereby an inner product is introduced on $\tilde{E}$. This inner product obviously extends the inner product given on E, and the corresponding norm $\|x\| = \sqrt{(x | x)} = \lim \sqrt{(x_n | x_n)} = \lim \|x_n\|$ coincides with the norm on $\tilde{E}$. Thus $\tilde{E}$ is a Hilbert space. ■

Exercises

1. The second assertion of Proposition 66.1 is false in non-complete inner product spaces. On such spaces there exist more continuous linear forms than are generated by the elements of the space.

2. An orthonormal basis of the inner product space E is also an orthonormal basis of its completion $\tilde{E}$. *Hint*: Use Exercise 1 in §65.

3. Every orthonormal sequence in a Hilbert space converges weakly to 0. *Hint*: Exercise 1 in §64.

$^+$4. Let E be an inner product space over $\mathbf{K}$. A map $(x, y) \mapsto s(x, y)$ from $E \times E$ into $\mathbf{K}$ is called a *sesquilinear form* if the following hold: $s(x_1 + x_2, y) = s(x_1, y) + s(x_2, y)$, $s(x, y_1 + y_2) = s(x, y_1) + s(x, y_2)$, $s(\alpha x, y) = \alpha s(x, y)$, $s(x, \alpha y) = \bar{\alpha} s(x, y)$. For every $A \in \mathscr{S}(E)$, e.g., $s(x, y) := (Ax | y)$ is sesquilinear. *Continuity* and *boundedness* of a sesquilinear form is defined as for bilinear forms

(Exercise 4 in §14); the bounded sesquilinear forms are precisely the continuous ones. Show the following:

(a) Let $A \in \mathscr{S}(E)$; then $s(x, y) := (Ax \mid y)$ is continuous if and only if A is continuous. In this case $\|A\| = \sup_{x, y \neq 0} |s(x, y)|/\|x\| \, \|y\|$.

(b) Let E be complete and $s(x, y)$ a continuous sesquilinear form. Then there exists an $A \in \mathscr{L}(E)$ such that $s(x, y) = (Ax \mid y)$ for all $x, y \in E$.

§67. The adjoint transformation

Let E and F be two Hilbert spaces over $\mathbf{K}$. The transformation A' dual to $A \in \mathscr{L}(E, F)$ maps F' into E'. Proposition 66.1 suggests, however, not to consider the map $g \mapsto f := A'g$ of *linear forms*, but the corresponding map $y_g \mapsto x_f$ of the *generating vectors* (one does not have to leave then the spaces E and F). The map $A^*: F \to E$ so defined is characterized by the equation

$$(Ax \mid y) = (x \mid A^*y) \qquad \text{for all} \quad x \in E, y \in F, \tag{67.1}$$

and is called the *transformation* (map, operator) *adjoint* to A or shortly the *adjoint* of A (a confusion with the algebraically dual transformation, which we also denote by A^*, is not to be feared: In the context of Hilbert spaces A^* will always denote the adjoint of A). One will expect that A^* has properties similar to A'; indeed the reader can easily convince himself of the truth of the following proposition in which A and B denote continuous linear maps of Hilbert spaces.

Proposition 67.1. *A^* is a continuous linear map with $\|A^*\| = \|A\|$. The following rules hold:*

$$(A + B)^* = A^* + B^*, \quad (\alpha A)^* = \bar{\alpha} A^*, \quad (AB)^* = B^*A^*,$$

$$I^* = I, \quad 0^* = 0;$$

together with A also A^* is bijective, and we have $(A^*)^{-1} = (A^{-1})^*$;

$$A^{**} = A.$$

In the assertion before the last one, observe that A^{-1} is continuous by Proposition 32.2. In the last assertion, A^{**} is a short way of writing $(A^*)^*$.

Also the simple proof of the following proposition is left to the reader (cf. Exercise 1 in §29).

Proposition 67.2. *If A is a continuous linear map from the Hilbert space E into the Hilbert space F, then the following identities hold:*

$$\overline{A(E)}^{\perp} = N(A^*), \qquad \overline{A(E)} = N(A^*)^{\perp}, \tag{67.2}$$

$$\overline{A^*(F)}^{\perp} = N(A), \qquad \overline{A^*(F)} = N(A)^{\perp}. \tag{67.3}$$

From Proposition 29.3 and the second equation in (67.2) we obtain immediately

Proposition 67.3. *Under the hypotheses of Proposition* 67.2 *the map A is normally solvable if and only if $A(E) = N(A^*)^\perp$ holds, i.e., if the equation $Ax = y$ has a solution exactly in the case*

$$\text{`}(y|z) = 0 \qquad \textit{for all} \quad z \in N(A^*)\text{'}.$$

Thus in the theory of normal solvability A^* takes over the role of A', the inner product the role of the canonical bilinear form.

We close this section with a useful identity concerning norms.

Proposition 67.4. *We have* $\|A^*A\| = \|AA^*\| = \|A\|^2$.

Proof. With the aid of Proposition 67.1 we get $\|A^*Ax\| \leqq \|A^*\|\,\|A\|\,\|x\| = \|A\|^2\|x\|$, hence $\|A^*A\| \leqq \|A\|^2$. On the other hand, $\|Ax\|^2 = (Ax|Ax) = (x|A^*Ax) \leqq \|A^*A\|\,\|x\|^2$, hence $\|A\|^2 \leqq \|A^*A\|$. Altogether we have $\|A^*A\| = \|A\|^2$. It follows further that $\|AA^*\| = \|A^{**}A^*\| = \|A^*\|^2 = \|A\|^2$. ∎

Exercises

+1. For $A \in \mathscr{L}(E)$ we have $\sigma(A^*) = \{\bar{\lambda}: \lambda \in \sigma(A)\}$ (cf. Proposition 44.2).

2. $N(A^*) = N(AA^*)$, $\overline{A(E)} = \overline{AA^*(F)}$.

3. A compact self-map of a Hilbert space has a compact adjoint (cf. Proposition 42.2).

4. If the endomorphism A of the space $l^2(n)$ or l^2 is generated by the matrix (α_{jk}) (i.e., if $(\eta_1, \eta_2, \ldots) := A(\xi_1, \xi_2, \ldots)$ with $\eta_j := \sum \alpha_{jk}\xi_k$, $j = 1, 2, \ldots$), then A^* is generated by the matrix $(\bar{\alpha}_{kj})$.

*5. A closed subspace F of a Hilbert space E is invariant under $A \in \mathscr{L}(E)$ if and only if $F^\perp$ is invariant under A^*; $(F, F^\perp)$ reduces A if and only if F is invariant under A and A^*.

X

Spectral theory in Hilbert spaces

§68. Symmetric operators

We remind the reader of the definition of a symmetric operator in §63: An endomorphism A of the inner product space E is said to be *symmetric* if

(68.1) $$(Ax \mid y) = (x \mid Ay) \qquad \text{for all} \quad x, y \text{ in } E.$$

We already saw an example in §58 Exercise 5.

Real multiples, sums and (pointwise) limits of symmetric operators are again symmetric. The product of the symmetric endomorphisms A, B is symmetric if and only if A and B commute.

Proposition 68.1. *For a symmetric operator A every eigenvalue and $(Ax \mid x)$ is* real. *Eigenvectors corresponding to distinct eigenvalues are* orthogonal *to each other.*

Proof. $(Ax \mid x)$ is real because of $(Ax \mid x) = (x \mid Ax) = \overline{(Ax \mid x)}$. For an eigenvalue λ we have $Ax = \lambda x$ with an appropriate $x \neq 0$ and thus $\lambda = (Ax \mid x)/(x \mid x)$ is real. If, furthermore, $Ay = \mu y$, $y \neq 0$ and $\mu \neq \lambda$, then $\lambda(x \mid y) = (\lambda x \mid y) = (Ax \mid y) = (x \mid Ay) = (x \mid \mu y) = \mu(x \mid y)$, hence $(\lambda - \mu)(x \mid y) = 0$ and therefore $(x \mid y) = 0$. ■

We say that the symmetric operator A is *positive*, and write $A \geqq 0$, if $(Ax \mid x) \geqq 0$ for every $x \in E$. In this case $[x \mid y] := (Ax \mid y)$ defines on E a so-called *semi-inner product*, i.e., an expression which has all the properties of an inner product with one possible exception: From $[x \mid x] = 0$ it does not have to follow that $x = 0$. *However, the Schwarz inequality* (57.4) *is valid for such semi-inner products,* as we can see by modifying slightly the proof. We can therefore state the following proposition:

Proposition 68.2. *If $A \geqq 0$, then the* generalized Schwarz inequality

(68.2) $$|(Ax \mid y)|^2 \leqq (Ax \mid x)(Ay \mid y) \qquad \textit{for all} \quad x, y \textit{ in } \quad E$$

holds.

From the generalized Schwarz inequality it follows that a symmetric endomorphism is uniquely determined by its *quadratic form* $(Ax|x)$; more precisely, we have:

Proposition 68.3. *If for the* symmetric *endomorphisms A and B one always has $(Ax|x) = (Bx|x)$, then they are equal.*

Indeed, in this case $T := A - B$ is symmetric and satisfies $(Tx|x) = 0$ for all x. From (68.2) it follows that $(Tx|y) = 0$ for all x, y, hence also $\|Tx\|^2 = (Tx|Tx) = 0$ for all x and thus $T = 0$. ■

For arbitrary endomorphisms this proposition is correct only if the space is complex. For the proof we use the representation of the *hermitian form* $(Ax|y)$ of A, which is valid for any endomorphism A of a *complex* inner product space:

$$\begin{aligned} 4(Ax|y) &= (A(x+y)|x+y) - (A(x-y)|x-y) \\ &\quad + i(A(x+iy)|x+iy) - i(A(x-iy)|x-iy). \end{aligned} \tag{68.3}$$

This formula expresses, stated briefly, the hermitian form of A by the corresponding quadratic form. Thus if the quadratic forms of A and B coincide, then $(Ax|y) = (Bx|y)$ for all x, y from where $A - B = 0$ follows as above. We consider along with (68.3) the analogous representation

$$\begin{aligned} 4(x|Ay) &= (x+y|A(x+y)) - (x-y|A(x-y)) \\ &\quad + i(x+iy|A(x+iy)) - i(x-iy|A(x-iy)). \end{aligned} \tag{68.4}$$

If $(Az|z)$ is always real, then it follows that $(Az|z) = (z|Az)$; from (68.3) and (68.4) we obtain then that $(Ax|y) = (x|Ay)$ for all x, y, hence A is symmetric. If we take into account Proposition 68.1, then we can state the following result:

Proposition 68.4. *Let A and B be endomorphisms of the* complex *inner product space E. Then the following assertions are valid:*

(a) *From $(Ax|x) = (Bx|x)$ for all x it follows that $A = B$.*
(b) *A is symmetric if and only if $(Ax|x)$ is real for all $x \in E$.*

The next proposition shows that the norm of a continuous symmetric operator can be determined with the help of its quadratic form.

Proposition 68.5. *For the continuous and symmetric operator A we have*

$$\|A\| = \sup_{\|x\|=1} |(Ax|x)|. \tag{68.5}$$

Proof. For every normalized x we have $|(Ax|x)| \leqq \|A\|\,\|x\|^2 = \|A\|$, hence

$$v(A) := \sup_{\|x\|=1} |(Ax|x)| \leqq \|A\|. \tag{68.6}$$

For an arbitrary $\lambda > 0$ we have

$$4\|Ax\|^2 = \left(A\left(\lambda x + \frac{1}{\lambda}Ax\right)\middle|\lambda x + \frac{1}{\lambda}Ax\right) - \left(A\left(\lambda x - \frac{1}{\lambda}Ax\right)\middle|\lambda x - \frac{1}{\lambda}Ax\right)$$

$$\leqq v(A)\left[\left\|\lambda x + \frac{1}{\lambda}Ax\right\|^2 + \left\|\lambda x - \frac{1}{\lambda}Ax\right\|^2\right]$$

$$= 2v(A)\left[\lambda^2\|x\|^2 + \frac{1}{\lambda^2}\|Ax\|^2\right]$$

(the last equality follows from the parallelogram law). If $\|Ax\| \neq 0$, and if one sets $\lambda^2 = \|Ax\|/\|x\|$, then it follows that $\|Ax\| \leqq v(A)\|x\|$. This inequality is trivially true also in the case $\|Ax\| = 0$ and yields the estimate $\|A\| \leqq v(A)$ for the norm of A, from where the assertion follows with the aid of (68.6). ■

If A is a symmetric operator, if $x_n \to x$ and $Ax_n \to y$, then for every z we have

$$(Ax_n|z) \to (y|z)$$

and

$$(Ax_n|z) = (x_n|Az) \to (x|Az) = (Ax|z),$$

so that $Ax = y$ and thus A is closed. With the aid of the closed graph theorem (Theorem 32.3) we obtain from here the important *Hellinger–Toeplitz Theorem* (see Exercise 6 in §63):

Proposition 68.6. *Every symmetric operator on a Hilbert space is* continuous.

For a symmetric operator A we have $(x|A^*y) = (Ax|y) = (x|Ay)$ for all x, y, hence $A = A^*$. Conversely, from $A = A^*$ the symmetry of A follows. Continuous endomorphisms A of a Hilbert space, which coincide with their adjoint A^*, are called *selfadjoint*. With this terminology *an endomorphism of a Hilbert space is symmetric if and only if it is selfadjoint.*

Because their quadratic form is real, symmetric operators have some properties which remind us of the situation in the real number field. Thus e.g., an *order* can be introduced: For two symmetric operators A, B the symbol $A \leqq B$ (or $B \geqq A$) means that $(Ax|x) \leqq (Bx|x)$ for all x (for the verification of the axioms of order one needs Proposition 68.3). Obviously $A \leqq B$ is equivalent to $B - A \geqq 0$. A sequence A_n of symmetric operators is said to be *monotone increasing* or *decreasing* if $A_1 \leqq A_2 \leqq \cdots$ or $A_1 \geqq A_2 \geqq \cdots$, respectively; it is said to be *bounded from above* (*below*) if there exists a symmetric operator B with $A_n \leqq B$ ($A_n \geqq B$) for all n. A sequence is said to be *bounded* if it is bounded from above and from below. Analogously to the well-known convergence theorem for monotone sequences of numbers, we have:

Proposition 68.7. *Every monotone and bounded sequence of symmetric operators on a Hilbert space converges pointwise to a symmetric operator.*

In the proof we consider a monotone increasing sequence: $A_1 \leqq A_2 \leqq \cdots \leqq B$. For $n > m$ we have $A_n - A_m \geqq 0$; with the aid of Proposition 68.5 we obtain first

$$\|A_n - A_m\| = \sup_{\|x\|=1} (A_n x - A_m x | x) \leqq \sup_{\|x\|=1} [(Bx|x) - (A_1 x|x)] =: \alpha,$$

and then with the aid of Proposition 68.2 the estimate

$$\begin{aligned} \|A_n x - A_m x\|^4 &= ((A_n - A_m)x|(A_n - A_m)x)^2 \\ &\leqq ((A_n - A_m)x|x)((A_n - A_m)^2 x|(A_n - A_m)x) \\ &\leqq [(A_n x|x) - (A_m x|x)]\alpha^3 \|x\|^2. \end{aligned}$$

The sequence of numbers $(A_n x|x)$ is monotone increasing and bounded, hence convergent. Our estimate shows therefore that $(A_n x)$ is a Cauchy sequence. Since the space is complete, this sequence converges, i.e., (A_n) tends pointwise to a (symmetric) operator A. ■

We return to the order relation between symmetric operators. If A, B, C are symmetric, then $A \leqq B$ implies that $A + C \leqq B + C$ and $\alpha A \leqq \alpha B$ if $\alpha \geqq 0$. The problem, whether the inequality $A \leqq B$ 'may' be multiplied by a positive operator, is more difficult. For its investigation we need *Reid's inequality* (see [132]):

Proposition 68.8. *If A, B are continuous operators, $A \geqq 0$ and AB symmetric, then for all x we have*

$$|(ABx|x)| \leqq \|B\|(Ax|x). \tag{68.7}$$

Proof. From (68.2) (generalized Schwarz inequality) and the inequality between the arithmetic and geometric mean we obtain

$$|(Ax|y)| \leqq \tfrac{1}{2}[(Ax|x) + (Ay|y)]. \tag{68.8}$$

Because of $(AB^n x|y) = (B^{n-1}x|ABy) = (AB^{n-1}x|By) = (B^{n-2}x|AB^2 y) = \cdots = (x|AB^n y)$, the operator AB^n is symmetric; from (68.8) we obtain for $n = 1, 2, \ldots$ the estimate

$$|(AB^n x|x)| = |(x|AB^n x)| = |(Ax|B^n x)| \leqq \tfrac{1}{2}[(Ax|x) + (AB^{2n}x|x)].$$

By mathematical induction we get now the inequality

$$|(ABx|x)| \leqq \left(\frac{1}{2} + \frac{1}{4} + \cdots + \frac{1}{2^n}\right)(Ax|x) + \frac{1}{2^n}(AB^{2^n}x|x). \tag{68.9}$$

If $\|B\| = 1$, then because of $|(AB^{2^n}x|x)| \leqq \|A\| \cdot \|x\|^2$ the last term in (68.9) tends to 0; thus for $n \to \infty$ we obtain $|(ABx|x)| \leqq (Ax|x)$. From this special case of Reid's inequality we obtain (68.7) if we replace B by $B/\|B\|$. ■

We can now show that the order relation between continuous symmetric operators is compatible with multiplication:

Proposition 68.9. *If A, B, C are continuous symmetric operators, then $A \leqq B$ and $C \geqq 0$ imply $AC \leqq BC$ whenever C commutes with A and B. In particular, the product of two continuous, positive and commuting operators is always positive.*

We prove first the last assertion. Let A, B be continuous, positive and commuting. We may assume $0 \leqq I - B \leqq I$; if this does not already hold, we replace B by βB with an appropriate positive factor β. By Proposition 68.5 we have then $\|I - B\| \leqq 1$; since A commutes with $I - B$, and since therefore $A(I - B)$ is symmetric, it follows from Proposition 68.8 that

$$(A[I - B]x|x) \leqq (Ax|x), \qquad \text{hence} \quad A - AB \leqq A$$

and thus $0 \leqq AB$. The first assertion now follows easily from what we have just proved; one only has to multiply the positive operator $B - A$ by C. ■

Exercises

*1. For every continuous positive operator we have $\|Ax\|^2 \leqq \|A\|(Ax|x)$.

$^+$2. If $[x|y]$ is a continuous semi-inner product on the Hilbert space E with the inner product $(x|y)$ (i.e., if $x_n \to x$, $y_n \to y$ implies that $[x_n|y_n] \to [x|y]$), then there exists a continuous, positive operator S with $[x|y] = (Sx|y)$ for all $x, y \in E$ (see §66 Exercise 4).

*3. On the vector space E let a semi-inner product $[x|y]$, and with it a semi-norm $|x| := \sqrt{[x|x]}$, be given. Furthermore let $A \in \mathscr{S}(E)$ be *symmetric*, i.e., let $[Ax|y] = [x|Ay]$ for all $x, y \in E$. Show the following:

(a) $[Ax|x] \in \mathbf{R}$ for all $x \in E$.

(b) If A is even *fully symmetric*, i.e., if $|u| \neq 0$ for all eigensolutions u of A corresponding to non-zero eigenvalues, then all the eigenvalues are real, and the eigensolutions corresponding to distinct eigenvalues are orthogonal to each other.

(c) From $\nu(A) := \sup_{|x|=1} |[Ax|x]| < \infty$ it follows that A is bounded and that its norm $|A|$ is equal to $\nu(A)$, provided that there exist at all elements x with $|x| \neq 0$; otherwise we have trivially $|A| = \nu(A) = 0$ (*boundedness* and *norm* of A are defined like in normed spaces).

$^+$4. The endomorphism A of the Hilbert space E is said to be *symmetrizable* if there exists a continuous semi-inner product $[x|y]$ on E, with respect to which A is symmetric; A is said to be *fully symmetrizable* if we have furthermore $|x| := \sqrt{[x|x]} \neq 0$ for all eigensolutions x corresponding to non-zero eigenvalues. Show the following:

(a) A is symmetrizable if and only if there exists an $H \geqq 0$ such that HA is symmetric with respect to the inner product of E.

(b) The eigenvalues of a fully symmetrizable operator are all real; for eigensolutions u, v corresponding to distinct eigenvalues we have $[u|v] = 0$.

(c) A symmetrizable and bounded operator A is also bounded with respect to the semi-norm $|x|$, and we have $|A| \leqq \|A\|$. *Hint*: Exercises 2, 3; Proposition 68.8.

$^{+}$5. The definition of symmetrizability and of full symmetrizability from Exercise 4 can be carried over word for word to the case when E is a *Banach space* (cf. [60]). Show the following:

(a) A semi-inner product $[x|y]$ on E is continuous if and only if there exists a $\gamma \geqq 0$ such that $|x| := \sqrt{[x|x]} \leqq \gamma\|x\|$ holds for all x.

(b) For a fully symmetrizable A assertion (b) in Exercise 4 holds.

(c) For a symmetrizable and bounded A assertion (c) of Exercise 4 holds. *Hint* for (c): From $|A^n x|^2 \leqq |x||A^{2n}x|$ it follows that $|Ax|/|x| \leqq (|A^{2^k}x|/|x|)^{1/2^k}$, if $|x| \neq 0$. Now use (a). See also the beginning of §74.

§69. Orthogonal projectors

In this section let E be a Hilbert space.

We have defined orthogonal projectors in §63; in Proposition 63.1 we asserted that precisely the symmetric projectors are the ones which are orthogonal.

The terminology introduced in §5 concerning reducing subspaces will be somewhat simplified in the context of Hilbert spaces where only orthogonal decompositions and orthogonal projectors play a role: We say that the closed subspace F of E *reduces* the operator $A \in \mathscr{S}(E)$ if F and $F^\perp$ are invariant under A (observe that $A = F \oplus F^\perp$ by Theorem 63.1). Proposition 5.4 takes now the following form:

Proposition 69.1. *Let P be the* orthogonal projector *from E onto the closed subspace F. Then the following assertions hold*:

(a) *F is invariant under the endomorphism A if and only if* $AP = PAP$.
(b) *F reduces A if and only if* $AP = PA$.

The following theorems deal with commuting projectors and order relations between them.

Proposition 69.2. *Let* P_1, P_2 *be the orthogonal projectors onto the closed subspaces* F_1, F_2 *of E. Then the following assertions hold*:

(a) *If* P_1 *and* P_2 *commute, then* P_1P_2 *is the orthogonal projector onto* $F_1 \cap F_2$.

(b) *If* $P_1P_2 = 0$, *then also* $P_2P_1 = 0$, *the subspaces* F_1, F_2 *are orthogonal to each other, and* $P_1 + P_2$ *is the orthogonal projector onto the subspace* $F_1 \oplus F_2$ *(which is closed by Exercise* 4 *in* §63).

Proof.

(a) From the assumption that P_1 and P_2 commute, it follows that P_1P_2 is idempotent and symmetric, hence an orthogonal projector. The image space of P_1P_2 is $\{x \in E: P_1P_2x = x\} = \{x \in E: P_2P_1x = x\} = F_1 \cap F_2$.

(b) We have $P_2P_1 = P_2^*P_1^* = (P_1P_2)^* = 0^* = 0$. For $x \in F_1$, $y \in F_2$ we have furthermore $(x|y) = (P_1x|P_2y) = (x|P_1P_2y) = 0$, i.e., F_1, F_2 are orthogonal to each other. $P_1 + P_2$ is symmetric and because of $(P_1 + P_2)^2 = P_1^2 + 2P_1P_2 + P_2^2 = P_1 + P_2$ idempotent, hence an orthogonal projector, and it is easy to see that its image space coincides with $F_1 \oplus F_2$. ■

For every orthogonal projector P we have $(Px|x) = (P^2x|x) = (Px|Px)$, hence

(69.1) $$(Px|x) = \|Px\|^2;$$

because of $0 \leqq \|Px\|^2 \leqq \|P\|^2\|x\|^2 \leqq (x|x)$ (we have $\|P\| \leqq 1$ by Proposition 63.1), it follows furthermore that

(69.2) $$0 \leqq P \leqq I.$$

A more general assertion concerning order is made by:

Proposition 69.3. *Under the hypotheses of Proposition* 69.2 *the following assertions are equivalent*:

(a) $P_1 \leqq P_2$.
(b) $\|P_1x\| \leqq \|P_2x\|$ *for all* $x \in E$.
(c) $P_1P_2 = P_2P_1 = P_1$.
(d) $F_1 \subset F_2$.
(e) $P_2 - P_1$ *is an orthogonal projector.*

Proof. Because of (69.1) we have (a) $\Leftrightarrow$ (b). From (a) it follows that $I - P_2 \leqq I - P_1$, thus because of (69.1) we have $\|(I - P_2)P_1x\|^2 = ((I - P_2)P_1x|P_1x) \leqq ((I - P_1)P_1x|P_1x) = 0$ for all $x \in E$ and therefore $(I - P_2)P_1 = 0$, hence $P_2P_1 = P_1$. But then also $P_1P_2 = P_1^*P_2^* = (P_2P_1)^* = P_1^* = P_1$. Thus we have proved (a) $\Rightarrow$ (c). (c) $\Rightarrow$ (d) since, because of Proposition 69.2a, we have $F_1 = P_1(E) = (P_1P_2)(E) = F_1 \cap F_2$. We now show (d) $\Rightarrow$ (b). With the aid of the orthogonal complement G of F_1 in F_2 we obtain in $E = F_1 + G + F_2^\perp$ a decomposition of E into pairwise orthogonal subspaces. If $x = x_1 + y + x_2$ is the corresponding decomposition of an arbitrary vector x, then $\|P_1x\|^2 = \|x_1\|^2 \leqq \|x_1\|^2 + \|y\|^2 = \|x_1 + y\|^2 = \|P_2x\|^2$, hence (b) holds. So far we have proved the equivalence of the assertions (a) through (d). (a) $\Rightarrow$ (e): $P_2 - P_1$ is symmetric, and since together with (a) also (c) holds, it follows that $(P_2 - P_1)^2 = P_2^2 - 2P_2P_1 + P_1^2 = P_2 - 2P_1 + P_1 = P_2 - P_1$, hence $P_2 - P_1$ is an idempotent, and thus (e) is true. With the help of (69.2) from (e) we get (a) immediately. ■

Exercises

*1. Under the hypotheses of Proposition 69.2 we have: $F_1 \perp F_2$ implies $P_1P_2 = P_2P_1 = 0$.

*2. Let $P_1, \ldots, P_n$ be the orthogonal projectors onto the closed subspaces $F_1, \ldots, F_n$ of E. Then the following holds: $P := P_1 + \cdots + P_n$ is an orthogonal projector if and only if $P_jP_k = 0$ for $j \neq k$. In this case $P(E) = F_1 + \cdots + F_n$.

3. If under the hypotheses of Proposition 69.2 P_1 and P_2 commute, then $P := P_1 + P_2 - P_1P_2$ is an orthogonal projector and $P(E) = F_1 + F_2$.

§70. Normal operators and their spectra

We consider first endomorphisms A of an n-dimensional Hilbert space E. On an eigenvector u the operator A acts as a multiplication by a number: $Au = \lambda u$. Consequently it will always be easy to describe A when E has a basis $u_1, \ldots, u_n$ consisting of eigenvectors u_k. Indeed, if $\lambda_1, \ldots, \lambda_n$ are the corresponding (not necessarily distinct) eigenvalues and if one represents x in the form $x = \sum_{k=1}^n \alpha_k u_k$, then A acts like a superposition of 'dilations' in the 'proper directions' (eigendirections): $Ax = \sum_{k=1}^n \lambda_k \alpha_k u_k$. The situation becomes even more transparent when the eigenvectors are pairwise orthogonal. In this case the occasionally troublesome determination of the expansion coefficients α_k is as simple as one can wish: One has $x = \sum_{k=1}^n (x|u_k)u_k$ (Example 58.1), hence

$$Ax = \sum_{k=1}^{n} \lambda_k(x|u_k)u_k. \tag{70.1}$$

In (70.1) we only have to take into account the eigenvectors corresponding to non-zero eigenvalues. The reader will verify without difficulty that the adjoint A^* of such an operator A has the representation

$$A^*x = \sum_{k=1}^{n} \bar{\lambda}_k(x|u_k)u_k \tag{70.2}$$

and that by the Theorem of Pythagoras one has thus

$$\|Ax\| = \|A^*x\| \qquad \text{for all} \quad x \in E. \tag{70.3}$$

In a certain sense (70.3) is the mildest weakening of the relation $Ax = A^*x$, characteristic for symmetric operators.

In the remainder of this section let E be a complex Hilbert space.

We say that a continuous endomorphism A of E is *normal* if it satisfies (70.3). Obviously every symmetric operator of E is normal. A useful characterization of normalcy is given by:

Proposition 70.1. *The operator A is normal if and only if it commutes with its adjoint.*

Proof. If A is normal, then $(Ax|Ax) = (A^*x|A^*x)$, hence $(A^*Ax|x) = (AA^*x|x)$. With the aid of Proposition 68.4a it follows that $A^*A = AA^*$. The implications can be reversed. ■

Obviously every scalar multiple αA, and even every polynomial $\alpha_0 I + \alpha_1 A + \cdots + \alpha_n A^n$ of a normal operator A is again normal. If B is also normal, and if A commutes with B^*, then $A + B$, AB and BA are again normal. If a sequence A_n of normal operators tends uniformly to an operator A, then because of

$\|A_n^* - A^*\| = \|A_n - A\|$ we also have $A_n^* \Rightarrow A^*$ and with the aid of (70.3) it follows immediately that A is normal.

If the closed subspace F reduces the normal operator A, then F is also invariant under A^* (Exercise 5 in §67), and obviously the restriction of A^* to F is the adjoint of the restriction of A to F. It follows immediately that this restriction is again normal: *Normalcy is preserved under restriction to reducing subspaces.*

In §54 we called an operator A on a complex Banach space *normaloid* if its spectral radius $r(A)$ is equal to $\|A\|$. The following proposition motivates this terminology.

Proposition 70.2. *Every normal operator is paranormal and thus also spectrally normaloid.*

Proof. For a normal A we have

$$\|Ax\|^2 = (Ax|Ax) = (A^*Ax|x) \leqq \|A^*Ax\|\,\|x\| = \|A^2x\|\,\|x\|, \tag{70.4}$$

hence A is paranormal. For the remainder of the assertion invoke Proposition 54.6. ∎

Since together with A also $\lambda I - A$ is normal, from (70.3) and Proposition 67.2 we obtain immediately

Proposition 70.3. *For a normal A and an arbitrary λ we have $N(\lambda I - A) = N(\bar{\lambda} I - A^*)$; consequently $\overline{(\lambda I - A)(E)} = N(\lambda - A)^\perp$, and we have the orthogonal decomposition $E = \overline{(\lambda I - A)(E)} \oplus N(\lambda I - A)$.*

From this proposition, together with Proposition 67.3, we obtain that *a normal operator A is normally solvable if and only if the equation $Ax = y$ has a solution exactly when $y \perp N(A)$.* Furthermore we get from it (see the proof of the corresponding assertion in Proposition 68.1):

Proposition 70.4. *Eigenvectors of a normal operator corresponding to distinct eigenvalues are orthogonal to each other.*

We now turn to the investigation of isolated points λ_0 in the spectrum of a normal operator A. The subspace M_σ invariant under A, belonging to the spectral set $\sigma := \{\lambda_0\}$ (see Theorem 49.1), coincides by Proposition 49.1 with the set $\{x: \lim \|(\lambda_0 I - A)^n x\|^{1/n} = 0\}$. Since A^* commutes with A and therefore also with $(\lambda_0 I - A)^n$ (Proposition 70.1), it follows from this representation that M_σ is also invariant under A^*, and from here we obtain, as above, that the restriction A_σ of A to M_σ is normal. The spectrum of A_σ is $\{\lambda_0\}$ (Theorem 49.1), the spectrum of the normal operator $\lambda_0 I_\sigma - A_\sigma$ is therefore $\{0\}$, hence we have $r(\lambda_0 I_\sigma - A_\sigma) = 0$ from where with the aid of Proposition 70.2 it follows that $\|\lambda_0 I_\sigma - A_\sigma\| = 0$, i.e., that $A_\sigma = \lambda_0 I_\sigma$. Consequently λ_0 is an eigenvalue of A and $M_\sigma = N(\lambda_0 I - A)$. From (50.9) and the remark following it we obtain now that λ_0 is a pole of first order of the resolvent R_λ. With the aid of Propositions 50.2 and 70.3 we obtain furthermore that $(\lambda_0 I - A)(E)$ is closed and that the spectral projector P_σ is an orthogonal projector. With this we have proved the following:

Proposition 70.5. *An* isolated *spectral point λ_0 of the normal operator A is an eigenvalue of A, even a pole of first order of the resolvent. $\lambda_0 I - A$ is normally solvable, and the spectral projector belonging to $\{\lambda_0\}$ is an orthogonal projector; it projects E onto $N(\lambda_0 I - A)$ along $(\lambda_0 I - A)(E)$.*

This proposition and the following lemma, which is an immediate consequence of (70.4), will enable us to give a simple characterisation of the essential spectrum of a normal operator.

Lemma 70.1. *The length of the nullchain of a normal operator A is at most* 1.

Proposition 70.6. *The essential spectrum $\sigma_e(A)$ of a normal operator A consists of the limit points of $\sigma(A)$ and of the isolated spectral points having infinite multiplicity.*

By the multiplicity of an isolated spectral point we mean, of course, its multiplicity as an eigenvalue. For the proof we assume first that λ_0 is an isolated spectral point with finite multiplicity. Then we have $E = (\lambda_0 I - A)(E) \oplus N(\lambda_0 I - A)$ by Proposition 70.5, hence $\operatorname{ind}(\lambda_0 I - A) = 0$: thus λ_0 does not lie in $\sigma_e(A)$. Now let conversely λ_0 be in $\sigma(A) \backslash \sigma_e(A)$. Then we have $0 < \alpha(\lambda_0 I - A) = \beta(\lambda_0 I - A) < \infty$ and by Lemma 70.1 the length of the nullchain is $p(\lambda_0 I - A) = 1$. With the aid of Proposition 38.6 it follows from these two assertions that also the length $q(\lambda_0 I - A)$ of the image chain is equal to 1. Proposition 50.2 now yields that λ_0 is an isolated spectral point (its multiplicity is finite by assumption). ■

We call λ an *approximate eigenvalue* of A if there exists a sequence of normalized vectors x_n with $\lambda x_n - A x_n \to 0$; the set of the approximate eigenvalues is the *approximative point spectrum.* It contains $\sigma_p(A)$ and lies in $\sigma(A)$. For normal operators we have furthermore

Proposition 70.7. *The approximate point spectrum of a normal operator A coincides with $\sigma(A)$.*

We only have to show that every spectral point λ of A, which is not an eigenvalue, belongs to the approximate point spectrum. For such a λ the subspace $(\lambda I - A)(E)$ is not closed (otherwise not only $\alpha(\lambda I - A)$, but because of Proposition 70.3 also $\beta(\lambda I - A)$, would be zero, hence λ would lie in $\rho(A)$). By Proposition 36.1 the minimal modulus is

$$\gamma(\lambda I - A) = \inf_{x \neq 0} \frac{\|(\lambda I - A)x\|}{\|x\|} = 0$$

from where the assertion immediately follows. ■

If the normal operator A has no eigenvector, then its spectrum has no isolated points (Proposition 70.5). Since it is also closed, it is a *perfect* set and has therefore the power of the continuum (see [181], p. 72. We assume, of course, that $E \neq \{0\}$, hence that $\sigma(A)$ is not empty). In this case one says that A possesses a *pure segment spectrum.* Consider now the other extreme case: let the set of all eigenvectors be so large that its closed linear hull is all of E. We want to show that then the spectrum of A consists of all eigenvalues of A and their limit points; we say that A has a *pure point spectrum.* For every eigenvalue λ we determine an orthonormal basis of the eigenspace $N(\lambda I - A)$—see Theorem 65.1—we combine these bases and obtain so an orthonormal system S (Proposition 70.4), which is obviously maximal, hence an orthonormal basis of E (Proposition 65.3). We now show—and with this our assertion concerning $\sigma(A)$ will be proved—that a point λ, which is neither an eigenvalue of A nor a limit point of eigenvalues, lies in $\rho(A)$. For this it is sufficient to prove that the equation

$$(\lambda I - A)x = y \tag{70.5}$$

has for every $y \in E$ a solution $x \in E$. We assume first that (70.5) has a solution x. Then there exists a sequence (u_k) in S such that

$$x = \sum_{k=1}^{\infty} (x\,|\,u_k)u_k \quad \text{and} \quad y = \sum_{k=1}^{\infty} (y\,|\,u_k)u_k.$$

If λ_k is the eigenvalue belonging to u_k, then

$$(\lambda I - A)x = \sum_{k=1}^{\infty} (\lambda - \lambda_k)(x\,|\,u_k)u_k = \sum_{k=1}^{\infty} (y\,|\,u_k)u_k,$$

hence $(x\,|\,u_k) = (y\,|\,u_k)/(\lambda - \lambda_k)$ for all k, and so

$$x = \sum_{k=1}^{\infty} \frac{(y\,|\,u_k)}{\lambda - \lambda_k} u_k. \tag{70.6}$$

In order to prove the solvability of (70.5) for a given $y \in E$, we write the tentative solution x in the form (70.6). According to our hypothesis concerning λ there exists $\varepsilon > 0$ such that $|\lambda - \lambda_k| \geqq \varepsilon$ for all k, consequently the series

$$\sum_{k=1}^{\infty} \left| \frac{(y\,|\,u_k)}{\lambda - \lambda_k} \right|^2$$

converges (Proposition 64.2), and by Proposition 64.1a the series

$$\sum_{k=1}^{\infty} \frac{(y\,|\,u_k)}{|\lambda - \lambda_k|} u_k$$

has a sum x in E. Obviously x is a solution of equation (70.5). ■

If A does have eigenvalues but not a pure point spectrum, then let F be the closed linear hull of the set of all eigenvectors of A. The subspaces F and $F^\perp$ are invariant under A and A^*, the restrictions A_p, A_s of A to F, $F^\perp$, respectively, are normal and $\sigma(A) = \sigma(A_p) \cup \sigma(A_s)$ (Exercise 2 in §49). According to construction, A_p has a pure point spectrum; A_s has a pure segment spectrum (an eigenvector x of A_s would lie in $F^\perp$, on the other hand—as eigenvector of A—also in F, thus we would have $(x\,|\,x) = 0$ in contradiction to $x \neq 0$). Thus $\sigma(A)$ *is obtained as the union of a pure point spectrum and of a pure segment spectrum.* $\sigma(A_s)$ is also called the *segment spectrum* of A.

The term 'segment spectrum' can be better understood if we consider symmetric operators, because the spectra of these lie in intervals of the real axis which are easily described. We now turn to the examination of these facts.

If A is a *symmetric* operator on E, then we call the numbers

$$m(A) := \inf_{\|x\|=1} (Ax|x) \quad \text{and} \quad M(A) := \sup_{\|x\|=1} (Ax|x) \tag{70.7}$$

the *lower* and the *upper bound* of A, respectively. It follows from Proposition 68.5 that

$$\|A\| = \max\{|m(A)|, |M(A)|\} \tag{70.8}$$

(observe in these considerations that $(Ax|x)$ is real according to Proposition 68.4 and that A is continuous according to Proposition 68.6). The bounds of A make it possible to describe more precisely the position of the spectrum $\sigma(A)$:

Proposition 70.8. *The spectrum of the symmetric operator A lies in the closed interval* $[m(A), M(A)]$ *of the real axis; the bounds* $m(A)$, $M(A)$ *belong to* $\sigma(A)$. *If* λ *is not real, then for the resolvent operator* $R_\lambda = (\lambda I - A)^{-1}$ *the estimate*

$$\|R_\lambda\| \leqq \frac{1}{|\text{Im } \lambda|} \tag{70.9}$$

is valid.

Proof. We first show that $\sigma(A)$ is real. For a non-real λ and for every vector $x \neq 0$ we have

$$\begin{aligned} 0 < |\lambda - \bar{\lambda}| \|x\|^2 &= |([\lambda I - A]x|x) - ([\bar{\lambda} I - A]x|x)| \\ &= |([\lambda I - A]x|x) - (x|[\lambda I - A]x)| \leqq 2\|(\lambda I - A)x\| \|x\|, \end{aligned}$$

thus λ does not lie in the approximate point spectrum of A, hence, because of Proposition 70.7, not in $\sigma(A)$. Also (70.9) follows immediately from the above estimate. Let us now assume temporarily that $m(A) > 0$. Then 0 lies in $\rho(A)$; otherwise there would exist because of Proposition 70.7 a sequence x_n of normalized vectors such that $Ax_n \to 0$ in contradiction to the inequality $0 < m(A) \leqq (Ax_n|x_n) \leqq \|Ax_n\|$. If we apply this result in the case $\lambda < m(A)$ or $\lambda > M(A)$ to the operators $A - \lambda I$ or $\lambda I - A$, respectively, we see that these values λ belong to $\rho(A)$. Thus $\sigma(A)$ lies in $[m(A), M(A)]$. Let now $\lambda = m(A)$. Then $A - \lambda I \geqq 0$, and from the generalized Schwarz inequality (68.2) we get

$$\begin{aligned} \|(A - \lambda I)x\|^4 &= ([A - \lambda I]x|[A - \lambda I]x)^2 \\ &\leqq ([A - \lambda I]x|x)([A - \lambda I]^2 x|[A - \lambda I]x) \\ &\leqq ([A - \lambda I]x|x)\|A - \lambda I\|^3 \|x\|^2. \end{aligned}$$

Since by the definition of $m(A)$ there exists a sequence x_n of normalized vectors such that $(Ax_n|x_n) \to m(A) = \lambda$, this estimate shows that λ belongs to the approximate point spectrum, hence, by Proposition 70.7, also to the spectrum of A. One sees analogously that $M(A)$ lies in $\sigma(A)$. ■

Exercises

1. Show with the help of the propositions of this section that for a normal operator A on the complex space $l^2(n)$ the expansion (70.1), with pairwise orthogonal eigenvectors u_k corresponding to the eigenvalues λ_k, is valid. This expansion, which motivated our definition of normal operators, is thus characteristic for normalcy in the finite-dimensional case.

2. A normal operator is normally solvable if and only if the length of its image chain is finite.

$^+$3. Every complex number λ can be written in the form $\lambda = \alpha + i\beta$ with $\alpha, \beta \in \mathbf{R}$; the real part α and the imaginary part β are uniquely determined: $\alpha = \frac{1}{2}(\lambda + \bar{\lambda})$, $\beta = (1/2i)(\lambda - \bar{\lambda})$. λ^{-1} exists if and only if $(\alpha^2 + \beta^2)^{-1}$ exists; in this case $\lambda^{-1} = \bar{\lambda}(\alpha^2 + \beta^2)^{-1}$. Show that analogous assertions hold for (in some cases normal) operators:

(a) $L \in \mathscr{L}(E)$ can be written in the form $L = A + iB$ with $A = A^*$, $B = B^*$; the (selfadjoint) operators A, B are uniquely determined: $A = \frac{1}{2}(L + L^*)$, $B = (1/2i)(L - L^*)$.

(b) L is normal if and only if the 'real part' A and the 'imaginary part' B commute.

(c) If L is normal, then L^{-1} exists on E if and only if $(A^2 + B^2)^{-1}$ exists on E; in this case $L^{-1} = L^*(A^2 + B^2)^{-1}$.

$^+$4. λ is an eigenvalue of the normal operator A if and only if $(\lambda I - A)(E)$ is not dense in E.

$^+$5. The definition of the approximate point spectrum applies to any continuous endomorphism A of a Banach space. Show that the approximate point spectrum of A is contained in $\sigma(A)$ and that it is closed.

6. The linear map of $L^2(a, b)$ defined by $(Ax)(t) := tx(t)$ is symmetric but has no eigenvalue.

$^+$7. The continuous endomorphism U of the complex Hilbert space E is said to be *unitary* if $UU^* = U^*U = I$. In what follows, let U be a unitary operator. Show the following:

(a) U is normal.

(b) The unitary operators form a multiplicative group.

(c) U preserves the inner product: $(Ux|Uy) = (x|y)$.

(d) U is isometric: $\|Ux\| = \|x\|$.

(e) $\sigma(U)$ lies on the boundary of the unit disk.

(f) For every symmetric operator A the operator e^{iA} is unitary.

§71. Normal meromorphic operators

In this section let A be a continuous endomorphism of the complex Hilbert space E.

If A is meromorphic and normal, hence also spectrally normaloid (Proposition 70.2), then the eigenvalues $\lambda_n \neq 0$ of A are simple poles of the resolvent R_λ (Proposition 54.5); consequently the spectral projector P_n belonging to λ_n

projects E along the (closed) image space $(\lambda_n I - A)(E)$ onto $N(\lambda_n I - A)$ (Proposition 50.2), and is an orthogonal projector because of Proposition 70.3. Since $P_n P_m$ vanishes for $n \neq m$, it follows from Exercise 2 in §69 that also $P_1 + \cdots + P_n$ is an orthogonal projector, and has therefore a norm $\leqq 1$ (Proposition 63.1). Thus A satisfies all the hypotheses and condition (a) of Theorem 54.1, so that we can state the following proposition:

Proposition 71.1. *If $A \neq 0$ is normal and meromorphic, if $\{\lambda_1, \lambda_2, \ldots\}$ is the set of $\neq 0$ eigenvalues of A ordered according to decreasing absolute values, and if P_n is the orthogonal projector of E onto the eigenspace $N(\lambda_n I - A)$, then for A the uniformly convergent expansion*

$$A = \sum \lambda_n P_n \tag{71.1}$$

is valid, hence because of Proposition 70.4 *we have*

$$\|Ax\|^2 = \sum |\lambda_n|^2 \|P_n x\|^2 \qquad \text{for every} \quad x \in E. \tag{71.2}$$

If one knows the eigenvalues and the eigensolutions (hence also the spectral projectors) of A, then Proposition 71.1 is very helpful in solving operator equations. We have namely:

Proposition 71.2. *Under the hypotheses of Proposition* 71.1 *the equation*

$$(\lambda I - A)x = y \qquad \text{with} \quad \lambda \neq 0 \tag{71.3}$$

is solvable if and only if y is orthogonal to $N(\lambda A - I)$; in this case a solution is given by

$$x := \frac{1}{\lambda} y + \frac{1}{\lambda} \sum_{\lambda_n \neq \lambda} \frac{\lambda_n}{\lambda - \lambda_n} P_n y. \tag{71.4}$$

Proof. The solvability criterion follows immediately from Proposition 70.3 since $(\lambda I - A)(E)$ is closed. In the case of solvability one first observes by means of (71.2) that

$$\sum_{\lambda_n \neq \lambda} \frac{1}{|\lambda_n - \lambda|^2} |\lambda_n|^2 \|P_n y\|^2$$

converges, hence by Proposition 64.1 the series in (71.4) has a sum x. We now obtain with the help of the expansion (71.1)

$$\begin{aligned}(\lambda I - A)x &= \frac{1}{\lambda}(\lambda I - A)y + \frac{1}{\lambda} \sum_{\lambda_n \neq \lambda} \frac{\lambda_n}{\lambda - \lambda_n} (\lambda I - A) P_n y \\ &= y - \frac{1}{\lambda} Ay + \frac{1}{\lambda} \sum_{\lambda_n \neq \lambda} \frac{\lambda_n}{\lambda - \lambda_n} (\lambda - \lambda_n) P_n y \\ &= y - \frac{1}{\lambda} Ay + \frac{1}{\lambda} \sum_{n=1}^{\infty} \lambda_n P_n y = y.\end{aligned}$$

■

Because of Proposition 71.1, *a normal Riesz operator* is the uniform limit of a sequence of compact operators, since its projectors P_n are finite-dimensional (Proposition 52.2), thus it is *compact* itself (Proposition 13.1). If for every eigenspace $N(\lambda_n I - A) = 0$ we determine an orthonormal basis $\{u_{n_1}, \ldots, u_{n_{k_n}}\}$, then we have $P_n x = (x|u_{n_1})u_{n_1} + \cdots + (x|u_{n_{k_n}})u_{n_{k_n}}$, hence $Ax = \sum_{n=1}^{\infty} \lambda_n[(x|u_{n_1})u_{n_1} + \cdots + (x|u_{n_{k_n}})u_{n_{k_n}}]$. In order to simplify this cumbersome notation, we agree on the following *convention*: Let $\{u_1, u_2, \ldots\}$ be the union of the above orthonormal bases of the eigenspaces corresponding to eigenvalues $\neq 0$, and let μ_n be the eigenvalue belonging to u_n. In the sequence $(\mu_1, \mu_2, \ldots)$ every eigenvalue $\neq 0$ will occur thus as many times as is indicated by its multiplicity, while in the set $\{\lambda_1, \lambda_2, \ldots\}$ the eigenvalues λ_n were pairwise distinct. If we take into account that $\mu_n(x|u_n) = (x|\bar{\mu}_n u_n) = (x|A^*u_n) = (Ax|u_n)$, then from Proposition 71.1 we obtain immediately:

Proposition 71.3. *The eigenvalues different from zero of a normal* compact *operator $A \neq 0$ form a nonempty sequence $\mu_1, \mu_2, \ldots$, if one lists every eigenvalue as many times as is given by its multiplicity; this sequence tends to zero if it is infinite. To this sequence there corresponds an orthonormal sequence $u_1, u_2, \ldots$ of eigensolutions (so that $Au_n = \mu_n u_n$ for all n), with the aid of which the expansion*

$$(71.5) \qquad Ax = \sum \mu_n(x|u_n)u_n = \sum (Ax|u_n)u_n \qquad \textit{for all} \quad x \in E$$

is valid.

If (μ_n) is a finite sequence, or a sequence tending to zero, and $\{u_1, u_2, \ldots\}$ is an orthonormal system, then (71.5) defines a continuous endomorphism A on E whose adjoint is given by $A^*x = \sum_{n=1}^{\infty} \bar{\mu}_n(x|u_n)u_n$. It follows that $\|Ax\| = \|A^*x\|$ for all x, thus A is normal. If we define the continuous, finite-dimensional operator A_n by $A_n x := \sum_{k=1}^{n} \mu_k(x|u_k)u_k$, then

$$\|(A - A_k)x\|^2 = \sum_{n>k} |\mu_n|^2 |(x|u_n)|^2 \leqq \max_{n>k} |\mu_n|^2 \|x\|^2,$$

consequently $A_n \Rightarrow A$; by Proposition 13.1 we obtain that A is compact. *The representation* (71.5)—with the stated properties of (μ_n) and (u_n)—*is therefore characteristic for normal compact operators.*

Under the hypotheses of Proposition 71.3, formula (71.4) for the solution of the equation $(\lambda I - A)x = y$ has the form

$$(71.6) \qquad x = \frac{1}{\lambda} y + \frac{1}{\lambda} \sum_{\mu_n \neq \lambda} \frac{\mu_n}{\lambda - \mu_n} (y|u_n)u_n;$$

here we assume that $y \perp N(\lambda I - A)$.

Exercises

+1. Under the hypotheses of Proposition 71.3 the eigensolutions $\{u_1, u_2, \ldots\}$ of A form an orthonormal basis of E if and only if 0 is not an eigenvalue of A.

*2. Assume that for an endomorphism A of the inner product space E over $\mathbf{K}$ one has $Ax = \sum \mu_n(x|u_n)u_n$ $(x \in E)$ with a sequence of numbers $\mu_n \neq 0$ and an orthonormal sequence (u_n). Show the following:

(a) $Au_n = \mu_n u_n$.

(b) A is bounded if and only if (μ_n) is.

(c) If A is bounded and $\tilde{A}$ is the extension of A to the Hilbert space completion $\tilde{E}$ of E, then $\tilde{A}x = \sum \mu_n(x|u_n)u_n$ for all $x \in \tilde{E}$; $\tilde{A}$ is normal.

(d) A is precompact if and only if (μ_n) is finite or converges to zero.

(e) A is symmetric if and only if all μ_n are real.

§72. Symmetric compact operators

In applications it is of considerable importance that for a symmetric compact operator $A \neq 0$ the expansion (71.5) is valid even if the space E is not complete. One can see this e.g., by extending A to the completion of E, where one will invoke Exercise 5 in §13. We want to expound here, however, another method which uses directly the compactness of A. E is a real or complex inner product space.

By Proposition 68.5 we have $\sup_{\|x\|=1} |(Ax|x)| = \|A\|$, consequently there exists a sequence (x_n) and a number μ with $|\mu| = \|A\| > 0$ such that

$$\|x_n\| = 1 \quad \text{and} \quad (Ax_n|x_n) \to \mu.$$

From $0 \leqq \|Ax_n - \mu x_n\|^2 = \|Ax_n\|^2 - 2\mu(Ax_n|x_n) + \mu^2\|x_n\|^2 \leqq \|A\|^2 - 2\mu(Ax_n|x_n) + \|A\|^2$ it follows now that

$$Ax_n - \mu x_n \to 0. \tag{72.1}$$

Because of the compactness of A the sequence (Ax_n) has a convergent subsequence (Ax_{n_k}); it follows from (72.1) that also (x_{n_k}) converges to a (normalized) element u and that $Au - \mu u = 0$, hence u is an eigensolution corresponding to the eigenvalue $\mu = \pm\|A\|$. Obviously

$$|(Au|u)| = \sup_{\|x\|=1} |(Ax|x)|, \tag{72.2}$$

and conversely, every vector u which satisfies (72.2) is an eigensolution of A corresponding to the eigenvalue $\pm\|A\|$ (choose $x_n = u$).

Let now $\mu_1 := \mu$, $u_1 := u$ and $E_1 := [u_1]^\perp$. The restriction A_1 of A to E_1 is a symmetric compact endomorphism of E_1; if it is $\neq 0$, then by what we have just proved it has an eigenvalue μ_2 such that $0 < |\mu_2| = \|A_1\| \leqq \|A\| = |\mu_1|$. Let u_2 be the corresponding normalized eigensolution. If A_2 is the restriction of A to $E_2 = [u_1, u_2]^\perp$, then the same arguments yield, if $A_2 \neq 0$, an eigenvalue μ_3 of A_2 such that $0 < |\mu_3| = \|A_2\| \leqq \|A_1\| = |\mu_2|$, and a corresponding eigensolution u_3; trivially u_2, u_3 are also eigensolutions of A corresponding to the eigenvalues μ_2, μ_3. It is clear now how the process continues. One obtains a possibly terminating sequence (μ_n) of eigenvalues such that $|\mu_1| \geqq |\mu_2| \geqq \cdots$

> 0, and an orthonormal sequence (u_n) of corresponding eigensolutions. (μ_n) terminates with μ_m precisely if A vanishes on $E_m := [u_1, \ldots, u_m]^\perp$; in this case $E = [u_1, \ldots, u_m] \oplus E_m$ (§63), hence $x = \sum_{k=1}^{m} (x|u_k)u_k + y$ with $y \in E_m$ and so

$$Ax = \sum_{k=1}^{m} \mu_k(x|u_k)u_k.$$

If (μ_n) does not terminate, then $\mu_n \to 0$ (Proposition 43.2). For an arbitrary $x \in E$ the vector $y_n := x - \sum_{k=1}^{n} (x|u_k)u_k$ lies in E_n, consequently $\|Ay_n\| \leqq \|A_n\| \|y_n\| = |\mu_{n+1}| \|y_n\|$ and $\|y_n\|^2 = \|x\|^2 - \sum_{k=1}^{n} |(x|u_k)|^2 \leqq \|x\|^2$. It follows that $Ay_n \to 0$, hence

$$Ax = \sum_{k=1}^{\infty} \mu_k(x|u_k)u_k.$$

In the sequence (μ_k) every eigenvalue $\neq 0$ of A occurs as many times as indicated by its multiplicity. Otherwise there would exist an eigensolution u with $Au \neq 0$ and $u \quad u_k$ $(k = 1, 2, \ldots)$, with which we would then have $Au = \sum \mu_k(u|u_k)u_k = 0$, which is a contradiction. We summarize:

Proposition 72.1. *If $A \neq 0$ is a symmetric compact endomorphism of the (real or complex, complete or non-complete) inner product space E, then one obtains an orthonormal sequence of eigenvectors u_n by determining first a solution u_1 of the variational problem*

$$|(Ax|x)| = \max \qquad \text{under the side condition } \|x\| = 1,$$

and then successively for $n = 2, 3, \ldots$ a solution of the problem

$$|(Ax|x)| = \max \qquad \text{under the side conditions } \|x\| = 1,$$
$$(x|u_k) = 0 \qquad \text{for} \quad k = 1, \ldots, n-1,$$

as long as this maximum is positive; the absolute value of the eigenvalue μ_n corresponding to u_n is equal to this maximum. The described procedure yields each eigenvalue $\neq 0$ of A as often as its multiplicity indicates, and one has the expansion

$$Ax = \sum \mu_n(x|u_n)u_n = \sum (Ax|u_n)u_n \qquad \text{for all} \quad x \in E. \tag{72.3}$$

Formula (71.6) is valid also under the more general hypotheses of Proposition 72.1, and we can make an even more general statement:

Proposition 72.2 *Assume that for the symmetric operator A of the inner product space E one has $\operatorname{ind}(\lambda I - A) = 0$ when $\lambda \neq 0$, and let the expansion (72.3) be valid with an orthonormal sequence of eigensolutions u_n and a sequence of corresponding eigenvalues μ_n, which can accumulate only at 0. Under these hypotheses the equation $(\lambda I - A)x = y$ is solvable if and only if $y \perp N(\lambda I - A)$; and in this case a solution x is given by*

$$x := \frac{1}{\lambda} y + \frac{1}{\lambda} \sum_{\mu_n \neq \lambda} \frac{\mu_n}{\lambda - \mu_n} (y|u_n)u_n. \tag{72.4}$$

Proof. Because of the symmetry of A we have $N(\lambda I - A) \perp (\lambda I - A)(E)$, thus the hypotheses concerning the index implies the orthogonal decomposition

$$E = N(\lambda I - A) \oplus (\lambda I - A)(E) \qquad \text{for} \quad \lambda \neq 0, \tag{72.5}$$

and so the solvability criterion of the proposition. From (72.3) it follows with the help of the Bessel inequality that

$$\|Ax\|^2 = \sum |\mu_n|^2 |(x|u_n)|^2 \leqq \left(\max_n |\mu_n|^2\right) \|x\|^2,$$

hence A is continuous. Consequently the continuous, linear and symmetric extension $\tilde{A}$ of A to the Hilbert space completion $\tilde{E}$ of E exists. Since the series

$$\sum_{\mu_n \neq \lambda} \left| \frac{\mu_n}{\lambda - \mu_n} \right|^2 |(y|u_n)|^2$$

converges, (72.4) defines a vector $x \in \tilde{E}$ for which $(\lambda \tilde{I} - \tilde{A})x = y$ (see the proof of Proposition 71.2); obviously x is perpendicular to $N(\lambda I - A)$. Because of (72.5) there exists a vector $z \in E$, which is also perpendicular to $N(\lambda I - A)$, and for which $(\lambda I - A)z = y$. It follows that $x - z$ belongs to $N(\lambda \tilde{I} - \tilde{A})$ and is perpendicular to $N(\lambda I - A)$. In the case $\operatorname{Im} \lambda \neq 0$ we obtain, with the aid of Proposition 68.1, from the first of these assertions that $x - z = 0$, i.e., x lies in E. Let now λ be real. We approximate $x - z$ by a sequence (v_n) from E, this sequence can obviously be chosen to be orthogonal to $N(\lambda I - A)$. Because of (72.5) it lies then in $(\lambda I - A)(E)$, thus $v_n = (\lambda I - A)w_n$. It follows that

$$0 = ((\lambda \tilde{I} - \tilde{A})(x - z)|w_n) = (x - z|(\lambda I - A)w_n) \to (x - z|x - z),$$

hence again $x - z = 0$ and so $x \in E$. ∎

Exercises

1. The lower bound $m(A)$ and the upper bound $M(A)$ of a symmetric compact operator in a complex Hilbert space is always an eigenvalue if it is $\neq 0$.

2. Let A be a continuous, symmetric endomorphism of the inner product space E with $\operatorname{ind}(\lambda I - A) = 0$ for all $\lambda \neq 0$. Let $\tilde{A}$ be its continuous, linear and symmetric extension to the Hilbert space completion $\tilde{E}$ of E. Show the following:
   (a) $N(\lambda I - A) = N(\lambda \tilde{I} - \tilde{A})$ for $\lambda \neq 0$.
   (b) If an equation $(\lambda \tilde{I} - \tilde{A})x = y$ $(\lambda \neq 0)$ holds, then either both x, y lie in E or both do not lie in E.

§73. The Sturm–Liouville eigenvalue problem

Let L be the differential operator defined by

$$Lx := (px')' + qx \tag{73.1}$$

and let R_1, R_2 be the boundary value operators given by

$$R_1 x := \alpha_1 x(a) + \alpha_2 x'(a), \qquad R_2 x := \beta_1 x(b) + \beta_2 x'(b). \tag{73.2}$$

The *Sturm–Liouville eigenvalue problem* consists in finding the solutions of

$$Lx - \lambda rx = 0, \qquad R_1 x = R_2 x = 0, \tag{73.3}$$

where the following hypotheses are to be satisfied:

$$\begin{aligned} &\text{all functions occurring are real and defined in } [a, b], \\ &p \in C^{(1)}[a, b] \quad \text{and} \quad q, r \in C[a, b], \\ &p(t) > 0 \ \text{and} \ \ r(t) > 0 \quad \text{on} \quad [a, b], \\ &(\alpha_1, \alpha_2) \quad \text{and} \quad (\beta_1, \beta_2) \text{ are real vectors} \neq (0, 0) \end{aligned} \tag{73.4}$$

(see (3.23) and the considerations made there).

$u \in C^{(2)}[a, b]$ is called a *comparison function* if $R_1 u = R_2 u = 0$. For two comparison functions u, v we have

$$\int_a^b vLu \, dt = \int_a^b uLv \, dt; \tag{73.5}$$

this equation follows directly from the relation $vLu - uLv = [p(vu' - uv')]'$, valid for all $u, v \in C^{(2)}[a, b]$.

We assume that $\lambda = 0$ *is not an eigenvalue of the problem* (73.3). Then the Green function $G(s, t)$ of the boundary value problem

$$Lx = y, \qquad R_1 x = R_2 x = 0 \tag{73.6}$$

exists; G is continuous on the square $a \leqq s, t \leqq b$, and for every $y \in C[a, b]$

$$x(s) = \int_a^b G(s, t) y(t) dt \tag{73.7}$$

is the uniquely determined solution of (73.6). We now show that G is *symmetric*: $G(s, t) = G(t, s)$. For arbitrary $y, z \in C[a, b]$ let

$$u(s) := \int_a^b G(s, t) y(t) dt, \qquad v(s) := \int_a^b G(s, t) z(t) dt. \tag{73.8}$$

By the remark just made we have

$$\begin{gathered} Lu = y, \qquad Lv = z, \\ R_1 u = R_2 u = R_1 v = R_2 v = 0; \end{gathered} \tag{73.9}$$

thus, in particular, u and v are comparison functions, so that we can apply (73.5) to them. If we take into account (73.8) and (73.9), then

$$\begin{aligned} \int_a^b \left[\int_a^b G(s, t) z(t) dt \right] y(s) ds &= \int_a^b \left[\int_a^b G(s, t) y(t) dt \right] z(s) ds \\ &= \int_a^b \left[\int_a^b G(t, s) y(s) ds \right] z(t) dt, \end{aligned}$$

hence

$$\int_a^b \int_a^b [G(s, t) - G(t, s)]y(s)z(t)\mathrm{d}t\, \mathrm{d}s = 0 \qquad \text{for all} \quad y, z \in C[a, b],$$

from where the symmetry of G immediately follows.

A function x is a solution of (73.3) if and only if it satisfies the integral equation

$$x(s) - \lambda \int_a^b k(s, t)x(t)\mathrm{d}t = 0 \quad \text{with} \quad k(s, t) := G(s, t)r(t)$$

(see (3.25) through (3.28)); with the aid of the corresponding integral operator $K: C[a, b] \to C[a, b]$ it can be written more briefly in the form

(73.10) $(I - \lambda K)x = 0$ or $(\mu I - K)x = 0$ with $\mu := \dfrac{1}{\lambda}$ for $\lambda \neq 0$.

From the symmetry of G it follows that K is symmetric with respect to the inner product

(73.11) $$(x \mid y) := \int_a^b r(t)x(t)y(t)\mathrm{d}t$$

on $C[a, b]$. We could now apply the eigenvalue theory of §72 if K were *compact* on the vector space $C[a, b]$ equipped with the inner product norm

(73.12) $$|x| = \left[\int_a^b r(t)x^2(t)\mathrm{d}t\right]^{1/2}.$$

This is indeed the case.

For the proof let (x_n) be a sequence from $C[a, b]$, let $y_n := Kx_n$ and $|x_n| \leqq \gamma$. Since $g(s, t) := G(s, t)\sqrt{r(t)}$ is bounded, $|g(s, t)| \leqq \alpha$, we obtain from the Cauchy-Schwarz inequality

$$|y_n(s)| = \left|\int_a^b k(s, t)x_n(t)\mathrm{d}t\right| \leqq \int_a^b \alpha|\sqrt{r(t)}x_n(t)|\mathrm{d}t$$
$$\leqq \left(\int_a^b \alpha^2\, \mathrm{d}t\right)^{1/2}\left(\int_a^b r(t)x_n^2(t)\mathrm{d}t\right)^{1/2} \leqq \sqrt{b - a}\,\alpha\gamma,$$

hence the image sequence (y_n) is bounded with respect to the maximum norm of $C[a, b]$. Since, furthermore, $g(s, t)$ is uniformly continuous on $a \leqq s, t \leqq b$, for any preassigned $\varepsilon > 0$ there exists $\rho > 0$ so that $|s_1 - s_2| \leqq \rho$ implies $|g(s_1, t) - g(s_2, t)| \leqq \varepsilon$. With the aid of the Cauchy–Schwarz inequality we obtain, as above,

$$|y_n(s_1) - y_n(s_2)| \leqq \int_a^b |g(s_1, t) - g(s_2, t)||\sqrt{r(t)}x_n(t)|\mathrm{d}t$$
$$\leqq \int_a^b \varepsilon|\sqrt{r(t)}x_n(t)|\mathrm{d}t \leqq \sqrt{b - a}\,\varepsilon\gamma,$$

hence (y_n) is equicontinuous. By the Arzelà–Ascoli theorem (see §13 Exercise 1), (y_n) contains therefore a subsequence which converges uniformly, and a fortiori in the sense of the norm (73.12), to a function in $C[a, b]$. Thus we have proved the compactness of K.

According to Proposition 72.1 there exists a sequence of eigenvalues $\mu_n \neq 0$ (either finite or converging to 0) of the operator K, and a corresponding orthonormal sequence u_n of eigensolutions such that

$$(73.13) \qquad Kz = \sum \mu_n(z|u_n)u_n = \sum (Kz|u_n)u_n \qquad \text{for all} \quad z \in C[a, b];$$

u_n is an eigensolution of the problem (73.3) corresponding to the eigenvalue $\lambda_n = 1/\mu_n$ (cf. (73.10)). Every comparison function x can be represented because of (73.7) in the form

$$(73.14) \qquad x(s) = \int_a^b G(s, t)y(t)\mathrm{d}t = \int_a^b k(s, t)z(t)\mathrm{d}t = (Kz)(s),$$

where we have set $y := Lx$, $z(t) := y(t)/r(t)$; according to (73.13) we have for such a function the expansion

$$(73.15) \qquad x = \sum \alpha_n u_n \quad \text{with} \quad \alpha_n := \int_a^b r(t)x(t)u_n(t)\mathrm{d}t$$

in the sense that

$$\int_a^b r(s)\left[x(s) - \sum_{k=1}^{n} \alpha_k u_k(s)\right]^2 \mathrm{d}s \to 0 \qquad \text{for} \quad n \to \infty.$$

We show now that $\sum \alpha_n u_n$ even converges absolutely and uniformly on $[a, b]$. Because of (73.15) we have $\alpha_n = (x|u_n) = (Kz|u_n) = (z|Ku_n) = \mu_n(z|u_n)$, hence

$$\left(\sum_{k=m}^{n} |\alpha_k u_k(s)|\right)^2 = \left(\sum_{k=m}^{n} |(z|u_k)||\mu_k u_k(s)|\right)^2 \leqq \sum_{k=m}^{n} (z|u_k)^2 \sum_{k=m}^{n} [\mu_k u_k(s)]^2;$$

the first sum on the right-hand side is arbitrarily small for sufficiently large m (Proposition 64.2), the second is bounded in $[a, b]$: indeed, writing $g_s(t) := G(s, t)$ we have

$$\mu_k u_k(s) = (Ku_k)(s) = \int_a^b r(t)G(s, t)u_k(t)\mathrm{d}t = (g_s|u_k),$$

hence (Bessel's inequality)

$$\sum_{k=m}^{n} [\mu_k u_k(s)]^2 \leqq |g_s|^2 = \int_a^b r(t)G^2(s, t)\mathrm{d}t \leqq (b - a) \max_{a \leqq s, t \leqq b} r(t)G^2(s, t).$$

Thus the assertion concerning convergence is proved. We summarize:

Every twice continuously differentiable function x, which satisfies the boundary conditions $R_1 x = R_2 x = 0$, can be expanded in a series (73.15) *with respect to the eigensolutions u_n of the Sturm–Liouville eigenvalue problem* (73.3). *The u_n can be chosen in such a way that they form an orthonormal sequence in the sense* *of the inner product* (73.11); *the expansion* (73.15) *converges then absolutely and uniformly on* $[a, b]$.

In §76 we shall discuss eigenvalue problems which are much more general than the Sturm–Liouville problem.

§74. Wielandt operators

In the preceding section we investigated an integral operator $K: C[a, b] \to C[a, b]$ with continuous kernel, which is symmetric with respect to a certain inner product on $C[a, b]$. We know that K is compact with respect to the maximum-norm of $C[a, b]$; in order to be able to apply the eigenvalue theory of §72, we had to prove, however, the compactness of K with respect to the inner product norm. The question arises whether one can get along without this 'second compactness', i.e., whether one can develop quite generally an eigenvalue theory for compact or even Riesz operators on a Banach space E, which are symmetric with respect to an inner product on E. We shall show first that this is indeed possible, second that only algebraic properties of Riesz operators are needed, and third that we obtain thereby the essential parts of the theory of symmetrizable operators (see [18] and the references given there). The decisive Propositions 74.1 through 74.3 were presented for the first time by H. Wielandt in a course at the University of Tübingen in 1952; cf. also [85].

In what follows we shall use concepts, notations and results from Exercise 3 in §68 without explicitly referring to it any more. Basic is:

Proposition 74.1. *On the complex vector space E let a semi-inner product* $[x|y]$ *with the semi-norm* $|x| := \sqrt{[x|x]}$ *and the nullspace* $N := \{w \in E: |w| = 0\}$ *be given, furthermore also a symmetric operator A such that* $\lambda I - A$ *is bijective whenever* $\operatorname{Im} \lambda > 0$. *Then the following holds: A vector which belongs to* $N + (\xi I - A)^2(E)$ *for all real* ξ *lies already in N.*

Proof. For a bijective $\lambda I - A$ let $R_\lambda := (\lambda I - A)^{-1}$. Exactly as in (44.7) and (70.9) one proves the equation

$$(74.1) \quad R_\lambda - R_\mu = (\mu - \lambda)R_\lambda R_\mu, \qquad \text{hence} \quad R_\lambda = R_\mu + (\mu - \lambda)R_\lambda R_\mu$$

and the estimate

$$(74.2) \qquad |R_\lambda| \leqq \frac{1}{\operatorname{Im} \lambda} \quad \text{for} \quad \operatorname{Im} \lambda > 0,$$

respectively. For a fixed $x \in E$ let

$$f(\lambda) := [R_\lambda x|x] = u(\lambda) + iv(\lambda) = u(\alpha, \beta) + iv(\alpha, \beta) \qquad (\lambda = \alpha + i\beta).$$

$f(\lambda)$ certainly exists for $\beta = \operatorname{Im} \lambda > 0$, and setting $y := R_\lambda x$, we have

$$f(\lambda) = [y|(\lambda I - A)y] = [(\alpha I - A)y|y] - i\beta[y|y],$$

hence

(74.3) $$v(\alpha, \beta) = -\beta[y|y] \leqq 0 \qquad \text{for} \quad \beta > 0.$$

From the second equation in (74.1) one obtains

$$R_\lambda = R_\mu + (\mu - \lambda)(R_\mu + (\mu - \lambda)R_\lambda R_\mu)R_\mu = R_\mu + (\mu - \lambda)R_\mu^2 + (\mu - \lambda)^2 R_\lambda R_\mu^2,$$

hence

$$\frac{f(\lambda) - f(\mu)}{\lambda - \mu} = -[R_\mu^2 x|x] + (\lambda - \mu)[R_\lambda R_\mu^2 x|x],$$

and since because of (74.2) the estimate

$$|(\lambda - \mu)[R_\lambda R_\mu^2 x|x]| \leqq |\lambda - \mu| \frac{1}{\operatorname{Im} \lambda} |R_\mu^2 x||x|$$

holds for $\operatorname{Im} \lambda > 0$, it follows that $\lim_{\lambda \to \mu} [(f(\lambda) - f(\mu))/(\lambda - \mu)]$ exists for $\operatorname{Im} \mu > 0$. Thus f is holomorphic in the upper half-plane, and so v is harmonic there.

Now assume that for a certain real ξ the vector x lies in $N + (\xi I - A)^2(E)$, i.e., that with appropriate $w \in N$ and $y \in E$ one has

$$x = w + (\xi I - A)^2 y;$$

furthermore let

$$z := (\xi I - A)y.$$

From the identity

$$(\xi I - A)^2 = (\xi I - A)(\lambda I - A) + (\xi - \lambda)(\lambda I - A) + (\xi - \lambda)^2 I$$

it follows then that

$$x = w + (\lambda I - A)z + (\xi - \lambda)(\lambda I - A)y + (\xi - \lambda)^2 y,$$

hence

$$\begin{aligned} f(\lambda) &= [R_\lambda x|x] \\ &= [R_\lambda w|x] + [z|x] + (\xi - \lambda)[y|x] + (\xi - \lambda)^2[R_\lambda y|x] \quad \text{for} \quad \operatorname{Im} \lambda > 0. \end{aligned}$$

Because of $|[R_\lambda w|x]| \leqq |R_\lambda||w||x| = 0$, the first term of this sum vanishes, the second is real ($[z|x] = [z|(\xi I - A)z]$), the third tends obviously to 0 as $\lambda \to \xi$ and the fourth tends also to 0 if λ tends to ξ in the angular domain $W := \{\lambda = \alpha + i\beta : |\alpha - \xi| \leqq \beta\}$; thus we have

(74.4) $$f(\lambda) \to [z|x] \in \mathbf{R}, \quad \text{i.e.,} \quad v(\lambda) \to 0 \qquad \text{for} \quad \lambda \to \xi \text{ in } W.$$

Since $-v$ is a positive harmonic function in the upper half-plane, it has a representation

(74.5) $$-v(\alpha, \beta) = \delta\beta + \int_{-\infty}^{\infty} \frac{\beta}{(t - \alpha)^2 + \beta^2} \, d\sigma(t)$$

with a constant $\delta \leqq 0$ and a monotone increasing function σ. From (74.4) it follows that $\sigma'(\xi)$ exists and is equal to zero (for these theorems of potential theory see [182]).

Next, assume that x lies in $N + (\xi I - A)^2(E)$ *for every* real ξ. By the result above, the integral in (74.5) will vanish and we will have $-v(\alpha, \beta) = \delta\beta$. Since for $\beta > 0$ we have by (74.2) the estimate

$$\delta\beta = |v(\alpha, \beta)| = |v(\lambda)| \leqq |f(\lambda)| \leqq |R_\lambda||x|^2 \leqq \frac{|x|^2}{\beta},$$

we must have $\delta = 0$ and thus $v(\alpha, \beta) \equiv 0$ in the upper half-plane.

It follows now from (74.3) that $|y| = 0$ ($y = R_\lambda x$) and so

$$\begin{aligned} |x|^2 &= [x|x] = [(\lambda I - A)y|x] \\ &= \lambda[y|x] - [y|Ax] \leqq |\lambda||y||x| + |y||Ax| = 0. \end{aligned}$$

Thus Proposition 74.1 is proved. From it we obtain now very easily an existence theorem for eigenvalues. Let it be preceded by a definition:

An endomorphism A of a complex vector space with a semi-inner product is called a *Wielandt operator* if it is fully symmetric and if $\operatorname{ind}(\lambda I - A)$ vanishes for all $\lambda \neq 0$.

Proposition 74.2. *The Wielandt operator A on E has a non-zero eigenvalue if and only if $|Ay|$ does not vanish for every $y \in E$.*

Proof. The condition is trivially necessary. Assume that it is satisfied. If A did not have a non-zero eigenvalue, then $\lambda I - A$ would be bijective for every $\lambda \neq 0$, consequently $x := A^2 y$ (y arbitrary in E) would lie for every real ξ in $(\xi I - A)^2(E)$, hence we would have $|x| = 0$ (Proposition 74.1) and thus $|Ay|^2 = [Ay|Ay] = [A^2 y|y] = [x|y] \leqq |x||y| = 0$. Since y was arbitrary, we obtain a contradiction to our assumption. ■

For a vector space E with semi-inner product $[x|y]$ the notion of an *orthonormal system* can be defined as usual. Since Bessel's identity and the results which follow from it do in reality not depend on the strict definiteness of the inner product, we have for an orthonormal system S in E the following assertions:

(74.6) $$\left| x - \sum_{\nu=1}^{n} [x|u_\nu]u_\nu \right|^2 = |x|^2 - \sum_{\nu=1}^{n} |[x|u_\nu]|^2, \qquad u_\nu \in S;$$

(74.7) $$\sum_{\nu=1}^{n} |[x|u_\nu]|^2 \leqq |x|^2, \qquad u_\nu \in S;$$

(74.8) for a fixed x at most countably many $[x|u]$ are $\neq 0$ ($u \in S$).

If S is an uncountable orthonormal system, then we list, as earlier, in a series in which Fourier coefficients $[x|u]$ occur, only those terms for which $[x|u] \neq 0$. It follows then from Bessel's inequality (74.7) that $\sum_{u \in S} |[x|u]|^2$ converges and is $\leqq |x|^2$, and with the aid of the Cauchy–Schwarz inequality one obtains that the expression

(74.9) $$\langle x|y\rangle := [x|y] - \sum_{u\in S} [x|u][u|y]$$

exists for all $x, y \in E$. It obviously is a semi-inner product on E. If $u_1, u_2, \ldots$ are the elements u from S with $[x|u] \neq 0$, then it follows from the Bessel identity (74.6) that

$$\langle x|x\rangle \quad \text{vanishes if and only if} \quad \left| x - \sum_{\nu=1}^{n} [x|u_\nu]u_\nu \right| \to 0.$$

In this case we write briefly

(74.10) $$x = \sum_{u\in S} [x|u]u,$$

but we keep in mind that the sum of an infinite series is not uniquely determined because of the lack of definiteness of our semi-norm. We state:

(74.11) (74.10) *holds if and only if* $\langle x|x\rangle$ *vanishes.*

The full symmetry of an operator A has as a consequence that on every eigenspace $N(\xi I - A)$ corresponding to a (real) eigenvalue $\xi \neq 0$ the semi-inner product $[\cdot|\cdot]$ is an inner product and that $N(\xi I - A) \cap (\xi I - A)(E) = \{0\}$. If A is a Wielandt operator, then one can therefore construct an orthonormal basis S_ξ for $N(\xi I - A)$; the union $S := \bigcup_{\xi\neq 0} S_\xi$ is an *orthonormal eigensystem* of A; furthermore we have the decomposition

(74.12) $$E = N(\xi I - A) \oplus (\xi I - A)(E)$$

so that $\xi I - A$ is chain-finite by Proposition 38.4 and that even the decomposition

(74.13) $$E = N(\xi I - A) \oplus (\xi I - A)^2(E)$$

is valid; the assertions from (74.12) on are valid trivially also for non-eigenvalues $\xi \neq 0$. With the aid of the above eigensystem S we introduce on E the semi-inner product (74.9) and the corresponding nullspace $N := \{w \in E: \langle w|w\rangle = 0\}$. As an endomorphism of the space $(E, \langle\cdot|\cdot\rangle)$, the operator A obviously satisfies the hypotheses of Proposition 74.1, and because of $N(\xi I - A) \subset N$, we have by (74.13) the decomposition $E = N + (\xi I - A)^2(E)$ for every $\xi \neq 0$. Thus every vector $y := A^2x$ belongs for every real ξ to $N + (\xi I - A)^2(E)$, hence by Proposition 74.1 it lies in N, and therefore $\langle Ax|Ax\rangle = \langle x|A^2x\rangle \leqq \langle x|x\rangle^{1/2}\langle A^2x|A^2x\rangle^{1/2} = 0$. It follows now from (74.11) that the expansion

(74.14) $$Ax = \sum_{u\in S} [Ax|u]u = \sum_{u\in S} \mu[x|u]u \qquad \text{for every} \quad x \in E$$

is valid in the following sense: If $u_1, u_2, \ldots$ are those vectors u from the eigensystem S of A for which $[x|u] \neq 0$, and if $\mu_1, \mu_2, \ldots$ are the corresponding eigenvalues (i.e., if $Au_n = \mu_n u_n$), then

$$(74.15) \quad \left|Ax - \sum_{k=1}^{n} [Ax|u_k]u_k\right| = \left|Ax - \sum_{k=1}^{n} \mu_k[x|u_k]u_k\right| \to 0 \quad \text{as} \quad n \to \infty.$$

We state this result:

Proposition 74.3. *For the Wielandt operator A the expansion* (74.14) *is valid in the sense of the limit assertion* (74.15); *here S is an orthonormal eigensystem of A.*

For Wielandt operators a formula analogous to (72.4) for the solution of the equation $(\lambda I - A)x = y$ can be given (see [106], Theorem 11 and its proof). We will not present it here but content ourselves with the assertion of Proposition 72.1.

The results obtained so far can be applied whenever a power of A is compact on the complex normed space E or if A is a Riesz operator on the Banach space E and if furthermore A can be made fully symmetric by the introduction of a semi-inner product on E (for the assertion concerning the compact power see Exercise 4 in §43). We still want to consider the more special case when the Riesz operator A on the complex Hilbert space E is *symmetrizable* by $H \geqq 0$ (see Exercise 4a in §68), furthermore let $Hu \neq 0$ for every eigensolution u corresponding to a non-zero eigenvalue of A. Then A is symmetric with respect to the semi-inner product $[x|y] := (x|Hy)$, and because of the inequality

$$(74.16) \qquad \|Hx\|^2 \leqq \|H\|(x|Hx) = \|H\|\,|x|^2$$

(Exercise 1 in §68) even fully symmetric. By Proposition 74.3 we have therefore (74.15), where μ_k runs through the sequence of all (real) non-zero eigenvalues (in which every eigenvalue occurs as often as its multiplicity indicates). If we set $s_n := \sum_{k=1}^{n} \mu_k[x|u_k]u_k$, then, because of $|[Ax|y] - [s_n|y]| = |[Ax - s_n|y]| \leqq |Ax - s_n||y|$, it follows from (74.15) immediately that

$$[Ax|y] = \sum \mu_k[x|u_k][u_k|y],$$

hence also

$$(74.17) \qquad (HAx|y) = \sum \mu_k(x|Hu_k)(Hu_k|y) \qquad \text{for} \quad x, y \in E.$$

With the aid of (74.16) we obtain the estimate

$$\left\|\sum_{k=m}^{n} \alpha_k Hu_k\right\|^2 \leqq \|H\| \left|\sum_{k=m}^{n} \alpha_k u_k\right|^2 = \|H\| \sum_{k=m}^{n} |\alpha_k|^2,$$

from where it follows immediately that $\sum \alpha_k Hu_k$ converges whenever $\sum |\alpha_k|^2$ is finite. We see from here that $Bx := \sum \mu_k[x|u_k]Hu_k = \sum \mu_k(x|Hu_k)Hu_k$ is defined for every $x \in E$, and with the help of (74.17) we obtain that $(Bx|y) = (HAx|y)$ for every $y \in E$, hence $Bx = HAx$ and thus

$$(74.18) \quad HAx = \sum \mu_k(x|Hu_k)Hu_k = \sum (HAx|u_k)Hu_k \qquad \text{for every} \quad x \in E.$$

Because of the relation $(u_j|Hu_k) = \delta_{jk}$ one says that the eigensolutions $u_1, u_2, \ldots$ are *H-orthogonal* to another. With these considerations we have obtained the most important assertions of the theory of symmetrizable Riesz operators.

Exercises

$^+$1. The operator A of the Hilbert space E is fully symmetric with respect to $[x|y] := (x|Hy)$ with $H \geqq 0$ if and only if $Au = \lambda u \neq 0$ implies $Hu \neq 0$; in this case we say that A is *fully symmetrizable* by H.

$^+$2. Let the Riesz operator A of the complex Hilbert space E be fully symmetrizable by $H \geqq 0$ (see Exercise 1). Then A has a non-zero eigenvalue if and only if $HA \neq 0$.

3. Obtain under the hypotheses of Exercise 2 non-zero eigenvalues of the Riesz operator A and the expansion (74.18) by an extremal procedure (see Proposition 72.1).

§75. Determination and estimation of eigenvalues

We investigate in this section operators A on a (real or complex) inner product space E for which an *expansion theorem* of the form

$$Ax = \sum \mu_n(x|u_n)u_n \tag{75.1}$$

holds, where the following hypotheses should be satisfied:

$$\begin{gathered}\text{all } \mu_n \text{ are real and } \neq 0,\\ \text{the } \mu_n \text{ can accumulate only at } 0,\\ (u_n) \text{ is an orthonormal sequence.}\end{gathered} \tag{75.2}$$

A is then a continuous symmetric endomorphism of E with eigensolutions u_n corresponding to the eigenvalues μ_n (Exercise 2 in §71); as in §72 (immediately before Proposition 72.1), we see that in the sequence (μ_n) every non-zero eigenvalue of A occurs as often as its (finite) multiplicity indicates. We say that *μ_n participates in x* if $(x|u_n) \neq 0$.

From (75.1) we obtain, by successive application of A, the expansion

$$A^k x = \sum \mu_n^k(x|u_n)u_n, \qquad k = 1, 2, \ldots. \tag{75.3}$$

The representation

$$(Ax|x) = \sum \mu_n|(x|u_n)|^2 \tag{75.4}$$

shows that *A is positive if and only if all $\mu_n > 0$*. Obviously together with A also every power A^k $(k = 1, 2, \ldots)$ is positive.

We now assume temporarily that A is positive and (μ_n) monotone decreasing, the last hypothesis is no restriction of the generality. For an x such that $Ax \neq 0$, the maximum $\mu(x)$ of all eigenvalues participating in x is then positive. If $\mu(x)$

is equal to the eigenvalues $\mu_m, \ldots, \mu_{m+r}$, and only to these, then writing $\mu := \mu(x)$ we obtain from (75.3) that

$$A^k x = \mu^k \sum_{n=m}^{m+r} (x|u_n)u_n + \sum_{n>m+r} \mu_n^k (x|u_n)u_n,$$

where the first sum does not vanish and $\mu_n < \mu$ for $n > m + r$. Consequently for $k \to \infty$ we have

$$\left(\frac{\|A^k x\|}{\mu^k}\right)^2 = \sum_{n=m}^{m+r} |(x|u_n)|^2 + \sum_{n>m+r} \left(\frac{\mu_n}{\mu}\right)^{2k} |(x|u_n)|^2 \to \sum_{n=m}^{m+r} |(x|u_n)|^2 > 0,$$

hence

$$\frac{\|A^{k+1}x\|}{\|A^k x\|} = \frac{\|A^{k+1}x\|}{\mu^{k+1}} \frac{\mu^k}{\|A^k x\|} \mu \to \mu.$$

One sees similarly that

$$\frac{A^k x}{\|A^k x\|} = \frac{A^k x}{\mu^k} \cdot \frac{\mu^k}{\|A^k x\|} \to \frac{\sum_{n=m}^{m+r} (x|u_n)u_n}{\|\sum_{n=m}^{m+r} (x|u_n)u_n\|};$$

the limit is a normalized eigensolution corresponding to the eigenvalue μ.

By similar considerations we see that the sequence of the *Schwarz quotients* converges:

$$\frac{(A^{k+1}x|x)}{(A^k x|x)} \to \mu.$$

This sequence has the advantage of being *monotone increasing*, since by Proposition 68.2 we have

$$(A^{k+1}x|x)^2 = (A^k(Ax)|x)^2 \leqq (A^k(Ax)|Ax)(A^k x|x) = (A^{k+2}x|x)(A^k x|x).$$

If A is not positive any more, then we can apply our results to the positive operator A^2 and obtain thus all the assertions of the following proposition, except the last one.

Proposition 75.1. *If* $Ax \neq 0$, *then*

$$\frac{\|A^{2k+2}x\|}{\|A^{2k}x\|} \to \alpha^2 > 0, \qquad \frac{(A^{2k+2}x|x)}{(A^{2k}x|x)} \nearrow \alpha^2,$$

$$\frac{A^{2k}x}{\|A^{2k}x\|} \to u, \qquad \text{where} \quad \|u\| = 1 \quad \text{and} \quad A^2 u = \alpha^2 u.$$

At least one of the vectors

$$v := u + \frac{1}{\alpha} Au \quad \text{or} \quad w := u - \frac{1}{\alpha} Au$$

is an eigensolution of A corresponding to the eigenvalue α or $-\alpha$, respectively.

The last assertion of the proposition follows from the equations $Av = \alpha v$, $Aw = -\alpha w$, which are easy to verify, and from the fact that because of $v + w = 2u \neq 0$, at least one of the vectors v, w does not vanish. ■

We now turn to estimating and comparing eigenvalues. We concentrate on positive eigenvalues. One obtains assertions concerning negative eigenvalues by applying Propositions 75.2 through 75.6 to the operator $-A$.

Proposition 75.2. *If α is a positive and r a natural number, then A has at least r eigenvalues $\geqq \alpha$ if and only if there exists an r-dimensional subspace F of E such that*

$$(Ax|x) \geqq \alpha(x|x) \qquad \textit{for all} \quad x \in F;$$

the eigenvalues are counted according to their multiplicity.

Proof. We first assume that A possesses r eigenvalues $\geqq \alpha$; by changing subscripts we may suppose that these are the eigenvalues $\mu_1, \ldots, \mu_r$. For every element $x = \sum_{\rho=1}^r \xi_\rho u_\rho$ of the r-dimensional subspace $F := [u_1, \ldots, u_r]$ we have then $(Ax|x) = \sum_{\rho=1}^r \mu_\rho |\xi_\rho|^2 \geqq \alpha \sum_{\rho=1}^r |\xi_\rho|^2 = \alpha(x|x)$. Now let conversely an r-dimensional subspace $F = [y_1, \ldots, y_r]$ exist on which $(Ax|x) \geqq \alpha(x|x)$. We assume that A has only $q < r$ eigenvalues $\geqq \alpha$; by changing subscripts we can obtain that $\mu_n \geqq \alpha$ for $n = 1, \ldots, q$ and $\mu_n < \alpha$ for $n > q$. We now determine a non-trivial solution $(\eta_1, \ldots, \eta_r)$ of the system of equations

$$\sum_{\rho=1}^{r} \xi_\rho (y_\rho | u_\sigma) = 0, \qquad \sigma = 1, \ldots, q$$

(this is possible because of $q < r$); for $z := \sum_{\rho=1}^r \eta_\rho y_\rho \in F$ we have then

$$z \neq 0 \quad \text{and} \quad (z|u_\sigma) = 0 \qquad \text{for} \quad \sigma = 1, \ldots, q,$$

hence

$$\alpha(z|z) \leqq (Az|z) = \sum_{n>q} \mu_n |(z|u_n)|^2 \leqq \alpha \sum_{n>q} |(z|u_n)|^2 \leqq \alpha(z|z)$$

(the last estimate follows from Bessel's inequality). Since the two extreme terms of this chain of inequalities agree, we obtain

$$\alpha(z|z) = \sum_{n>q} \mu_n |(z|u_n)|^2 = \alpha \sum_{n>q} |(z|u_n)|^2 \tag{75.5}$$

hence

$$\sum_{n>q} (\alpha - \mu_n)|(z|u_n)|^2 = 0,$$

from where because of $\alpha - \mu_n > 0$ $(n > q)$ we obtain $(z|u_n) = 0$ for $n > q$. From (75.5) we obtain therefore that $(z|z) = 0$, in contradiction to $z \neq 0$. Thus A must indeed have at least r eigenvalues $\geqq \alpha$. ■

We now decompose (μ_n) into a monotone decreasing sequence of positive and a monotone increasing sequence of negative eigenvalues:

$$\mu_1^+ \geqq \mu_2^+ \geqq \cdots > 0, \qquad \mu_1^- \leqq \mu_2^- \leqq \cdots < 0$$

(each of the two sequences can be finite and one of them even empty); let u_n^+, u_n^- be the corresponding eigensolutions. From Proposition 75.2 we obtain immediately:

Proposition 75.3. *The eigenvalue μ_r^+ exists and is $\geqq \alpha > 0$ if and only if there exists an r-dimensional subspace of E on which we have $(Ax|x) \geqq \alpha(x|x)$.*

If for every non-zero vector x of an r-dimensional subspace F we have $(Ax|x) > 0$, then also $\alpha(F) := \min\{(Ax|x): x \in F, \|x\| = 1\} > 0$ (the minimum exists because the function $x \mapsto (Ax|x)$ is continuous on the set $\{x \in F: \|x\| = 1\}$, which according to Proposition 10.6 is compact); then μ_r^+ exists and is $\geqq \alpha(F)$ by Proposition 75.3. For $F_0 := [u_1^+, \ldots, u_r^+]$ we have $\mu_r^+ = \alpha(F_0)$, since on the one hand for every $x = \sum_{\rho=1}^r \xi_\rho u_\rho^+$ with $\|x\| = \sqrt{\sum_{\rho=1}^r |\xi_\rho|^2} = 1$ one has $(Ax|x) = \sum_{\rho=1}^r \mu_\rho^+ |\xi_\rho|^2 \geqq \mu_r^+$, and on the other $(Au_r^+|u_r^+) = \mu_r^+$. Through these considerations we have proved the *Courant maximum-minimum principle*:

Proposition 75.4. *We have*

$$\mu_r^+ = \max_F \min_{0 \neq x \in F} \frac{(Ax|x)}{(x|x)},$$

where F runs through all r-dimensional subspaces of E on which $(Ax|x) > 0$ for $x \neq 0$. The maximum is attained for $F = [u_1^+, \ldots, u_r^+]$. In particular, μ_1^+ exists and is given by

$$\mu_1^+ = \max_{x \neq 0} \frac{(Ax|x)}{(x|x)}$$

provided that $(Ax|x)$ is positive for at least one $x \in E$.

$\dfrac{(Ax|x)}{(x|x)}$ is called the *Rayleigh quotient.*

The following proposition, which is complementary to Proposition 75.4 in so far as finite-codimensional subspaces occur in it instead of finite-dimensional ones, is called the *Courant minimum-supremum principle.* Its proof can be left to the reader.

Proposition 75.5. *If the right-hand side of the following equation is positive, then we have*

$$\mu_r^+ = \min_F \sup_{0 \neq x \in F^\perp} \frac{(Ax|x)}{(x|x)},$$

where F runs through all $(r-1)$-dimensional subspaces of E. The minimum is attained for $F = [u_1^+, \ldots, u_{r-1}^+]$.

From the minimum-supremum principle we obtain in a simple fashion the *Weyl comparison theorem*:

Proposition 75.6. *If the operators A, B and C satisfy the hypotheses* (75.1) *and* (75.2), *if α_n^+, β_n^+ and γ_n^+ are their positive eigenvalues arranged in a monotone decreasing order, and if*

$$A = B + C,$$

then the estimates

$$\alpha_{r+s-1}^+ \leqq \beta_r^+ + \gamma_s^+$$

are valid.

For the proof let v_n^+, w_n^+ be the eigensolutions of B, C corresponding to β_n^+, γ_n^+ and

$$F := [v_1^+, \ldots, v_{r-1}^+, w_1^+, \ldots, w_{s-1}^+]. \qquad G := [v_1^+, \ldots, v_{r-1}^+],$$
$$H := [w_1^+, \ldots, w_{s-1}^+].$$

Then we obtain with the help of Proposition 75.5 the following chain of inequalities, in which x should satisfy the condition $\|x\| = 1$:

$$\begin{aligned}\alpha_{r+s-1}^+ &\leqq \sup_{x\in F^\perp} (Ax|x) \leqq \sup_{x\in F^\perp} (Bx|x) + \sup_{x\in F^\perp} (Cx|x)\\ &\leqq \sup_{x\in G^\perp} (Bx|x) + \sup_{x\in H^\perp} (Cx|x) = \beta_r^+ + \gamma_s^+.\end{aligned}$$ ■

To conclude we present an *inclusion theorem*:

Proposition 75.7. *If for an $x \in E$ with $\|x\|^2 = \sum_{n=1}^{\infty} |(x|u_n)|^2 > 0$ and for a real polynomial $p(t) := \alpha_0 + \alpha_1 t + \alpha_2 t^2$ we have*

$$(p(A)x|x) = \alpha_0(x|x) + \alpha_1(Ax|x) + \alpha_2(A^2x|x) \geqq 0,$$

then the set $\{t \in \mathbf{R}: p(t) \geqq 0\}$ contains at least one non-zero eigenvalue of A.

Proof. With the help of (75.3) we obtain from our hypotheses the estimate

$$\begin{aligned}\sum_{n=1}^{\infty} p(\mu_n)|(x|u_n)|^2 &= \sum_{n=1}^{\infty} [\alpha_0|(x|u_n)|^2 + \alpha_1\mu_n|(x|u_n)|^2 + \alpha_2\mu_n^2|(x|u_n)|^2]\\ &= (p(A)x|x) \geqq 0,\end{aligned}$$

and since by assumption at least one $|(x|u_n)|^2$ does not vanish, not all $p(\mu_n)$ can be negative. ■

Exercises

Let the operator A satisfy the assumptions (75.1), (75.2).

1. State the propositions corresponding to Propositions 75.2 through 75.6 for the negative eigenvalues μ_n^- of A.
2. The number of eigenvalues $\geqq \alpha > 0$ of A is equal to $\sup\{\dim F : F \subset E, (Ax|x) \geqq \alpha(x|x)$ for all $x \in F\}$.
3. A is precompact.

§76. General eigenvalue problems for differential operators

We study in this section a generalization of the boundary value problem (3.23). Let two differential operators L, M be defined by

$$(76.1) \qquad (Lx)(t) := \sum_{\nu=0}^{l} f_\nu(t)x^{(\nu)}(t) \qquad \text{for} \quad x \in E_l := C^{(l)}[a, b],$$

$$(76.2) \qquad (Mx)(t) := \sum_{\mu=0}^{m} g_\mu(t)x^{(\mu)}(t) \qquad \text{for} \quad x \in E_m := C^{(m)}[a, b].$$

We assume that all functions have values in $\mathbf{K}$, and that f_ν, g_μ are continuous on $[a, b]$. Under these circumstances the image spaces of L and M lie in $E := C[a, b]$.

Besides the two differential operators, let two linear maps

$$(76.3) \qquad P: E_l \to \mathbf{K}^l, \qquad Q: E_m \to \mathbf{K}^l$$

be given; e.g., P could be a boundary value operator defined by

$$(76.4) \quad Px := (R_1x, \ldots, R_lx) \quad \text{with} \quad R_\mu x := \sum_{\nu=0}^{l-1} [\alpha_{\mu\nu}x^{(\nu)}(a) + \beta_{\mu\nu}x^{(\nu)}(b)]$$

(cf. (3.19) and (73.2)). We now consider the problem to find non-trivial solutions x of the system of equations

$$(76.5) \qquad Lx = \lambda Mx, \qquad Px = \lambda Qx.$$

Every $\lambda \in \mathbf{K}$ for which such a solution x exists is called an *eigenvalue* of the problem (76.5), x itself is called an *eigensolution* corresponding to the eigenvalue λ. Again one has to distinguish between the eigenvalue of a *problem* and that of an *operator*.

In order to have something definite before our eyes, we assume

$$(76.6) \qquad l > m \quad \text{and} \quad f_l(t) \neq 0 \qquad \text{for all} \quad t \in [a, b].$$

Under these assumptions we have

(a) $E_l \subset E_m \subset E$,
(b) $(L - \lambda M)(E_l) = E$ for every $\lambda \in \mathbf{K}$,
(c) $N_\lambda := \{x \in E_l : (L - \lambda M)x = 0\}$ has for every $\lambda \in \mathbf{K}$ the same dimension l.

The last two assertions result from the theory of linear differential equations. The following considerations are based uniquely on the properties (a) through (c). We will therefore assume, more generally than up to now, that E_l, E_m, E are arbitrary vector spaces over $\mathbf{K}$ and that

$$L: E_l \to E, \qquad M: E_m \to E, \qquad P: E_l \to \mathbf{K}^l, \qquad Q: E_m \to \mathbf{K}^l$$

are linear maps; furthermore let assertions (a) through (c) be valid. Under these hypotheses we study the eigenvalue problem (76.5) which with the help of the linear operators

$$\text{(76.7)} \qquad \mathbf{L}: E_l \to E \times \mathbf{K}^l \qquad \mathbf{L}x := (Lx, Px)$$

$$\text{(76.8)} \qquad \mathbf{M}: E_m \to E \times \mathbf{K}^l, \qquad \mathbf{M}x := (Mx, Qx)$$

can be transformed into the equivalent eigenvalue problem

$$\text{(76.9)} \qquad \mathbf{L}x = \lambda \mathbf{M}x.$$

We start our investigations with two lemmas. Let

$$\mathbf{N}_\lambda := \{x \in E_l: (\mathbf{L} - \lambda \mathbf{M})x = 0\}.$$

Lemma 76.1. $\dim \mathbf{N}_\lambda = \operatorname{codim}(P - \lambda Q)(N_\lambda)$.

Proof. Let $\{x_{\lambda 1}, \ldots, x_{\lambda l}\}$ be a basis of N_λ and

$$\text{(76.10)} \qquad \mathfrak{a}_{\lambda\nu} := (P - \lambda Q)x_{\lambda\nu} \qquad \text{for} \quad \nu = 1, \ldots, l.$$

x lies in $\mathbf{N}_\lambda$ if and only if

$$x = \sum_{\nu=1}^{l} \alpha_\nu x_{\lambda\nu} \quad \text{and} \quad (P - \lambda Q)x = \sum_{\nu=1}^{l} \alpha_\nu \mathfrak{a}_{\lambda\nu} = 0.$$

The maximal number of linearly independent elements in $\mathbf{N}_\lambda$ is thus equal to the maximal number of linearly independent vectors $(\alpha_1, \ldots, \alpha_l)$ which satisfy the system of equations $\sum_{\nu=1}^{l} \alpha_\nu \mathfrak{a}_{\lambda\nu} = 0$. It is well-known that the latter number is $l - \dim[\mathfrak{a}_{\lambda 1}, \ldots, \mathfrak{a}_{\lambda l}] = l - \dim(P - \lambda Q)(N_\lambda) = \operatorname{codim}(P - \lambda Q)(N_\lambda)$. ■

Lemma 76.2. *If $\lambda = 0$ is not an eigenvalue of the problem* (76.9), *then* $\mathbf{L}^{-1}$ *exists on* $E \times \mathbf{K}^l$.

Proof. Obviously we only have to show that for an arbitrary element $(y, \mathfrak{y})$ from $E \times \mathbf{K}^l$ there exists an $x \in E_l$ with $\mathbf{L}x = (y, \mathfrak{y})$. Because of (b) there exists in the first place an $x_1 \in E_l$ such that $Lx_1 = y$, and since $\operatorname{codim} P(N_0) = 0$ (Lemma 76.1), there exists an $x_0 \in N_0 = N(L)$ such that $Px_0 = \mathfrak{y} - Px_1$. Consequently $L(x_1 + x_0) = y$, $P(x_1 + x_0) = \mathfrak{y}$, hence $\mathbf{L}(x_1 + x_0) = (y, \mathfrak{y})$. ■

If $\lambda = 0$ is not an eigenvalue of the problem (76.5), then by Lemma 76.2 the linear map

$$G := \mathbf{L}^{-1}\mathbf{M}: E_m \to E_l$$

exists; we call it the *Green operator* of problem (76.5). *The eigenvalues of G are precisely the reciprocals of the eigenvalues of the problem* (76.5); *the corresponding eigenspaces are the same.* The eigenvalue problem (76.5) amounts thus to determining the eigenvalues and the eigensolutions of the Green operator. The system of equations

$$(L - \lambda M)x = y, \qquad (P - \lambda Q)x = \mathfrak{y} \qquad (y \in E, \mathfrak{y} \in \mathbf{K}^l)$$

is equivalent to the operator equation

$$(76.11) \qquad (I - \lambda G)x = z, \qquad z := \mathbf{L}^{-1}(y, \mathfrak{y}).$$

The most important result of this section is:

Proposition 76.1. *For all λ we have* $\operatorname{ind}(I - \lambda G) = 0$.

Proof. Because of $G(E_m) \subset E_l \subset E_m$, it is sufficient by Proposition 23.4 to prove the asserted equation for the restriction of G to E_l. The subspace N_λ contains a basis $\{u_1, \ldots, u_d\}$ of $\mathbf{N}_\lambda$, where $d := \operatorname{codim}(P - \lambda Q)(N_\lambda)$ (Lemma 76.1); if $d = 0$, let $\{u_1, \ldots, u_d\} = \emptyset$. We complete this basis of $\mathbf{N}_\lambda$ by $r := l - d$ elements $z_1, \ldots, z_r$ to a basis $u_1, \ldots, u_d, z_1, \ldots, z_r$ of N_λ. We then have

$$(P - \lambda Q)u_\nu = 0 \qquad \text{for} \quad \nu = 1, \ldots, d,$$

$$\mathfrak{z}_\nu := (P - \lambda Q)z_\nu \neq 0 \qquad \text{for} \quad \nu = 1, \ldots, r.$$

The vectors $\mathfrak{z}_1, \ldots, \mathfrak{z}_r$ are linearly independent: From $\alpha_1\mathfrak{z}_1 + \cdots + \alpha_r\mathfrak{z}_r = 0$ it follows namely that $(P - \lambda Q)(\alpha_1 z_1 + \cdots + \alpha_r z_r) = 0$, hence $\alpha_1 z_1 + \cdots + \alpha_r z_r \in \mathbf{N}_\lambda$; according to the construction of the z_ν all α_ν must vanish. We complete the set $\{\mathfrak{z}_1, \ldots, \mathfrak{z}_r\}$ by $d = l - r$ vectors $\mathfrak{y}_1, \ldots, \mathfrak{y}_d$ to a basis $\{\mathfrak{y}_1, \ldots, \mathfrak{y}_d, \mathfrak{z}_1, \ldots, \mathfrak{z}_r\}$ of $\mathbf{K}^l$. By Lemma 76.2 for $\mathfrak{y}_\nu$ there exists exactly one $y_\nu \in E_l$ such that $\mathbf{L}y_\nu = (0, \mathfrak{y}_\nu)$, $\nu = 1, \ldots, d$. Furthermore by Proposition 16.1 there exist d linear forms $x_1^*, \ldots, x_d^*$ on E_l such that $\langle u_i, x_k^* \rangle = \delta_{ik}$ and $\langle z_\rho, x_k^* \rangle = 0$ for $i, k = 1, \ldots, d$, $\rho = 1, \ldots, r$. With these y_ν and x_k^* we define the finite-dimensional operator S on E_l by

$$Sx := \sum_{\nu=1}^{d} \langle x, x_\nu^* \rangle y_\nu$$

(if $d = 0$ we set $S = 0$; in this case the remainder of the proof becomes simpler). Finally, let $\bar{G}$ be the restriction of G to E_l and

$$R := I - \lambda\bar{G} - S, \qquad \text{hence} \quad I - \lambda\bar{G} = R + S.$$

We first show that R is injective. From $Rx = 0$ it follows that $(I - \lambda\bar{G})x = Sx$, hence

$$(76.12) \quad (\mathbf{L} - \lambda\mathbf{M})x = \mathbf{L}Sx = \sum_{\nu=1}^{d} \langle x, x_\nu^* \rangle (0, \mathfrak{y}_\nu) = \left(0, \sum_{\nu=1}^{d} \langle x, x_\nu^* \rangle \mathfrak{y}_\nu\right);$$

consequently x lies in N_λ and has therefore the form

$$x = \alpha_1 z_1 + \cdots + \alpha_r z_r + \beta_1 u_1 + \cdots + \beta_d u_d$$

so that

$$(P - \lambda Q)x = \alpha_1 \mathfrak{z}_1 + \cdots + \alpha_r \mathfrak{z}_r \quad \text{and} \quad \langle x, x_\nu^* \rangle = \beta_\nu.$$

It thus follows from (76.12) that

$$\alpha_1 \mathfrak{z}_1 + \cdots + \alpha_r \mathfrak{z}_r = \beta_1 \mathfrak{y}_1 + \cdots + \beta_d \mathfrak{y}_d,$$

hence $\alpha_1 = \cdots = \alpha_r = \beta_1 = \cdots = \beta_d = 0$, and so $x = 0$. Next we show that $R(E_l) = E_l$, i.e., that for an arbitrary $y \in E_l$ the equation

(76.13) $$Rx = y \qquad \text{i.e.,} \quad (I - \lambda \bar{G})x = y + Sx$$

can be solved by an $x \in E_l$. (76.13) is equivalent to the equation

$$(\mathbf{L} - \lambda \mathbf{M})x = \mathbf{L}(y + Sx) = \mathbf{L}y + \left(0, \sum_{\nu=1}^{d} \langle x, x_\nu^* \rangle \mathfrak{y}_\nu\right),$$

which decomposes into the two equations

(76.14) $$(L - \lambda M)x = z, \qquad (P - \lambda Q)x = \mathfrak{a} + \sum_{\nu=1}^{d} \langle x, x_\nu^* \rangle \mathfrak{y}_\nu,$$

if we set $\mathbf{L}y = (z, \mathfrak{a})$. Because of (b) there exists a solution x_0 to the first equation; then *all* its solutions can be represented in the form

(76.15) $$x = x_0 + \sum_{\rho=1}^{r} \alpha_\rho z_\rho + \sum_{\nu=1}^{d} \beta_\nu u_\nu.$$

If we write $\mathfrak{x}_0 := (P - \lambda Q)x_0$, then for such an x we have

$$(P - \lambda Q)x = \mathfrak{x}_0 + \sum_{\rho=1}^{r} \alpha_\rho \mathfrak{z}_\rho, \qquad \langle x, x_\nu^* \rangle = \langle x_0, x_\nu^* \rangle + \beta_\nu,$$

so that the second equation in (76.14) takes the form

$$\mathfrak{x}_0 + \sum_{\rho=1}^{r} \alpha_\rho \mathfrak{z}_\rho = \mathfrak{a} + \sum_{\nu=1}^{d} [\langle x_0, x_\nu^* \rangle + \beta_\nu] \mathfrak{y}_\nu$$

or

$$\sum_{\rho=1}^{r} \alpha_\rho \mathfrak{z}_\rho - \sum_{\nu=1}^{d} \beta_\nu \mathfrak{y}_\nu = \mathfrak{a} - \mathfrak{x}_0 + \sum_{\nu=1}^{d} \langle x_0, x_\nu^* \rangle \mathfrak{y}_\nu.$$

Since $\{\mathfrak{z}_1, \ldots, \mathfrak{z}_r, \mathfrak{y}_1, \ldots, \mathfrak{y}_d\}$ is a basis of $\mathbf{K}^l$, this equation can be satisfied by appropriate numbers α_ρ, β_ν. The element x formed with these α_ρ, β_ν according to (76.15) then solves (76.13). Taking all together R is thus bijective.

We finish the proof with the observation that by Proposition 23.3 we have $\operatorname{ind}(I - \lambda \bar{G}) = \operatorname{ind}(R + S) = \operatorname{ind}(R) = 0$. ■

The applicability of the Green operator G is improved by the following fact which is obtained immediately from the results up to now and from Proposition 23.4:

Proposition 76.2. *Let the hypotheses* (a), (b), (c) *be fulfilled for the eigenvalue problem* (76.5). *Assume that $\lambda = 0$ is not an eigenvalue, and let G be the Green operator of problem* (76.5). *Furthermore let $\tilde{E}$ be an arbitrary vector space between $G(E_m)$ and E_m, and let $\tilde{G}$ be the restriction of G to $\tilde{E}$. Then for all λ the relation* $\operatorname{ind}(I - \lambda\tilde{G}) = 0$ *is valid, and x is an eigensolution of problem* (76.5) *corresponding to the eigenvalue λ if and only if x is an eigensolution of $\tilde{G}$ corresponding to the eigenvalue* $1/\lambda$.

Owing to this proposition, the eigenvalue problem (76.5) amounts to—expressed briefly—finding the eigenvalues $\neq 0$ of any $\tilde{G}$ and the corresponding eigensolutions. The greater flexibility we have gained thereby is felt particularly pleasantly when we want to make the connection with the eigenvalue theory of symmetric operators: In order to make this theory fruitful for dealing with problem (76.5), it is sufficient to introduce an inner or semi-inner product on *any* vector space $\tilde{E}$ between $G(E_m)$ and E_m in such a way that $\tilde{G}$ becomes symmetric or fully symmetric, respectively. In particular we have the following:

Proposition 76.3. *If the hypotheses of Proposition* 76.2 *are satisfied and if $\tilde{G}$ is fully symmetric with respect to a semi-inner product $[x|y]$ on $\tilde{E}$, then $\tilde{G}$ is a* Wielandt operator. *Consequently* (*Proposition* 74.3) *every vector $x = Gy$ with $y \in \tilde{E}$ can be expanded into a series of the form*

$$x = \sum_{u \in S} \frac{1}{\lambda}[y|u]u = \sum_{u \in S} [x|u]u$$

with respect to an orthonormal system S of eigensolutions of the problem (76.5) *corresponding to its* (*real*) *eigenvalues λ. The expansion is to be understood in the sense of the limit assertion* (74.15).

We want to investigate more precisely the distribution of eigenvalues in the case that L, M are the differential operators (76.1), (76.2) which satisfy hypothesis (76.6); let $\mathbf{K}$ be the complex number field, furthermore let P and Q be boundary value operators, i.e., of the form (76.4). It is well known that the functions $x_{\lambda 1}(t), \ldots, x_{\lambda l}(t)$ which satisfy the equations

$$(L - \lambda M)x_{\lambda\nu} = 0, \qquad x_{\lambda\nu}^{(\mu-1)}(a) = \delta_{\mu\nu} \qquad (\nu, \mu = 1, \ldots, l)$$

form a basis of the nullspace N_λ of $L - \lambda M$, and for every fixed t depend together with their derivatives differentiably on the complex parameter λ. Consequently the determinant $D(\lambda)$, whose rows are the vectors $\mathfrak{a}_{\lambda\nu} := (P - \lambda Q)x_{\lambda\nu}$, is a holomorphic function of λ on $\mathbf{C}$. Since because of Lemma 76.1 exactly the zeros of $D(\lambda)$ are eigenvalues of problem (76.5), we have the following situation: Either every complex number is an eigenvalue, or the eigenvalues form an at most countable—possibly empty—set without finite limit points. If λ is not an eigenvalue—this is the situation considered from Lemma 76.2 on, in which the Green operator exists—then the second case of this alternative holds and we can apply Proposition 76.3, Proposition 74.3 and the propositions of §75, provided that an inner product can be found on the space $\tilde{E}$ which makes $\tilde{G}$ symmetric.

The investigations of this section present in an abstract and generalized form considerations which figured in the course on eigenvalue theory at Tübingen by H. Wielandt in 1952 we already mentioned (§74). For further developments of these ideas we refer to [157].

Exercise

Let L, M be the differential operators (76.1), (76.2) which satisfy the hypotheses (76.6). Let P be the boundary value operator (76.4) and $Q = 0$. Furthermore assume that 0 is not an eigenvalue of the problem (76.5), and let $G(s, t)$ be the Green function of the boundary value problem $Lx = y$, $Px = 0$ (see the considerations from (3.20) on). Then the Green operator G is given by $(Gy)(s) = \int_a^b G(s, t)[My](t)dt$, and (76.5) is equivalent to finding a non-trivial solution of the equation $x(s) - \int_a^b G(s, t)[Mx](t)\,dt = 0$ (cf. (3.27), (3.28) for $(Mx)(t) := r(t)x(t)$). In the case $m \geqq 1$ this equation is a so-called *integro-differential equation.*

§77. Preliminary remarks concerning the spectral theorem for symmetric operators

A symmetric operator A of $\mathbf{C}^n$ (in other words: a hermitian (n, n)-matrix) can be represented in the form

$$(77.1)\qquad A = \sum_{\rho=1}^{r} \lambda_\rho P_\rho;$$

here $\lambda_1, \ldots, \lambda_r$ are the (real and pairwise different) eigenvalues of A, and P_ρ is the orthogonal projector of $\mathbf{C}^n$ onto the eigenspace $N(\lambda_\rho I - A)$ (cf. §54). Since these eigenspaces are pairwise perpendicular to each other, we have

$$(77.2)\qquad P_\rho P_\sigma = 0 \quad \text{for} \quad \rho \neq \sigma$$

(Exercise 1 in §69), furthermore

$$P_1 + \cdots + P_r = I.$$

For normal meromorphic, in particular symmetric compact operators on Hilbert spaces we have found expansions which are analogous to (77.1). In the last analysis this was possible because the said operators are richly endowed with eigenvectors. There exist, however, symmetric operators which have no eigenvalues at all. Naturally the question arises, whether also in such cases an analog of (77.1) exists, and in case of a positive answer, what it looks like.

To approach the answer to this question, we recall that the natural generalization of a (finite or infinite) sum is the Stieltjes integral. We will write (77.1) therefore in the form of such an integral and examine then whether a corresponding representation is possible for arbitrary symmetric operators.

For this purpose we consider the λ_ρ arranged in monotone increasing order: $\lambda_1 < \lambda_2 < \cdots < \lambda_r$. By Proposition 70.8 we have $\lambda_1 = m(A)$, $\lambda_r = M(A)$ ($m(A)$ and $M(A)$ are the *bounds* of A (see (70.7)); observe that $\sigma(A) = \{\lambda_1, \ldots, \lambda_r\}$). We set $m := m(A)$, $M := M(A)$, and define a family of endomorphisms E_λ for $-\infty < \lambda < \infty$ by

$$(77.3)\qquad E_\lambda := \begin{cases} 0 & \text{for } \lambda < \lambda_1 = m \\ P_1 & \text{for } \lambda_1 \leqq \lambda < \lambda_2 \\ P_1 + P_2 & \text{for } \lambda_2 \leqq \lambda < \lambda_3 \\ \vdots & \vdots \\ P_1 + \cdots + P_{r-1} & \text{for } \lambda_{r-1} \leqq \lambda < \lambda_r \\ P_1 + \cdots + P_r = I & \text{for } \lambda \geqq \lambda_r = M. \end{cases}$$

Now let $Z_n := \{\mu_0, \ldots, \mu_n\}$ be a (generalized) partition of the interval $[m, M]$, i.e., let

$$(77.4)\qquad \mu_0 < m < \mu_1 < \cdots < \mu_{n-1} < \mu_n = M.$$

Then $E_{\mu_k} - E_{\mu_{k-1}} = \sum P_\rho$, where the sum is taken with respect to those subscripts ρ for which $\mu_{k-1} < \lambda_\rho \leqq \mu_k$. With the aid of (77.1) we obtain immediately, that for any choice of intermediate points $\lambda_k' \in [\mu_{k-1}, \mu_k]$ and for any sequence of partitions Z_n with $\max(\mu_k - \mu_{k-1}) \to 0$ we have

$$\sum_{k=1}^{n} \lambda_k'(E_{\mu_k} - E_{\mu_{k-1}}) \Rightarrow A.$$

Postponing a rigorous definition of the Stieltjes integral with an operator-valued integrating function, we write for this limit relation more briefly

$$(77.5)\qquad A = \int_{m-0}^{M} \lambda\, \mathrm{d}E_\lambda.$$

We also state some properties of the 'spectral family' (E_λ):

(77.6)

E_λ is an orthogonal projector,

(E_λ) is monotone increasing: $E_\lambda \leqq E_\mu$ for $\lambda \leqq \mu$,

(E_λ) is continuous to the right: $E_{\lambda+\varepsilon} \to E_\lambda$ for $\varepsilon \to +0$,

$E_\lambda = 0$ for $\lambda < m$, $E_\lambda = I$ for $\lambda \geqq M$.

The first property follows because of (77.2) from Exercise 2 in §69, the second follows from Proposition 69.3, the third is trivial and the fourth is already listed in (77.3). We shall see that for every symmetric operator A of a complex Hilbert space a spectral family (E_λ) can be found, which has all the properties (77.6), and which with the representation (77.5) is valid.

In the case of the operator A in (77.1), one can represent E_λ very easily as a function of A. Since for every polynomial $\varphi(\lambda) := \alpha_0 + \alpha_1\lambda + \cdots + \alpha_n\lambda^n$ one has obviously $\varphi(A) = \sum_{\rho=1}^{r} \varphi(\lambda_\rho)P_\rho$, for a polynomial $\varphi_\mu(\lambda)$ such that

$$\varphi_\mu(\lambda_\rho) = \begin{cases} 1 & \text{for} \quad \lambda_\rho \leqq \mu \\ 0 & \text{for} \quad \lambda_\rho > \mu \end{cases}$$

we get immediately $E_\mu = \varphi_\mu(A)$. In the general case we cannot expect to obtain E_λ as a polynomial in A. The considerations so far suggest, however, the following procedure: Define $e_\mu(A)$ for functions

$$(77.7) \qquad e_\mu(\lambda) := \begin{cases} 1 & \text{for} \quad \lambda \leqq \mu \\ 0 & \text{for} \quad \lambda > \mu \end{cases}, \qquad -\infty < \mu < +\infty,$$

set $E_\mu := e_\mu(A)$ and check whether (E_μ) has the properties (77.6) and whether A can be represented in the form (77.5). This method of proof for equation (77.5) can be found in [10]; its seems to yield the most elementary approach to the 'spectral theorem' (77.5) and therefore we shall make use of it.

§78. Functional calculus for symmetric operators

In this section let A be a symmetric operator of the complex Hilbert space H. A is continuous by Proposition 68.6. Let $m := m(A)$ and $M := M(A)$. In what follows, all polynomials $p(\lambda)$ will have *real* coefficients; then $p(A)$ is always symmetric. Basic is the following:

Lemma 78.1. *If two polynomials p, q satisfy the inequality $p(\lambda) \geqq q(\lambda)$ for all $\lambda \in [m, M]$, then $p(A) \geqq q(A)$.*

We obviously only have to show that $p(\lambda) \geqq 0$ for $\lambda \in [m, M]$ implies $p(A) \geqq 0$. For this we represent $p(\lambda)$ in the form $p(\lambda) = \alpha \prod (\lambda - \lambda_\nu)$, where every zero λ_ν occurs as often as its multiplicity indicates; to avoid the trivial case, we assume $\alpha \neq 0$. We introduce:

α_j the zeros $\leqq m$,

β_k the zeros $\geqq M$,

γ_l the zeros in the interval (m, M),

$\delta_n = \zeta_n + i\eta_n$ the non-real zeros.

Every γ_l has even multiplicity, and together with δ_n also $\bar{\delta}_n$ is a zero. With appropriate indexing we have

$$(\lambda - \gamma_l)(\lambda - \gamma_{l+1}) = (\lambda - \gamma_l)^2 \qquad \text{for} \quad l = 1, 3, \ldots,$$

$$(\lambda - \delta_n)(\lambda - \delta_{n+1}) = (\lambda - \delta_n)(\lambda - \bar{\delta}_n) = (\lambda - \zeta_n)^2 + \eta_n^2 \qquad \text{for} \quad n = 1, 3, \ldots$$

and so

$$p(\lambda) = \alpha' \prod_j (\lambda - \alpha_j) \prod_k (\beta_k - \lambda) \prod_l (\lambda - \gamma_l)^2 \prod_n [(\lambda - \zeta_n)^2 + \eta_n^2] \quad \text{with} \quad \alpha' > 0$$

hence also

$$p(A) = \alpha' \prod_j (A - \alpha_j I) \prod_k (\beta_k I - A) \prod_l (A - \gamma_l I)^2 \prod_n [(A - \zeta_n I)^2 + \eta_n^2 I], \quad \alpha' > 0.$$

We have thus represented $p(A)$ as a product of positive, commuting operators; by Proposition 68.9 we have $p(A) \geqq 0$. ■

Let now K_1 be the class of all functions $f: [m, M] \to \mathbf{R}$ for which the following holds: There exists a sequence of continuous functions f_n with $f_n(\lambda) \geqq f_{n+1}(\lambda) \geqq 0$ and $f_n(\lambda) \to f(\lambda)$—all this for $\lambda \in [m, M]$. If one approximates the functions $g_n(\lambda) := f_n(\lambda) + 1/n$ sufficiently closely by polynomials according to the Weierstrass approximation theorem, then one sees that in this definition 'sequence of continuous functions' can be replaced by 'sequence of polynomials', while everything else remains unchanged. Every function in K_1 is non-negative, and all non-negative continuous functions on $[m, M]$, but also the step functions e_μ defined in (77.7), lie in K_1.

We leave to the reader the simple proof of the following lemma by means of the Heine–Borel covering theorem.

Lemma 78.2. *If the functions $f, g \in K_1$ are approximated according to definition by continuous functions f_n, g_n, and if $f(\lambda) \leqq g(\lambda)$ for all $\lambda \in [m, M]$, then for every natural number k there exists an index n such that*

$$f_n(\lambda) \leqq g_k(\lambda) + \frac{1}{k} \quad \text{for} \quad \lambda \in [m, M].$$

If f is from K_1, and (f_n) is a monotone decreasing sequence of polynomials converging to f, then we have by Lemma 78.1

$$f_1(A) \geqq f_2(A) \geqq \cdots \geqq 0,$$

consequently $(f_n(A))$ converges pointwise to a symmetric operator B (Proposition 68.7). With the aid of Lemma 78.2 one sees that B does not depend on the choice of the approximating sequence of polynomials. We have thus the right to denote the limit operator B by $f(A)$. The correspondence $f \mapsto f(A)$ has the following properties, which follow from the two lemmas or immediately from the definition of $f(A)$, respectively:

From $f(\lambda) \leqq g(\lambda)$ for all $\lambda \in [m, M]$ it follows that $f(A) \leqq g(A)$;

$$\alpha f \mapsto \alpha f(A) \text{ for } \alpha \geqq 0, \qquad f + g \mapsto f(A) + g(A), \qquad fg \mapsto f(A)g(A);$$

$f(A)$ commutes with $g(A)$ and even with every continuous endomorphism which commutes with A;

if f is approximated according to its definition by continuous functions f_n, then $f_n(A) \to f(A)$.

Let the class K_2 consist of all differences $h = f - g$ of functions $f, g \in K_1$. We set $h(A) := f(A) - g(A)$ and ask the reader to convince himself that this definition is unique, and that K_2 and the correspondence $h \mapsto h(A)$ have the following properties (for the proof use the assertions listed above concerning K_1):

K_2 is a real algebra, containing $C[m, M]$;

from $g(\lambda) \leqq h(\lambda)$ for all $\lambda \in [m, M]$ it follows that $g(A) \leqq h(A)$;

$\alpha h \mapsto \alpha h(A)$ *for all* $\alpha \in \mathbf{R}$, $g + h \mapsto g(A) + h(A)$, $gh \mapsto g(A)h(A)$;

$h(A)$ commutes with $g(A)$ and even with every continuous endomorphism which commutes with A.

We already observed that the functions

$$e_\mu(\lambda) := \begin{cases} 1 & \text{for } \lambda \leqq \mu \\ 0 & \text{for } \lambda > \mu \end{cases}$$

lie in K_1. Consequently

$$E_\mu := e_\mu(A)$$

exists for every real μ. The operator E_μ is symmetric, and because of $e_\mu^2 = e_\mu$ it is idempotent, thus it is an orthogonal projector (Proposition 63.1). It is just as easy to see that the family (E_μ) possesses all the other properties listed in (77.6); one only has to use the corresponding properties of the basic functions e_μ. We call (E_μ) the *spectral family* of the operator A.

Exercise

$^+$For a positive operator A there exists exactly one positive operator B such that $B^2 = A$. B is called the *square root* of A and is denoted by $A^{1/2}$.

§79. The spectral theorem for symmetric operators on Hilbert spaces

In this section let again A be a symmetric endomorphism of the complex Hilbert space H, and let $m := m(A)$, $M := M(A)$.

We consider a (generalized) partition Z_n of the interval $[m, M]$:

$$\mu_0 < m < \mu_1 < \mu_2 < \cdots < \mu_{n-1} < \mu_n = M;$$

its *mesh* is

$$\delta(Z_n) := \max_{k=1}^{n}(\mu_k - \mu_{k-1}).$$

By λ_k we denote intermediary points:

$$\mu_{k-1} \leqq \lambda_k \leqq \mu_k, \qquad \text{where we require that } \lambda_1 \geqq m.$$

Let now f be a real-valued function on $[m, M]$, let (E_λ) be the spectral family of A and assume that there exists a (symmetric) operator B such that the following holds: For an arbitrary $\varepsilon > 0$ there exists an $\eta > 0$ such that

$$\left\| B - \sum_{k=1}^{n} f(\lambda_k)(E_{\mu_k} - E_{\mu_{k-1}}) \right\| \leqq \varepsilon$$

for *every* partition Z_n with $\delta(Z_n) \leqq \eta$ and *every* choice of intermediary points λ_k. We then say that the *Stieltjes integral*

$$\int_{m-0}^{M} f(\lambda) \mathrm{d}E_\lambda \qquad \text{exists and is equal to } B.$$

Let us now consider in particular the function $f(\lambda) = \lambda$. From the inequalities

$$\mu_{k-1}[e_{\mu_k}(\lambda) - e_{\mu_{k-1}}(\lambda)] \leqq \lambda_k[e_{\mu_k}(\lambda) - e_{\mu_{k-1}}(\lambda)] \leqq \mu_k[e_{\mu_k}(\lambda) - e_{\mu_{k-1}}(\lambda)],$$

$$\mu_{k-1}[e_{\mu_k}(\lambda) - e_{\mu_{k-1}}(\lambda)] \leqq \lambda[e_{\mu_k}(\lambda) - e_{\mu_{k-1}}(\lambda)] \leqq \mu_k[e_{\mu_k}(\lambda) - e_{\mu_{k-1}}(\lambda)],$$

which are valid for all $\lambda \in \mathbf{R}$, we obtain

$$\mu_{k-1}(E_{\mu_k} - E_{\mu_{k-1}}) \leqq \lambda_k(E_{\mu_k} - E_{\mu_{k-1}}) \leqq \mu_k(E_{\mu_k} - E_{\mu_{k-1}}),$$

$$\mu_{k-1}(E_{\mu_k} - E_{\mu_{k-1}}) \leqq A(E_{\mu_k} - E_{\mu_{k-1}}) \leqq \mu_k(E_{\mu_k} - E_{\mu_{k-1}}).$$

With the aid of the abbreviations

$$R := \sum_{k=1}^{n} \mu_{k-1}(E_{\mu_k} - E_{\mu_{k-1}}), \quad S := \sum_{k=1}^{n} \lambda_k(E_{\mu_k} - E_{\mu_{k-1}}), \quad T := \sum_{k=1}^{n} \mu_k(E_{\mu_k} - E_{\mu_{k-1}})$$

we obtain from these inequalities by summation:

(79.1) $$R \leqq S \leqq T,$$

(79.2) $$R \leqq A \leqq T.$$

Let now $\varepsilon > 0$ be given arbitrarily and set $\eta := \varepsilon/2$. For every partition Z_n with $\delta(Z_n) \leqq \eta$ we have $0 \leqq T - R \leqq \eta I$; because of (79.1), (79.2) we obtain from here

$$0 \leqq T - S \leqq \eta I \quad \text{and} \quad 0 \leqq T - A \leqq \eta I,$$

hence

$$\|T - S\| \leqq \eta \quad \text{and} \quad \|T - A\| \leqq \eta$$

(see (70.8)). Consequently $\|A - S\| \leqq \|A - T\| + \|T - S\| \leqq \eta + \eta = \varepsilon$, and so $A = \int_{m-0}^{M} \lambda \, dE_\lambda$. We have thus proved the *spectral theorem*:

Theorem 79.1. *For every symmetric operator A of the complex Hilbert space H there exists a family of orthogonal projectors E_λ of H, namely the spectral family of A, so that the following hold*:

(a) $E_\lambda \leqq E_\mu$ *for* $\lambda \leqq \mu$,
(b) $E_{\lambda+\varepsilon} \to E_\lambda$ *for* $\varepsilon \to +0$,
(c) $E_\lambda = 0$ *for* $\lambda < m$, $E_\lambda = I$ *for* $\lambda \geqq M$ *and*
(d) $A = \int_{m-0}^{M} \lambda \, dE_\lambda$.

It will be shown in Exercises 3 and 4 how the resolvent set and the point spectrum of a symmetric operator can be described with the help of its spectral family.

For *non-continuous* self-adjoint operators, which because of the Hellinger–Toeplitz theorem cannot be defined on the whole Hilbert space H, one can also construct a family E_λ of orthogonal projectors, which is defined for all real λ, monotone increasing and continuous to the right, tends to 0 for $\lambda \to -\infty$ and to I for $\lambda \to +\infty$, and with which the spectral decomposition

$$A = \int_{-\infty}^{+\infty} \lambda \, dE_\lambda$$

holds. We refer the reader to [10].

Exercises

$^+$1. For $x, y \in H$ the function $(E_\lambda x | x) = \|E_\lambda x\|^2$ is monotone increasing, $(E_\lambda x | y)$ is of bounded variation and

$$(Ax|x) = \int_{m-0}^{M} \lambda \, d(E_\lambda x | x) = \int_{m-0}^{M} \lambda \, d\|E_\lambda x\|^2, \qquad (Ax|y) = \int_{m-0}^{M} \lambda \, d(E_\lambda x | y).$$

$^+$2. Show successively the following:

(a) $(E_{\mu_k} - E_{\mu_{k-1}})(E_{\mu_l} - E_{\mu_{l-1}}) = \delta_{kl}(E_{\mu_k} - E_{\mu_{k-1}})$.
(b) $p(A) = \int_{m-0}^{M} p(\lambda) dE_\lambda$ for every polynomial p with real coefficients.
(c) $f(A) = \int_{m-0}^{M} f(\lambda) dE_\lambda$ for every real-valued function f continuous on $[m, M]$.
(d) If f, g are real-valued, continuous functions on $[m, M]$ and if $f(\lambda) \geqq g(\lambda)$ holds for all $\lambda \in \sigma(A)$, then $f(A) \geqq g(A)$.
(e) Assertion (d) also holds for functions f, g from the class K_2. In particular $f(A) = g(A)$ if f and g agree on $\sigma(A)$.
(f) If $f \in K_2$ and $Ax = \mu x$, $x \neq 0$, then $f(A)x = f(\mu)x$.

$^+$3. The point $\mu \in \mathbf{R}$ belongs to $\rho(A)$ if and only if it is an interior point of an interval of constancy of the spectral family (E_λ).

+4. For every $\mu \in \mathbf{R}$ the limit in the sense of pointwise convergence $E_{\mu-0} := \lim_{\varepsilon \to +0} E_{\mu-\varepsilon}$ exists. μ is an eigenvalue if and only if (E_λ) has a jump at μ, i.e., if $S_\mu := E_\mu - E_{\mu-0} \neq 0$. In this case S_μ is an orthogonal projector onto the eigenspace $N(\mu I - A)$. Setting

$$\delta_\mu(\lambda) := \begin{cases} 1 & \text{for} \quad \lambda = \mu \\ 0 & \text{for} \quad \lambda \neq \mu \end{cases},$$

one has $S_\mu = \delta_\mu(A)$.

XI

Topological vector spaces

§80. Metric vector spaces

In this book we have considered so far almost exclusively normed spaces. This type of space is, however, not sufficient for the purposes of the applications. We want to document this by examples of spaces which occur in a natural manner in analysis and are not normed. We ask the reader to remember the concept of a valued space, introduced in §6.

Example 80.1. Our first example is the space (s) of all numerical sequences $x = (\xi_n)$ with the absolute value

$$|x| = \sum_{n=1}^{\infty} \frac{1}{2^n} \frac{|\xi_n|}{1 + |\xi_n|}$$

(see Example 1.5). We saw in §1 that convergence in the sense of the absolute value is equivalent to componentwise convergence. Since $|x| \leqq 1$ for all x, the absolute value cannot be a norm; *but one cannot even introduce any norm on* (s) *so that convergence in the sense of that norm is equivalent to the convergence in the sense of the absolute value.* Otherwise the linear forms f_k defined by $f_k(\xi_1, \xi_2, \ldots) := \xi_k$ would be continuous, not only in the sense of the absolute value (which they are, because convergence in the sense of the norm implies componentwise convergence), but also in the sense of the norm, i.e., they would be bounded. Consequently the sequence of kth components $(f_k(x_1), f_k(x_2), \ldots)$ of a norm-bounded sequence $(x_1, x_2, \ldots)$ would be bounded. With the help of the Bolzano–Weierstrass theorem and the diagonal process, one could choose a subsequence $(y_1, y_2, \ldots)$ such that $f_k(y_n) \to \eta_k$ as $n \to \infty$ $(k = 1, 2, \ldots)$. The sequence (y_n) would therefore converge componentwise, hence in the sense of the absolute value, and so, by assumption, also in the sense of the norm. By Theorem 10.1 the space (s) would then be finite-dimensional, which is obviously not the case.

Example 80.2. On the vector space $C^{(\infty)}[a, b]$ of the infinitely differentiable functions $x: [a, b] \to \mathbf{K}$ we define the semi-norms p_n and an absolute value by

$$p_n(x) := \max_{a \leq t \leq b} |x^{(n-1)}(t)|, \qquad |x| := \sum_{n=1}^{\infty} \frac{1}{2^n} \frac{p_n(x)}{1 + p_n(x)}.$$

The triangle inequality follows from the estimate

$$\text{(80.1)} \qquad \frac{p(x+y)}{1 + p(x+y)} \leq \frac{p(x)}{1 + p(x)} + \frac{p(y)}{1 + p(y)},$$

valid for every semi-norm p on a vector space E (see the proof of (1.8)); the other properties of an absolute value are trivial. Similarly as in the case of the space (s), one proves the assertion:

(80.2) $\quad |x_k - x| \to 0$ if and only if $p_n(x_k - x) \to 0$ for $n = 1, 2, \ldots$.

Thus convergence $x_k \to x$ with respect to absolute value is equivalent to

$$x_k^{(n)}(t) \to x^{(n)}(t) \textit{ uniformly} \text{ on } [a, b] \text{ for } n = 0, 1, \ldots.$$

Example 80.3. In the vector space $C(a, b)$ of all continuous functions $x: (a, b) \to \mathbf{K}$ the 'uniform convergence on all compact subsets of (a, b)' ($x_k(t) \to x(t)$ uniformly on every compact $K \subset (a, b)$) plays usually a more important role than the uniform convergence on the whole open interval (a, b). Since each such K lies in a closed, bounded (hence compact) interval $[\alpha(K), \beta(K)] \subset (a, b)$, one sees that this convergence is equivalent to uniform convergence on all intervals

$$I_n := \left[a + \frac{1}{n+p}, b - \frac{1}{n+p}\right], \qquad n = 1, 2, \ldots,$$

where the fixed $p > 0$ is chosen so large that $a + (1/p) < b - (1/p)$. We define the semi-norms p_n on $C(a, b)$ by

$$p_n(x) := \max_{t \in I_n} |x(t)|$$

and see that $x \to x_k$ in the sense of 'uniform convergence on all compact subsets of (a, b)' is equivalent to $p_n(x_k - x) \to 0$ for $n = 1, 2, \ldots$. If we define an absolute value on $C(a, b)$ by

$$|x| := \sum_{n=1}^{\infty} \frac{1}{2^n} \frac{p_n(x)}{1 + p_n(x)},$$

then assertion (80.2) is again valid, so that uniform convergence on all compact subsets of (a, b) is finally nothing but convergence in the sense of the absolute value.

We close this series of examples by emphasizing what is common in them:

Example 80.4. If $(p_1, p_2, \ldots)$ is a *total* (finite or infinite) sequence of semi-norms on a vector space E, i.e., $p_1(x) = p_2(x) = \cdots = 0$ implies $x = 0$, then

$$|x| := \sum \frac{1}{2^n} \frac{p_n(x)}{1 + p_n(x)} \tag{80.3}$$

defines an absolute value on E, and $x_k \to x$ in the sense of the absolute value is equivalent to $p_n(x_k - x) \to 0$ for $n = 1, 2, \ldots$.

In a valued space the triangle inequality implies the continuity of addition: from $x_n \to x$, $y_n \to y$ it follows that $x_n + y_n \to x + y$. The discrete absolute value (6.2) shows that multiplication by a scalar does not have to be continuous. We call a valued space, i.e., a vector space with a translation invariant metric, a *metric vector space* if the scalar multiplication is also continuous. *The valued space in Example* 80.4, *and with it each of the spaces presented above, is a metric vector space.* This follows immediately from the meaning of convergence together with the estimate $p_n(\alpha_k x_k - \alpha x) \leqq |\alpha_k| p_n(x_k - x) + |\alpha_k - \alpha| p_n(x)$. Because of Proposition 6.1 normed spaces are always metric vector spaces. The continuity of the norm asserted there is also, by the way, true for the absolute value.

It is a significant fact that the open mapping theorem and the theorem about the continuous inverse, as well as the closed graph theorem, are also valid in the framework of metric vector spaces. Let the reader go through the proofs and make the necessary modifications; we only indicate that the equation $E = \bigcup_{n=1}^{\infty} nK$, $K := K_r(0)$, at the beginning of (b) in the proof of Principle 32.1, now follows from the continuity of multiplication by scalars: For a fixed $x \in E$ we have $(1/n)x \to 0$, thus $|(1/n)x| < r$ for sufficiently large n, so $(1/n) \in K$ or $x \in nK$. We formulate the results:

Theorem 80.1. *Every continuous linear map from a complete metric vector space E* onto *the complete metric vector space F is open; if it is, moreover, injective, then its inverse is also continuous. A closed linear map from E into F is always continuous.*

The Hahn–Banach theorem does not need to hold, however, on metric vector spaces.

Exercises

1. We recall that the spaces (s) and $C^{(\infty)}[a, b]$ are complete (§1 and Exercise 1 of §6). Show that $C(a, b)$ equipped with the metric given by the absolute value from Example 80.3 is also complete (see also Exercise 2).

$^+$2. For the completeness of the space E of Example 80.4 a criterion can be given which uses only the sequence of semi-norms $P := (p_1, p_2, \ldots)$. A sequence

(x_k) from E is called a P-Cauchy sequence if $p_n(x_k - x_l) \to 0$ as $k, l \to \infty$, for every $n \in \mathbf{N}$. Show that E is complete if and only if for every P-Cauchy sequence (x_k) there exists an $x \in E$ such that $p_n(x_k - x) \to 0$ as $k \to \infty$, for every $n \in \mathbf{N}$.

+3. Every continuous linear form on (s) has the form

$$(80.4) \qquad f(x) = \sum_{k=1}^{\infty} \alpha_k \xi_k, \qquad \alpha_k = 0 \text{ with finitely many exceptions.}$$

Conversely, (80.4) defines always a continuous linear form on (s). If we define f_k: $(s) \to \mathbf{K}$ by $f_k(x) := \xi_k$, then every continuous linear form f on (s) can be represented in the form $f = \sum_{k=1}^{\infty} \alpha_k f_k$, $\alpha_k = 0$ with finitely many exceptions.

4. Let Δ be a simply connected domain in $\mathbf{C}$, and $H(\Delta)$ the complex vector space of holomorphic functions on Δ. Show that $H(\Delta)$ can be made into a complete metric vector space with the help of a sequence of semi-norms in such a fashion that metric convergence $x_n \to x$ is equivalent to $x_n(\xi) \to x(\xi)$ *uniformly* on all compact subsets of Δ.

§81. Basic notions from topology

In normed spaces and metric vector spaces addition and multiplication by a scalar are *continuous* operations, and the significance of this circumstance manifested itself again and again. It is therefore indicated to investigate in its purest form the situation here present by considering vector spaces which are equipped with a topology in such a way that for it addition and multiplication by a scalar are continuous. Topologies are namely the best conceptual means to study questions of continuity separately from any concrete particularities. We first recall the simplest topological concepts and facts. The proofs can be found in any text on topology; see e.g., [179].

A non-empty set E is called a *topological space* if with every $x \in E$ a system $\mathfrak{U}(x)$ of subsets $U \subset E$ is associated so that the following *neighborhood axioms* are satisfied:

(U1) x lies in every $U \in \mathfrak{U}(x)$.
(U2) If $U \in \mathfrak{U}(x)$, then every set containing U also belongs to $\mathfrak{U}(x)$.
(U3) The intersection of two sets in $\mathfrak{U}(x)$ belongs to $\mathfrak{U}(x)$.
(U4) For every $U \in \mathfrak{U}(x)$ there exists a $V \in \mathfrak{U}(x)$ such that $U \in \mathfrak{U}(y)$ for every $y \in V$.

Every $U \in \mathfrak{U}(x)$ is called a *neighborhood* of x, and $\mathfrak{U}(x)$ is called the *filter of neighborhoods* of x. The system of filters of neighborhoods defines the *topology* τ of E. For greater clarity we occasionally write (E, τ) for a topological space E with topology τ.

A metric space E becomes a topological space in a canonical way if for $\mathfrak{U}(x)$ we take the collection of all sets containing an open ball around x. The topology so introduced in called the *metric* topology of E or the topology of E *generated by the metric*.

On a set $E \neq \emptyset$ one can always introduce two extreme topologies:

(a) The *discrete* topology: $\mathfrak{U}(x)$ consists of all sets $U \subset E$ which contain x;
(b) The *chaotic* topology: $\mathfrak{U}(x)$ contains only E.

The discrete metric generates the discrete topology. The chaotic topology is not a metric topology if the space contains more than one point.

$\mathfrak{B}(x) \subset \mathfrak{U}(x)$ is called a *basis of neighborhoods* of the point x if every $U \in \mathfrak{U}(x)$ contains a $V \in \mathfrak{B}(x)$. Then $\mathfrak{U}(x)$ is the collection of all supersets of the sets from $\mathfrak{B}(x)$. In a metric space the family of all open balls around x, and also the at most countable collection $\{K_{1/n}(x): n = 1, 2, \ldots\}$ is a basis of neighborhoods of x. Bases of neighborhoods simplify the handling of topologies when their elements have an easily discernible structure (as for instance the open balls in the metric case).

A topology τ on E, or the topological space (E, τ) itself, is said to be *separated* or *Hausdorff* if any two distinct points x, y always have disjoint neighborhoods $U \in \mathfrak{U}(x)$, $V \in \mathfrak{U}(y)$. A separated topological space is also called a *Hausdorff space. Metric topologies are always separated.*

A subset M of a topological space E is said to be *open* if every $x \in M$ has a neighborhood which lies completely in M. The system $\mathfrak{O}$ of all open subsets of E has the following properties:

(O1) $\emptyset$ and E belong to $\mathfrak{O}$.
(O2) The union of arbitrarily many sets from $\mathfrak{O}$ belongs to $\mathfrak{O}$.
(O3) The intersection of finitely many sets from $\mathfrak{O}$ belongs to $\mathfrak{O}$.

In the discrete topology *all* subsets of E are open, while in the chaotic topology *only* $\emptyset$ and E are open.

The open sets containing x form a basis of neighborhoods of x. Thus the topology is determined already by the open sets. This fact suggests to invert the procedure used up to now which consisted in defining open sets by means of neighborhoods, and to introduce neighborhoods, and thus topologies, starting out from 'open sets'. For this let $\mathfrak{O}$ be a collection of subsets of the set $E \neq \emptyset$ which has the properties (O1) through (O3). *Then there exists exactly one topology on E whose collection of open sets is precisely $\mathfrak{O}$.*

Let two topologies τ_1, τ_2 be defined on E, and let $\mathfrak{O}_1, \mathfrak{O}_2$ be the corresponding collections of open sets. We say that τ_1 is *coarser* than τ_2 (in symbols: $\tau_1 \prec \tau_2$) if $\mathfrak{O}_1 \subset \mathfrak{O}_2$; we then also say that τ_2 is *finer* than τ_1. The finer topology possesses, roughly speaking, more open sets. The relation '$\prec$' is an order in the set of all topologies on E. *The discrete topology is the finest, the chaotic topology the coarsest topology.* A family $(\tau_\iota: \iota \in J)$ of topologies on E with the corresponding collections of open sets $\mathfrak{O}_\iota$ has a *greatest lower bound* (infimum) τ, i.e., there exists a topology τ with the following properties:

(a) $\tau \prec \tau_\iota$ for all $\iota \in J$;
(b) from $\bar{\tau} \prec \tau_\iota$ for all $\iota \in J$ it follows that $\bar{\tau} \prec \tau$.

The collection of open sets corresponding to τ is $\mathfrak{D} := \bigcap_{\iota \in J} \mathfrak{D}_\iota$. For this reason τ is also called the *intersection* of the topologies τ_ι.

Analogously $(\tau_\iota: \iota \in J)$ also has a *least upper bound* (supremum) ω, i.e., there exists a topology ω with the following properties:

(a) $\tau_\iota \prec \omega$ for all $\iota \in J$;
(b) from $\tau_\iota \prec \bar{\tau}$ for all $\iota \in J$ it follows that $\omega \prec \bar{\tau}$.

ω is the intersection of all topologies which are finer than every τ_ι (such topologies exist, e.g., the discrete one).

When comparing two topologies τ_1, τ_2 on E with bases of neighborhoods $\mathfrak{B}_1(x)$, $\mathfrak{B}_2(x)$, the *Hausdorff criterion* is useful:

We have $\tau_1 < \tau_2$ *if and only if for all* $x \in E$ *the following assertion holds: For every* $V_1 \in \mathfrak{B}_1(x)$ *there exists a* $V_2 \in \mathfrak{B}_2(x)$ *so that* $V_2 \subset V_1$ *(i.e., grosso modo, every basic* τ_1*-neighborhood can be minorized by a basic* τ_2*-neighborhood).*

$x \in E$ is called a *closure point* of $G \subset E$ if every neighborhood of x—or if every neighborhood belonging to a basis of neighborhoods $\mathfrak{B}(x)$—has points in common with G. The *closure* or *closed hull* $\bar{G}$ of G is the set of all closure points of G. We always have $G \subset \bar{G}$; if $G = \bar{G}$ then we say that G is *closed. The closure* $\bar{G}$ *is always closed.* $G \subset E$ *is closed if and only if its complement* $E \setminus G$ *is open.* We therefore obtain from the properties (O1) through (O3) of open sets, by the known rules of complementation, immediately the following properties of closed sets:

(C1) $\varnothing$ and E are closed.
(C2) The intersection of arbitrarily many closed sets is closed.
(C3) The union of finitely many closed sets is closed.

The closure $\bar{G}$ is the intersection of all closed sets containing G. In a Hausdorff space every finite set is closed.

We say that a sequence (x_n) from E *converges* to $x \in E$, or that x is the limit of (x_n), (in symbols: $x_n \to x$ or $\lim x_n = x$) if every neighborhood of x—or any neighborhood belonging to a basis of neighborhoods $\mathfrak{B}(x)$—contains almost all terms of the sequence, i.e., all but finitely many. *In Hausdorff spaces a sequence has at most one limit.* The limit of a sequence from $G \subset E$ is always a closure point of G; in opposition to the situation in metric spaces, the converse of this is not true. This is one of the reasons why convergent sequences do not play a great role in general topological spaces.

The map $f: E \to F$ of the topological spaces E, F is said to be *continuous at the point* $x_0 \in E$ if for every neighborhood V of $f(x_0)$—or for every neighborhood belonging to a basis of neighborhoods $\mathfrak{B}(f(x_0))$—there exists a neighborhood U of x_0 such that $f(U) \subset V$; it is called *sequentially continuous* at x_0 if $x_n \to x_0$ implies $f(x_n) \to f(x_0)$. A function continuous at x_0 is also sequentially continuous there; if E, F are metric spaces, then the converse is also true.

A function which is continuous or sequentially continuous at every point of E is simply said to be *continuous* or *sequentially continuous*, respectively. Continuous functions are by far more important than sequentially continuous ones.

The map $f: E \to F$ is continuous at x_0 *if and only if for every V from a basis of neighborhoods of $f(x_0)$, the preimage $f^{-1}(V)$ is a neighborhood of x_0; it is* continuous *if and only if the preimages of open (closed) sets in F are open (closed) in E.*

The composition $g \circ f$ of the continuous functions $f: E \to F$, $g: F \to G$ is continuous.

A bijective map $f: E \to F$ is said to be *homeomorphic* or a *homeomorphism* if both f and f^{-1} are continuous. Such a homeomorphism establishes a one-to-one correspondence between the open sets of E and the open sets of F, so that E cannot be distinguished from F topologically. In this case we say that the spaces E, F are *homeomorphic* (to each other); the symmetric terminology is justified by the fact that with f also f^{-1} is homeomorphic.

If E is made by the topologies τ_1 and τ_2 into the topological spaces E_1 and E_2, respectively, then τ_1 is finer than τ_2 if and only if $I: E_1 \to E_2$ is continuous (I is the identity mapping of E).

We now discuss the *transfer of topologies*:

Let E be an arbitrary non-empty set, $(F_\iota: \iota \in J)$ a family of topological spaces, and $(f_\iota: \iota \in J)$ a family of maps $f_\iota: E \to F_\iota$. All f_ι become continuous if we introduce on E the discrete topology. The intersection of all topologies on E, for which all the f_ι are continuous, is then the *coarsest* topology with this property; it is called the *initial topology* on E for the family (f_ι).

If a family $(E_\iota: \iota \in J)$ of topological spaces, a non-empty set F and for each $\iota \in J$ a map $f_\iota: E_\iota \to F$ are given, then there exists a *finest* topology on F for which every f_ι is continuous; it is called the *final topology* on F for the family (f_ι). A subset M of F is open in this topology if and only if $f_\iota^{-1}(M)$ for all $\iota \in J$ is open in E_ι.

The initial and final topologies are also called the *topologies generated by the family* (f_ι).

We now present some examples of the transfer of topologies. The first three concern the initial, the last one the final topology.

Example 81.1. Let F be a topological space and $f: E \to F$ a bijective map. The collection of open sets in E for the topology generated by f is $\{f^{-1}(M): M$ open subset of $F\}$. E and F are homeomorphic.

Example 81.2. Let E be a non-empty subset of the topological space F and let $f: E \to F$ be the (canonical) injection i of E into F: $i(x) = x$ for all $x \in E$. The topology generated by i is called the *relative* topology or also the topology *induced* by F (or by the topology of F). The open sets of this topology are the sets $M \cap E$, where M runs through the open sets of F. Unless something else is explicitly said, *we always equip a subset $E \neq \varnothing$ of a topological space F with the relative topology* and call then E a *subspace* of F. If F is a metric space, then the induced metric on E generates precisely the induced topology.

The set $M \subset E$ is said to be *relatively open* if it is open in the relative topology of E; the *relatively closed* sets and *relative neighborhoods* are defined analogously.

The relative neighborhoods of $x \in E$ are precisely the intersections of the neighborhoods of x with E; if $\mathfrak{B}(x)$ is a basis of neighborhoods of x, then $\{V \cap E: V \in \mathfrak{B}(x)\}$ is a basis of the relative neighborhoods of x.

$f: G \to H$ (G, H topological spaces) is said to be *open* if the image of every open subset of G is open in the relative topology of $f(G)$.

Example 81.3. On the cartesian product $E := \prod_{\iota \in J} F_\iota$ of the topological spaces F_ι the family of projectors $\pi_\iota: E \to F_\iota$ onto the components generates the so-called *product topology*; E itself, equipped with the product topology, is called the *topological product* of the spaces F_ι. By definition, the product topology is the coarsest topology for which all the π_ι are continuous.

If $\mathfrak{B}(x_\iota)$ is a basis of neighborhoods of $x_\iota \in F_\iota$, then a basis of neighborhoods of $x = (x_\iota) \in E$ is given by the sets $V := \prod_{\iota \in J} V_\iota$, where $V_\iota = F_\iota$ with the exception of *finitely* many indices and $V_\iota \in \mathfrak{B}(x_\iota)$ for these exceptional indices.

Example 81.4. Let $\{R_\lambda: \lambda \in \Lambda\}$ be a *partition* of the set $E \neq \emptyset$, i.e., let $\bigcup_{\lambda \in \Lambda} R_\lambda = E$ and $R_\lambda \cap R_\mu = \emptyset$ for $\lambda \neq \mu$. If we call two points x, y of E equivalent provided they lie in the same set R_λ, then we define thereby an equivalence relation in E, and R_λ is the equivalence or residue class of each of its elements. We already encountered partitions when forming the quotient space of a vector space with respect to a subspace. Like we did there, we denote by $\hat{x}$ the residue class of x and by $\hat{E}$ the set of all residue classes; thus $\hat{E} := \{R_\lambda: \lambda \in \Lambda\}$. The map $h: E \to \hat{E}$ which associates with every $x \in E$ its equivalence class $\hat{x}$ is called the canonical surjection from E onto $\hat{E}$. If now E is a topological space, then h generates a topology on $\hat{E}$ called the *quotient topology*; $\hat{E}$ itself, equipped with this topology, is called the *quotient space. A subset M of $\hat{E}$ is open if and only if $h^{-1}(M)$ is open.* By definition, the quotient topology is the finest topology on $\hat{E}$ for which h is continuous.

A map $g: \hat{E} \to G$ (G a topological space) is continuous if and only if $g \circ h: E \to G$ is continuous.

We close this section with the definition of a filter and of a filter basis.

A collection $\mathfrak{F}$ of subsets of the non-empty set X is called a *filter* on X, if the following conditions are satisfied:

(F1) $\mathfrak{F} \neq \emptyset$, $\emptyset \notin \mathfrak{F}$.
(F2) Every superset of a set from $\mathfrak{F}$ lies again in $\mathfrak{F}$.
(F3) The intersection of any two sets from $\mathfrak{F}$ lies in $\mathfrak{F}$.

A system of subsets $\mathfrak{B}$ is called a *filter basis* on X if the following holds:

(FB1) $\mathfrak{B} \neq \emptyset$, $\emptyset \notin \mathfrak{B}$.
(FB2) The intersection of any two sets from $\mathfrak{B}$ contains a set from $\mathfrak{B}$.

The filter of neighborhoods of a point is a filter, a basis of neighborhoods is a filter basis.

Every filter is a filter basis. The collection of all supersets of sets of a filter basis $\mathfrak{B}$ is a filter $\mathfrak{F}$. It is called the *filter generated* by $\mathfrak{B}$, while $\mathfrak{B}$ is called a *basis* of $\mathfrak{F}$.

Exercises

1. Let E be a subspace of the topological space F. Show the following:
(a) The relatively closed subsets of E are the intersections $M \cap E$, where M runs through the closed subsets of F.
(b) If E is itself closed, then a relatively closed set $M \subset E$ is also closed (as a subset of F). (*Example*: $F = \mathbf{R}^2$, $E = \mathbf{R}$, $M = [a, b]$). The assertion is not necessarily true if E is open.
*2. With a map $f: E \to F$ we associate the map $f_0: E \to f(E)$ which is defined by $f_0(x) = f(x)$ for $x \in E$. If we equip $f(E)$ with the relative topology, then the following is true: f is continuous if and only if f_0 is continuous. Thus when considering questions of continuity, we may restrict ourselves to relative neighborhoods of the image points.
*3. The restriction f_1 of a continuous map $f: E \to F$ to the subspace G of E is continuous.
*4. Let $f: E \to F$ be injective, so that the map $f_0: E \to f(E)$ defined in Exercise 2 is bijective. $f_0^{-1}: f(E) \to E$ is continuous if and only if f is open. Thus a bijective map f is homeomorphic if and only if it is continuous and open.
*5. Every subspace of a Hausdorff space is Hausdorff.
*6. The topological product of Hausdorff spaces is Hausdorff.
7. Let F be a topological space and $\varnothing \neq G \subset E \subset F$. Then both F and the subspace E induce topologies on G. Show that they coincide.
8. Like in Exercises 6 and 7 of §1, we equip the cartesian products $E := \prod_{k=1}^{n} F_k$ and $E := \prod_{k=1}^{\infty} F_k$ with the metric

$$d(x, y) := \sum_{k=1}^{n} d_k(x_k, y_k) \quad \text{and} \quad d(x, y) := \sum_{k=1}^{\infty} \frac{1}{2^k} \frac{d_k(x_k, y_k)}{1 + d_k(x_k, y_k)},$$

respectively; here $x = (x_1, x_2, \ldots)$, $y = (y_1, y_2, \ldots)$. Show that d generates the product topology. Thus the product of at most countably many metric spaces is again a metric space. The metric topology ('absolute value topology') of $(s) = \prod_{k=1}^{\infty} F_k$ ($F_k := \mathbf{K}$ for all k) coincides with the product topology.
*9. Let E, F, G be metric spaces, $f: E \times F \to G$ a 'function of two variables'. f is continuous at $(x, y) \in E \times F$ if and only if $x_n \to x$, $y_n \to y$ implies $f(x_n, y_n) \to f(x, y)$. An analogous assertion is valid for a map $f: E_1 \times \cdots \times E_k \to G$ if $E_1, \ldots, E_k, G$ are metric spaces. This fact justifies that we defined the continuity of addition and multiplication by a scalar in metric vector spaces and of the inner product in prehilbert spaces by means of sequences.
10. Let $E := \prod_{\iota \in J} F_\iota$ be the product of the topological spaces F_ι. A sequence $(x_\iota^{(n)})$ in E converges to $(x_\iota) \in E$ if and only if $x_\iota^{(n)} \to x_\iota$ as $n \to \infty$ for every $\iota \in J$ (convergence in the topological product is equivalent to componentwise convergence).

11. After the study of this section it is advisable to read again Exercises 11 through 15 of §1 and Exercises 9 through 10 of §7.

§82. The weak topology

Let (E, E^+) be a bilinear system. Every $x^+ \in E^+$ generates, according to the definition

$$f_{x^+}(x) := \langle x, x^+ \rangle,$$

a linear form f_{x^+} on E. The initial topology for the family of functions $f_{x^+}: E \to \mathbf{K}$, i.e., *the* coarsest *topology on E for which all linear forms f_{x^+} are continuous*, is called the *weak topology* generated by E^+ on E, and is denoted by $\sigma(E, E^+)$.

The preimage of an ε-neighborhood of the image point $f_{x^+}(x_0)$, i.e., the set

$$U_{x^+;\varepsilon}(x_0) := \{x \in E: |f_{x^+}(x) - f_{x^+}(x_0)| \leqq \varepsilon\} = \{x \in E: |\langle x - x_0, x^+ \rangle| \leqq \varepsilon\}$$

must be a $\sigma(E, E^+)$-neighborhood of x_0. The same is true for the finite intersections

$$\text{(82.1)} \quad U_{x_1^+, \ldots, x_n^+;\,\varepsilon}(x_0) := \bigcap_{\nu=1}^{n} U_{x^+_\nu;\,\varepsilon}(x_0) = \left\{x \in E: \max_{\nu=1}^{n} |\langle x - x_0, x_\nu^+ \rangle| \leqq \varepsilon\right\}$$

and their supersets, the collection of which we will call $\mathfrak{U}(x_0)$. The reader will verify easily that $\mathfrak{U}(x_0)$ satisfies the neighborhood axioms (U1) through (U4). The topology defined by the systems of neighborhoods $\mathfrak{U}(x)$, $x \in E$, is coarser than $\sigma(E, E^+)$ by the Hausdorff criterion; since on the other hand every f_{x^+} is trivially continuous for it, it must also be finer than $\sigma(E, E^+)$, and so it must coincide with $\sigma(E, E^+)$. *The neighborhoods of the form* (82.1) *form a basis of neighborhoods $\mathfrak{B}(x_0)$ of the point x_0 in the topology $\sigma(E, E^+)$ if $\{x_1^+, \ldots, x_n^+\}$ runs through all finite subsets of E^+ and ε through all positive numbers.* With the help of the *neighborhoods of zero*

$$\text{(82.2)} \qquad U_{x_1^+, \ldots, x_n^+;\,\varepsilon} := \left\{x \in E: \max_{\nu=1}^{n} |\langle x, x_\nu^+ \rangle| \leqq \varepsilon\right\}$$

the neighborhoods belonging to the basis $\mathfrak{B}(x_0)$ can also be written in the form

$$\text{(82.3)} \qquad x_0 + U_{x_1^+, \ldots, x_n^+;\,\varepsilon}.$$

Let now f be a $\sigma(E, E^+)$-continuous linear form on E. Then, since in particular f is continuous at the origin, there exist elements $x_1^+, \ldots, x_n^+$ and an $\varepsilon > 0$ such that

$$\text{(82.4)} \qquad p(x) := \max_{\nu=1}^{n} |\langle x, x_\nu^+ \rangle| \leqq \varepsilon \quad \text{implies} \quad |f(x)| \leqq 1.$$

We now show that we have

$$\text{(82.5)} \qquad |f(x)| \leqq \frac{1}{\varepsilon} p(x) \quad \text{for all} \quad x \in E.$$

Indeed, if $p(x) = 0$, i.e., $p(\lambda x) = |\lambda| p(x) = 0$ for all $\lambda \in \mathbf{K}$, then it follows from (82.4) that $|\lambda||f(x)| = |f(\lambda x)| \leqq 1$ for all λ, hence $f(x) = 0$ and so (82.5) is valid in this case. If $p(x) \neq 0$, then $p(\varepsilon x/p(x)) = (\varepsilon/p(x))p(x) = \varepsilon$, and so we obtain from (82.4) that

$$\frac{\varepsilon}{p(x)} |f(x)| = \left| f\left(\frac{\varepsilon x}{p(x)}\right)\right| \leqq 1,$$

and therefore (82.5) is valid also in this case.

(82.5) shows that $f_{x_\nu^+}(x) = \langle x, x_\nu^+ \rangle = 0$ for $\nu = 1, \ldots, n$ implies that $f(x) = 0$. With the aid of Lemma 15.1 we obtain from here that $f = \sum_{\nu=1}^n \alpha_\nu f_{x_\nu^+}$. If (E, E^+) is even a left dual system, i.e., if we can identify the linear form f_{x^+} with its generating vector x^+ (see §15), then the above representation of f becomes the equation $f = \sum_{\nu=1}^n \alpha_\nu x_\nu^+$. Thus f lies in E^+, so that we can say: In the case of a left dual system (E, E^+), the space E^+ is the set of *all* $\sigma(E, E^+)$-continuous linear forms on E.

The weak topology $\sigma(E, E^+)$ is Hausdorff if and only if (E, E^+) is a right dual system. Indeed, if it is Hausdorff and if we have $\langle x, x^+ \rangle = 0$ for all $x^+ \in E^+$, then x lies in every neighborhood of zero and must therefore vanish, since otherwise there would exist a neighborhood U of 0 (containing x) and a neighborhood V of x disjoint from U. Conversely, if (E, E^+) is a right dual system, then there exists to any two distinct points $x_1, x_2 \in E$ an x^+ with $\langle x_1 - x_2, x^+ \rangle \neq 0$. One immediately sees that the neighborhoods $U_{x^+;\varepsilon/3}(x_1)$ and $U_{x^+;\varepsilon/3}(x_2)$ are disjoint if one sets $\varepsilon = |\langle x_1 - x_2, x^+ \rangle|$.

If we apply our results to the bilinear system (E^+, E) with the bilinear form $\langle x^+, x \rangle := \langle x, x^+ \rangle$, then we obtain the weak topology $\sigma(E^+, E)$ on E^+ generated by E, and the assertions concerning it, which are analogous to those made above. We summarize and add a few evident supplementary remarks:

Theorem 82.1. *Let (E, E^+) be a bilinear system with respect to the bilinear form $\langle x, x^+ \rangle$. Then the sets*

$$U_{x_1^+, \ldots, x_n^+; \varepsilon} := \left\{ x \in E : \max_{\nu=1}^{n} |\langle x, x_\nu^+ \rangle| \leqq \varepsilon \right\}$$

form a basis of neighborhoods of 0 and the sets $x_0 + U_{x_1^+, \ldots, x_n^+; \varepsilon}$ a basis of neighborhoods of x_0 for the weak topology $\sigma(E, E^+)$. This is the coarsest topology on E for which all the linear forms $x \mapsto \langle x, x^+ \rangle$ are continuous. Conversely, for every $\sigma(E, E^+)$-continuous linear form f there exists an x^+ such that $f(x) = \langle x, x^+ \rangle$ for all $x \in E$. If (E, E^+) is a left *dual system, then E^+ is the vector space of all weakly continuous linear forms on E. The topology $\sigma(E, E^+)$ is Hausdorff if and only if (E, E^+) is a* right *dual system. The E^+-weak convergence $x_n \rightharpoonup x$ (§46) is equivalent to $x_n \to x$ in the sense of the weak topology $\sigma(E, E^+)$. Analogous assertions hold about the weak topology $\sigma(E^+, E)$ on E^+. If (E, E^+) is a* dual system, *then the weak topologies are Hausdorff and each of the spaces E, E^+ is the vector space of all weakly continuous linear forms on the other one.*

The weak topologies make it possible to describe the components E, E^+ of a bilinear system *topologically*. The next proposition shows that also the conjugability of an operator has a topological characterization. We first introduce the following definition:

Let (E, E^+) and (F, F^+) be bilinear systems. A linear map $A: E \to F$ is said to be *weakly continuous* if it is continuous with respect to the weak topologies $\sigma(E, E^+)$, $\sigma(F, F^+)$.

Proposition 82.1. *If (E, E^+) is a left dual system and (F, F^+) a bilinear system, then the linear map $A: E \to F$ is conjugable if and only if it is weakly continuous.*

Proof. If A is weakly continuous, then $\langle Ax, y^+ \rangle$ defines for a fixed $y^+ \in F^+$ a linear form A^+y^+ on E, which is $\sigma(E, E^+)$-continuous and therefore lies in E^+ (Theorem 82.1). Thus A^+ is a map from F^+ into E^+ such that $\langle Ax, y^+ \rangle = \langle x, A^+y^+ \rangle$ for all $x \in E$, $y^+ \in F^+$ and so it is conjugate to A. Now let A be conjugable and

$$Ax_0 + V := Ax_0 + \left\{ y \in F : \max_{\nu=1}^{n} |\langle y, y_\nu^+ \rangle| \leqq \varepsilon \right\}$$

an arbitrary (weak) basic neighborhood of Ax_0. Then

$$x_0 + U := x_0 + \left\{ x \in E : \max_{\nu=1}^{n} |\langle x, A^+ y_\nu^+ \rangle| \leqq \varepsilon \right\}$$

is a (weak) neighborhood of x_0 and $A(x_0 + U) = Ax_0 + A(U) \subset Ax_0 + V$. Thus A is weakly continuous at the arbitrary point x_0. ■

Since A is conjugate to A^+ with respect to the bilinear systems (F^+, F), (E^+, E), from Proposition 82.1 we obtain immediately:

Proposition 82.2. *If (E, E^+) is a left dual system and (F, F^+) a right dual system, then the weak continuity of $A: E \to F$ implies the weak continuity of $A^+: F^+ \to E^+$.*

By weak continuity of A^+ we mean continuity with respect to the topologies $\sigma(F^+, F)$ and $\sigma(E^+, E)$.

For every subspace G^+ of E^+ the restriction of the bilinear form of (E, E^+) establishes (E, G^+) as a bilinear system, and by the Hausdorff criterion $\sigma(E, G^+) \prec \sigma(E, E^+)$. If (E, E^+)—and with it also (E, G^+)—is a left dual system, then the set of all $\sigma(E, G^+)$-continuous linear forms coincides with G^+, and the set of all $\sigma(E, E^+)$-continuous linear forms coincides with E^+. This observation yields:

Proposition 82.3. *If (E, E^+) is a left dual system and G^+ a proper subspace of E^+, then the topology $\sigma(E, E^+)$ is strictly finer than $\sigma(E, G^+)$.*

Exercises

1. If (E, E^+) is a dual system, then the finite-dimensional endomorphism K of E is weakly continuous if and only if it can be represented in the form

$$Kx = \sum_{\nu=1}^{n} \langle x, x_\nu^+ \rangle x_\nu \qquad (x_\nu \in E, x_\nu^+ \in E^+).$$

Hint: Proposition 16.5.

$^+$2. If (E, E^+) is a left dual system and $\mathscr{A}(E)$ an E^+-saturated algebra of operators, then every $A \in \mathscr{A}(E)$ is weakly continuous. *Hint*: Proposition 25.5.

$^+$3. If (E, E^+) is a dual system, then the algebra of operators $\mathscr{A}(E)$ is E^+-saturated if and only if I lies in $\mathscr{A}(E)$, every $A \in \mathscr{A}(E)$ is weakly continuous, and every weakly continuous endomorphism of finite rank of E belongs to $\mathscr{A}(E)$. *Hint*: Proposition 25.6.

§83. The concept of a topological vector space. Examples

A topology τ on a vector space E over $\mathbf{K}$ is said to be a *vector space topology* for E if addition and multiplication by scalars, i.e., the maps

$$(x, y) \mapsto x + y \qquad \text{from} \quad E \times E \qquad \text{into} \quad E$$

and

$$(\alpha, x) \mapsto \alpha x \qquad \text{from} \quad \mathbf{K} \times E \qquad \text{into} \quad E$$

are continuous, where $E \times E$ and $\mathbf{K} \times E$ are equipped with the respective product topologies; E itself, or more precisely (E, τ), is then called a *topological vector space*. Continuity of addition means: For every neighborhood W of $x_0 + y_0$ there exist neighborhoods U of x_0 and V of y_0 so that

$$x + y \in W \quad \text{for} \quad x \in U, y \in V, \qquad \text{briefly:} \quad U + V \subset W.$$

Continuity of multiplication by scalars means: For every neighborhood W of $\alpha_0 x_0$ there exists a $\delta > 0$ and a neighborhood U of x_0 so that

$$\alpha x \in W \quad \text{for} \quad |\alpha - \alpha_0| \leqq \delta \quad \text{and} \quad x \in U.$$

Example 83.1. Normed spaces and metric vector spaces are topological vector spaces when they are equipped with their metric topology (consider Exercise 9 in §81).

Example 83.2. The weak topology $\sigma(E, E^+)$ is a vector space topology on E. We first observe that

$$p(x) := |\langle x, x^+ \rangle| \qquad \text{for a fixed} \quad x^+ \in E^+$$

defines a semi-norm p on E. The typical basic neighborhood of the origin is then given by

(83.1) $U_{p_1,\ldots,p_n;\varepsilon} := \left\{x \in E : \max_{\nu=1}^{n} p_\nu(x) \leqq \varepsilon\right\}$ where $p_\nu(x) := |\langle x, x_\nu^+\rangle|$

(Theorem 82.1). We shall handle it similarly as a ball $K_\varepsilon[0]$ in a normed space. Let now $W := z_0 + U_{p_1,\ldots,p_n;\varepsilon}$ be an arbitrary basic neighborhood of $z_0 := x_0 + y_0$ (Theorem 82.1). The neighborhoods $U := x_0 + U_{p_1,\ldots,p_n;\varepsilon/2}$ of x_0 and $V := y_0 + U_{p_1,\ldots,p_n;\varepsilon/2}$ of y_0 obviously satisfy $U + V \subset W$ and therefore addition is continuous. Now let

$$W := \alpha_0 x_0 + U \quad \text{with} \quad U := U_{p_1,\ldots,p_n;\varepsilon}$$

be an arbitrary basic neighborhood of $\alpha_0 x_0$. We will have proved the continuity of multiplication by scalars if we can exhibit a $\delta > 0$ and a neighborhood of zero V such that

(83.2) $\alpha x - \alpha_0 x_0 \in U$ whenever $|\alpha - \alpha_0| \leqq \delta$ and $x - x_0 \in V$.

For this we write

(83.3) $\alpha x - \alpha_0 x_0 = \alpha_0(x - x_0) + (\alpha - \alpha_0)x_0 + (\alpha - \alpha_0)(x - x_0)$

and try to make each of the three summands 'sufficiently small'. Setting $V_1 := U_{p_1,\ldots,p_n;\varepsilon/3}$ we have

(83.4) $$V_1 + V_1 + V_1 \subset U,$$

and setting

$$V_2 := \begin{cases} V_1 & \text{if } \alpha_0 = 0 \\ U_{p_1,\ldots,p_n;\varepsilon/3|\alpha_0|} & \text{if } \alpha_0 \neq 0 \end{cases}$$

we have

(83.5) $$\alpha_0 V_2 \subset V_1.$$

Clearly, either V_1 is contained in V_2 or V_2 is contained in V_1, consequently

(83.6) $$V := V_1 \cap V_2$$

is the smaller one of the neighborhoods V_1, V_2; because of (83.5) we have thus

(83.7) $$\alpha_0 V \subset V_1.$$

Finally, there exists a $\delta > 0$ so that $|\alpha - \alpha_0| \leqq \delta$ implies $p_\nu((\alpha - \alpha_0)x_0) = |\alpha - \alpha_0| p_\nu(x_0) \leqq \varepsilon/3$ for $\nu = 1, \ldots, n$, hence

(83.8) $$(\alpha - \alpha_0)x_0 \in V_1.$$

We may obviously choose $\delta \leqq 1$, but then

(83.9) $$(\alpha - \alpha_0)V \subset V \subset V_1.$$

From (83.3) we now obtain with the aid of (83.7) through (83.9) and of (83.4) that $\alpha x - \alpha_0 x_0 \in V_1 + V_1 + V_1 \subset U$ whenever $|\alpha - \alpha_0| \leqq \delta$ and $x - x_0 \in V$; thus we have proved the continuity of multiplication by scalars. ■

Example 83.3. The proofs of continuity in the last example were based solely on the *semi-norm properties* of the p_ν. This suggests the following generalization: Let a family P of semi-norms p be given on a vector space E. For every finite subset $\{p_1, \ldots, p_n\}$ of P and every $\varepsilon > 0$ let

$$U_{p_1, \ldots, p_n; \varepsilon} := \left\{x \in E: \max_{\nu=1}^{n} p_\nu(x) \leqq \varepsilon\right\}$$

(cf. (83.1)). The collection $\mathfrak{U}(y)$, which consists of all sets of the form

$$y + U_{p_1, \ldots, p_n; \varepsilon} = \left\{x \in E: \max_{\nu=1}^{n} p_\nu(x - y) \leqq \varepsilon\right\}$$

and their supersets, satisfies the neighborhood axioms (U1) through (U4). The topology defined on E by the neighborhood filters $\mathfrak{U}(y)$ is called the *topology generated by the family P. It is a vector space topology for E*, and since for semi-norms p we always have

$$|p(x) - p(y)| \leqq p(x - y), \tag{83.10}$$

all $p \in P$ are continuous with respect to it. One sees, as in the case of weak topologies, that *it is Hausdorff if and only if from the fact that $p(x) = 0$ for all $p \in P$, it follows that $x = 0$*; in this case we say that P is *total.* The convergence $x_n \to y$ is equivalent to $p(x_n - y) \to 0$ for all $p \in P$. Topologies of normed spaces are included in our construction: We obtain them when P consists only of the given norm.

A total *sequence* $P = (p_1, p_2, \ldots)$ of semi-norms on E generates, besides the vector space topology τ_1 just described, also a *metric* vector space topology τ_2 by means of the absolute value

$$|x| := \sum \frac{1}{2^\nu} \frac{p_\nu(x)}{1 + p_\nu(x)} \tag{83.11}$$

(see Example 80.4, and the remarks made there). τ_1 coincides with τ_2: Indeed, if for the ball $K_r[y]$ we first choose a natural n such that $\sum_{\nu=n+1}^{\infty} (1/2^\nu) \leqq r/2$, and then an $\varepsilon > 0$ so that $t/(1 + t) \leqq r/2n$ for $0 < t \leqq \varepsilon$, then for all $x \in V := y + U_{p_1, \ldots, p_n; \varepsilon}$ we have the estimate

$$|x - y| = \sum_{\nu=1}^{n} \frac{1}{2^\nu} \frac{p_\nu(x - y)}{1 + p_\nu(x - y)} + \sum_{\nu=n+1}^{\infty} \frac{1}{2^\nu} \frac{p_\nu(x - y)}{1 + p_\nu(x - y)} \leqq n \frac{r}{2n} + \frac{r}{2} = r,$$

thus V lies in $K_r[y]$ and so $\tau_2 \prec \tau_1$. If now a basic neighborhood $V := y + U_{p_1, \ldots, p_n; \varepsilon}$ of the point y with respect to the topology τ_1 is given, then it first follows from $|x - y| \leqq (1/2^n)(\varepsilon/(1 + \varepsilon))$ that $p_\nu(x - y)/(1 + p_\nu(x - y)) \leqq \varepsilon/(1 + \varepsilon)$ for $\nu = 1, \ldots, n$ and then, since the function $t \mapsto t/(1 + t)$ is monotone increasing for $t > -1$, that $p_\nu(x - y) \leqq \varepsilon$ for $\nu = 1, \ldots, n$, hence

$x \in V$. The ball $K_r[y]$ with $r := (1/2^n)(\varepsilon/(1+\varepsilon))$ thus lies in V, and therefore also $\tau_1 \prec \tau_2$. ■

In general, we say that the topology of a topological space (or the space itself) is *metrizable* if it coincides with a metric topology. In this sense a vector space topology, which is generated by a total sequence of semi-norms, is metrizable. Let us now assume that the family P of semi-norms generates a metrizable topology on E. Then the balls $K_{1/n}[0]$ $(n = 1, 2, \ldots)$ form a basis of neighborhoods of the origin, consequently there also exists a countable basis of neighborhoods $U^{(1)}, U^{(2)}, \ldots$ of 0, where each $U^{(k)}$ is determined the usual way by finitely many semi-norms from P. The totality $\tilde{P}$ of semi-norms which thus occur is a sequence taken from P, and obviously $\tilde{P}$ generates the same topology as P. Since the latter is metrizable, therefore Hausdorff, $\tilde{P}$ must be total. In summary, we proved the following proposition:

Proposition 83.1. *A topology defined by a family of semi-norms P is metrizable if and only if it can be generated by a total sequence* $(p_1, p_2, \ldots)$ *from P. In this case the metric is obtained from the absolute value* (83.11).

Example 83.4. Let F be a vector space of functions $x: T \to \mathbf{K}$. We say that a non-empty subset S of T is *F-bounded* if every $x \in F$ is bounded on S. Finite subsets of T are always F-bounded; compact subsets of $T \subset \mathbf{R}$ are $C(T)$-bounded. An F-bounded subset S generates by means of

(83.12) $$p_S(x) := \sup_{t \in S} |x(t)|$$

a semi-norm p_S on F. Let now $\mathfrak{S}$ be a non-empty system of F-bounded subsets of T. Then the family of semi-norms $(p_S: S \in \mathfrak{S})$ generates a vector space topology on F (Example 83.3) which is called the *topology generated by* $\mathfrak{S}$. *The sequence* x_k *converges to x in this topology if and only if* $x_k(t) \to x(t)$ *for* $t \in \bigcup_{S \in \mathfrak{S}} S$ *and this convergence is* uniform *on* every $S \in \mathfrak{S}$; because of this, the topology is also called the *topology of uniform convergence on all sets from* $\mathfrak{S}$. We encountered a special case already in Example 80.3, the topology of uniform convergence (of continuous functions) on all compact subsets of the interval (a, b). If F is the vector space of all functions on T, then the system of all subsets with one point of T generates the *topology of pointwise convergence*. Thus the classical pointwise convergence ($x_n(t) \to x(t)$ for all $t \in T$) is represented as a vector space topology. This is not metrizable if T is not countable (Exercise 3), which shows again the necessity to go beyond metric vector spaces. The topology of those function spaces on T which have a supremum-norm topology (e.g., $B(T)$, l^∞, (c), etc.) is generated by $\mathfrak{S} := \{T\}$.

Example 83.5. We undertake a slight generalization of Example 83.4. Let T be a non-empty set and F a vector space over $\mathbf{K}$. With every $(t, x) \in T \times F$ let a scalar $\langle t | x \rangle$ be associated, and assume that this correspondence is linear in the second term:

$$\langle t | \alpha x \rangle = \alpha \langle t | x \rangle, \qquad \langle t | x + y \rangle = \langle t | x \rangle + \langle t | y \rangle.$$

If F is a linear function space, then e.g., $\langle t|x\rangle := x(t)$ defines a correspondence of this kind; here we have by definition $x = y \Leftrightarrow \langle t|x\rangle = \langle t|y\rangle$ for all $t \in T$. In the general case, we can have, however, $\langle t|x\rangle = \langle t|y\rangle$ for all $t \in T$ without having $x = y$. We proceed as in the preceding example: We say that $S \subset T$ is *F-bounded* if $\sup_{t \in S} |\langle t|x\rangle| < \infty$ for all $x \in F$, i.e., if every function $t \mapsto \langle t|x\rangle$ is bounded on S, and define for an F-bounded set S a semi-norm p_S on F by

$$p_S(x) := \sup_{t \in S} \langle t|x\rangle.$$

If $\mathfrak{S}$ is a system of F-bounded sets, then the family $(p_S : S \in \mathfrak{S})$ defines, according to Example 83.3, a vector space topology on F which is again called the *topology generated by* $\mathfrak{S}$. *A sequence* (x_k) *converges to* x *for this topology if and only if* $\langle t|x_k\rangle \to \langle t|x\rangle$ *for* $t \in \bigcup_{S \in \mathfrak{S}} S$, *and* uniformly *on* every $S \in \mathfrak{S}$.

Example 83.6. Let (E, E^+) be a bilinear system with the bilinear form $\langle x, x^+\rangle$. If we set $T := E$, $F := E^+$ in Example 83.5, and $\langle x|x^+\rangle := \langle x, x^+\rangle$, then every system $\mathfrak{S}$ of E^+-bounded sets S of E generates a vector space topology on E^+ by means of the semi-norms

$$p_S(x^+) := \sup_{x \in S} |\langle x, x^+\rangle|, \tag{83.13}$$

which is called briefly the $\mathfrak{S}$-*topology*. It is the *topology of uniform convergence on the sets of* $\mathfrak{S}$. The weak topology $\sigma(E^+, E)$ is generated by the system of all sets having one point, the so-called *strong* topology, which is denoted by $\beta(E^+, E)$, by the system of all E^+-bounded subsets of E. Obviously $\beta(E^+, E)$ *is the finest* $\mathfrak{S}$-*topology on* E^+.

Given the bilinear system (E, E') consisting of a normed space E and its dual E', because of Proposition 41.4, the E'-bounded subsets of E are precisely those which are bounded for the norm. For every E'-bounded set S there exists therefore a ball $K := K_\rho[0] \supset S$, consequently we have

$$U_{p_S;\varepsilon} = \left\{x' \in E' : \sup_{x \in S} |\langle x, x'\rangle| \leqq \varepsilon\right\} \supset U_{p_K;\varepsilon} = \left\{x' \in E' : \sup_{x \in K} |\langle x, x'\rangle| \leqq \varepsilon\right\}.$$

Because of $\sup_{x \in K} |\langle x, x'\rangle| = \rho\|x'\|$ the last set is the ball $K'_{\varepsilon/\rho}[0]$ in the normed space E'. It follows that the balls around 0 form in E' a basis of neighborhoods of the origin with respect to $\beta(E', E)$ and that consequently $\beta(E', E)$ *coincides with the norm topology on* E'.

Exercises

1. The discrete topology is never a vector space topology, the chaotic topology always (we assume that the vector space is not trivial).

$^+$2. The product topology on $F(T) := \prod_{t \in T} \mathbf{K}_t$, $\mathbf{K}_t = \mathbf{K}$ for all $t \in T$ ($F(T)$ is the vector space of all functions $x\colon T \to \mathbf{K}$) coincides with the topology of pointwise convergence of Example 83.4, and is therefore a vector space topology.

3. The topology of pointwise convergence on the vector space $F(T)$ of all functions $x: T \to \mathbf{K}$ is not metrizable if T is not countable. *Hint*: No sequence of semi-norms $p_n(x) := |x(t_n)|$ is total under the stated hypothesis.

*4. (a) A semi-norm on a topological vector space E is continuous if and only if it is continuous at the origin. *Hint*: (83.10).

(b) If the topology of E is generated by the family of semi-norms P, then the semi-norm q on E is continuous if and only if there exist $\gamma > 0$ and $p_1, \ldots, p_n$ in P so that $q(x) \leqq \gamma \max_{\nu=1}^{n} p_\nu(x)$ for all $x \in E$. *Hint*: Proof of (82.5).

(c) Under the hypotheses of (b), the family of all continuous semi-norms also generates the topology of E.

5. If $P = \{p_1, \ldots, p_n\}$ is a finite total set of semi-norms on E, then $\|x\| := \sum_{\nu=1}^{n} p_\nu(x)$ defines a norm on E which generates the same topology as P.

6. No system $\mathfrak{S}$ of subsets $S \subset \mathbf{N}$ generates the normed space topology of l^p, $1 \leqq p < \infty$.

$^+$7. Let E be a normed space. On $\mathscr{L}(E)$ we define semi-norms by

(a) $p_S(A) := \sup_{x \in S} \|Ax\|$ for every bounded subset S of E,
(b) $q_x(A) := \|Ax\|$ for every $x \in E$,
(c) $r_{x,x'}(A) := |\langle Ax, x' \rangle|$ for every $x \in E$ and $x' \in E'$.

The family (a) generates the normed space topology (the topology of uniform convergence on all bounded subsets of E), the family (b) generates the topology of pointwise convergence, and the family (c) the topology of weak operator-convergence $A_n \rightharpoonup A$ (see Exercise 7 in §46).

$^+$8. Let $(\alpha_{\nu\mu})$ be an infinite matrix of non-negative numbers and $1 \leqq p < \infty$. We denote by $k^p(\alpha_{\nu\mu})$ the set of all sequences $x = (\xi_1, \xi_2, \ldots)$ for which all the series $\sum_{\mu=1}^{\infty} \alpha_{\nu\mu}|\xi_\mu|^p$ $(\nu = 1, 2, \ldots)$ converge. $k^p(\alpha_{\nu\mu})$ is a vector space of sequences, and the functions $p_\nu(x) := (\sum_{\nu=1}^{\infty} \alpha_{\nu\mu}|\xi_\mu|^p)^{1/p}$ are semi-norms on $k^p(\alpha_{\nu\mu})$. The linear space $k^p(\alpha_{\nu\mu})$ equipped with the topology generated by $(p_1, p_2, \ldots)$ is called a *Köthe space*. For $\alpha_{\nu\mu} = 1$ $(\nu, \mu = 1, 2, \ldots)$ we have $k^p(\alpha_{\nu\mu}) = l^p$. The sequence $(p_1, p_2, \ldots)$ is total if and only if in every column of the matrix $(\alpha_{\nu\mu})$ there exists at least one positive element. In this case $k^p(\alpha_{\nu\mu})$ is metrizable and complete.

9. Let (E, E^+) be a bilinear system. The set $S \subset E$ is E^+-bounded if and only if for every $\sigma(E, E^+)$-neighborhood U of the origin there exists a $\rho > 0$ such that $S \subset \rho U$.

10. Let the topology τ on E be generated by the family of semi-norms P. Show that $U := U_{p_1, \ldots, p_n; \varepsilon}$ has the following properties:

(a) For every $x \in E$ there exists a $\rho > 0$ so that $x \in \alpha U$ whenever $|\alpha| \geqq \rho$.
(b) If x lies in U, so does αx for $|\alpha| \leqq 1$.
(c) U is τ-closed.
(d) U is convex.

Simplest case: Let E be a normed space, $U := K_\varepsilon[0]$.

§84. The neighborhoods of zero in topological vector spaces

A vector space topology τ on E is related to the linear structure of E: addition and multiplication by scalars are continuous with respect to τ. This fact will have an influence on the form of the bases of neighborhoods. Our investigations in this direction will be considerably simplified by the fact that we need only to know a basis of neighborhoods of the origin; one obtains from them by translation bases of neighborhoods of all points. This is the main content of the next proposition.

Proposition 84.1. *The self-map f of a topological vector space E defined by $f(x) := \alpha_0 x + x_0$ is continuous, and in the case that $\alpha_0 \neq 0$, even a homeomorphism of E onto itself. From here it follows in particular:*

(a) *$f(M) = \alpha_0 M + x_0$ is open or closed for every open or closed set M, respectively, if $\alpha_0 \neq 0$.*

(b) *If U is a neighborhood of 0 and $\alpha_0 \neq 0$, then $\alpha_0 U$ is a neighborhood of 0 and $\alpha_0 U + x_0$ is a neighborhood of x_0.*

(c) *If $\mathfrak{N}$ is a basis of neighborhoods of 0, then the sets $x_0 + U$ $(U \in \mathfrak{N})$ form a basis of neighborhoods of x_0.*

Proof. The continuity of the map f and of its inverse map $y \mapsto (1/\alpha_0)(y - x_0)$, which is defined on E in the case $\alpha_0 \neq 0$, is an immediate consequence of the continuity of the linear operations. (a) is now trivial. (b) follows from (a) if we take into consideration that the open neighborhoods of a point form a basis of neighborhoods. (c) is a very simple consequence of (a) and (b). ■

We saw in §83 Exercise 10 that the neighborhoods $U_{p_1, \ldots, p_n; \varepsilon}$ of zero in a vector space, whose topology is generated by semi-norms, have some pleasant geometric properties. We ask whether we can find in an arbitrary topological vector space a basis of neighborhoods of the origin consisting of similarly structured sets. In order to be able to express ourselves easily, we first introduce the following definitions (cf. parts (a) and (b) of the Exercise just mentioned):

A set M in an arbitrary vector space E is said to be *absorbing* if for every $x \in E$ there exists a $\rho > 0$ so that $x \in \alpha M$ whenever $|\alpha| \geqq \rho$. It is said to be *balanced* if whenever x lies in M, also αx lies in M for any α such that $|\alpha| \leqq 1$.

Supersets, finite intersections and arbitrary unions of absorbing sets are absorbing. Arbitrary intersections and unions of balanced sets are balanced. Since every $M \subset E$ is contained in a balanced set, namely in E, there exists a smallest balanced set which contains M (the intersection of all balanced supersets of M); it is called the *balanced hull* of M. The set $\{0\}$ is balanced; if M contains the zero vector, then there exists therefore a largest balanced subset $\neq \varnothing$ of M (the union of all balanced $N \subset M$); it is called the *balanced core* of M.

Lemma 84.1. (a) *A balanced set $D \neq \varnothing$ contains* 0 *and is* symmetric, *i.e., if $x \in D$ then also $-x \in D$.*

(b) *If D is balanced and $|\alpha| \leqq |\beta|$, then $\alpha D \subset \beta D$.*

(c) *A balanced set is already absorbing if for every $x \in E$ there exists $\mu \neq 0$ such that $x \in \mu D$; μ can be chosen positive.*

(d) *The balanced core of a set $M \supset \{0\}$ is $\bigcap_{|\alpha| \geqq 1} \alpha M$.*

(e) *The balanced hull of M is $\bigcup_{|\alpha| \leqq 1} \alpha M$.*

(f) *Let M be a subset of a topological vector space. If M is open, then also its balanced hull is open. If M is closed and contains $\{0\}$, then also its balanced core is closed.*

Proof. (a) and (b) are trivial. (c) follows from (b). (d) Let $D := \bigcap_{|\alpha| \geqq 1} \alpha M$ and K the balanced core of M. Obviously D is a balanced subset of M, hence $D \subset K$. If now $x \in K$, then $\lambda x \in M$ for all $|\lambda| \leqq 1$, hence $x \in \alpha M$ for all $|\alpha| \geqq 1$. Thus also $K \subset D$. (e) Let $V := \bigcup_{|\alpha| \leqq 1} \alpha M$ and H the balanced hull of M. Obviously V is a balanced superset of M, hence $H \subset V$. On the other hand, every balanced superset of M, in particular H, must contain all sets αM ($|\alpha| \leqq 1$), and thus also their union V. (f) follows from (d), (e) and Proposition 84.1. ■

Proposition 84.2. *For every neighborhood of zero U in a topological vector space the following assertions hold:*

(a) *U is absorbing.*
(b) *There exists a neighborhood of zero V such that $V + V \subset U$.*
(c) *U contains a balanced neighborhood of zero (e.g., the balanced core of U).*
(d) *U contains a closed neighborhood of zero.*

Proof. (a) Because of $0 \cdot x = 0 \in U$, and the continuity of multiplication by scalars, there exists a $\delta > 0$ so that $\alpha x \in U$ for $|\alpha| \leqq \delta$, hence $x \in \lambda U$ for $|\lambda| \geqq 1/\delta$.

(b) Because of $0 + 0 = 0 \in U$, and the continuity of addition, there exist neighborhoods V_1, V_2 of zero such that $V_1 + V_2 \subset U$. Then $V := V_1 \cap V_2$ satisfies the requirements.

(c) Because of $0 \cdot 0 = 0 \in U$, and the continuity of multiplication by scalars, there exists $\delta > 0$ and a neighborhood V of 0 so that $\alpha V \subset U$ for $|\alpha| \leqq \delta$. It follows that $\delta V \subset \lambda U$ for $|\lambda| \geqq 1$, the neighborhood δV of zero lies therefore in the balanced core of U (Lemma 84.1d), which is consequently itself a neighborhood of 0.

(d) According to (b) there exists a neighborhood V of zero such that $V + V \subset U$. Because of (c) V contains a balanced neighborhood of zero W for which a fortiori $W + W \subset U$. Now let $x \in \overline{W}$. Then the neighborhood $x + W$ of x (Proposition 84.1) intersects the set W, therefore there exists a $y \in W$ of the form $y = x + w$, $w \in W$. Together with w also $-w$ lies in W (Lemma 84.1a), hence $x = y + (-w) \in W + W \subset U$. Thus the closed set $\overline{W}$ lies in U, and as a superset of W it is a neighborhood of zero. ■

The proposition just proved makes it possible to guarantee the existence of bases of neighborhoods of zero with pleasant properties.

Theorem 84.1. *In every topological vector space there exists a basis $\mathfrak{N}$ of neighborhoods of zero with the following properties:*

(a) *Every $U \in \mathfrak{N}$ is absorbing, balanced and closed;*
or alternatively:
(a′) *Every $U \in \mathfrak{N}$ is absorbing, balanced and open.*
(b) *For every $U \in \mathfrak{N}$ there exists a $V \in \mathfrak{N}$ such that $V + V \subset U$.*

Proof. (a) Because of Proposition 84.2 the closed neighborhoods of zero form a basis $\mathfrak{N}'$ of neighborhoods of zero, and their balanced cores form a further basis $\mathfrak{N}$. By Lemma 84.1f every $U \in \mathfrak{N}$ is closed, by Proposition 84.2a it is absorbing.

(a′) Because of Proposition 84.2 the set of all balanced neighborhoods of zero form a basis $\mathfrak{N}''$ of neighborhoods of zero. $W \in \mathfrak{N}''$ contains an open neighborhood of zero V; its balanced hull is open (Lemma 84.1f) and contained in W. Thus the set of all open and balanced neighborhoods of zero forms a basis $\mathfrak{N}$ of neighborhoods of zero. Of course, each $U \in \mathfrak{N}$ is again absorbing.

(b) now follows immediately from Proposition 84.2b. ■

If the vector space topology on E is generated by a family of semi-norms, then the sets $U_{p_1, \dots, p_n; \varepsilon}$ form a basis of neighborhoods of zero with the properties (a) and (b) in Theorem 84.1 (cf. §83 Exercise 10). These sets are furthermore convex. We want to observe here already that one cannot find in every topological vector space a basis of neighborhoods of zero, which consists of convex sets.

Since a basis of neighborhoods of zero determines the topology of a topological vector space completely, one must be able to decide any topological question with their help. The next two propositions give examples for this.

Proposition 84.3. *In a topological vector space E with the basis $\mathfrak{N}$ of neighborhoods of zero, the following assertions are equivalent:*

(a) *E is Hausdorff.*
(b) *For every $x \neq 0$ there exists a neighborhood $U \in \mathfrak{N}$ which does not contain x.*
(c) $\bigcap_{U \in \mathfrak{N}} U = \{0\}$.

Proof. The chain of implications (a) ⇒ (b) ⇒ (c) is trivial. If (c) is valid and if $x \neq y$, then there exist a $U_0 \in \mathfrak{N}$ which does not contain $x - y$. Because of Theorem 84.1 there exists a balanced, hence also symmetric, neighborhood of zero V such that $V + V \subset U_0$. The neighborhoods $x + V$ and $y + V$ of x and y, respectively, are disjoint; otherwise there would exist vectors $v_1, v_2 \in V$ so that $x + v_1 = y + v_2$, hence $x - y = v_2 + (-v_1) \in V + V \subset U_0$, in contradiction to the choice of U_0. Thus we have deduced (a) from (c). ■

From Proposition 84.1 one obtains, with the help of the Hausdorff criterion, immediately:

Proposition 84.4. *Let two vector space topologies τ_1, τ_2 with the respective bases $\mathfrak{N}_1, \mathfrak{N}_2$ of neighborhoods of zero be given on E. We have $\tau_1 \prec \tau_2$ if and only if for every $U_1 \in \mathfrak{N}_1$ there exists a $U_2 \in \mathfrak{N}_2$ such that $U_2 \subset U_1$. In particular, τ_1 and τ_2 coincide if they have the same basis of neighborhoods of zero.*

Exercises

1. The closure of a balanced set is balanced.
2. Let $A: E \to F$ be linear and $M \subset E$, $N \subset F$ balanced. Then the sets $A(M)$ and $A^{-1}(N)$ are also balanced.
3. A topological vector space is Hausdorff if and only if $\{0\}$ is closed.
4. For a vector space topology generated by semi-norms, determine a basis of neighborhoods of zero which has properties (a′) and (b) in Theorem 84.1.

$^+$5. We use the notation of Example 83.5. Show the following:
(a) The topology on F generated by $\mathfrak{S}$ is Hausdorff if and only if from $\langle t|x\rangle = 0$ for all $t \in \bigcup_{S \in \mathfrak{S}} S$ it follows that $x = 0$.
(b) A subsystem of $\mathfrak{S}$ generates a coarser topology than $\mathfrak{S}$ does.

§85. The generation of vector space topologies

Theorem 84.1 does essentially have a converse. More precisely, the following assertion holds, for the understanding of which it is useful to remember that every basis of neighborhoods is a filter basis:

Theorem 85.1. *Let $\mathfrak{N}$ be a filter basis on the vector space E over* **K** *which satisfies the following conditions:*

(NU1) *Every $U \in \mathfrak{N}$ is absorbing.*
(NU2) *Every $U \in \mathfrak{N}$ is balanced.*
(NU3) *For every $U \in \mathfrak{N}$ there exists $V \in \mathfrak{N}$ such that $V + V \subset U$.*

Then there exists one and only one vector space topology on E for which $\mathfrak{N}$ is a basis of neighborhoods of zero.

Proof. We first define a topology on E for which $\mathfrak{N}$ is a basis of neighborhoods of zero. To do this, we call a set $W \subset E$ a neighborhood of x if W contains a set $x + U$, $U \in \mathfrak{N}$. One sees immediately that these neighborhoods satisfy axioms (U1) through (U3) of §81. Also (U4) is satisfied: Let W be a neighborhood of x, $x + U \subset W$ $(U \in \mathfrak{N})$ and let $V \in \mathfrak{N}$ be such that $V + V \subset U$. Then also $W_0 := x + V$ is a neighborhood of x, and for every $y \in W_0$ we have $y + V \subset x + V + V \subset x + U \subset W$, i.e., W is a neighborhood of y. The neighborhoods so introduced define thus a topology τ on E for which $\mathfrak{N}$ is a basis of neighborhoods of zero, and $\{x + U: U \in \mathfrak{N}\}$ is a basis of neighborhoods of x.

We now show that τ is a vector space topology. For every neighborhood $x + y + U$ ($U \in \mathfrak{N}$) of $x + y$ there exists $V \in \mathfrak{N}$ such that $V + V \subset U$; $x + V$ and $y + V$ are neighborhoods of x and y, respectively, and since $(x + V) + (y + V) = x + y + V + V \subset x + y + U$, the continuity of addition has been verified. The continuity of multiplication by scalars at $\alpha_0 x_0$ can be proved just like in the case of the weak topology (Example 83.2); we only have to show that for $U \in \mathfrak{N}$ there exists a $V_1 \in \mathfrak{N}$ with $V_1 + V_1 + V_1 \subset U$, that for V_1 there exists a $V_2 \in \mathfrak{N}$ with $\alpha_0 V_2 \subset V_1$, that for V_1, V_2 there exists $V \in \mathfrak{N}$ with $V \subset V_1 \cap V_2$, and that $(\alpha - \alpha_0)x_0 \in V_1$ for $|\alpha - \alpha_0| \leqq \delta$ provided that δ is sufficiently small (see (83.2) through (83.6) and (83.8)). The existence of V_1 and V_2 is guaranteed by the subsequent lemma, V exists because $\mathfrak{N}$ is a filter basis, and because of (NU1) there also exists an appropriate δ. Thus τ is indeed a vector space topology on E which has $\mathfrak{N}$ for a basis of neighborhoods of zero, and it is the only such topology (Proposition 84.4). ■

We now supply the announced lemma:

Lemma 85.1. *Under the hypotheses of Theorem* 85.1 *let an arbitrary* $U \in \mathfrak{N}$ *be given. Then there exists for every* $n \in \mathbf{N}$ *and* $\alpha \in \mathbf{K}$ *a* $V \in \mathfrak{N}$ *so that the sum with* n *terms* $V + V + \cdots + V$ *and the product* αV*, respectively, is a part of* U.

Proof. If we apply (NU3) successively, we obtain for every $m \in \mathbf{N}$ a $V_m \in \mathfrak{N}$ such that

$$\underbrace{V_m + \cdots + V_m}_{2^m \text{ terms}} \subset U.$$

If $2^m \geqq n$ and $V := V_m$, then

$$\underbrace{V + \cdots + V}_{n \text{ terms}} \subset \underbrace{V + \cdots + V}_{2^m \text{ terms}} \subset U$$

(the first inclusion holds because $0 \in V$). For a given α we choose an $n \geqq |\alpha|$ and for n, according to what we have just proved, a $V \in \mathfrak{N}$ such that $V + \cdots + V \subset U$ (the sum has n terms). Because of $nV \subset V + \cdots + V$ we have a fortiori $nV \subset U$. By (NU2) it follows that $\alpha V = (\alpha/n)nV \subset (\alpha/n)U \subset U$. ■

Example 85.1. Let P be a family of semi-norms p on the vector space E. The sets $U_{p;\varepsilon} := \{x \in E: p(x) \leqq \varepsilon\}$ satisfy (NU1) through (NU3) but do not form in general a filter basis. Therefore one considers the collection $\mathfrak{N}$ of all finite intersections $U_{p_1, \ldots, p_n; \varepsilon} := \bigcap_{\nu=1}^{n} U_{p_\nu; \varepsilon} = \{x \in E: \max_{\nu=1}^{n} p_\nu(x) \leqq \varepsilon\}$. $\mathfrak{N}$ is a filter basis which satisfies the conditions of Theorem 85.1. It defines therefore a vector space topology on E, which we already met in Example 83.3: the topology generated by P.

Example 85.2. We can construct the $\mathfrak{S}$-topologies considered in Example 83.5 also without the explicit use of semi-norms. We employ the notation introduced there, set $U_{S;\varepsilon} := \{x \in F: \sup_{t \in S} |\langle t|x\rangle| \leqq \varepsilon\}$ and observe that it is exactly the

F-bounded sets $S \subset T$ which yield absorbing sets $U_{S,\varepsilon}$. Since the collection of all sets $U_{S;\varepsilon}$, which belong to a system $\mathfrak{S}$ of F-bounded subsets of T, does in general not form a filter basis, we consider the set $\mathfrak{N}$ of all finite intersections

$$U_{S_1,\ldots,S_n;\varepsilon} := \bigcap_{\nu=1}^{n} U_{S_\nu;\varepsilon} = \left\{x \in F: \sup_{t \in S_\nu} |\langle t|x\rangle| \leqq \varepsilon \text{ for } \nu = 1, \ldots, n\right\}.$$

$\mathfrak{N}$ satisfies all conditions of Theorem 85.1, defines therefore a vector space topology on F, which is, of course, nothing but the $\mathfrak{S}$-topology.

Exercises

1. Let the collection $\mathfrak{M}$ of subsets of the vector space E have the following properties:
(a) Every $U \in \mathfrak{M}$ is absorbing.
(b) Every $U \in \mathfrak{M}$ is balanced.
(c) For every $U \in \mathfrak{M}$ there exists a $V \in \mathfrak{M}$ such that $V + V \subset U$.

Then the finite intersections $\bigcap_{\nu=1}^{n} U_\nu$ $(U_\nu \in \mathfrak{M})$ form a basis of neighborhoods of zero for a uniquely determined vector space topology on E (cf. Examples 85.1, 85.2).

2. To Example 85.2: If for every $S_1, S_2 \in \mathfrak{S}$ there exists an $S_3 \in \mathfrak{S}$ with $S_1 \cup S_2 \subset S_3$, then already the sets $U_{S;\varepsilon}$ $(S \in \mathfrak{S}, \varepsilon > 0)$ form a basis of neighborhoods of zero for the topology generated by $\mathfrak{S}$.

3. Again to Example 85.2: Let $\mathfrak{S}_1$ be the collection of all subsets of the $S \in \mathfrak{S}$, and let $\mathfrak{S}_2$ be the collection of all finite unions of sets from $\mathfrak{S}$. Then $\mathfrak{S}$, $\mathfrak{S}_1$ and $\mathfrak{S}_2$ generate the same topology. If $T := E$, $F := E^+$ and (E, E^+) is a bilinear system, then let $\mathfrak{S}_3$ be the collection of all sets αS $(\alpha \in \mathbf{K}, S \in \mathfrak{S})$, and let $\mathfrak{S}_4$ be the collection of all balanced hulls of the $S \in \mathfrak{S}$. Show that $\mathfrak{S}, \mathfrak{S}_3, \mathfrak{S}_4$ generate the same topologies on E^+.

4. Let (E, E^+) be a bilinear system. Among all the $\mathfrak{S}$-topologies on E^+ which satisfy $\bigcup_{S \in \mathfrak{S}} S = E$, the topology $\sigma(E^+, E)$ is the coarsest. *Hint*: §84 Exercise 5b and the above Exercise 3.

§86. Subspaces, product spaces and quotient spaces

In this section we deal with the question whether certain algebraic operations in and with topological vector spaces (the formation of linear subspaces, product spaces and quotient spaces) again yield topological vector spaces. The proofs of the first two propositions are left to the reader as simple exercises.

Proposition 86.1. *For a linear subspace F of a topological vector space E the following assertions hold*:

(a) *F is a topological vector space with respect to the induced topology. A basis of neighborhoods of zero $\mathfrak{N}$ in E yields the basis of neighborhoods of zero $\{U \cap F : U \in \mathfrak{N}\}$ in F.*

(b) *The closure $\overline{F}$ is a linear subspace of E.*

Proposition 86.2. *The product of an arbitrary family of topological vector spaces, equipped with the product topology, is a topological vector space.*

Proposition 86.3. *If the quotient space E/F of a topological vector space E with respect to a linear subspace F is equipped with the quotient topology, then the following assertions are valid:*

(a) *E/F is a topological vector space.*
(b) *E/F is Hausdorff if and only if F is closed.*
(c) *The canonical homomorphism $h: E \to E/F$ is continuous and open.*
(d) *A neighborhood V of $x \in E$ is mapped by h onto a neighborhood $h(V)$ of $h(x)$. A basis of neighborhoods of zero $\mathfrak{N}$ in E generates the basis of neighborhoods of zero $\{h(U): U \in \mathfrak{N}\}$ in E/F.*

We first prove (c). By Example 81.4 the map h is continuous. If $M \subset E$ is open then $h^{-1}(h(M)) = M + F = \bigcup_{x \in F} (M + x)$ is also open, from where it follows that $h(M)$ is open; thus h is an open map. (d) follows from (c).

(a) Let W be a neighborhood of $h(x) + h(y) = h(x + y)$. Then $h^{-1}(W)$ is a neighborhood of $x + y$, hence there exist neighborhoods U of x, V of y so that $U + V \subset h^{-1}(W)$. It follows that $h(U) + h(V) = h(U + V) \subset h(h^{-1}(W)) \subset W$. Because of (d) we obtain from here the continuity of addition. The continuity of multiplication by scalars can be seen just as easily.

(b) If E/F is Hausdorff, then $\{h(0)\}$ and thus also $F = h^{-1}(h(0))$ is closed. Let, conversely, F be closed and $h(x) \neq 0$. Then x does not lie in F, hence there exists a neighborhood V of x such that $V \cap F = \emptyset$. For every $y \in V$ we have therefore $y \notin F$ and so $h(y) \neq 0$. Thus $h(V)$ is a neighborhood of $h(x)$ (see (d)) which does not contain 0. Then $h(x) - h(V)$ is a neighborhood of zero which does not contain $h(x)$, and it follows now from Proposition 84.3 that E/F is Hausdorff. ∎

Unless something else is said explicitly, we *always* equip subspaces, products and quotients of topological vector spaces with the topologies we gave them in this section: the relative-topology, the product topology and the quotient topology.

Exercises

+1. The product of at most countably many metric vector spaces is again a metric vector space.

+2. The product of infinitely many normed spaces is not normable (i.e., the product topology does not derive from a norm). *Hint*: Otherwise there would exist a neighborhood of zero $U := \prod U_\iota \subset \{x: \|x\| \leqq 1\}$. Choose now an $x \neq 0$ with $\alpha x \in U$ for all α.

+3. Let F be a closed subspace of the normed space E. Then the quotient norm on E/F generates the quotient topology.

§87. Continuous linear maps of topological vector spaces

If E, F are topological vector spaces over $\mathbf{K}$, then we define, as in the case of normed spaces:

$$\mathscr{L}(E, F) \text{ the set of all continuous linear maps } A: E \to F,$$

$$\mathscr{L}(E) := \mathscr{L}(E, F),$$

$$E' := \mathscr{L}(E, \mathbf{K}).$$

E', the set of all continuous linear forms on E, is also called the (topological) *dual* of E.

Proposition 87.1. (a) *A linear map is everywhere continuous if it is continuous at the origin.*

(b) *Sums, scalar multiples and products of continuous linear maps are again continuous. In particular, $\mathscr{L}(E, F)$ is a vector space and $\mathscr{L}(E)$ is an algebra.*

(c) *The nullspace of $A \in \mathscr{L}(E, F)$ is closed whenever F is a Hausdorff space. Thus the nullspaces of continuous linear forms are always closed.*

Proof. (a) Let $W = Ax + V$ (V a neighborhood of zero in F) be an arbitrary neighborhood of Ax. Since A is continuous at 0, with V we can associate a neighborhood of zero $U \subset E$ so that $A(U) \subset V$. For the neighborhood $x + U$ of x we have then $A(x + U) \subset Ax + V = W$.

(b) Let $A, B \in \mathscr{L}(E, F)$ and let W be a neighborhood of zero in F. For W there exists a neighborhood V of zero in F such that $V + V \subset W$, and for V there exist neighborhoods U_1, U_2 of zero in E so that $A(U_1) \subset V$, $B(U_2) \subset V$. Then $U := U_1 \cap U_2$ is a neighborhood of zero in E for which $(A + B)(U) \subset A(U) + B(U) \subset V + V \subset W$. It follows that $A + B$ is continuous at 0, and because of (a), everywhere. We leave the proof of the remaining assertions in (b) to the reader.

(c) $\{0\} \subset F$ is closed if F is Hausdorff, the same is then valid for $A^{-1}(\{0\}) = N(A)$. ■

Continuity and openness of a linear map A can be recognized by means of its canonical injection $\hat{A}$ (cf. Proposition 21.3):

Proposition 87.2. *$A \in \mathscr{S}(E, F)$ is continuous or open if and only if the corresponding canonical injection $\hat{A}: E/N(A) \to F$ is continuous or open, respectively.*

Proof. With the aid of the canonical homomorphism h from E onto $E/N(A)$ we have

$$A = \hat{A} \circ h. \tag{87.1}$$

The statement about continuity follows now from the proposition following Example 81.4, and the assertion concerning openness by means of Proposition 86.3c.

Continuously projectable subspaces and *topological complementary spaces* (or *complements*) are defined as in §24 and just like there, it is true that *exactly the continuously projectable subspaces possess topological complements.* The proof of Proposition 24.1 can be taken over word for word. When doing this, one sees that it also covers the generalization in Exercise 1 of §24; thus we have:

Proposition 87.3. *The continuous linear map $A: E \to F$ is* relatively regular, *i.e., there exists $B \in \mathscr{L}(F, E)$ such that $ABA = A$, if and only if it is open and its nullspace and image space are continuously projectable.*

Through an evident generalization of Proposition 26.1, we obtain from here immediately:

Proposition 87.4. *Given $A \in \mathscr{L}(E, F)$, there exists*

(a) *$B \in \mathscr{L}(F, E)$ such that $BA = I_E$ if and only if A is injective, open and $A(E)$ is continuously projectable,*
(b) *$C \in \mathscr{L}(F, E)$ such that $AC = I_F$ if and only if A is surjective, open and $N(A)$ is continuously projectable,*
(c) *$B \in \mathscr{L}(F, E)$ such that $BA = I_E$ and $AB = I_F$ if and only if A is bijective and open.*

We now turn to linear forms on a topological vector space. The basis of our investigations is the simple

Lemma 87.1. *Every non-trivial linear form on a one-dimensional topological vector space E is open. If the topology is Hausdorff, then it is also continuous.*

Proof. We can represent f in the form $f(\alpha x_0) = \alpha$ where $x_0 \neq 0$. The map f is bijective, and the inverse map $\alpha \mapsto \alpha x_0$ is trivially continuous. Therefore f must be open. Now let E be Hausdorff. Then for every $\varepsilon > 0$ there exists a balanced neighborhood U of zero in E which does not contain εx_0 (Proposition 4.3 and Theorem 84.1). If αx_0 lies in U, then $|f(\alpha x)| = |\alpha| \leqq \varepsilon$ (which proves the continuity of f according to Proposition 87.1a); otherwise we would have $|\varepsilon/\alpha| < 1$, hence $(\varepsilon/\alpha)U \subset U$ and so $\varepsilon x_0 = (\varepsilon/\alpha)\alpha x_0 \in U$ in contradiction to the choice of U. ■

Proposition 87.5. *A non-trivial linear form f on the topological vector space E is open. It is continuous if and only if it has a closed nullspace.*

Proof. By §15 Exercise 7 the space $E/N(A)$ is one-dimensional. The assertion concerning the openness of f now follows immediately from Lemma 87.1 in combination with Proposition 87.2. To see the correctness of the statement about continuity, invoke also Propositions 87.1c and 86.3b. ■

With the help of Proposition 87.1b the reader immediately sees that for every $A \in \mathscr{L}(E, F)$ there exists a map $A': F' \to E'$ such that

$$(87.2) \qquad \langle Ax, y' \rangle = \langle x, A'y' \rangle \quad \text{for all} \quad x \in E \quad \text{and} \quad y' \in F'.$$

A' is linear and it is uniquely determined by (87.2) because (E, E') is a left dual system (see end of §16). *Thus A is conjugable.* By the *conjugate* of A we understand always, unless something else is explicitly said, the map A' just defined.

Exercises

$^+$1. If E is a complete metric vector space and if the subspaces E_1, E_2 are closed and algebraically complementary, then they are also topologically complementary. *Hint*: Closed graph theorem (last assertion in Theorem 80.1).

2. A continuous projector of a topological vector space is always open.

$^+$3. Every finite-dimensional subspace of a Hausdorff topological vector space E has a topological complement if and only if (E, E') is a dual system. *Hint*: Proof of Proposition 25.4b.

§88. Finite-dimensional topological vector spaces

Basic for this section is:

Theorem 88.1. *On a finite-dimensional linear space there exists only one separated vector space topology; it is generated by a norm.*

Proof. If $\{x_1, \ldots, x_n\}$ is a basis of the linear space E, then $\|\xi_1 x_1 + \cdots + \xi_n x_n\| := \max_{\nu=1}^{n} |\xi_\nu|$ defines a norm on E; let τ_N be the topology it generates. A basis of τ_N-neighborhoods of zero is formed by the balls $K_r := \{x \in E: \|x\| \leqq r\}$. Now let τ be a second Hausdorff vector space topology on E, and let $\mathfrak{N}$ be a basis of τ-neighborhoods of zero with the properties (a) and (b) listed in Theorem 84.1. By Lemma 85.1 for every $U \in \mathfrak{N}$ there exists a $V \in \mathfrak{N}$ such that

$$(88.1) \qquad \underbrace{V + \cdots + V}_{n \text{ terms}} \subset U,$$

and for this (absorbing and balanced) V there exists $r > 0$ so that $rx_\nu \in V$ for $\nu = 1, \ldots, n$. It follows that K_r lies in U; indeed, if $\xi_1 x_1 + \cdots + \xi_n x_n \in K_r$, then $(|\xi_\nu|/r) \leqq 1$, hence $\xi_\nu x_\nu = (\xi_\nu/r) r x_\nu \in V$ for $\nu = 1, \ldots, n$ and thus $\xi_1 x_1 + \cdots + \xi_n x_n \in U$ (see (88.1)). With the aid of Proposition 84.4 it follows that $\tau \prec \tau_N$. We now show that conversely $\tau_N \prec \tau$, i.e., that an arbitrary K_r is always a τ-neighborhood of zero. For this it is sufficient to prove that there exists a $V_0 \in \mathfrak{N}$ which is bounded with respect to the norm, i.e., lies in a K_{r_0}, since in this case $K_r = (r/r_0) K_{r_0}$ contains the τ-neighborhood of zero $(r/r_0)V_0$, and is thus itself a τ-neighborhood of zero. Let us assume that $W \in \mathfrak{N}$ lies in no ball around 0

Then for every natural number k there exists a $w_k \in W$ such that $\|w_k\| \geqq k$. Because W is balanced, also $v_k := (k/\|w_k\|)w_k$ lies in W and thus $u_k := (1/k)v_k$ lies in $(1/k)W$. Obviously $\|u_k\| = 1$; by Proposition 10.4 there exists a subsequence (u_{k_m}) which converges with respect to τ_N, and a fortiori with respect to the coarser topology τ, to an element $u_0 \neq 0$. Because W is balanced, we have $W \supset \frac{1}{2}W \supset \frac{1}{3}W \supset \cdots$; it follows therefore from $u_{k_m} \in (1/k_m)W$ that u_{k_m} lies in $(1/k)W$ for all sufficiently large m. Since $(1/k)W$ is closed, also u_0 lies in $(1/k)W$, hence $ku_0 \in W$ for $k = 1, 2, \ldots$. From here it follows, if we use again that W is balanced, that W contains the one-dimensional subspace $[u_0]$. The existence of a neighborhood $V_0 \in \mathfrak{N}$ which is bounded with respect to the norm will therefore certainly be proved, if we can find a τ-neighborhood of zero in which *no* one-dimensional subspaces lie. We now show that such a neighborhood does indeed exist.

Since τ is Hausdorff, there exists a $U_0 \in \mathfrak{N}$ which does not coincide with E. Let m be the largest dimension of the subspaces contained in U_0; obviously $0 \leqq m \leqq n - 1$. If $m = 0$, then we are ready. If, however, $m \geqq 1$, then we choose $V \in \mathfrak{N}$ so that $V + V \subset U_0$. Either V contains only subspaces of dimension $\leqq m - 1$, or it contains exactly one m-dimensional subspace F_m. In the second case there exists an $x \neq 0$ in F_m, and a $W \in \mathfrak{N}$ which does not contain x (Proposition 84.3). $V \cap W$ contains a $U_1 \in \mathfrak{N}$, and this U_1 can contain only subspaces of dimension $\leqq m - 1$; indeed, an m-dimensional subspace lying in U_1 and thus also in V would have to coincide with F_m, so that U_1 would contain the vector x in contradiction to the construction of U_1. Thus starting with U_0 we have found in $\mathfrak{N}$ a neighborhood of zero, namely V or U_1, which contains only subspaces of dimension $\leqq m - 1$. We only have to continue this procedure in order to obtain finally a $V_0 \in \mathfrak{N}$ in which no one-dimensional subspaces lie. ■

Proposition 88.1. *Every finite-dimensional subspace F of a Hausdorff topological vector space E is closed.*

Proof. Let $x_0 \in \overline{F}$ and let F_0 be the linear hull of $F \cup \{x_0\}$. Then F_0 is a finite-dimensional Hausdorff topological vector space for the relative topology τ_0; by Theorem 88.1 this topology τ_0 is generated by a norm (on F_0). Since obviously x_0 is also a τ_0-closure point of F, the assertion of Proposition 88.1 follows immediately from Proposition 10.5. ■

Proposition 88.2. *If F is a closed and G a finite-dimensional subspace of the topological vector space E, then $F + G$ is closed.*

Proof. Let h be the canonical homomorphism from E onto the separated topological vector space E/F (Proposition 86.3b). $h(G)$ is finite-dimensional, hence, because of Proposition 88.1, closed in E/F. Thus $F + G = h^{-1}[h(G)]$ is also closed (Proposition 86.3c). ■

Proposition 88.3. *Every linear map A from a finite-dimensional Hausdorff topological vector space E into an arbitrary topological vector space is continuous. In particular $E' = E^*$.*

Proof. Let $\{x_1, \ldots, x_n\}$ be a basis of E. The topology of E is generated by the norm $\|\xi_1 x_1 + \cdots + \xi_n x_n\| := \max_{\nu=1}^n |\xi_\nu|$ (Theorem 88.1). If W is an arbitrary neighborhood of zero in F and V a neighborhood of zero such that $V + \cdots + V \subset W$ (sum with n terms; see Lemma 85.1), then there exists a $\delta > 0$ such that $\xi_\nu A x_\nu \in V$ whenever $|\xi_\nu| \leqq \delta$, hence

$$A\left(\sum_{\nu=1}^{n} \xi_\nu x_\nu\right) = \sum_{\nu=1}^{n} \xi_\nu A x_\nu \in W \qquad \text{whenever} \quad \left\|\sum_{\nu=1}^{n} \xi_\nu x_\nu\right\| \leqq \delta.$$

Thus A is continuous at 0 and therefore on E (Proposition 87.1a). ∎

Proposition 88.4. *If F is a closed, finite-codimensional subspace of the topological vector space E, then every algebraic complement G of F in E is also a topological complement.*

Proof. Let P project E onto G along F and let $\hat{P}: E/F \to G$ be the injection associated with P. Then E/F is a finite-dimensional separated topological vector space (Proposition 86.3b), because of Proposition 88.3 the map $\hat{P}$ is thus continuous, from where the continuity of P follows with the aid of Proposition 87.2. ∎

Exercises

1. If E is a finite-dimensional Hausdorff topological vector space with the basis $\{x_1, \ldots, x_n\}$, then the map A defined by $A(\sum_{\nu=1}^n \xi_\nu x_\nu) := (\xi_1, \ldots, \xi_n)$ is an isomorphism from E onto $l^p(n)$, $1 \leqq p \leqq \infty$, which is continuous in both directions.

2. Let E be a topological vector space with a 'large' dual space, i.e., let (E, E') be a dual system. Show the following:

(a) E is Hausdorff.

(b) A continuous endomorphism of E with finite deficiency is relatively regular if and only if it is open and its image space is closed. *Hint*: Exercise 3 in §87.

§89. Fredholm operators on topological vector spaces

In this section let E be a topological vector space. A Fredholm operator in $\mathscr{L}(E)$ is also called a *Fredholm operator on E*. Because of Proposition 87.3, *the endomorphism A of E is a Fredholm operator on E if and only if it has finite deficiency, it is continuous and open, and has a continuously projectable nullspace and image space.* Basic for the theory of Fredholm operators on E is

Proposition 89.1. *If E is* Hausdorff, *then the algebra $\mathscr{L}(E)$ is E'-saturated.*

Proof. Let the finite-dimensional endomorphism A of E be represented in the form $Ax = \sum_{\nu=1}^{n} \langle x, x_\nu^* \rangle y_\nu$ with the help of linearly independent vectors $y_1, \ldots, y_n$ from E and certain linear forms $x_1^*, \ldots, x_n^*$. If all x_ν^* are continuous, then every map $x \mapsto \langle x, x_\nu^* \rangle y_\nu$ and therefore also their sum A is continuous. Now let conversely A be continuous. The linear form f_ν on $F := [y_1, \ldots, y_n]$ defined by $f(\sum_{\mu=1}^{n} \xi_\mu y_\mu) := \xi_\nu$ is continuous by Proposition 88.3, thus $x_\nu^* = f_\nu \circ A$ is also continuous. ■

Proposition 89.1 permits to make use of the results of §25, 26, 27 and 39 for Fredholm operators on Hausdorff topological spaces:

Theorem 89.1. *Let E be a separated topological vector space. Then we obtain assertions concerning Fredholm operators on E if in the Propositions and Theorems of* §25, 26, 27 *and* 39 *we replace $\mathscr{A}(E)$ by $\mathscr{L}(E)$, $\mathscr{F}(\mathscr{A}(E))$ by the ideal $\mathscr{F}(E)$ of all continuous, finite-dimensional endomorphisms, E^+ by E', and A^+ by A'. Where (E, E^+) is supposed to be a dual system, we must require that (E, E') is a right dual system.*

In the latter case—(E, E') a right dual system—$A \in \mathscr{L}(E)$ is a Fredholm operator if and only if it has finite deficiency, it is open and has a closed image space (see §88 Exercise 2).

Exercise

Formulate the theorems, propositions and exercises of §25, 26, 27 and 39 for the case that E is a separated topological vector space, $\mathscr{A}(E) = \mathscr{L}(E)$, $\mathscr{F}(\mathscr{A}(E)) = \mathscr{F}(E)$ and $E^+ = E'$.

XII

Locally convex vector spaces

§90. Bases of neighborhoods of zero in locally convex vector spaces

In the most important examples of topological vector spaces we have investigated so far, the topology was generated in the sense of Example 83.3 by a family P of semi-norms on the vector space E. The canonical basis of neighborhoods of zero in such a topology consists of the sets

$$U_{p_1, \ldots, p_n; \varepsilon} := \left\{ x \in E : \max_{\nu=1}^{n} p_\nu(x) \leqq \varepsilon \right\}, \tag{90.1}$$

where $\{p_1, \ldots, p_n\}$ runs through all finite subsets of P, and ε through all positive numbers. The basic neighborhoods (90.1) have a property whose fundamental importance will become increasingly clear in the following sections: they are *convex*. Topological vector spaces, which have a basis of neighborhoods of zero consisting of convex sets, are called *locally convex*. They are the most important spaces of functional analysis. The topology of a locally convex space is also said to be locally convex.

Topologies which are generated by semi-norms, in particular normed topologies, weak topologies and $\mathfrak{S}$-topologies, are locally convex.

For the following investigations we need three lemmas.

Lemma 90.1. *For convex sets K in a vector space E over* **K** *the following assertions hold:*

(a) *$x + K$ and αK are convex for every $x \in E$ and $\alpha \in$* **K**.
(b) *The intersection of convex sets is convex.*
(c) *If K contains the point zero, then the balanced core of K is convex.*
(d) *For $\alpha, \beta \geqq 0$ we have $(\alpha + \beta)K = \alpha K + \beta K$.*

Proof. (a) and (b) are trivial. (c) follows from (a) and (b) with the aid of Lemma 84.1d. To prove (d) we may assume that $\alpha, \beta > 0$. Obviously $(\alpha + \beta)K \subset \alpha K + \beta K$. But because of the convexity of K we also have $(\alpha/(\alpha + \beta))K + (\beta/(\alpha + \beta))K \subset K$, hence $\alpha K + \beta K \subset (\alpha + \beta)K$. ■

Because of the above assertion (b), there exists for every $M \subset E$ a smallest convex set containing M, namely the intersection of all convex sets $K \supset M$. It is called the *convex hull* of M, and is denoted by $\text{co}(M)$. A sum of the form $\alpha_1 x_1 + \cdots + \alpha_n x_n$ with $\alpha_\nu \geqq 0$ $(\nu = 1, \ldots, n)$, $\alpha_1 + \cdots + \alpha_n = 1$, is called a *convex combination* of the vectors $x_1, \ldots, x_n$. We leave the simple proof of the next lemma to the reader.

Lemma 90.2. *The convex hull of a set M consists of all convex combinations of the elements of M. The convex hull of an open set in a topological vector space is open.*

A balanced and convex set is called *absolutely convex* (see Exercise 1). By Lemma 90.1c, *the balanced core of a convex set containing* 0 *is absolutely convex.*

Lemma 90.3. *The closure of a convex set K in a topological vector space is convex.*

Proof. Let $x, y \in \overline{K}$, $\alpha, \beta > 0$ and $\alpha + \beta = 1$. We must show that every neighborhood W of $\alpha x + \beta y$ intersects the set K. For this we choose neighborhoods U, V of x, y, respectively, so that $\alpha U + \beta V \subset W$. Since x, y are closure points of K, there exist in $U \cap K$, $V \cap K$ points u, v, respectively. Obviously $\alpha u + \beta v$ lies in $W \cap K$, hence this intersection is not empty. ■

Proposition 90.1. *In a locally convex space the absolutely convex, closed, and the absolutely convex, open neighborhoods of zero form bases of neighborhoods of zero.*

Proof. According to Proposition 84.2d, every neighborhood W of zero contains a closed neighborhood V of zero, and by hypothesis V contains a convex neighborhood U of zero. The closure $\overline{U}$ lies in V, and is convex according to Lemma 90.3. The balanced core of $\overline{U}$ is closed by Lemma 84.1f, convex by Lemma 90.1c, and by Proposition 84.2c it is a neighborhood of zero which lies in $\overline{U}$, hence a fortiori in W. According to what we have just proved, every neighborhood W of zero contains an absolutely convex neighborhood V of zero, and V contains by Theorem 84.1a′ an open neighborhood U of zero. The balanced hull D of U is open by Lemma 84.1f and lies in V. By Lemma 90.2 the set $\text{co}(D)$ is also open and is a subset of V, hence a fortiori of W. Since $\text{co}(D)$ is obviously balanced and a neighborhood of zero, we have proved also the second assertion. ■

With the aid of Lemma 90.1 we obtain from Proposition 90.1 immediately

Proposition 90.2. *Every point of a locally convex space has a basis of neighborhoods which consists of convex closed sets, and one which consists of convex open sets.*

Exercises

1. A subset K of a vector space E is absolutely convex if and only if $x, y \in K$ and $|\alpha| + |\beta| \leqq 1$ imply that $\alpha x + \beta y \in K$.

*2. Let A be a linear map from E into F, let M be a convex set in E and N a convex set in F. Then the sets $A(M)$ and $A^{-1}(N)$ are also convex. The corresponding statements are true for absolutely convex sets M, N.

*3. If $K_1, \ldots, K_n$ are convex subsets of the vector space E, then

$$\operatorname{co}\left(\bigcup_{\nu=1}^{n} K_\nu\right) = \left\{\alpha_1 x_1 + \cdots + \alpha_n x_n : x_\nu \in K_\nu, \alpha_\nu \geqq 0\ (\nu = 1, \ldots, n),\ \sum_{\nu=1}^{n} \alpha_\nu = 1\right\}.$$

4. Let the filter basis $\mathfrak{B}$ on the vector space E consist of absorbing, absolutely convex sets. Then there exists exactly one locally convex topology on E for which $\mathfrak{N} := \{\varepsilon U : \varepsilon > 0, U \in \mathfrak{B}\}$ is a basis of neighborhoods of zero. Example: P a family of semi-norms,

$$\mathfrak{B} := \{U_{p_1, \ldots, p_n; 1}\}, \quad \mathfrak{N} := \{\varepsilon U_{p_1, \ldots, p_n; 1}\} = \{U_{p_1, \ldots, p_n; \varepsilon}\}$$

(see (90.1)).

§91. The generation of locally convex topologies by semi-norms

We know that every family P of semi-norms on the vector space E generates a locally convex topology for which all the $p \in P$ are continuous (Example 83.3). We will show in this section that conversely every locally convex topology on E is generated by a family of semi-norms.

With every absorbing subset U of the vector space E we associate its *Minkowski functional* p, which is defined by

(91.1) $$p(x) := \inf\{\alpha > 0 : x \in \alpha U\}.$$

p is also called the *distance function* or *gauge* of U and will be denoted occasionally more precisely by p_U. Obviously $0 \leqq p(x) < \infty$. The (open or closed) unit ball U of a normed space generates the Minkowski functional $p(x) = \|x\|$. Every neighborhood of zero of a topological vector space has a Minkowski functional.

Proposition 91.1. *The Minkowski functional p of an absorbing and absolutely convex set $U \subset E$ is a semi-norm on E, and we have*

(91.2) $$\{x : p(x) < 1\} \subset U \subset \{x : p(x) \leqq 1\}.$$

Proof. Given two vectors x, y and an arbitrary $\varepsilon > 0$, there exist positive numbers α, β such that

$$p(x) \leqq \alpha \leqq p(x) + \varepsilon \quad \text{and} \quad x \in \alpha U,$$

$$p(y) \leqq \beta \leqq p(y) + \varepsilon \quad \text{and} \quad y \in \beta U.$$

With the aid of Lemma 90.1d it follows that $x + y \in (\alpha + \beta)U$, hence $p(x + y) \leqq \alpha + \beta \leqq p(x) + p(y) + 2\varepsilon$. Since ε was arbitrary, we obtain $p(x + y) \leqq$

$p(x) + p(y)$. Next we prove the homogeneity property $p(\lambda x) = |\lambda| p(x)$. Because of $0 \in U$ we have

$$p(0 \cdot x) = p(0) = \inf\{\alpha > 0 : 0 \in \alpha U\} = 0 = 0 \cdot p(x).$$

For $\lambda > 0$ we obtain

$$p(\lambda x) = \inf\{\alpha > 0 : \lambda x \in \alpha U\} = \inf\left\{\alpha > 0 : x \in \frac{\alpha}{\lambda} U\right\};$$

setting $\beta := \alpha/\lambda$ we have thus $p(\lambda x) = \lambda \inf\{\beta > 0 : x \in \beta U\} = \lambda p(x)$. Now let $|\lambda| = 1$. Since U is balanced, we have in this case $\lambda x \in \alpha U \Leftrightarrow x \in \alpha U$, from where $p(\lambda x) = p(x)$ follows. For an arbitrary $\lambda \neq 0$ we have, according to what has been shown so far,

$$p(\lambda x) = p\left(|\lambda| \frac{\lambda}{|\lambda|} x\right) = |\lambda| p\left(\frac{\lambda}{|\lambda|} x\right) = |\lambda| p(x). \tag{91.3}$$

Now we prove (91.2). From $p(x) < 1$ it follows that there exists a positive α with $\alpha < 1$ such that $x \in \alpha U$. Since U is balanced, x lies also in U. The second inclusion in (91.2) is trivial because of $U = 1 \cdot U$. ∎

Proposition 91.2. *In a topological vector space E the following assertions are valid:*

(a) *An absorbing, absolutely convex set $U \subset E$ is a neighborhood of zero if and only if the semi-norm p_U is continuous.*

(b) *For any absolutely convex neighborhood of zero U we have $U = \{x : p_U(x) \leqq 1\}$ if U is closed, and $U = \{x : p_U(x) < 1\}$ if U is open.*

Proof. (a) Let U be a neighborhood of zero, $\varepsilon > 0$ and $p := p_U$. The εU is also a neighborhood of zero, and $x \in \varepsilon U$ implies $(1/\varepsilon)x \in U$, hence $p((1/\varepsilon)x) \leqq 1$, (Proposition 91.1) and so $p(x) \leqq \varepsilon$. Hence p is continuous at 0 and therefore on E (Exercise 4 in §83). Now let p be continuous. Then $\{x : p(x) < 1\}$ is an open neighborhood of zero, which lies in U because of Proposition 91.1. Consequently U must also be a neighborhood of zero.

(b) Let U be an absolutely convex, closed neighborhood of zero and $p := p_U$. By Proposition 91.1 we have $U \subset \{x : p(x) \leqq 1\}$. Now let $p(x) \leqq 1$. Then for $0 < \alpha < 1$ we have $p(\alpha x) = \alpha p(x) < 1$, hence

$$\alpha x \in U \quad \text{for all} \quad \alpha \in (0, 1) \tag{91.4}$$

(Proposition 91.1). If x were not in U, then there would exist, since U is closed, a neighborhood V of x such that $U \cap V = \varnothing$. Since the vectors αx lie in V for all numbers α sufficiently close to 1, these vectors could not lie in U, in contradiction to (91.4). Thus $p(x) \leqq 1$ implies indeed $x \in U$. Now let the absolutely convex neighborhood U of zero be open. By Proposition 91.1 we have $\{x : p(x) < 1\} \subset U$. If x lies in U, then so does a neighborhood $x + V$ of x, therefore there exists a $\beta > 0$ such that $(1 + \beta)x \in U$, from where $p(x) \leqq 1/(1 + \beta) < 1$ follows. ∎

From the propositions of this section we obtain easily the following theorem which makes the locally convex topologies in a certain sense analytically accessible.

Theorem 91.1. *The topology of a locally convex space E can always be generated by a family P of continuous semi-norms on E (e.g., by the family of all continuous semi-norms). E is Hausdorff if and only if P is total.*

Proof. Let P be the family of all continuous semi-norms on E, let $\mathfrak{N}$ be the canonical basis of neighborhoods of zero for the topology generated by P (i.e., the collection of sets $U_{p_1,\ldots,p_n;\varepsilon} := \{x: \max_{\nu=1}^{n} p_\nu(x) \leqq \varepsilon\}$, where $p_\nu \in P$ and $\varepsilon > 0$), and $\mathfrak{M}$ the set of all absolutely convex, closed neighborhoods of zero in E. By Proposition 90.1 the collection $\mathfrak{M}$ is a basis of neighborhoods of zero in E. Obviously $\mathfrak{N} \subset \mathfrak{M}$. But the opposite inclusion is also valid: indeed, for every $U \in \mathfrak{M}$ we have $p_U \in P$ by Proposition 91.2, and $U = \{x: p_U(x) \leqq 1\} \in \mathfrak{N}$. It follows now with the aid of Proposition 84.4 that P generates the topology of E. The condition for E to be Hausdorff has already been proved in Example 83.3. ■

It is very well possible that not every continuous semi-norm is needed to generate the topology of E. In normed spaces, for instance, it is sufficient to consider the single semi-norm $p(x) := \|x\|$. The canonical basis of neighborhoods of zero $\mathfrak{N}$ can be described particularly easily if P is *saturated*, i.e., if for any $p_1, \ldots, p_n$ in P also the semi-norm $x \mapsto \max_{\nu=1}^{n} p_\nu(x)$ is in P. Then $\mathfrak{N}$ consists of all sets $U_{p;\varepsilon} = \{x: p(x) \leqq \varepsilon\}$, where $p \in P$, $\varepsilon > 0$. The family of all continuous semi-norms on E is saturated; thus *every locally convex topology can be generated by a* saturated *family of semi-norms.* As the first application of Theorem 91.1, we show that the continuity of linear maps on locally convex spaces can be described by means of semi-norms.

Proposition 91.3. *Let the topologies of the locally convex spaces E, F be generated by the families of semi-norms P, Q. In this case the linear map $A: E \to F$ is continuous if and only if for every $q \in Q$ there exists a $\gamma > 0$ and finitely many $p_1, \ldots, p_n$ from P such that*

$$q(Ax) \leqq \gamma \max_{\nu=1}^{n} p_\nu(x) \qquad \textit{for all} \quad x \in E. \tag{91.5}$$

Proof. Let (91.5) be satisfied, and let $V := \{y \in F: \max_{\mu=1}^{m} q_\mu(x) \leqq \varepsilon\}$ be a basic neighborhood of zero in F. For every q_μ there exists a $\gamma_\mu > 0$ and a continuous semi-norm $p^{(\mu)}$ on E such that $q_\mu(Ax) \leqq \gamma_\mu p^{(\mu)}(x)$ for all $x \in E$. It follows that the neighborhood of zero $U := \bigcap_{\mu=1}^{m} \{x \in E: p^{(\mu)}(x) \leqq \varepsilon/\gamma_\mu\}$ is mapped by A into V, so that A is continuous at 0 and therefore on E. Now let A be continuous and $q \in Q$. Then there exist $p_1, \ldots, p_n \in P$ and $\varepsilon > 0$ so that $q(Ax) \leqq 1$ whenever $p(x) := \max_{\nu=1}^{n} p_\nu(x) \leqq \varepsilon$. From here one obtains (91.5) with $\gamma := 1/\varepsilon$ by the same argument which led from (82.4) to (82.5). ■

From Proposition 91.3 we obtain immediately:

Proposition 91.4. *If the topology on E is generated by the family of semi-norms P, then a linear form f on E is continuous if and only if there exist $\gamma > 0$ and finitely many $p_1, \ldots, p_n$ from P such that*

$$|f(x)| \leqq \gamma \max_{\nu=1}^{n} p_\nu(x) \quad \text{for all} \quad x \in E. \tag{91.6}$$

The estimates (91.5), (91.6) become simpler when P is *saturated*: in this case the semi-norm $x \mapsto \max_{\nu=1}^{n} p_\nu(x)$ can be replaced by a $p \in P$.

Exercises

+1. The set of all semi-norms on the vector space E generates a locally convex topology τ on E for which the collection of all absorbing, absolutely convex subsets of E is a basis of neighborhoods of zero. τ is the *finest locally convex topology* on E. It is Hausdorff.

+2. Let the families of semi-norms P, Q generate the topologies τ_P, τ_Q on E. We have $\tau_Q \prec \tau_P$ if and only if for every $q \in Q$ there exists a $\gamma > 0$ and finitely many $p_1, \ldots, p_n$ from P so that $q(x) \leqq \gamma \max_{\nu=1}^{n} p_\nu(x)$ on E.

3. Let P be the set of all continuous semi-norms on the locally convex space E. Then the collection of all neighborhoods $U_p := \{x : p(x) \leqq 1\}$ $(p \in P)$ is a basis of neighborhoods of zero of E.

+4. A sequence (x_n) in a topological vector space E is called a *Cauchy sequence*, if for every neighborhood U of zero—or for every neighborhood from a basis of neighborhoods of zero—there exists an index $n_0(U)$ such that $x_n - x_m \in U$ for $n, m \geqq n_0(U)$. Every convergent sequence is a Cauchy sequence. E is said to be *sequentially complete* if every Cauchy sequence in E converges to an element of E. If the topology of E is generated by the family of semi-norms P, then the following hold:

(a) (x_n) is a Cauchy sequence if and only if $p(x_n - x_m) \to 0$ as $n, m \to \infty$ for every $p \in P$.

(b) $x_n \to x$ holds if and only if $p(x_n - x) \to 0$ for every $p \in P$.

(c) E is sequentially complete if and only if from $p(x_n - x_m) \to 0$ as $n, m \to \infty$ for every $p \in P$, it follows that there exists an $x \in E$ such that $p(x_n - x) \to 0$ for every $p \in P$.

§92. Subspaces, products and quotients of locally convex spaces

From Proposition 86.1a one obtains immediately:

Proposition 92.1. *Let the topology of the locally convex space E be generated by the family P of semi-norms. Let F be a linear subspace of E and let $\bar{P}$ be the family of restrictions $\bar{p}$ of $p \in P$ to F (i.e., $\bar{p}(x) = p(x)$ for $x \in F$). Then the relative topology on F is generated by $\bar{P}$, it is thus locally convex.*

If we consider that a product $\prod_{\iota \in J} K_\iota$ of convex sets $K_\iota \subset E_\iota$ is convex in the product space $\prod_{\iota \in J} E_\iota$, then we obtain:

Proposition 92.2. *The product of locally convex spaces is locally convex.*

If p is a semi-norm on the vector space E and if F is a linear subspace of E, then

$$\hat{p}(\hat{x}) := \inf_{y \in \hat{x}} p(y) \qquad \text{for} \quad \hat{x} \in E/F$$

defines a semi-norm $\hat{p}$ on E/F, called the *quotient semi-norm.* The image of $U := \{x : p(x) < \varepsilon\}$ under the canonical homomorphism $h: E \to E/F$ is then $h(U) = \{\hat{x} : \hat{p}(\hat{x}) < \varepsilon\}$. From here follows, with the aid of Proposition 86.3d:

Proposition 92.3. *Let E be a locally convex space whose topology is generated by a* saturated *family P of semi-norms, and let F be a linear subspace. Then the quotient topology on E/F is locally convex, and is generated by the family of quotient semi-norms $\hat{p}$ $(p \in P)$.*

Exercise

$^+$The product of countably many normed spaces is not a normed space but it is a metric locally convex vector space (see Exercise 1 and 2 in §86).

§93. Normable locally convex spaces. Bounded sets

A topological vector space is said to be *normable* if its topology arises from a norm. A normable space is locally convex and possesses a neighborhood of zero which is bounded (for the norm), namely the unit ball V; the collection $\{\varepsilon V : \varepsilon > 0\}$ is a basis of neighborhoods of zero. In this section we shall define the concept of a bounded set in topological vector spaces and show that a separated locally convex space, in which there exists a bounded neighborhood of zero, is always normable.

A subset M of the topological vector space E is said to be *bounded* if for every neighborhood U of zero—or for every neighborhood from a basis of neighborhoods of zero—there exists an $\alpha > 0$ such that $M \subset \alpha U$.

In a metric vector space this *topological* notion of boundedness does not have to coincide with the *metric* notion of boundedness (being contained in a ball) we define earlier. Thus e.g., (s) is bounded metrically but not topologically. In a normed space, however, the two concepts of boundedness are trivially identical. The following proposition is usually called the *Kolmogorov normability theorem.*

Proposition 93.1. *A locally convex space E is normable if and only if it is Hausdorff and has a bounded neighborhood of zero.*

Proof. Since the necessity of the conditions is clear after the above considerations, we turn to the proof of their sufficiency. Let U be a bounded neighborhood of zero and V an absolutely convex, closed neighborhood of zero contained in U (Proposition 90.1); by Proposition 91.2 we have $V = \{x: p_V(x) \leqq 1\}$. Since V is obviously bounded, for every neighborhood of zero W there exists an $\varepsilon > 0$ so that $\varepsilon V \subset W$. Thus the collection of sets $\varepsilon V = \{x: p_V(x) \leqq \varepsilon\}$ is a basis of neighborhoods of zero, and the topology of E is therefore generated by the semi-norm p_V. Because E is Hausdorff, p_V is even a norm (Theorem 91.1), and the proof is complete. ■

The simple proofs of the next two propositions are left to the reader.

Proposition 93.2. *In a topological vector space the following sets are always bounded:*

(a) *finite sets and subsets of bounded sets;*
(b) *scalar multiples, unions and sums of finitely many bounded sets;*
(c) *balanced and closed hulls of bounded sets;*
(d) *continuous linear images of bounded sets.*

Proposition 93.3. *Let the topology of the locally convex space E be generated by the family of semi-norms P. The subset $M \subset E$ is bounded if and only if every $p \in P$ is bounded on M. From here it follows that if (E, E^+) is a bilinear system, then $M \subset E$ is $\sigma(E, E^+)$-bounded if and only if every linear form $x \mapsto \langle x, x^+ \rangle$, $x^+ \in E^+$, is bounded on M (i.e., if M is E^+-bounded in the sense of Example* 83.6).

Exercises

1. We keep the notation and the hypotheses of Example 83.5. Let the collection $\mathfrak{S}$ of F-bounded subsets of T generate the locally convex topology τ on F. The set $M \subset F$ is τ-bounded if and only if the family of functions $t \mapsto \langle t | x \rangle$ $(x \in M)$ is uniformly bounded on every $S \in \mathfrak{S}$.

$^+$2. Every Cauchy sequence, in particular every convergent sequence, in a topological vector space is bounded.

3. A subset M of the topological vector space E is said to be *B-bounded* if for every sequence (x_n) in M and every sequence (α_n) converging to zero, one has $\alpha_n x_n \to 0$. Show that a set is bounded if and only if it is B-bounded.

$^+$4. Let E, F be topological vector spaces. $A \in \mathscr{S}(E, F)$ is said to be *bounded* if in E there exists a neighborhood U of zero with bounded image $A(U)$. Show the following:

(a) Bounded maps are continuous.

(b) The identity transformation I of a Hausdorff locally convex space E is bounded if and only if E is normable.

(c) A continuous finite-dimensional operator $A: E \to F$ is bounded whenever F is Hausdorff.

(d) Sums and scalar multiples of bounded operators are bounded.

(e) The product of a bounded operator by a continuous operator is bounded.

+5. Let the topology of the locally convex space E be generated by the family $P = (p_\iota : \iota \in J)$ of semi-norms. Show that $A \in \mathscr{S}(E)$ is bounded (see Exercise 4) if and only if there exists a $p \in P$ and numbers $\gamma_\iota > 0$ so that $p_\iota(Ax) \leqq \gamma_\iota p(x)$ for all $\iota \in J$ and all $x \in E$. In this case A is also said to be p-bounded.

+6. Let the hypotheses of Exercise 5 be given and let A be p-bounded. Set

$$|A|_\iota := \inf\{\gamma_\iota > 0 : p_\iota(Ax) \leqq \gamma_\iota p(x)\}$$

and

$$|A| := \inf\{\gamma > 0 : p(Ax) \leqq \gamma p(x)\}.$$

of course, $|A|_\iota$ and $|A|$ depend also on p; we fix, however, p in this discussion. Show that the set of p-bounded endomorphisms of E forms an algebra $\mathscr{B}_p(E)$, and for $A, B \in \mathscr{B}_p(E)$ we have $|A + B|_\iota \leqq |A|_\iota + |B|_\iota$, $|\alpha A|_\iota = |\alpha||A|_\iota$, $|AB|_\iota \leqq |A|_\iota |B|$, in particular $|A^n|_\iota \leqq |A|_\iota |A|^{n-1}$ for $n = 1, 2, \ldots$. If E is even sequentially complete (Exercise 4 in §91) and $|A| < 1$, then $I - A$ has a continuous inverse defined on E. *Hint*: For every $y \in E$ the sequence $x_n := y + Ay + \cdots + A^n y$ converges; it appears that $I - A$ is bijective and that $(I - A)^{-1} - I$ is p-bounded.

+7. Let A be a bounded endomorphism of the sequentially complete locally convex space E. Then there exist an $r > 0$ such that $\lambda I - A$ has for every $|\lambda| > r$ a continuous inverse defined on E. *Hint*: Exercise 6.

XIII

Duality and compactness

§94. The Hahn–Banach theorem

The following *extension theorem of Hahn–Banach* is a generalization of Theorem 28.1. It is of basic importance for our further work.

Theorem 94.1. *For every continuous linear form f on the subspace F of the locally convex space E there exists a continuous linear form g on E such that $g(x) = f(x)$ for all $x \in F$.*

The proof consists in generating the topology of E and F by semi-norms (Theorem 91.1 and Proposition 92.1) and then applying Proposition 91.4 and Principle 28.1. ∎

A counterpart to Theorem 28.2 is:

Theorem 94.2. *Let F be a closed subspace of the locally convex space E and $x_0 \notin F$. Then there exists a continuous linear form f on E such that*

$$f(x) = 0 \quad \text{for} \quad x \in F \quad \text{and} \quad f(x_0) = 1.$$

Proof. On the linear hull H of $\{x_0\} \cup F$ we define a linear form h by

$$h(\alpha x_0 + x) := \alpha \qquad (\alpha \in \mathbf{K}, x \in F)$$

cf. the beginning of the proof of Theorem 28.2). Obviously

$$h(x) = 0 \quad \text{for} \quad x \in F \quad \text{and} \quad h(x_0) = 1.$$

We now prove that h is continuous. $x_0 + F$ is closed and does not contain 0. Consequently there exists an absolutely convex, closed neighborhood of zero U which does not intersect $x_0 + F$, and since $U = \{x : p(x) \leqq 1\}$ (Proposition 91.2 with $p := p_U$), we must have $p(x_0 + x) > 1$ for $x \in F$. In the case $\alpha \neq 0$, for all $x \in F$ we obtain

$$|h(\alpha x_0 + x)| = |\alpha| \leqq |\alpha| p\left(x_0 + \frac{x}{\alpha}\right) = p(\alpha x_0 + x),$$

and this inequality holds obviously also for $\alpha = 0$. By Proposition 91.4 the continuity of h follows since p, and with it also its restriction to H, is continuous (Proposition 91.2). We now only need to extend h to E according to Theorem 94.1. ■

In a separated space E the subspace $F := \{0\}$ is closed. Therefore we immediately obtain from Theorem 94.2:

Theorem 94.3. *A* separated *locally convex space E forms, together with its dual E′, a* dual system (E, E').

There exist separated, non-locally convex spaces E which have a trivial dual $E' = \{0\}$; see [8], vol. I, p. 158.

Exercise

$^{+}$A continuous endomorphism of a separated locally convex space E is a Fredholm operator on E if and only if it has finite deficiency, it is open and has a closed image space.

§95. The topological characterization of normal solvability

The continuous endomorphism A of the topological vector space E is said to be *normally solvable* if it is E'-normally solvable in the sense of §29 (observe that the E'-conjugate A' always exists).

An (E, E')-orthogonally closed subspace of E is obviously always closed. In the case of a locally convex E also the converse holds (Lemma 29.2 and Theorem 94.2). Because of Proposition 29.1 we can state:

Proposition 95.1. *The continuous endomorphism A of the locally convex space E is normally solvable if and only if its image space is closed.*

With the aid of Proposition 82.1 we obtain from Proposition 95.1 immediately

Proposition 95.2. *Let (E, E^+) be a left dual system. The E^+-conjugable endomorphism A of E is E^+-normally solvable if and only if its image space is* weakly *(i.e., for the topology $\sigma(E, E^+)$) closed.*

Propositions 82.1 and 95.2 describe the conjugability and normal solvability, i.e., the central concepts of the theory of solving equations, with the help of the weak topology.

Exercises

1. Let E be a separated locally convex space and A a continuous, open endomorphism of E with finite deficiency, whose image space is closed. Then $\alpha(A) = \beta(A')$, $\beta(A) = \alpha(A')$, A is E'-normally solvable, A' is E-normally solvable *Hint*: Theorem 27.1, Exercise in §94.

2. Let E be a topological vector space. $A \in \mathscr{L}(E)$ is normally solvable whenever one of the following conditions is satisfied:
(a) $\beta(A)$ is finite and $A(E)$ closed;
(b) (E, E') is a right dual system and $A(E)$ is continuously projectable.
$^{+}$3. If E is a complete metric vector space, $A \in \mathscr{L}(E)$ and $\beta(A)$ finite, then $A(E)$ is closed and A is normally solvable. *Hint*: $E = A(E) \oplus C$. Then $E \times C$ is a complete metric vector space, $P(x, c) := x$, $Q(x, c) := c$ define continuous linear maps P, Q from $E \times C$ onto E, C. The map $B := AP + Q$ is continuous. Show with the aid of Theorem 80.1 that $E \backslash A(E) = B((E \times C)\backslash(E \times \{0\}))$ is open.

§96. Separation theorems

In this section let E be a topological vector space. We recall that every neighborhood of zero U generates a Minkowski functional p_U. The concept of *sublinear functional* occurring in the next lemma has been defined in §28 Exercise 2.

Lemma 96.1. *The Minkowski functional p of a convex neighborhood of zero U in E is sublinear. If U is furthermore open, then we have*

(96.1) $\varepsilon U = \{x : p(x) < \varepsilon\}$ *for all* $\varepsilon > 0$, *in particular* $U = \{x : p(x) < 1\}$.

Proof. The first assertion is obtained by repeating word for word the beginning of the proof of Proposition 91.1. We need to prove the second assertion obviously only in the indicated special case ($\varepsilon = 1$). If $p(x) < 1$, then there exists an α such that $p(x) \leqq \alpha < 1$ and $x = \alpha y$ for some $y \in U$, consequently also $x = \alpha y + (1 - \alpha) \cdot 0$ lies in U; thus we have $\{x : p(x) < 1\} \subset U$. The converse inclusion is obtained as in the proof of Proposition 91.2. ■

A set $M := x_0 + F$, where F is a linear subspace of E, is called a *linear manifold*. It is called a *hyperplane* (through x_0) if codim $F = 1$; in this terminology F itself is a hyperplane through 0. It follows immediately from Exercise 7 in §15 that *if f is a linear form* $\neq 0$ *on E and* $f(x_0) = \alpha$, *then*

$$x_0 + N(f) = \{x : f(x) = \alpha\}$$

represents a hyperplane through x_0, *and conversely every hyperplane through* x_0 *can be described in this fashion.* $f(x) = \alpha$ is called the *equation* of H. The hyperplane H is closed if and only if f is continuous (Proposition 87.5).

Theorem 96.1. *For every convex, open set* $K \neq \varnothing$, *and every linear manifold M in E which does not intersect K, there exists a closed hyperplane H which contains M and does not meet K.*

In the first part of the proof let E be *real*. We may assume without restricting the generality that 0 lies in K, hence that K is a neighborhood of zero. Let p

be its Minkowski functional. Furthermore let $M = x_0 + F$, where F is a linear subspace of E. It follows from $K \cap M = \varnothing$ that

$$(96.2) \qquad p(x_0 + y) \geqq 1 \qquad \text{for all} \quad y \in F$$

(Lemma 96.1) and $x_0 \notin F$. Because of the second assertion the scalar α in the representation $x = \alpha x_0 + y$ of a vector $x \in F_0 := \{\alpha x_0 + y : \alpha \in \mathbf{R}, y \in F\}$ is uniquely determined, hence we can define a linear form f on F_0 by

$$f(\alpha x_0 + y) := \alpha.$$

We have

$$(96.3) \qquad f(\alpha x_0 + y) \leqq p(\alpha x_0 + y) \qquad (\alpha \in \mathbf{R};\ y \in F);$$

this is trivial for $\alpha \leqq 0$, while for $\alpha > 0$ it follows from (96.2) that

$$f(\alpha x_0 + y) = \alpha \leqq \alpha p\left(x_0 + \frac{y}{\alpha}\right) = p(\alpha x_0 + y).$$

By Exercise 2 in §28 there exists a linear form g on E such that

$$(96.4) \quad g(x) = f(x) \qquad \text{for} \quad x \in F_0 \quad \text{and} \quad g(x) \leqq p(x) \qquad \text{for} \quad x \in E.$$

We now show that g is continuous. For this, let K_ε be the balanced core of εK ($\varepsilon > 0$). The set K_ε is a neighborhood of zero (Proposition 84.2) such that $K_\varepsilon \subset \varepsilon K$ and $K_\varepsilon = -K_\varepsilon$. From $x \in K_\varepsilon$ it follows because of Lemma 96.1 that

$$|g(x)| = \begin{cases} g(x) \leqq p(x) < \varepsilon & \text{if} \quad g(x) \geqq 0 \\ -g(x) = g(-x) \leqq p(-x) < \varepsilon & \text{if} \quad g(x) < 0, \end{cases}$$

thus g is indeed continuous, so that $H := \{x : g(x) = 1\}$ is a closed hyperplane. From (96.1) and (96.4) it follows that $K \subset \{x : g(x) < 1\}$, and since $g(x) = 1$ whenever $x \in M$, we have $M \subset H$. The hyperplane H satisfies therefore all the requirements of the theorem. Now let E be *complex* and let E_r be the real space corresponding to E (see Exercise 2 in §4). E_r with the topology of E is a topological vector space, M is a linear manifold and K a convex, open set in E_r. We assume without restricting the generality that 0 lies in M, i.e., that M is a linear subspace. By what we have proved already, there exists a continuous linear form g on E_r such that $\{x : g(x) = 0\}$ contains M and does not intersect K. If we now define a continuous linear form f on E by $f(x) := g(x) - ig(ix)$ (cf Part II of the proof of Principle 28.1), then $H := \{x : f(x) = 0\}$ is a closed hyperplane which contains $M = M \cap (iM)$ and does not meet K. ■

Theorem 96.2. *Let E be locally convex and $K \subset E$ not empty, closed and convex. Then for every $y \notin K$ there exists a continuous linear form f on E and an $\alpha \in \mathbf{R}$ such that*

$$(96.5) \qquad \operatorname{Re} f(y) < \alpha < \operatorname{Re} f(x) \qquad \textit{for all} \quad x \in K.$$

For the proof we assume first that E is *real*. Because of $y \notin K$, there exists, according to Proposition 90.2, an open, convex neighborhood W of y such that $W \cap K = \varnothing$. The non-empty, open and convex set $K - W$ does not intersect the subspace $M := \{0\}$, consequently there exists, because of Theorem 96.1, a continuous linear form g on E so that $H := \{x : g(x) = 0\}$ does not intersect the set $K - W$, hence $g(x) \neq 0$ for all $x \in K - W$. If there would exist vectors x_1, x_2 in $K - W$ with $g(x_1) < 0$ and $g(x_2) > 0$, then we would have

$$g(\lambda x_1 + (1 - \lambda)x_2) = \lambda g(x_1) + (1 - \lambda)g(x_2) = 0$$

for an appropriate $\lambda \in (0, 1)$, consequently $\lambda x_1 + (1 - \lambda)x_2$ would lie in $(K - W) \cap H = \varnothing$. This contradiction shows that $g(x)$ has a constant sign in $K - W$; without restricting the generality we may assume that it is positive. It follows that $g(x) > g(w)$ for all $x \in K$ and $w \in W$, hence $\beta := \inf_{x \in K} g(x) \geqq g(w)$ for $w \in W$, and since $g(W)$ is open by Proposition 87.5, one sees easily that even $\beta > g(w)$ for $w \in W$, and that in particular $\beta > g(y)$. Choosing $\alpha := (g(y) + \beta)/2$, we then have $g(y) < \alpha < g(x)$ for $x \in K$. If E is *complex*, then we again pass to the real space E_r, which is locally convex with the topology of E, we construct g and α according to the first part of this proof and define the continuous linear form f by $f(x) := g(x) - ig(ix)$. It is now trivial that (96.5) holds. ■

Exercises

1. Let K_1, K_2 be disjoint, non-empty, convex sets in the locally convex space E, furthermore let K_1 be open. Then there exists an $f \in E'$ such that

$$f(K_1) \cap f(K_2) = \varnothing.$$

2. Under the hypotheses of Exercise 1, there exists an $f \in E'$ and an $\alpha \in \mathbf{R}$ such that $\operatorname{Re} f(x_1) < \alpha \leqq \operatorname{Re} f(x_2)$ for $x_1 \in K_1$, $x_2 \in K_2$.

§97. Three applications to normed spaces

The first two applications are concerned with the conjugates of continuous operators on a Banach space. We start with a lemma.

Lemma 97.1. *Let E be a complete, F an arbitrary normed space,*

$$K_r := \{x \in E : \|x\| \leqq r\},\ V_\rho := \{y \in F : \|y\| < \rho\} \qquad \text{and} \qquad A \in \mathscr{L}(E, F).$$

Then it follows from $V_\rho \subset \overline{A(K_1)}$ that $V_\rho \subset A(K_1)$.

Proof. As it can be easily seen, it is sufficient to prove that $V_\rho \subset A(K_r)$ for every $r > 1$. From the hypothesis $V_\rho \subset \overline{A(K_1)}$ it follows that $V_{\rho\varepsilon^n} \subset \overline{A(K_{\varepsilon^n})}$ for every $\varepsilon \in (0, 1)$ and $n = 0, 1, \ldots$. One now sees as in the proof of Principle 32.1 (from (32.2) on) that for every $y \in V_\rho$ there exists an $x \in K_{1/(1-\varepsilon)}$ such that $Ax = y$, which proves the assertion. ■

Proposition 97.1. *The continuous linear map A from the Banach space E into the Banach space F is surjective if and only if A' possesses a continuous inverse.*

Proof. We assume that A' has a continuous inverse and show first that (with the notation of the above lemma)

$$V_\rho \subset \overline{A(K_1)} \quad \text{for} \quad \rho := \frac{1}{\|(A')^{-1}\|}. \tag{97.1}$$

To do this, we consider a y which lies in V_ρ but not in $\overline{A(K_1)}$. The set $A(K_1)$ is convex, being the linear image of a convex set (Exercise 2 in §90), hence by Lemma 90.3 also $\overline{A(K_1)}$ is convex. Consequently there exists a $y' \in F'$ such that $\operatorname{Re}\langle y, y'\rangle > \operatorname{Re}\langle z, y'\rangle$ for all $z \in \overline{A(K_1)}$ (Theorem 96.2), in particular $\operatorname{Re}\langle y, y'\rangle > \operatorname{Re}\langle Ax, y'\rangle$ for all $x \in K_1$. Let $\langle Ax, y'\rangle = re^{i\varphi}$, $r \geqq 0$. Since with x also $e^{-i\varphi}x$ lies in K_1, it follows that

$$\operatorname{Re}\langle y, y'\rangle > \operatorname{Re}\langle Ae^{-i\varphi}x, y'\rangle = \operatorname{Re} e^{-i\varphi}\langle Ax, y'\rangle = r = |\langle Ax, y'\rangle|,$$

from where we obtain

$$\begin{aligned} \rho\|y'\| &= \rho\|(A')^{-1}A'y'\| \leqq \rho\|(A')^{-1}\|\,\|A'y'\| = \|A'y'\| \\ &= \sup_{x \in K_1} |\langle x, A'y'\rangle| = \sup_{x \in K_1} |\langle Ax, y'\rangle| \leqq \operatorname{Re}\langle y, y'\rangle \leqq |\langle y, y'\rangle| \leqq \|y\|\,\|y'\|. \end{aligned}$$

Thus we have $\rho \leqq \|y\|$ in contradiction to our assumption. With the aid of Lemma 97.1 it follows from (97.1) that $V_\rho \subset A(K_1)$, from where we obtain immediately that $A(E) = F$. Now let A be surjective. If A' would not have a continuous inverse on $A'(F')$, then by Proposition 7.5 there would exist a sequence $(y_n') \subset F'$ such that $\|y_n'\| = 1$ and $\|A'y_n'\| \to 0$. If we set $\alpha_n := \max\{\sqrt{\|A'y_n'\|}, 1/\sqrt{n}\}$ and $z_n' := y_n'/\alpha_n$, then

$$\|z_n'\| \to \infty, \qquad \|A'z_n'\| \to 0, \tag{97.2}$$

consequently $\langle Ax, z_n'\rangle = \langle x, A'z_n'\rangle \to 0$ for every $x \in E$. Because of the surjectivity of A, we obtain from here, with the aid of Proposition 33.1, that $(\|z_n\|)$ is bounded, in contradiction to (97.2). ■

The counterpart to the proposition just proved is Proposition 41.3. Compare also with the proof of Proposition 36.5 and with Exercise 3 in §36.

Let A be a continuous endomorphism of the Banach space E. If $A(E)$ is closed, then it follows from Proposition 36.4 that also $A'(E')$ is closed. We now show that *the converse is also valid*. Let $A'(E')$ be closed. We set $F := \overline{A(E)}$, define $B \in \mathscr{L}(E, F)$ by $Bx := Ax$ for $x \in E$, and show that $B(E) = F$, which will prove that $A(E)$ is closed. Because of Proposition 97.1, we only have to verify that $B' \colon F' \to E'$ has a continuous inverse. From $\overline{B(E)} = F$ the injectivity of B' follows in a very simple fashion (see Exercise 1 in §29). Furthermore we obtain from the Hahn–Banach extension theorem that $B'(F') = A'(E')$, hence B' has a closed image space. With the aid of Lemma 36.1 we now see that $(B')^{-1}$ is continuous. ■

We can summarize this proposition and Propositions 29.3, 36.4 in:

Theorem 97.1. *Let A be a continuous endomorphism of the Banach space E. Then the following assertions are equivalent:*

(a) *$A(E)$ is closed.*
(b) *$A'(E')$ is closed.*
(c) *A is E'-normally solvable: $A(E) = \{y: \langle y, x'\rangle = 0$ for all $x' \in N(A')\}$.*
(d) *A' is E-normally solvable: $A'(E') = \{y': \langle x, y'\rangle = 0$ for all $x \in N(A)\}$.*

As the last application, we supply the *proof of Proposition* 62.3 for the case the closed subset $F \neq \varnothing$ of the reflexive Banach space E, which occurs there, is not linear but only convex. Obviously, the following hint is sufficient: The minimizing sequence $(y_n) \subset F$, which was introduced in the first part of the proof of Proposition 62.3, has a subsequence (y_{n_k}) which converges weakly to a vector y_0 of E; the assumption that y_0 does not lie in F leads immediately to a contradiction because of Theorem 96.2. ■

Exercises

1. Let E, F be Banach spaces. The image space of the compact operator $K: E \to F$ is closed if and only if K has finite rank. *Hint*: Consider the compact operator $K_1: E \to K(E)$ defined by $K_1 x := Kx$ for $x \in E$, and use the Propositions 97.1 and 42.2.
2. A compact endomorphism of a Banach space is relatively regular if and only if it has finite rank. *Hint*: Exercise 1.

§98. Admissible topologies

A locally convex topology τ on the vector space E is said to be *admissible with respect to the left dual system* (E, E^+) if E^+ is the space of all τ-continuous linear forms of E. Obviously $\sigma(E, E^+)$ is the coarsest admissible topology. The topology of a locally convex space E is, by definition, admissible with respect to (E, E').

The importance of admissible topologies is based on the fact that sets M in a locally convex space E may have topological properties which can be described *with the help of continuous linear forms alone*. If M has such a property in *one* admissible topology, for instance in the weak one, then it has the same property in *all of them*. We now give an example.

Proposition 98.1. *Let (E, E^+) be a left dual system. Then a convex set $K \subset E$ is closed either for all admissible topologies or for none of them.*

Proof. We may assume that E and E^+ are real, because convexity and closedness are not influenced when we pass to the real space E_r. Let K be non-empty and closed with respect to an admissible topology τ. A set of the form

$$\{x \in E\colon f(x) \leqq \alpha\}$$

or

$$\{x \in E\colon f(x) \geqq \alpha\},$$

where $f \neq 0$ is a τ-continuous linear form, is called a *closed half-space* determined by f and α. It follows from Theorem 96.2 that K is the intersection of all closed half-spaces $H \supset K$. From this representation of K it follows immediately that K is closed for all admissible topologies. ■

The aim of the following sections is to characterize the admissible topologies and to construct them as $\mathfrak{S}$-topologies. Our most important tool in these investigations will be the bipolar theorem to which we now turn.

Exercise

A closed linear manifold M in a locally convex space is the intersection of all closed hyperplanes $H \supset M$.

§99. The bipolar theorem

Let (E, E^+) be a linear system, $\mathfrak{S}$ a family of $\sigma(E, E^+)$-bounded (i.e., E^+-bounded) subsets of E (see Proposition 93.3). The typical neighborhood of zero U for the $\mathfrak{S}$-topology on E^+ (Example 83.6) can be written with the aid of certain sets $S_1, \ldots, S_n$ from $\mathfrak{S}$ in the following form:

$$\begin{aligned} U &= \left\{x^+ \in E^+ : \sup_{x \in S_\nu} |\langle x, x^+\rangle| \leqq \varepsilon,\ \nu = 1, \ldots, n\right\} \\ &= \varepsilon \bigcap_{\nu=1}^{n} \left\{x^+ \in E^+ : \sup_{x \in S_\nu} |\langle x, x^+\rangle| \leqq 1\right\}. \end{aligned}$$

With every subset M of E we associate its *polar*

$$M^0 := \left\{x^+ \in E^+ : \sup_{x \in M} |\langle x, x^+\rangle| \leqq 1\right\}$$

in E^+ (and correspondingly with every subset N of E^+ its *polar*

$$N^0 := \left\{x \in E\colon \sup_{x^+ \in N} |\langle x, x^+\rangle| \leqq 1\right\}$$

in E). With this, U assumes the form

$$U = \varepsilon \bigcap_{\nu=1}^{n} S_\nu^0.$$

If E is a normed space, $E^+ = E'$ and $S = K_1[0]$, then $S^0 = \{x' \in E' : \|x'\| \leqq 1\}$: *The polar of the closed unit ball of E is the closed unit ball of E'*. We now turn to the study of polar sets. We emphasize that the concept of polarity makes sense only in connection with a bilinear system; this must in principle always be given in advance.

$M^{00} := (M^0)^0$ is called the *bipolar* of M. For $M^{000} := (M^{00})^0$ we do not introduce a special name.

Proposition 99.1. *Let (E, E^+) be a bilinear system and M, M_ι subsets of E. Then the following assertions hold:*

(a) *From $M_1 \subset M_2$ it follows that $M_1^0 \supset M_2^0$.*
(b) $M \subset M^{00}$.
(c) $M^0 = M^{000}$.
(d) $(\alpha M)^0 = (1/\alpha)M^0$ for $\alpha \neq 0$.
(e) $(\bigcup_{\iota \in J} M_\iota)^0 = \bigcap_{\iota \in J} M_\iota^0$.
(f) *M^0 is absolutely convex.*
(g) *For the balanced hull D of M we have $D^0 = M^0$.*
(h) *M^0 is $\sigma(E^+, E)$-closed.*
(i) *M^0 is absorbing if and only if M is $\sigma(E, E^+)$-bounded.*
(j) *For linear subspaces M we have $M^0 = M^\perp$.*

Proof. (a), (b), (d), (e), (f) and (j) are easy to see; (c) is proved similarly as the equation $M^\perp = M^{\perp\perp\perp}$ (see Lemma 29.3). (g) Because of $M \subset D$, we have $D^0 \subset M^0$ by (a). The converse inclusion follows from Lemma 84.1e with the aid of (d) and (e). (h) For every $x \in E$ the linear form $x^+ \mapsto \langle x, x^+ \rangle$ is weakly continuous on E^+, hence $N_x := \{x^+ \in E^+ : |\langle x, x^+ \rangle| \leqq 1\}$ and thus also $M^0 = \bigcap_{x \in M} N_x$ is weakly closed. (i) Let M be weakly bounded. Then by Proposition 93.3 for every $x^+ \in E^+$ there exists an $\alpha > 0$ such that $\sup_{x \in M} |\langle x, x^+ \rangle| \leqq \alpha$, from where it follows immediately that $x^+ \in \alpha M^0$: the set M^0 is absorbing. The implications are reversible. ■

The following principle, the so-called *bipolar theorem*, is the central theorem of the theory of polarity. We first introduce a definition:

If M is a subset of the vector space E, then the intersection of all absolutely convex sets $N \subset E$ which contain M is itself absolutely convex and is called the *absolutely convex hull* of M. If E is a topological vector space, then the intersection of all absolutely convex, closed sets $N \subset E$ which contain M is called the *absolutely convex, closed hull* of M; it is absolutely convex and closed.

Principle 99.1. *Let (E, E^+) be a bilinear system. Then the bipolar M^{00} of a non-empty subset M of E is the* absolutely convex, $\sigma(E, E^+)$-closed hull *of M.*

Proof. Let H be this hull. From Proposition 99.1b, f and h, it follows that $H \subset M^{00}$. Thus we only have to prove that $y \notin H$ implies $y \notin M^{00}$. By Theorem

96.2 there exists a $\sigma(E, E^+)$-continuous linear form f and a real α such that $\operatorname{Re} f(x) < \alpha < \operatorname{Re} f(y)$ for all $x \in H$. Because of $0 \in H$ we have $0 = f(0) < \alpha$; for the $\sigma(E, E^+)$-continuous linear form $g := (1/\alpha)f$ we have therefore

$$\text{(99.1)} \qquad \operatorname{Re} g(x) < 1 < \operatorname{Re} g(y), \qquad x \in H.$$

If we set $g(x) = re^{i\varphi}$ $(r \geqq 0)$ and observe that together with x also $e^{-i\varphi}x$ lies in H, then it follows from (99.1) that

$$\text{(99.2)} \qquad |g(x)| = r = g(e^{-i\varphi}x) < 1 \qquad \text{for all} \quad x \in H.$$

By Theorem 82.1 there exists an $x^+ \in E^+$ such that $g(x) = \langle x, x^+ \rangle$ for every $x \in E$. From (99.2) it now follows that $x^+ \in H^0$, and since because of $M \subset H$ by Proposition 99.1a we have $H^0 \subset M^0$, a fortiori $x^+ \in M^0$, and thus $|\langle x, x^+ \rangle| \leqq 1$ for all $x \in M^{00}$. Because of $|\langle y, x^+ \rangle| \geqq \operatorname{Re} \langle y, x^+ \rangle = \operatorname{Re} g(y) > 1$ (see (99.1)), we obtain from here that $y \notin M^{00}$. ∎

The following frequently used result is only a special case of the bipolar theorem:

Proposition 99.2. *Let (E, E^+) be a bilinear system and M a linear subspace or only a non-empty,* absolutely convex *subset of E. Then M^{00} is the $\sigma(E, E^+)$-*closure *of M.*

Exercises

$^+$1. Let (E, E^+) be a bilinear system and F a linear subspace of E. Then $F^\perp$ is $\sigma(E^+, E)$-closed, and $F^{\perp\perp}$ is the $\sigma(E, E^+)$-closure of F.

2. Let (E, E^+) be a bilinear system and $(M_\iota : \iota \in J)$ a family of non-empty, absolutely convex and $\sigma(E, E^+)$-closed subsets of E. Then $(\bigcap_{\iota \in J} M_\iota)^0$ is the absolutely convex, $\sigma(E^+, E)$-closed hull of $\bigcup_{\iota \in J} M_\iota^0$.

3. The absolutely convex hull of a set M is the convex hull of the balanced hull of M.

4. Let (E, E^+) be a dual system and A an E^+-conjugable endomorphism of E. Then for $M \subset E$, $N \subset E^+$ we have

$$(A(M))^0 = (A^+)^{-1}(M^0) \qquad \text{and} \qquad (A^+(N))^0 = A^{-1}(N^0).$$

§100. Locally convex topologies are $\mathfrak{S}$-topologies

Let (E, E^+) be a bilinear system and $\mathfrak{S}$ a collection of $\sigma(E^+, E)$-bounded subsets of E^+. Then the sets U of the form

$$\text{(100.1)} \qquad U := \varepsilon \bigcap_{\nu=1}^{n} S_\nu^0, \qquad \text{where} \quad \varepsilon > 0, S_\nu \in \mathfrak{S} \quad \text{for} \quad \nu = 1, \ldots, n$$

form a basis of neighborhoods of zero for the $\mathfrak{S}$-topology on E (see beginning of §99). We now raise the following question: Can a given locally convex topology τ on E be represented as an $\mathfrak{S}$-topology, where $\mathfrak{S}$ is a system of $\sigma(E', E)$-bounded subsets of the τ-dual E' of E?

We assume that such a system $\mathfrak{S}$ exists. Then for $\varepsilon > 0$ and $S \in \mathfrak{S}$ the set

$$U_\varepsilon := \varepsilon S^0 = \left\{x \in E : \sup_{x' \in S} |\langle x, x'\rangle| \leqq \varepsilon\right\}$$

is a τ-neighborhood of zero in E. For an arbitrary $\varepsilon > 0$ there exists thus a neighborhood of zero U (namely $U := U_\varepsilon$) such that

(100.2) $\quad |x'(x)| = |\langle x, x'\rangle| \leqq \varepsilon \quad$ for all $\quad x \in U \quad$ and *all* $\quad x' \in S$.

This situation will remind the reader of the concept of equicontinuity. As is well known, a family $\mathscr{F}$ of functions $f: [a, b] \to \mathbf{K}$ is *equicontinuous at the point* $x_0 \in [a, b]$ if for every $\varepsilon > 0$ there exists a $\delta > 0$ such that

$$|f(x) - f(x_0)| \leqq \varepsilon \quad \text{for} \quad |x - x_0| \leqq \delta \quad \text{and } all \quad f \in \mathscr{F}.$$

If E, F are topological vector spaces, then a family $\mathscr{F}$ of maps $A: E \to F$ is said to be *equicontinuous at the point* $x_0 \in E$ if for every neighborhood of zero (or only for every neighborhood of zero belonging to a basis of neighborhoods of zero) V in F there exists a neighborhood of zero U in E such that

$$Ax - Ax_0 \in V \quad \text{for} \quad x - x_0 \in U \quad \text{and } all \quad A \in \mathscr{F}.$$

The family will be called shortly *equicontinuous* if it is equicontinuous at every point of E. *Obviously a family of linear maps is equicontinuous if and only if it is equicontinuous at the point zero.* Thus we obtain from (100.2) the assertion: The system $\mathfrak{G}$, which generates the topology τ, consists of equicontinuous subsets of E'. We will now prove the following theorem:

Theorem 100.1. *The topology τ of the locally convex space E is generated by the system $\mathfrak{G}$ of* all *equicontinuous subsets of E'.*

The proof will be preceded by two lemmas. The first one shows essentially that $\mathfrak{G}$ generates a topology, because the sets in $\mathfrak{G}$ are all $\sigma(E', E)$-bounded.

Lemma 100.1. *Let E be locally convex. $S \subset E'$ is equicontinuous if and only if there exists a continuous semi-norm p on E such that*

(100.3) $\quad |f(x)| \leqq p(x) \quad$ *for all* $\quad x \in E \quad$ *and all* $\quad f \in S$.

In this case S is also $\sigma(E', E)$-bounded.

The first assertion is proved similarly as Proposition 91.4. The second assertion follows from (100.3) in combination with Proposition 93.3, since (100.3) shows that S is E-bounded. ■

Lemma 100.2. *Let E be a topological vector space. The set $S \subset E'$ is equicontinuous if and only if there exists a neighborhood of zero U in E such that $S \subset U^0$.*

Proof. If S is equicontinuous, then there exists a neighborhood of zero U in E such that $|\langle x, x'\rangle| \leqq 1$ for all $x \in U$ and all $x' \in S$, thus we have $S \subset U^0$. Now let $S \subset U^0$ for a certain neighborhood of zero U in E. For an arbitrary $\varepsilon > 0$ we have then $|\langle \varepsilon x, x'\rangle| \leqq \varepsilon$ for all $x \in U$ and $x' \in S$, or expressed differently: We have $|\langle y, x'\rangle| \leqq \varepsilon$ for all y in the neighborhood of zero εU and all $x' \in S$. Consequently S is equicontinuous. ■

We now turn to the proof of Theorem 100.1.

By Proposition 90.1 the set $\mathfrak{N}$ of all absolutely convex, closed neighborhoods of zero of E is a basis of τ-neighborhoods of zero. Because of Proposition 98.1, every $U \in \mathfrak{N}$ is also $\sigma(E, E')$-closed, hence we have $U^{00} = U$ (Proposition 99.2). Because of the equicontinuity of U^0 (Lemma 100.2) it follows that U is a neighborhood of zero in the $\mathfrak{G}$-topology γ and so $\tau \prec \gamma$. Now let $V := \varepsilon \bigcap_{\nu=1}^{n} S_\nu^0$ ($\varepsilon > 0$, $S_\nu \subset E'$ equicontinuous) be a γ-neighborhood of zero belonging to the canonical basis. By Lemma 100.2 and Proposition 90.1 there exists τ-neighborhoods of zero W_ν and absolutely convex, closed τ-neighborhoods of zero U_ν such that $S_\nu \subset W_\nu^0$ and $U_\nu \subset W_\nu$. Because of $U_\nu^{00} = U_\nu$ (see the first part of the proof), it follows first, with the aid of Proposition 99.1a, that $S_\nu \subset U_\nu^0$, then $S_\nu^0 \supset U_\nu$, and finally that $V \supset \varepsilon \bigcap_{\nu=1}^{n} U_\nu$. Consequently $\gamma \prec \tau$. ■

Exercises

1. Let $(f_\iota : \iota \in J)$ be an equicontinuous family of linear forms on the subspace F of the locally convex space E. Then there exists an equicontinuous family $(g_\iota : \iota \in J)$ of linear forms on E such that $g_\iota(x) = f_\iota(x)$ for all $x \in F$ and all $\iota \in J$.

*2. Let E, F be normed spaces, $A_1 \in \mathscr{L}(E, F)$ for $\iota \in J$. The family $(A_\iota : \iota \in J)$ is equicontinuous if and only if the family $(\|A_\iota\| : \iota \in J)$ is bounded.

§101. Compact sets

We first recall some concepts and facts from general topology.

Let M be a subset of the topological space E. Every system $\mathfrak{G}$ of open sets $G \subset E$ such that $M \subset \bigcup_{G \in \mathfrak{G}} G$ is called an *open cover* of M; every subsystem $\mathfrak{G}' \subset \mathfrak{G}$, which also covers M, is called a *subcover* of $\mathfrak{G}$. The set M is said to be *compact* if every open cover of M contains a finite subcover. Every subset of a compact set is said to be *relatively compact*.

The subset M of a *metric* space is compact or relatively compact, respectively, in the sense just defined if and only if it is (sequentially) compact or (sequentially) relatively compact, respectively, according to the definition given earlier in §10.

Finite sets and finite unions of compact sets are compact. Every closed subset of a compact set is compact. Compact subsets of *separated* topological spaces are closed, in such spaces a subset M is relatively compact if and only if $\overline{M}$ is compact. Compactness is an 'intrinsic property', more precisely: The set $M \subset E$ is compact if and only if it is compact for its relative topology.

Let $\mathfrak{B}$ be a system of non-empty, closed subsets of a compact set, which is totally ordered with respect to set-theoretical inclusion. Then $\bigcap_{M \in \mathfrak{B}} M \neq \emptyset$.

Of basic importance is the *theorem of Tikhonov*: *The topological product of compact topological spaces is compact.*

Finally, we shall use the fact that *continuous images of compact sets are compact*, that consequently compactness is preserved when the topology becomes coarser, and a *real-valued, continuous function on a compact topological space has a maximum and a minimum.* We now turn to compact and relatively compact sets in topological vector spaces.

Proposition 101.1. *Every compact subset, hence a fortiori every relatively compact subset of a topological vector space E is bounded.*

Proof. Let $M \subset E$ be compact, U a neighborhood of zero, and V a balanced, open neighborhood of zero such that $V + V \subset U$ (Theorem 84.1). Since $x + V$ is open and $M \subset \bigcup_{x \in M} (x + V)$, there exist finitely many vectors $x_1, \ldots, x_n$ in M such that $M \subset \bigcup_{\nu=1}^{n} (x_\nu + V)$. Furthermore, since V is absorbing, there exists an $\alpha \geqq 1$ such that $x_\nu \in \alpha V$ for $\nu = 1, \ldots, n$. Because V is balanced, we have $(1/\alpha)V \subset V$, and so we obtain finally

$$M \subset \bigcup_{\nu=1}^{n} (x_\nu + V) \subset \alpha V + V = \alpha\left\{v_1 + \frac{v_2}{\alpha} : v_1, v_2 \in V\right\} \subset \alpha(V + V) \subset \alpha U.$$

■

Proposition 101.1 cannot be reversed. We have, however:

Proposition 101.2. *If we equip the algebraic dual E^* of the vector space E with the weak topology $\sigma(E^*, E)$, then a subset of E^* is relatively compact if and only if it is bounded.*

Because of Proposition 101.1, we only have to show that a bounded set $M \subset E^*$ is relatively compact. If $\{x_\iota : \iota \in J\}$ is an algebraic basis of E, then the spaces E^* and $\mathbf{K}^J := \prod_{\iota \in J} \mathbf{K}_\iota$ ($\mathbf{K}_\iota := \mathbf{K}$ for all $\iota \in J$) are isomorphic to each other by means of the map $x^* \mapsto (\langle x_\iota, x^* \rangle : \iota \in J)$ (Exercise 9 in §15), hence they can be identified. Thus E^* carries in a natural way two locally convex topologies, namely the weak topology $\sigma := \sigma(E^*, E)$ and the product topology τ of $\mathbf{K}^J$. We now show that these two topologies coincide. For this purpose, we first prove the relation $\sigma \prec \tau$ by showing that every $x \in E$ is a τ-continuous linear form on E^*. Let $\varepsilon > 0$ be given arbitrarily. We represent x in the form $x = \sum_{\iota \in J_0} \xi_\iota x_\iota$, where J_0 is a subset of J having n elements and $\xi_\iota \neq 0$ for $\iota \in J_0$. Then

$$U := \left\{(\alpha_\iota : \iota \in J) : |\alpha_\iota| \leqq \frac{\varepsilon}{n|\xi_\iota|} \quad \text{for} \quad \iota \in J_0\right\}$$

is a τ-neighborhood of zero. For every linear form $x^* = (\alpha_\iota : \iota \in J)$ in U we have

$$|x(x^*)| = |\langle x, x^* \rangle| = \left| \sum_{\iota \in J_0} \alpha_\iota \xi_\iota \right| \leqq \sum_{\iota \in J_0} |\alpha_\iota||\xi_\iota| \leqq \varepsilon,$$

which proves the τ-continuity of x. Now, in order to show that $\tau \prec \sigma$, let $U := \{(\alpha_\iota): |\alpha_{\iota_\nu}| \leqq \varepsilon$ for $\nu = 1, \ldots, n\}$ be a τ-neighborhood of zero in E^*. Because of $U_{x_{\iota_1}, \ldots, x_{\iota_n}; \varepsilon} = \{x^* \in E^*: |\langle x_{\iota_\nu}, x^* \rangle| \leqq \varepsilon$ for $\nu = 1, \ldots, n\} = U$, we see that U is also a σ-neighborhood of zero, thus τ is indeed coarser than σ. Because of the σ-boundedness of M, and by Proposition 93.3, the sets $M_\iota := \{x_\iota(x^*): x^* \in M\} \subset \mathbf{K}$ are bounded, hence their closures $\overline{M}_\iota$ are compact. Tikhonov's theorem now shows that $\prod_{\iota \in J} \overline{M}_\iota$ is τ-compact, hence by what has been proved above, also σ-compact. Thus M as a subset of $\prod_{\iota \in J} \overline{M}_\iota$ is relatively compact. ∎

Proposition 101.3. *If $K_1, \ldots, K_n$ are compact and convex (absolutely convex) subsets of a topological vector space, then also the convex (absolutely convex) hull of $\bigcup_{\nu=1}^n K_\nu$ is compact.*

Proof. Let the sets K_ν be convex. The collection A of all vectors $a := (\alpha_1, \ldots, \alpha_n) \in \mathbf{K}^n$ with $\alpha_\nu \geqq 0$ $(\nu = 1, \ldots, n)$, $\alpha_1 + \cdots + \alpha_n = 1$ is compact; thus by Tikhonov's theorem the product $A \times K_1 \times \cdots \times K_n$ is compact, hence also its image under the continuous map $(a, x_1, \ldots, x_n) \mapsto \sum_{\nu=1}^n \alpha_\nu x_\nu$. But this image is $\operatorname{co}(\bigcup_{\nu=1}^n K_\nu)$ (Exercise 3 in §90). Now let the K_ν be even absolutely convex. Then $\operatorname{co}(\bigcup_{\nu=1}^n K_\nu)$ is balanced, hence it is already the absolutely convex hull of $\bigcup_{\nu=1}^n K_\nu$ and so it is compact by what has been just proved. ∎

Exercises

1. The balanced hull of a compact set in a topological vector space is compact.

2. Let K be a compact, A a closed subset of a topological vector space and let $K \cap A = \varnothing$. Then there exists a neighborhood of zero V such that $(K + V) \cap (A + V) = \varnothing$.

3. If K is a compact and A a closed subset of a topological vector space, then $K + A$ is closed. *Hint*: Exercise 2.

+4. The subset M of the topological vector space E is said to be *totally bounded* or *precompact* if for every neighborhood of zero U there exist finitely many points $x_1, \ldots, x_n$ in M so that $M \subset \bigcup_{i=1}^n (x_i + U)$ (cf. Exercise 3 in §42). Show the following:
(a) Every compact subset of E is precompact.
(b) Precompact subsets of E are bounded.
(c) Subsets and finite unions of precompact sets are precompact.
(d) The continuous linear image of a precompact set is precompact.

5. A subset $M := \{(\xi_1^{(\iota)}, \xi_2^{(\iota)}, \ldots): \iota \in J\}$ of (s) is relatively compact if and only if for every index k the set $\{\xi_k^{(\iota)}: \iota \in J\}$ is bounded. *Hint*: Diagonal process. The map $(\xi_1, \xi_2, \ldots) \mapsto \xi_k$ is continuous.

*6. Let E be a topological space. $y \in E$ is called a *limit point* of the sequence $(x_n) \subset E$ if in every neighborhood U of y there lie *infinitely* many terms of the sequence (i.e., if for every neighborhood U of y there exist infinitely many indices $n_1 < n_2 < \cdots$ such that $x_{n_k} \in U$ for $k = 1, 2, \ldots$). Show that every sequence from a compact set $K \subset E$ has at least one limit point in K. *Hint*: Proof by contradiction.

§102. The Alaoglu–Bourbaki theorem

It reads as follows:

Theorem 102.1. *Every equicontinuous subset of the dual E' of a topological vector space E is relatively $\sigma(E', E)$-compact.*

Because of Lemma 100.2, this theorem is proved when we show that the following assertion is valid:

Theorem 102.2. *Let E be a topological vector space. Then the polar U^0 formed in E' of a neighborhood of zero $U \subset E$ is $\sigma(E', E)$-compact.*

Proof. Together with the bilinear system (E, E'), we consider also the dual system (E, E^*), and denote by U^p the polar of U formed in E^*. Since U is absorbing and $U \subset U^{pp}$ (Proposition 99.1b), also U^{pp} is absorbing. Thus because of Proposition 99.1i the set U^p is certainly $\sigma(E^*, E)$-bounded and thus even relatively $\sigma(E^*, E)$-compact (Proposition 101.2). But since U^p is also $\sigma(E^*, E)$-closed by Proposition 99.1h, and since $\sigma(E^*, E)$ is furthermore separated, we now obtain the $\sigma(E^*, E)$-compactness of U^p. Obviously $U^0 \subset U^p$; but we also have $U^p \subset U^0$. Indeed, for $x^* \in U^p$ it follows from $x \in \varepsilon U$ that $|\langle x, x^* \rangle| \leqq \varepsilon$, whatever $\varepsilon > 0$ is; consequently x^* is continuous and lies therefore in U^0. Thus the polar U^0, since it coincides with U^p, is $\sigma(E^*, E)$-compact. Now $\sigma(E', E)$ is the topology induced by $\sigma(E^*, E)$ on E', so that U^0 must be also $\sigma(E', E)$-compact. ■

The polar in E' of the closed unit ball in the normed space E is the closed unit ball in E' (§99). Thus we obtain from Theorem 102.2:

Proposition 102.1. *The closed unit ball in the normed dual E' of the normed space E is $\sigma(E', E)$-compact.*

This proposition is the basis of the following *representation theorem*:

Proposition 102.2. *Let E be a* normed *space over* **K**. *We equip the closed unit ball K' of E' with the topology induced by $\sigma(E', E)$; by Proposition* 102.1

the set K' is compact. Let $C(K')$ be the vector space of all continuous functions $f: K' \to \mathbf{K}$ equipped with the norm $\|f\| := \sup_{x' \in K'} |f(x')|$. Then E is isomorphic as a normed space to a subspace of $C(K')$.

Proof. For every $x \in E$ we define the $\sigma(E', E)$-continuous linear form F_x on E' by $F_x(x') := \langle x, x' \rangle$. The restriction f_x of F_x to K' belongs then to $C(K')$. The map $x \mapsto f_x$ is obviously linear and preserves the norm:

$$\|f_x\| = \sup_{x' \in K'} |f_x(x')| = \sup_{\|x'\| \leq 1} |F_x(x')| = \|F_x\| = \|x\| \text{ (see (41.3))}. \qquad \blacksquare$$

In an abbreviated form, Proposition 102.2 states that every normed space can be considered as a subspace of a space of continuous functions, which are defined on a compact set and equipped with the supremum-norm.

§103. The characterization of the admissible topologies

Let τ be an admissible topology on E with respect to the *left dual system* (E, E^+). By Theorem 100.1 the sets of the form $\varepsilon \bigcap_{\nu=1}^{n} S_\nu^0$ ($\varepsilon > 0$, S_ν a τ-equicontinuous subset of E^+) form a basis of neighborhoods of zero for τ. With the aid of Lemma 100.2 and Proposition 99.1a, b, one sees immediately that already the neighborhoods $\varepsilon \bigcap_{\nu=1}^{n} U_\nu^{00}$ ($\varepsilon > 0$, U_ν a τ-neighborhood of zero) form a basis of neighborhoods of zero, hence that τ is generated by the system $\mathfrak{G}_0 := \{U^0 : U$ is a τ-neighborhood of zero$\}$. U^0 is absolutely convex by Proposition 99.1f and by Theorem 102.2 it is $\sigma(E^+, E)$-compact. It follows that the system $\mathfrak{M}$ of *all* absolutely convex and $\sigma(E^+, E)$-compact subsets of E^+ generates a topology on E which is finer than any topology admissible with respect to (E, E^+). It is called the *Mackey topology* and will be denoted by $\tau(E, E^+)$.

We now show that $\tau(E, E^+)$ is admissible with respect to (E, E^+), whereby it will be proved that the *Mackey topology is the* finest *admissible topology* (while the weak topology $\sigma(E, E^+)$ is the *coarsest* admissible topology). For this purpose we prove first

Lemma 103.1. *The polars of the absolutely convex, $\sigma(E^+, E)$-compact subsets of E^+ form a basis of neighborhoods of zero for $\tau(E, E^+)$.*

Proof. The typical neighborhood of zero, belonging to the basis by which the Mackey topology has been defined, is $\varepsilon \bigcap_{\nu=1}^{n} K_\nu^0$ ($\varepsilon > 0$, $K_\nu \subset E^+$ absolutely convex and $\sigma(E^+, E)$-compact). The absolutely convex hull H of $\bigcup_{\nu=1}^{n} K_\nu$ is by Proposition 101.3 again $\sigma(E^+, E)$-compact, hence the same is true also for the absolutely convex set $K := (1/\varepsilon)H$. The assertion follows now from the inclusions obtained from Proposition 99.1:

$$\varepsilon \bigcap_{\nu=1}^{n} K_\nu^0 = \varepsilon\left(\bigcup_{\nu=1}^{n} K_\nu\right)^0 \supset \varepsilon H^0 = \left(\frac{1}{\varepsilon} H\right)^0 = K^0. \qquad \blacksquare$$

For the proof of the admissibility of $\tau(E, E^+)$ we remind the reader that E^+ is a linear subspace of E^*, hence every subset of E^+ is also a subset of E^*. Polars will be formed in the dual system (E, E^*). Let now x'_0 be a $\tau(E, E^+)$-continuous linear form on E. By Lemma 103.1 there exists an absolutely convex, $\sigma(E^+, E)$-compact subset K of E^+ so that $|\langle x, x'_0\rangle| \leqq 1$ for $x \in K^0$. Consequently x'_0 lies in K^{00}. Since $\sigma(E^+, E)$ is the topology induced by $\sigma(E^*, E)$, we can assert the $\sigma(E^*, E)$-compactness, hence also the $\sigma(E^*, E)$-closedness of the absolutely convex subset K of E^*. From Proposition 99.2 it follows that $K^{00} = K$, consequently x'_0 lies in K, hence also in E^+, i.e., the $\tau(E, E^+)$-dual of E is contained in E^+. The converse inclusion follows very simply from the fact that an $x^+ \in E^+$ is always $\sigma(E, E^+)$-continuous, hence a fortiori $\tau(E, E^+)$-continuous. ■

We summarize our essential results concerning admissible topologies in the *Mackey–Arens theorem*:

Theorem 103.1. *A locally convex topology on E is admissible with respect to the left dual system (E, E^+) if and only if it is finer than $\sigma(E, E^+)$ and coarser than $\tau(E, E^+)$.*

The *strong topology* $\beta(E, E^+)$ as the finest $\mathfrak{S}$-topology on E is finer than $\tau(E, E^+)$ (see Example 83.6 in combination with Proposition 93.3). *It is in general not admissible.* If for instance F is a normed space, then $\beta(F', F)$ is the topology v defined by the norm on F' (see end of Example 83.6); the dual of (F', v)—i.e., the bidual F''—does in general not coincide with F.

Exercise

$^+$Let (E, E^+) be a dual system. Then every E^+-conjugable endomorphism of E is $\tau(E, E^+)$-continuous. *Hint*: Let V be an absolutely convex, closed neighborhood of zero in E, where E is equipped with the topology $\tau(E, E^+)$. Then V^0 is $\sigma(E^+, E)$-compact. Apply now Propositions 82.1, 82.2, the definition of the Mackey topology, and Exercise 4 in §99.

§104. Bounded sets in admissible topologies

It follows from Propositions 41.4 and 93.3 that the weakly bounded subsets of a normed space are even bounded. We shall now generalize this assertion to separated locally convex spaces. We begin with a preliminary consideration.

Let V be a non-empty, absolutely convex subset of a vector space E. It follows from Lemmas 84.1b and 90.1d that $[V] = \bigcup_{n=1}^{\infty} nV$. Consequently V is absorbing in $[V]$. By Proposition 91.1 the Minkowski functional p of V defined on $[V]$ is a semi-norm. Let E_V be the vector space $[V]$ equipped with the locally convex topology $\tau(p)$ generated by p. By Example 83.3 the semi-norm p is continuous with respect to $\tau(p)$, consequently V is a $\tau(p)$-neighborhood of zero (Proposition 91.2) and with the aid of Proposition 91.1 it follows now that $\{\varepsilon V : \varepsilon > 0\}$ is a basis of the $\tau(p)$-neighborhoods of zero in E_V.

Now let E be equipped with an additional separated vector space topology τ and let V be τ-bounded. Then for every τ-neighborhood of zero W in E there exists an $\varepsilon > 0$ such that $\varepsilon V \subset W$. It follows that $\tau(p)$ is finer than the topology induced by τ on $[V]$. Since the latter is Hausdorff (Exercise 5 in 81), a fortiori $\tau(p)$ must be Hausdorff, hence p is a norm (Theorem 91.1).

Now let us assume that V is even τ-compact. Then the normed space E_V is complete. To see this, let us consider a Cauchy sequence (x_n) in E_V. For every $\varepsilon > 0$ there exists an index $n_0 = n_0(\varepsilon)$ such that

$$(104.1) \qquad x_n \in x_m + \varepsilon V \qquad \text{for all} \quad n, m \geqq n_0.$$

Without restricting the generality we assume that $(x_n) \subset V$. Because of the τ-compactness of V the sequence (x_n) has a τ-limit point $x_0 \in V$ (Exercise 6 in §101). Since $x_m + \varepsilon V$ is τ-closed, it follows from (104.1) that $x_0 \in x_m + \varepsilon V$ for all $m \geqq n_0$, and from here we obtain that $x_m \to x_0$ for the topology defined by the norm on E_V. We state this result:

Lemma 104.1. *Every non-empty, absolutely convex and compact subset V of a separated topological vector space E generates a Banach space E_V, whose construction is described above. The topology given by the norm on E_V is finer than the topology induced by E.*

We can now prove the theorem announced at the outset:

Theorem 104.1. (Theorem of Mackey).

(a) *If (E, E^+) is a dual system, then all admissible topologies on E yield the same bounded sets.*

(b) *A subset of a separated locally convex space E is bounded if and only if it is $\sigma(E, E')$-bounded.*

Proof. Since the assertions (a) and (b) are equivalent, it is sufficient to prove (b). For this purpose we only have to show that every $\sigma(E, E')$-bounded set is also bounded for the topology τ of E. Let U be a τ-closed, absolutely convex neighborhood of zero in E. Then U is also $\sigma(E, E')$-closed (Proposition 98.1). Thus we obtain from Proposition 99.2 that $U = U^{00}$. Furthermore U^0 is absolutely convex and $\sigma(E', E)$-compact (Proposition 99.1 and Theorem 102.2), so that $(E')_{U^0}$ is a Banach space (Lemma 104.1).

Now let $M \subset E$ be $\sigma(E, E')$-bounded. By Proposition 93.3 every $x' \in E'$ is bounded on M, in particular we have

$$(104.2) \qquad \sup_{x \in M} |\langle x, x' \rangle| < \infty \qquad \text{for all} \quad x' \in (E')_{U^0}.$$

The linear form x—i.e., the map $x' \mapsto \langle x, x' \rangle$—is $\sigma(E', E)$-continuous; its restriction φ_x to $(E')_{U^0}$ is therefore continuous in the relative weak topology and a fortiori for the finer topology on $(E')_{U^0}$ defined by the norm (Lemma 104.1). Hence, for every $x \in E$ the linear form φ_x belongs to the dual of $(E')_{U^0}$. We conclude from (104.2) that the family $(\varphi_x : x \in M)$ is pointwise bounded on the Banach space $(E')_{U^0}$. Principle 33.2 now ensures the existence of a positive constant γ so that

$$\|\varphi_x\| = \sup_{x' \in U^0} |\varphi_x(x')| = \sup_{x' \in U^0} |\langle x, x' \rangle| \leqq \gamma \qquad \text{for all} \quad x \in M.$$

From here it follows immediately that $M \subset \gamma U^{00} = \gamma U$, i.e., that M is indeed τ-bounded. ∎

Because of the theorem of Mackey, we are entitled to speak simply of the *bounded* sets of a separated locally convex space without specifying the (admissible) topology.

A simple consequence of Theorem 104.1 is:

Proposition 104.1. *The topology of a metric locally convex vector space E is the Mackey topology $\tau(E, E')$.*

Proof. The balls $U_n := K_{1/n}(0)$ form a basis of neighborhoods of zero for the metric topology τ of E. Because of the Mackey–Arens theorem (Theorem 103.1) we only need to show that τ is finer than $\tau(E, E')$, i.e., that every absolutely convex $\tau(E, E')$-neighborhood of zero is also a τ-neighborhood of zero (Proposition 84.4). Let us assume that this is not the case. Then there exists an absolutely convex $\tau(E, E')$-neighborhood of zero V, which is not a τ-neighborhood of zero. Consequently no nV $(n = 1, 2, \ldots)$ is a τ-neighborhood of zero, hence for every n there exists an x_n which lies in U_n but not in nV. Thus (x_n) converges to zero and is thus (topologically) bounded. By Theorem 104.1 the sequence (x_n) is also $\tau(E, E')$-bounded, hence there exists an $\alpha > 0$ such that $x_n \in \alpha V$ for $n = 1, 2, \ldots$. It follows that $x_n \in nV$ for all $n \geqq \alpha$ (Lemma 84.1b), in contradiction to the construction of (x_n). ∎

Exercises

1. Where have we used in the proof of Lemma 104.1 that E is Hausdorff?
2. All admissible topologies on (s) are identical.

§105. Barrelled spaces. Reflexivity

The strong topology $\beta(E, E')$ on a locally convex space E is an $\mathfrak{S}$-topology, where $\mathfrak{S}$ consists of all $\sigma(E', E)$-bounded subsets of E' (Example 83.6 in combination with Proposition 93.3). It is finer than the Mackey topology $\tau(E, E')$ and can be different from it, hence it might not be admissible with respect to the dual system (E, E') (see end of §103). In this section we shall give, among others, conditions under which the strong topology is nevertheless admissible.

An absolutely convex, closed and absorbing subset T of E is called a *barrel*; because of Proposition 98.1 this concept is independent of the particular admissible topology. E possesses a basis of neighborhoods of zero which consists of barrels (Propositions 90.1 and 84.2a). If every barrel is a neighborhood of zero, i.e., if the system of all barrels forms a basis of neighborhoods of zero in E, then E is said to be *barrelled.*

Lemma 105.1. *The barrels of a locally convex space E have $\sigma(E', E)$-bounded polars. They are themselves exactly the polars of the $\sigma(E', E)$-bounded subsets of E'.*

Proof. For a barrel $T \subset E$ we have $T^{00} = T$ (Proposition 99.2), and since T is absorbing, it follows with the aid of Proposition 99.1i that T^0 is $\sigma(E', E)$-bounded. Thus one direction of the second assertion is proved. Now let M be a $\sigma(E', E)$-bounded subset of E'. Then M^0 is according to Proposition 99.1f, h, i a barrel in E. ∎

From this Lemma and from the Mackey–Arens theorem (Theorem 103.1) we obtain immediately:

Proposition 105.1. *The barrels of a locally convex space E form a basis of neighborhoods of zero for the* strong *topology $\beta(E, E')$. Thus the strong topology coincides with the topology τ of E if and only if E is barrelled. In this case $\tau = \tau(E, E') = \beta(E, E')$, in particular $\beta(E, E')$ is* admissible.

A complete, metric vector space, whose topology is locally convex, is called a *Fréchet space* (*F-space*). Banach spaces are Fréchet spaces. The following important result holds:

Theorem 105.1. *Every Fréchet space is barrelled.*

Proof. Let T be a barrel in the Fréchet space E. Since T is absorbing and closed, we have the representation $E = \bigcup_{n=1}^{\infty} nT$, where the sets nT are closed. By the Baire category theorem (Principle 31.1) at least one nT, say mT, contains a ball; let its center be mx_0, $x_0 \in T$. It follows that a neighborhood of x_0 lies in T, hence there exists a balanced neighborhood U of zero so that $x_0 + U \subset T$. Since also T is balanced, this inclusion implies $-x_0 + U \subset T$. For every $x \in U$ we then have $x = (x_0 + x)/2 + (-x_0 + x)/2 \in \frac{1}{2}T + \frac{1}{2}T = T$ (Lemma 90.1d) and thus $U \subset T$. Consequently T is a neighborhood of zero. ∎

Because of Exercise 2 in §100, we can formulate the principle of uniform boundedness (Principle 33.2) as follows: *Every* pointwise *bounded family of continuous linear maps from a Banach space into a normed space is* equicontinuous. For barrelled spaces the same assertion is valid, more precisely we have:

Theorem 105.2. *Every* pointwise *bounded family of continuous linear maps A_ι, $\iota \in J$, from a barrelled space E into a locally convex space is* equicontinuous.

Pointwise boundedness means of course that the set $\{A_\iota x : \iota \in J\}$ is bounded in F for every $x \in E$.

Proof. Let V be a closed, absolutely convex neighborhood of zero in F. Then $T := \bigcap_{\iota \in J} A_\iota^{-1}(V)$ is closed and absolutely convex (Exercise 2 in §90). Furthermore T is absorbing: for every $x \in E$ there exists an $\alpha > 0$ such that $\{A_\iota x : \iota \in J\} \subset \alpha V$, hence $(1/\alpha)x \in A_\iota^{-1}(V)$ for all $\iota \in J$ and thus $x \in \alpha T$. We conclude that T is a barrel and therefore a neighborhood of zero in E. For every A_ι we now have $A_\iota(T) \subset A_\iota(A_\iota^{-1}(V)) \subset V$, hence the family A_ι is equicontinuous. ■

From this theorem we obtain an analogue of Proposition 33.1, namely:

Proposition 105.2. *Let (A_n) be a sequence of continuous linear maps from a barrelled space E into a separated locally convex space F. If (A_n) converges pointwise to $A: E \to F$, i.e., $A_n x \to Ax$ for every $x \in E$, then A is linear and continuous.*

Proof. The linearity of A is trivial. (A_n) is equicontinuous according to Theorem 105.2. Hence for every closed neighborhood of zero V in F there exists a neighborhood of zero U in E such that $A_n(U) \subset V$ for $n = 1, 2, \ldots$. For every $x \in U$ we have therefore $Ax = \lim A_n x \in \overline{V} = V$, hence $A(U) \subset V$. Thus the continuity of A is proved. ■

Since Fréchet spaces are complete metric and at the same time barrelled vector spaces, we can say summarizing in an abbreviated form that *for them all the fundamental principles of functional analysis are valid*: The open mapping theorem and with it the theorem concerning the continuous inverse and the closed graph theorem (Theorem 80.1), the Hahn–Banach extension theorem (Theorem 94.1), and the uniform boundedness principle in the version of Theorem 105.2. We shall not enter here into the discussion of the important question, for what nonmetrizable topological vector spaces the open mapping theorem remains valid. We refer the reader to [8], vol. II; [43], [50], [56].

Let E be a normed space. The dual E'' of the dual E', equipped with the topology defined by the norm, has been called the bidual of E in §41. At the end of §83 we saw that the topology defined by the norm on E' coincides with the strong topology $\beta(E', E)$. If E is an arbitrary topological vector space, then it is reasonable to equip the dual E' with the strong topology $\beta(E', E)$—we then denote it by E'_β—and to define the *bidual* E'' of E as the dual of E'_β, i.e., to set $E'' := (E'_\beta)'$.

We now assume that E is a Hausdorff locally convex space. Then (E, E') is a dual system, and the canonical imbedding J from E into $(E')^*$, which associates with every $x \in E$ the linear form F_x on E' defined by $F_x(x') := \langle x, x' \rangle$, is consequently an algebraic isomorphism. F_x is continuous when E' is equipped with the weak topology $\sigma(E', E)$; thus F_x is a fortiori continuous if E' carries the finer topology $\beta(E', E)$. Hence $J(E)$ lies in E''. Thus if we identify, as before, F_x with x, then $E \subset E''$.

We say that E is *semi-reflexive* if $E = E''$. A semi-reflexive normed space is reflexive in the sense of §41, and the topology defined by the norm on E'', i.e., the strong topology $\beta(E'', E')$, coincides with the topology defined by the norm on E. Correspondingly we call a Hausdorff locally convex space E *reflexive* if it is semi-reflexive ($E = E''$), and the strong topology $\beta(E'', E')$ of E'' coincides with the topology τ of E.

We shall now investigate the reflexive spaces more thoroughly and we begin with:

Lemma 105.2. *If E is a Hausdorff barrelled space, then $\beta(E'', E')$ induces on E the initial topology τ of E.*

Proof. We denote by γ the topology induced by $\beta(E'', E')$ on E. Let U be an absolutely convex, closed neighborhood of zero in E and M a bounded subset of E. Then there exists an $\alpha > 0$ such that $M \subset \alpha U$, from where it follows that $U^0 \subset \alpha M^0$. Since M^0 is the typical neighborhood of zero for the strong topology $\beta(E', E)$, we see that U^0 is bounded with respect to $\beta(E', E)$, hence also with respect to $\sigma(E', E'')$. Consequently the polar $(U^0)^p$ of U^0, formed with respect to (E', E''), is a $\beta(E'', E')$-neighborhood of zero. Since U is also $\sigma(E, E')$-closed (Proposition 98.1), it follows now with the aid of Proposition 99.2 that

$$(U^0)^p \cap E = \{x \in E : |\langle x, x' \rangle| \leqq 1 \text{ for all } x' \in U^0\} = U^{00} = U.$$

We conclude that U is a γ-neighborhood of zero and thus γ is finer than τ. We now show that conversely τ is finer than γ (whereby the lemma will be proven completely). The typical $\beta(E'', E')$-neighborhood of zero is the polar M^p, formed with respect to (E', E''), of a $\sigma(E', E'')$-bounded subset M of E'. The set M is also $\sigma(E', E)$-bounded; hence $U := M^0$ is, because of Lemma 105.1, a barrel in E. But E is barrelled, hence U must be a τ-neighborhood of zero, and because of

$$M^p \cap E = \{x \in E : |\langle x, x' \rangle| \leqq 1 \text{ for all } x' \in M\} = M^0 = U,$$

it follows that $\tau \succ \gamma$. ■

The main theorem of the theory of reflexive spaces is:

Theorem 105.3. *A Hausdorff locally convex space is reflexive if and only if it is barrelled and every* bounded *subset of E is even relatively $\sigma(E, E')$-*compact.

Proof. (a) Let E be reflexive. Then we have $\tau = \beta(E, E')$ by definition, hence E is barrelled by Proposition 105.1. Because of $E = E'' = (E'_\beta)'$ the topology $\beta(E', E)$ is admissible with respect to (E', E), hence coincides with the Mackey topology $\tau(E', E)$ (Theorem 103.1 and the remark following it). Now let M be a bounded subset of E. By definition M^0 is a $\beta(E', E)$-neighborhood of zero, hence also a $\tau(E', E)$-neighborhood of zero in E'. By Lemma 103.1 there exists an absolutely convex, $\sigma(E, E')$-compact subset K of E such that $K^0 \subset M^0$.

It follows that $M \subset M^{00} \subset K^{00} = K$, hence M is indeed relatively $\sigma(E, E')$-compact.

(b) Now let E be barrelled and every bounded subset M of E even relatively $\sigma(E, E')$-compact. We have then $\beta(E', E) = \tau(E', E)$, hence the strong topology $\beta(E', E)$ is admissible on E' with respect to the dual system (E', E). It follows that $E'' = E$, i.e., E is semi-reflexive. Since $\beta(E'', E') = \beta(E, E')$ coincides with the topology of E because of Proposition 105.1, E is even reflexive. ■

The reader can convince himself without effort that the following theorem holds:

Theorem 105.4. *A Banach space E is reflexive if and only if its closed unit ball is $\sigma(E, E')$-compact.*

Proposition 10.6 and the last theorem say that compactness properties of the closed unit ball K of a Banach space E determine to a certain extent its destiny:

K is compact for the norm $\Leftrightarrow$ E is finite-dimensional,

K is weakly compact $\Leftrightarrow$ E is reflexive.

The reader should recall that the closed unit ball in the dual E' is *always* weakly compact, i.e., $\sigma(E', E)$-compact (Proposition 102.1).

We obtain from Proposition 62.2 that the closed unit ball K of a reflexive Banach space is *weakly sequentially compact*, i.e., that from every sequence $(x_n) \subset K$ one can select a subsequence which converges weakly to an $x \in K$. We want to mention without proof that the converse also holds. Thus *reflexive Banach spaces can also be characterized by the weak* sequential *compactness of their closed unit ball*; see [8], vol. I, p. 315.

Exercises

+1. (s) is a Fréchet space. On (s) all $\mathfrak{S}$-topologies formed with respect to the dual system $((s), (s)')$ coincide.

2. Find examples of Fréchet spaces. *Hint*: Proposition 83.1 and the Examples and Exercises of §83.

+3. Every E'-conjugable endomorphism of a separated barrelled space E (e.g., a Fréchet or Banach space) is continuous (cf. Proposition 41.5). *Hint*: Exercise in §103.

4. Let T be a barrel and K an absolutely convex, compact subset of a Hausdorff locally convex space E. Then there exists an $\alpha > 0$ such that $K \subset \alpha T$. *Hint*: T^0 is $\tau(E', E)$-bounded, K^0 is a $\tau(E', E)$-neighborhood of zero.

+5. A continuous endomorphism A of a Fréchet space is a Fredholm operator if and only if the deficiencies $\alpha(A)$, $\beta(A)$ are finite. *Hint*: Exercise in §94, Exercise 3 in §95.

6. By $A(\xi_1, \xi_2, \ldots) := (\xi_2, \xi_3, \ldots)$ we define a continuous endomorphism of the Fréchet space (s). For all $\lambda \neq 0$ we have $\alpha(\lambda I - A) = 1$ and $\beta(\lambda I - A) = 0$. It follows from Exercise 5 that $\lambda I - A$ is for all $\lambda \neq 0$ a Fredholm operator with $\operatorname{ind}(\lambda I - A) = 1$; cf. Proposition 52.1 and Exercise 8.

+7. Let A be a *bounded* Fredholm operator on a Fréchet space E. Then for all sufficiently small λ also $A - \lambda I$ is a Fredholm operator on E with

$$\operatorname{ind}(A - \lambda I) = \operatorname{ind}(A).$$

Hint: Exercise 6 in §93, proof of Proposition 37.2.

+8. Let A be a *bounded* endomorphism of a Fréchet space, and let $\alpha(\lambda I - A)$, $\beta(\lambda I - A)$ be finite for all $\lambda \neq 0$. Then $\lambda I - A$ is for all $\lambda \neq 0$ a Fredholm operator with vanishing index. Investigate how far the theory of the Fredholm domain (§51) can be carried over to the present case. *Hint*: Exercise 7 in §93 and the above Exercise 7.

9. A non-complete metric and locally convex vector space does not have to be barrelled. *Hint*: Exercise 1 in §33.

10. We consider the complete metric vector space $H(\Delta)$ of all holomorphic functions on the simply connected domain $\Delta \subset \mathbf{C}$ (see Exercise 4 in §80). Show the following:

(a) $H(\Delta)$ is a Fréchet space, hence also barrelled.

(b) $M \subset H(\Delta)$ is (topologically) bounded if and only if the family of functions M is uniformly bounded on every compact subset of Δ.

(c) Every bounded and closed subset of $H(\Delta)$ is compact (this is precisely the assertion of the famous *theorem of Montel* concerning normal families of holomorphic functions; observe that in a metric space exactly the sequentially compact sets are compact).

+11. Exercise 10 is the motivation for the following definition: A separated locally convex space E is called a *Montel space* if it is barrelled and if every closed, bounded subset is compact. Show the following:

(a) $H(\Delta)$ (see Exercise 10) is a Montel space.

(b) Normed Montel spaces are finite-dimensional.

(c) A Montel space is reflexive.

(d) Together with E also E'_β is a Montel space.

§106. Convex, compact sets: The theorems of Krein–Milman and Schauder

In this section we shall be concerned first with so-called extremal points and at the end we shall return to the problem of fixed points.

If x, y are elements of the vector space E, then

$$S(x, y) := \{\alpha x + (1 - \alpha)y : 0 < \alpha < 1\}$$

is called the *open* and

$$S[x, y] := \{\alpha x + (1 - \alpha)y : 0 \leqq \alpha \leqq 1\}$$

the *closed segment* with *end-points* x, y. The end-points are allowed to coincide; if this is, however, not the case, we speak of a *proper segment*. Let now M be a subset of E. A point $x_0 \in M$ is called an *extremal point* of M, if it lies on no proper open segment whose end-points belong to M, i.e., if the assumptions $x_0 \in S(x, y)$ and $x, y \in M$ imply that $x = y = x_0$. This formulation suggests the following generalization: $N \subset M$ is called an *extremal subset* of M if N is not empty and if the assumptions $N \cap S(x, y) \neq \varnothing$ and $x, y \in M$ imply that $x, y \in N$. Obviously x_0 is an extremal point of M if and only if $\{x_0\}$ is an extremal subset of M.

The relation 'N is an extremal subset of M' creates an order relation '$<$' in the set of all non-empty subsets of E, in particular we have the transitivity: From $P < Q$ and $Q < R$ it follows that $P < R$. It follows that an extremal point of an extremal subset of M is also an extremal point of M.

Lemma 106.1. *Let $K \neq \varnothing$ be a compact subset of the topological vector space E, f a continuous linear form on E, and $\mu := \min_{x \in K} \operatorname{Re} f(x)$. Then $K_1 := \{x \in K : \operatorname{Re} f(x) = \mu\}$ is a closed (hence also compact) extremal subset of K.*

Proof. K_1 is trivially closed and non-empty. For $x, y \in K$ and an $\alpha \in (0, 1)$ let now $\alpha x + (1 - \alpha)y$ lie in K_1. Then on the one hand

$$\operatorname{Re} f(x) \geqq \mu \qquad \text{and} \qquad \operatorname{Re} f(y) \geqq \mu,$$

and on the other $\alpha \operatorname{Re} f(x) + (1 - \alpha)\operatorname{Re} f(y) = \operatorname{Re} f(\alpha x + (1 - \alpha)y) = \mu$, thus $\operatorname{Re} f(x) = \operatorname{Re} f(y) = \mu$, i.e., $x, y \in K_1$. ■

Proposition 106.1. *A non-empty compact subset K of the separated locally convex space E possesses extremal points.*

Proof. On the non-empty set $\mathfrak{M}$ of all closed extremal subsets of K we define an order '$\prec$' by $F \prec G \Leftrightarrow G \subset F$. If $\mathfrak{M}_0 \subset \mathfrak{M}$ is totally ordered, then $D := \bigcap_{M \in \mathfrak{M}_0} M$ is not empty (§101), and one sees easily that D is an upper bound for $\mathfrak{M}_0$ in $\mathfrak{M}$. Thus by Zorn's lemma there exists an $M_0 \in \mathfrak{M}$ which does not contain properly any closed extremal subset of K. If M_0 would contain two distinct points x_0, y_0, then there would exist by Theorem 96.2 a continuous linear form f on E such that $\operatorname{Re} f(x_0) \neq \operatorname{Re} f(y_0)$, consequently

$$M_1 := \left\{x \in M_0 : \operatorname{Re} f(x) = \min_{y \in M_0} f(y)\right\}$$

would be a proper subset of M_0. Since, however, M_1 is by Lemma 106.1 a closed extremal subset of M_0 and therefore also a closed extremal subset of K, we obtain a contradiction to the Zorn minimal property of M_0. Thus this set contains exactly one point z, and z is an extremal point of K. ■

Theorem 106.1 (Theorem of Krein–Milman). *Let K be a non-empty, convex and compact subset of the separated locally convex space E and K_{ex} the set of its extremal points. Then*

$$K = \overline{\mathrm{co}(K_{ex})}.$$

Proof. Since obviously $\overline{\mathrm{co}(K_{ex})} \subset K$, we only have to prove the converse inclusion. For this we assume that y lies in K but not in $K_0 := \overline{\mathrm{co}(K_{ex})}$. Since K_0 is not empty because of Proposition 106.1 and since by Lemma 90.3 it is convex, there exists an $f \in E'$ and an $\alpha \in \mathbf{R}$ so that $\mathrm{Re}\, f(y) < \alpha < \mathrm{Re}\, f(x)$ for all $x \in K_0$ (Theorem 96.2). Consequently $K_1 := \{x \in K: \mathrm{Re}\, f(x) = \min_{z \in K} \mathrm{Re}\, f(z)\}$ does not intersect the set K_0, hence a fortiori

$$K_1 \cap K_{ex} = \varnothing. \tag{106.1}$$

By Lemma 106.1 the set K_1 is a compact extremal subset of K and contains thus an extremal point (Proposition 106.1), which must at the same time be an extremal point of K—in contradiction to (106.1). ∎

A further property of convex, compact sets, which is particularly important for applications, is given by the *first Schauder fixed-point theorem* (Proposition 106.2). We obtain it from the *second Schauder fixed-point theorem* (Theorem 106.3), which in turn will be deduced from the *Brouwer fixed-point theorem* (Theorem 106.2). We emphasize that the maps which occur in these fixed-point theorems do *not* have to be linear.

Theorem 106.2. *Every continuous self-map of the unit ball of $l^2(n)$ has at least one fixed point.*

Proof. It is easy to see that it is sufficient to prove the theorem for the case that $l^2(n)$ is $\mathbf{R}^n$ equipped with the euclidean norm. We assume this, and denote by A a continuous self-map of the unit ball $K := \{x \in \mathbf{R}^n: \|x\| \leqq 1\}$ of $\mathbf{R}^n$. The components of the image vector $(\eta_1, \ldots, \eta_n) := A(\xi_1, \ldots, \xi_n)$ can be represented in the form $\eta_i = \varphi_i(\xi_1, \ldots, \xi_n)$ with continuous functions $\varphi_i: K \to \mathbf{R}$, and we shall first assume that φ_i has even continuous partial derivatives with respect to all variables ξ_k. We will prove the theorem by contradiction, and assume for this purpose that $Ax \neq x$ for every $x \in K$. For a given $x \in K$ the set

$$g := \{x + \lambda(x - Ax): \lambda \in \mathbf{R}\}$$

is then a straight line through x with the direction vector $x - Ax$, and we study first the points of intersection of g with the surface of K, i.e., we consider the equation $\|x + \lambda(x - Ax)\| = 1$. If we square and express the square of the norm by the inner product, we obtain the equation

$$\|x - Ax\|^2\lambda^2 + 2\lambda(x|x - Ax) + \|x\|^2 = 1, \tag{106.2}$$

which can be written also in the form

$$(\|x - Ax\|^2\lambda + (x|x - Ax))^2 = (x|x - Ax)^2 + (1 - \|x\|^2)\|x - Ax\|^2. \tag{106.3}$$

The right-hand side of (106.3) is always positive: This is trivial for points in the interior of K and follows for the points of the surface from the estimate

$$(x|x - Ax) > 0 \quad \text{for} \quad \|x\| = 1 \tag{106.4}$$

(on the one hand $(x|x - Ax) = 1 - (x|Ax) \geqq 1 - \|x\|\,\|Ax\| \geqq 0$, on the other hand $(x|x - Ax) \neq 0$ since otherwise we would have $(x|Ax) = 1$ which because of $\|x\| = 1$, $\|Ax\| \leqq 1$ can occur only in the excluded case $x = Ax$). It follows that equation (106.2) has two distinct real solutions. We shall denote the larger one of these solutions by $\lambda(x)$; it is given by

(106.5)

$$\lambda(x) = \frac{1}{\|x - Ax\|^2}\left[(x|Ax - x) + \sqrt{(x|x - Ax)^2 + (1 - \|x\|^2)\|x - Ax\|^2}\right].$$

We obtain from this representation that the function $x \mapsto \lambda(x)$ has continuous partial derivatives with respect to all the variables ξ_k.

We now define on $[0, 1] \times K$ a map B by

$$B(t, x) := x + t\lambda(x)(x - Ax).$$

B has continuous partial derivatives with respect to $\xi_1, \ldots, \xi_n$; furthermore

$$B(0, x) = x, \ \|B(1, x)\| = 1 \quad \text{for all} \quad x \in K \tag{106.6}$$

$$\|B(t, x)\| < 1 \quad \text{for} \quad \|x\| < 1, \qquad B(t, x) = x \quad \text{for} \quad \|x\| = 1; \tag{106.7}$$

the last equality holds because by (106.4), (106.5) one has $\lambda(x) = 0$ whenever $\|x\| = 1$. Hence for a fixed t the map $x \mapsto B(t, x)$ maps the boundary

$$\partial K = \{x: \|x\| = 1\}$$

of K in a one-to-one way onto itself, and the interior $\mathring{K} = \{x: \|x\| < 1\}$ of K into itself. We now show that $x \mapsto B(t, x)$ is a bijection of K for small t.

Let $B_k(t, x)$ be the kth component of $B(t, x)$ and

$$\frac{\partial B}{\partial x} := \begin{pmatrix} \dfrac{\partial B_1}{\partial \xi_1}, \ldots, \dfrac{\partial B_1}{\partial \xi_n} \\ \cdots\cdots\cdots \\ \dfrac{\partial B_n}{\partial \xi_1}, \ldots, \dfrac{\partial B_n}{\partial \xi_n} \end{pmatrix} = \begin{pmatrix} \operatorname{grad} B_1 \\ \vdots \\ \operatorname{grad} B_n \end{pmatrix}$$

the functional matrix of B with respect to x. For every $x \in K$ the Jacobian $|(\partial B/\partial x)(0, x)|$ is equal to 1. Consequently there exists a $\delta > 0$ such that

$$\begin{vmatrix} \operatorname{grad} B_1(t, x_1) \\ \vdots \\ \operatorname{grad} B_n(t, x_n) \end{vmatrix} > 0 \quad \text{for} \quad 0 \leqq t \leqq \delta \quad \text{and} \quad \|x_j - x_k\| \leqq \delta. \tag{106.8}$$

Because of (106.6) there exists an ε, $0 < \varepsilon \leqq \delta$, such that

$$\|B(t, x) - x\| \leqq \tfrac{1}{2}\delta \qquad \text{for} \quad 0 \leqq t \leqq \varepsilon \qquad \text{and all} \quad x \in K.$$

If we now have $B(t, x) = B(t, y)$ for such a t, then $\|x - y\| \leqq \|x - B(t, x)\| + \|y - B(t, y)\| \leqq \delta$. For points $x_1, \ldots, x_n$ on the segment joining x and y we have a fortiori $\|x_j - x_k\| \leqq \delta$ $(j, k = 1, \ldots, n)$. For these points and for $t \in [0, \varepsilon]$ the determinant (106.8) is positive and thus its matrix non-singular. If according to the mean value theorem of differential calculus we choose the x_k so that $B_k(t, x) - B_k(t, y) = (\text{grad } B_k(t, x_k) | x - y)$, then

$$0 = B(t, x) - B(t, y) = \begin{pmatrix} \text{grad } B_1(t, x_1) \\ \vdots \\ \text{grad } B_n(t, x_n) \end{pmatrix} (x - y),$$

from where it follows that $x - y = 0$ and so that the map $x \mapsto B(t, x)$ is injective for $0 \leqq t \leqq \varepsilon$. For the proof of its surjectivity we observe that $|(\partial B/\partial x)(t, x)|$ is positive for $0 \leqq t \leqq \varepsilon$ and all $x \in K$ because of (106.8), and thus the image $G := \{B(t, x) : x \in \mathring{K}\}$ of the open set $\mathring{K}$ is itself again open. The compact set $\{B(t, x) : x \in K\}$ is therefore the union of ∂K and of the open set $G \subset \mathring{K}$, from where $G = \mathring{K}$ and thus the asserted surjectivity follows.

According to what has been proved so far,

$$V(t) := \int_K \left| \frac{\partial B}{\partial x} \right| \mathrm{d}v$$

is for $0 \leqq t \leqq \varepsilon$ the volume of the set $\{B(t, x) : x \in K\} = K$, hence for these values of t the function $V(t)$ is constant. But since $V(t)$ is obviously a polynomial in t, it follows that

$$(106.9) \qquad V(t) = V(0) > 0 \qquad \text{for all} \quad t \in [0, 1]$$

Now we obtain from the equality $\|B(1, x)\|^2 = B_1(1, x)^2 + \cdots + B_n(1, x)^2 = 1$ (see (106.6)) by differentiation with respect to $\xi_1, \ldots, \xi_n$ that the rows of $(\partial B/\partial x)(1, x)$ are linearly dependent and that so $|(\partial B/\partial x)(1, x)|$ vanishes. Consequently $V(1) = 0$. This contradiction to (106.9) shows that there exists in K at least one point x such that $Ax = x$, provided that the φ_i have continuous partial derivatives. The general case can, however, be reduced to the 'differentiable' one with the help of the Weierstrass approximation theorem. We leave the details to the reader. ■

A simple consequence of the Brouwer fixed point theorem is:

Lemma 106.2. *Let E be an n-dimensional normed space. Then every continuous self-map A of a convex, compact subset $C \neq \emptyset$ of E has at least one fixed point.*

Proof. Since E is homeomorphic to $l^2(n)$, it is sufficient to prove the lemma for the case $E = l^2(n)$. The set C lies in a positive multiple ρK of the unit ball K of $l^2(n)$. We now associate with every $x \in \rho K$ its uniquely determined best approximation Bx in C (see Theorem 60.1). B is continuous, and for $x \in C$ we have $Bx = x$. The continuous self-map $A \circ B$ of ρK has, according to Theorem 106.2, a fixed point x. Because of $x = A(Bx)$ the point x lies in C, and consequently even $x = Ax$. ∎

We now come to the announced *Schauder fixed-point theorems.*

Theorem 106.3. *Let E be a normed space, $K \subset E$* convex, *and $C \subset K$* compact *and non-empty. Then every continuous map $A: K \to C$ has at least one fixed point.*

Proof. By Lemma 42.1 the set C is totally bounded; thus for any preassigned $\varepsilon > 0$ one can find finitely many points $x_1, \ldots, x_n$ in C so that the following is valid:

(106.10) For every $x \in C$ there exists an x_ν so that $\|x - x_\nu\| < \varepsilon$.

We now define functions $f_1, \ldots, f_n$ on C by

$$f_\nu(x) := \begin{cases} 0, & \text{if} \quad \|x - x_\nu\| \geqq \varepsilon \\ \varepsilon - \|x - x_\nu\| & \text{if} \quad \|x - x_\nu\| \leqq \varepsilon. \end{cases}$$

Obviously f_ν is continuous, $f_\nu(x) \geqq 0$ and $\sum_{\nu=1}^n f_\nu(x) > 0$ (observe (106.10)). Thus

$$f(x) := \frac{\sum_{\nu=1}^n f_\nu(x) x_\nu}{\sum_{\nu=1}^n f_\nu(x)}$$

defines a continuous map f from C into $K_0 := \operatorname{co}(x_1, \ldots, x_n)$; we have

(106.11) $$\|f(x) - x\| \leqq \varepsilon \qquad \text{for all} \quad x \in C.$$

$B := f \circ A$ is a continuous map from K into $K_0 \subset K$, hence the restriction $\hat{B}$ of B to K_0 is a continuous self-map of the convex, compact subset K_0 of the finite-dimensional normed space $[x_1, \ldots, x_n]$. By Lemma 106.2 there exists now a $z \in K_0$ such that $Bz = \hat{B}z = z$, hence $f(Az) = z$. Because of (106.11) we have $\|Az - z\| = \|Az - f(Az)\| \leqq \varepsilon$. We summarize this intermediary result: For every $\varepsilon > 0$ there exists a $z = z(\varepsilon) \in K$ such that $\|Az - z\| \leqq \varepsilon$. We now determine according to this assertion for every $m \in \mathbf{N}$ a point $z_m \in K$ such that

(106.12) $$\|Az_m - z_m\| \leqq \frac{1}{m}.$$

Since Az_m lies in the compact set C, there exists a subsequence (z_{m_k}) and an $x \in C$ so that $Az_{m_k} \to x$. With the aid of (106.12) it follows that $z_{m_k} \to x$, and since A

is continuous, (Az_{m_k}) converges to Ax. The limits x and Ax of (Az_{m_k}) must coincide: x is a fixed point of A. ∎

From Theorem 106.3 we obtain immediately:

Proposition 106.2. *Let A be a continuous self-map of the non-empty convex subset K of the normed space E. The map A has a fixed point whenever one of the following conditions is satisfied*:

(a) *K is compact.*
(b) *K is closed and $A(K)$ is relatively compact.*

In contrast to Banach's fixed-point theorem, the fixed-point theorems of this section do not require that A shall be a *contracting* map. They do, however, not yield uniqueness of the fixed point, nor do they give a constructive procedure for obtaining it. Tikhonov has shown that Proposition 106.2 (with condition a) is valid also for locally convex spaces. An easily accessible proof can be found in [4].

It is shown in Exercise 1 that one can obtain the *Peano existence theorem* from Schauder's Proposition 106.2. We shall prepare a completely different application of the fixed-point theorem by the following considerations.

Every eigenspace of an endomorphism A of E is obviously invariant under A. If A has no eigenvalue, then one can still ask whether there exists at least one subspace $\neq E, \{0\}$, invariant under A, more briefly: whether A has a non-trivial invariant subspace. For a compact A this question got a positive answer in 1954 in [121]. An eigenspace of A has even the property to be invariant under any endomorphism which commutes with A. The question immediately arises whether every compact A has a non-trivial subspace, which is even invariant under all continuous endomorphisms which commute with A. The answer was given by Lomonosov in 1973 in [131]:

Proposition 106.3. *For every compact endomorphism $A \neq 0$ of a complex normed space E there exists a non-trivial, closed subspace of E which is invariant under all continuous endomorphisms of E which commute with A.*

Proof. $\mathscr{R} := \{B \in \mathscr{L}(E): AB = BA\}$ is obviously an algebra over $\mathbf{C}$. We shall prove by contradiction, and assume therefore that the assertion is false. Then A has certainly no eigenvalue, in particular $Ay \neq 0$ for all $y \neq 0$. It follows that $F_y := \overline{\{By: B \in \mathscr{R}\}}$ is for every $y \neq 0$ a closed subspace $\neq \{0\}$, which is invariant under all $B \in \mathscr{R}$.

For the points $x = x_0 + y$, $\|y\| \leqq 1$ of the closed unit ball K around x_0 we have $\|Ax\| \geqq \|Ax_0\| - \|Ay\| \geqq \|Ax_0\| - \|A\|$; because of $A \neq 0$ we can choose x_0 so that the zero vector does not lie in $\overline{A(K)}$ and thus $F_y \neq \{0\}$ for every $y \in \overline{A(K)}$. We fix x_0 in such a way. For every $y \in \overline{A(K)}$ there exists a $B \in \mathscr{R}$ such that $\|By - x_0\| < 1$: If we had, namely, $\|By - x_0\| \geqq 1$ for all $B \in \mathscr{R}$, then we would also have $\|z - x_0\| \geqq 1$ for all $z \in F_y$, thus x_0 would not lie in

F_y and so F_y would be a non-trivial, closed subspace, invariant under all $B \in \mathscr{R}$, in contradiction to our assumption that such subspaces do not exist. If we also invoke the continuity of B, we can say: For every $y \in \overline{A(K)}$ there exists an open ball $K(y)$ around y and a $B_y \in \mathscr{R}$ so that $\|B_y z - x_0\| < 1$ for $z \in K(y)$. Because of the compactness of A the set $\overline{A(K)}$ is sequentially compact, hence also compact; consequently $\overline{A(K)}$ is covered already by finitely many among the balls $K(y)$. It follows that there exist in $\mathscr{R}$ finitely many operators $B_1, \ldots, B_n$ with the following property:

(106.13) For every $y \in \overline{A(K)}$ there exists a B_i such that $\|B_i y - x_0\| < 1$.

If we introduce the non-negative function

$$\varphi(t) := \begin{cases} 1 - t & \text{for} \quad 0 \leqq t < 1 \\ 0 & \text{for} \quad t \geqq 1, \end{cases}$$

then $\sum_{i=1}^n \varphi(\|B_i y - x_0\|) > 0$ whenever $y \in \overline{A(K)}$, because of (106.13). Consequently we can define a map $f: \overline{A(K)} \to E$ by

$$f(y) := \frac{\sum_{i=1}^n \varphi(\|B_i y - x_0\|) B_i y}{\sum_{i=1}^n \varphi(\|B_i y - x_0\|)}. \tag{106.14}$$

f is continuous, hence $f(\overline{A(K)})$ is compact. Furthermore since $f(y)$ is a convex combination of the $B_1 y, \ldots, B_n y$, and since $B_i y$ lies in K because of (106.13), we have $f(A(K)) \subset f(\overline{A(K)}) \subset K$, so that $f \circ A$ maps the convex set K continuously into a compact subset of K. By Theorem 106.3 there exists an $x \in K$ such that $f(Ax) = x$. Thus because of (106.14) we have $\sum_{i=1}^n \alpha_i B_i Ax = x$ with certain numbers α_i, and since x, as an element of K, is distinct from the vector zero, this equation means that $B := \sum_{i=1}^n \alpha_i B_i A$ has the eigenvalue 1. If we take into account that B is compact, we can assert that dim $N(I - B)$ is positive and finite. Since $N(I - B)$ is obviously invariant under A, the restriction of A to $N(I - B)$, and thus also A itself, has an eigenvalue, in contradiction to our hypothesis. ■

Exercises

1. Let $T := [t_0 - \alpha, t_0 + \alpha]$, $X := [\xi_0 - \beta, \xi_0 + \beta]$ and $f: T \times X \to \mathbf{R}$ continuous. Then the *initial value problem* $\mathrm{d}x/\mathrm{d}t = f(t, x)$, $x(t_0) = \xi_0$ has at least one solution defined in a neighborhood of t_0 (existence theorem of Peano). *Hint*: The problem is equivalent to solving the (non-linear) integral equation

$$x(t) = \xi_0 + \int_{t_0}^{t} f(\tau, x(\tau)) \mathrm{d}\tau.$$

Set $\mu := \max_{T \times X} |f(t, \xi)|$, $\delta := \min(\beta/\mu, \alpha)$, $E := C[t_0 - \delta, t_0 + \delta]$,

$$K := \{x \in E : |x(t) - \xi_0| \leqq \beta\}, \qquad (Ax)(t) := \xi_0 + \int_{t_0}^{t} f(\tau, x(\tau)) \mathrm{d}\tau.$$

K is convex and closed, A a continuous self-map of K. With the help of the Arzelà–Ascoli theorem we see that $A(K)$ is relatively compact.

2. We use the notations and hypotheses of Exercise 1 (with the exception of δ). Let f satisfy additionally a *Lipschitz condition* with respect to the second variable, i.e., let there exist a constant $\lambda > 0$ so that $|f(t, \xi) - f(t, \eta)| \leqq \lambda|\xi - \eta|$. Let $0 < \delta < \min(\beta/\mu, \alpha, 1/\lambda)$. Show that A is a contracting map of the complete metric space K. Banach's fixed-point theorem shows now that the initial value problem of Exercise 1 can be solved *uniquely* and *constructively* (existence and uniqueness theorem of Picard–Lindelöf).

3. Let F be a closed subset of the metric space E and A a continuous map from F into E with a relatively compact image $A(F)$. Furthermore for every $\varepsilon > 0$ let there exist an $x = x(\varepsilon) \in F$ such that $d(Ax, x) \leqq \varepsilon$. Then A has a fixed-point.

XIV

The representation of commutative Banach algebras

§107. Preliminary remarks on the representation problem

We have seen in Proposition 102.2 that a normed space E is nothing but a subspace of the normed space $C(T)$ of all continuous functions on a certain compact set T. The question arises, when does E coincide with the whole space $C(T)$? But since $C(T)$ is not only a normed space, but even a commutative Banach algebra (Exercise 1), E can be made in this case in a natural way into a commutative Banach algebra, which differs from $C(T)$ only in the notation of its elements. We shall therefore pose the problem when is $E = C(T)$ only for commutative Banach algebras E. First we make some general comments on the *representation problem*, which will motivate our procedure and also throw new light on Proposition 102.2.

The general representation problem consists in mapping a set E which is equipped with some structure Σ (e.g., a group, a normed space, a Banach algebra) onto a set C of concrete, familiar objects (which must of course also carry the structure Σ, i.e., which must also be e.g., a group, a normed space, a Banach algebra) in such a way that the structure be preserved, i.e., so that the map is a homomorphism. If such a map is possible, then one says that E is represented by C, or that one has found a *representation* for E. The representation is *faithful* if the representing homomorphism from E onto C is injective, i.e., if it is even an isomorphism. Only in this case will one have a complete grasp of E by means of C and consider the representation problem as solved.

Which set C one chooses is, within certain bounds, arbitrary. In the representation theory of groups, C is frequently a set of matrices. To the analyst (real or complex-valued) functions are the most familiar 'concrete' objects, and therefore he will try to represent his structures E by *sets of functions* C, whenever possible. The decisive question will then be how one can obtain a function from an element $x \in E$, or somewhat more pointedly, how one can convert x into a function.

This conversion is always possible if on E there is defined an—at first entirely arbitrary—non-empty set Φ of **K**-valued functions. The conversion procedure, considered canonical hereafter, consists in associating with every $x \in E$ the function $F_x: \Phi \to \mathbf{K}$ whose 'table of values' is given by the family $(f(x): f \in \Phi)$ of numbers. In other words, F_x is defined by

$$F_x(f) := f(x) \qquad \text{for all} \quad f \in \Phi. \tag{107.1}$$

The correspondence $x \mapsto F_x$, which we will denote henceforth by H, is obviously injective if and only if from $f(x) = f(y)$ for all $f \in \Phi$ it follows that $x = y$, or equivalently: if for any two distinct elements x, y there always exists an $f \in \Phi$ such that $f(x) \neq f(y)$. In this case one says that Φ *separates* the points of E. Thus the injectivity of H is assured only if Φ is 'sufficiently large' (i.e., separates points).

Let now E be a vector space over $\mathbf{K}$. The map H is a homomorphism if and only if for x, y in E and $\alpha \in \mathbf{K}$ we have

$$F_{x+y} = F_x + F_y, \qquad F_{\alpha x} = \alpha F_x, \tag{107.2}$$

that is, if for all $f \in \Phi$ we have $f(x + y) = f(x) + f(y), f(\alpha x) = \alpha f(x)$. Therefore H will be a homomorphism if and only if Φ is a subset of E^*. In this case $\tilde{E} := \{F_x: x \in E\}$ is itself a vector space.

Now let E be even an algebra over $\mathbf{K}$. Then the map H is a homomorphism if and only if in addition to (107.2) also

$$F_{xy} = F_x F_y$$

is valid, i.e., if for every $f \in \Phi$ we have $f(xy) = f(x)f(y)$. Thus H is a homomorphism if and only if Φ is an (arbitrary) set of *multiplicative linear forms* on E. In this case $\tilde{E}$ is itself an algebra.

A set Φ of linear forms on E is said to be *total* if from $f(x) = 0$ for all $f \in \Phi$ it follows that $x = 0$. The set Φ separates points of E if and only if it is total. *The homomorphism H of vector spaces or algebras is an isomorphism of vector spaces or algebras, respectively, if and only if Φ is total.*

So far we have only considered *algebraic* structures on E. Now we add a *topological* structure: let E be a normed space. We will now try to equip the representing functions F_x with a norm in such a way that H preserves norms, i.e., that we have $\|x\| = \|F_x\|$ for every $x \in E$. Because of (41.3) this is possible if Φ is the set of all continuous linear forms f on E with $\|f\| \leq 1$, i.e., the closed unit ball of E'; indeed, in this case we have

$$\|x\| = \sup_{f \in \Phi} |F_x(f)|, \tag{107.3}$$

so that H preserves norms, or is isometric, if on $\tilde{E}$ we introduce a norm by

$$\|F_x\| := \sup_{f \in \Phi} |F_x(f)| \tag{107.4}$$

(observe that, as opposed to the situation in (41.3), the function F_x considered here is defined only on Φ, and is thus the restriction to Φ of the function F_x figuring in (41.3)).

It is unsatisfactory in this consideration that we represent E by a function space $\tilde{E}$ which to a certain extent has been tailored artificially for E and which is not defined by inner properties like boundedness or continuity of its elements: It is not defined as the space of *all* bounded or *all* continuous functions on Φ; besides, the second concept (continuity) makes sense only if Φ has a topology. Now Φ carries in a natural way the topology v induced by the topology defined by the norm of E', and for v every F_x is continuous so that $\tilde{E}$ becomes a linear subspace of the space of all v-continuous functions on Φ. We will certainly be able to equip this space with the supremum-norm—and if H has to be norm-preserving, then because of (107.3) there is no question of any other norm—if Φ is v-compact. This is, because of Proposition 10.6, the case only if E', and with it also E, is finite-dimensional. Thus v is eliminated for our purposes.

Loosely formulated, a topology has the more compact sets the coarser it is. If there exists at all a topology on Φ for which all the F_x are continuous, and for which at the same time Φ is compact, then certainly the initial topology for the set $\{F_x : x \in E\}$ must have these properties. This topology σ on Φ is, as one sees immediately, the topology induced by $\sigma(E', E)$ on Φ (Exercise 2), *and for it Φ is indeed compact* (Proposition 102.1).

If we equip Φ with the topology σ, then we can introduce thus on the vector space $C(\Phi)$ of all σ-continuous functions $F: \Phi \to \mathbf{K}$ the supremum-norm $\|F\| := \sup_{f \in \Phi} |F(f)|$, and in this way H becomes an isomorphism of normed spaces from E onto a subspace of $C(\Phi)$.

If E is a commutative Banach algebra, then several particular difficulties arise. For Φ one will now choose the set of all multiplicative linear forms $f \in E'$ such that $\|f\| \leqq 1$. Do there exist at all non-trivial linear forms of this kind? And if yes, do they form a σ-compact set Φ so that $C(\Phi)$ can be equipped with the supremum-norm (see Exercise 3)? And if all these questions have a satisfactory answer, then the problem arises whether—or under what additional conditions—H preserves norms; indeed, the supremum of the $|F_x(f)|$ can become smaller if it is not taken over *all* continuous linear forms f with $\|f\| \leqq 1$ but only over the *multiplicative* ones; it is, however, only about the supremum formed in the first way that we know that it coincides with $\|x\|$. And finally we will have to ask whether—or under what conditions—Φ is total, i.e., H is an isomorphism. The investigation and solution of these problems is the content of the representation theory of commutative Banach algebras created by Gelfand to which we now turn.

Exercises

$^{+}$1. The set $C(T)$ of all $\mathbf{K}$-valued continuous functions on a compact topological space T is an algebra over $\mathbf{K}$ if we define multiplication by scalars αx, addition $x + y$ and multiplication xy pointwise. $C(T)$ becomes a (commutative) Banach algebra through $\|x\| := \sup_{t \in T} |x(t)|$.

2. Let a family of maps $F_\iota: \Psi \to T_\iota$ be given, where Ψ is a non-empty set and T_ι is for every $\iota \in J$ a topological space. Let the initial topology for $(F_\iota : \iota \in J)$

be denoted by τ. Furthermore let Φ be a non-empty subset of Ψ, φ_ι the restriction of F_ι to Φ and τ_0 the initial topology for $(\varphi_\iota : \iota \in J)$. Show that τ_0 coincides with the topology induced by τ on Φ.

3. We ask for compactness with respect to σ instead of compactness with respect to the topology generated on Φ by the family $F_x : \Phi \to \mathbf{K}$ (which at first would seem more natural), because the two notions of compactness coincide. *Hint*: Exercise 2.

§108. Multiplicative linear forms and maximal ideals

Let E be first a commutative (real or complex) Banach algebra with the (normalized) unit element e, and $E^\times$ the set of all multiplicative linear forms $f \neq 0$ on E. For $f \in E^\times$ we have $f(x) = f(xe) = f(x)f(e)$, from where it follows immediately that $f(e) = 1$. *If for f, g in $E^\times$ the nullspaces coincide, then* $f = g$; for by Lemma 15.1 we have $f = \alpha g$, whence $\alpha = \alpha g(e) = f(e) = 1$ and so $f = g$.

The nullspace $N(f)$ of $f \in E^\times$ is obviously an ideal with codim $N(f) = 1$ (Exercise 7 in §15). If conversely N is an ideal of codimension 1 in E, then e does not lie in N (otherwise, because of $x = xe$, every $x \in E$ would lie in N), consequently for every $x \in E$ the representation $x = \alpha e + y$ holds with uniquely determined $\alpha \in \mathbf{K}$ and $y \in N$. It follows that $f(x) := \alpha$ determines a non-trivial multiplicative linear form f with $N(f) = N$. Consequently, *there is a one-to-one correspondence between $E^\times$ and the set of all ideals in E with codimension* 1.

An ideal J in E is said to be *proper* if it is distinct from E. A proper ideal is said to be *maximal* if it is not contained properly in any ideal except E. Obviously every ideal of codimension 1 is maximal. For a complex E also the converse is true, as we shall see shortly. First we prove, however, the basic theorem concerning maximal ideals:

Proposition 108.1. *A commutative Banach algebra E possesses maximal ideals. All maximal ideals are closed. An element of E is non-invertible if and only if it belongs to a maximal ideal. Every proper ideal in E is contained in a maximal ideal.*

Proof. Let J be a proper ideal in E. In the set $\mathfrak{M}$ of all proper ideals which contain J set-theoretical inclusion defines an order. The union of the sets of a totally ordered subset $\mathfrak{K}$ of $\mathfrak{M}$ is an ideal which contains J, and it is even proper (it does not contain e), hence it is an upper bound of $\mathfrak{K}$ in $\mathfrak{M}$. By Zorn's lemma $\mathfrak{M}$ possesses a maximal element M. Obviously M is a maximal ideal containing J. Since $\{0\}$ is a proper ideal, it follows from what has just been proved that E possesses maximal ideals. Every element x of an arbitrary proper ideal J is non-invertible; otherwise because of $y = (yx^{-1})x$ all E would lie in J. If conversely x is any non-invertible element of E, then $\{yx : y \in E\}$ is a proper ideal of E containing x. By what has already been proved, x lies in a maximal ideal. The closure $\overline{M}$ of a maximal ideal M is an ideal containing M (Exercise 1 in §22).

Since all the elements of M are non-invertible, it follows with the aid of Proposition 9.4 that the same holds also for the elements of $\overline{M}$. Thus $\overline{M}$ is a proper ideal (it does not contain e) and coincides therefore with M. ■

From the proposition just proved we obtain the surprising

Proposition 108.2. *Every multiplicative linear form f on the commutative Banach algebra E is* continuous. *For a non-trivial f we have $\|f\| = 1$. Thus $E^\times$ lies on the surface of the unit ball of E'.*

In the proof let $f \neq 0$. As we have seen, $N(f)$ is a maximal ideal in E. It follows from Proposition 108.1 that $N(f)$ is closed, which by Proposition 87.5 (or Exercise 1 in §21) ensures the continuity of f. From $|f(x)|^2 = |f(x^2)| \leqq \|f\| \|x\|^2$ it follows that $\|f\| \leqq 1$; because of $|f(e)| = 1 = \|e\|$ we must even have $\|f\| = 1$. ■

In order to prove the existence of non-trivial multiplicative linear forms, we need to suppose that E is complex:

Proposition 108.3. *Let E be a* complex, *commutative Banach algebra. Every maximal ideal has codimension* 1, *$E^\times$ is not empty, and the map $f \mapsto N(f)$ is a bijection between $E^\times$ and the set of all maximal ideals in E.*

Because of the results we have just established, it is sufficient to prove the first assertion of the proposition. So let M be a maximal ideal. Then $\check{E} := E/M$ is a complex Banach algebras (§22). E is even a division algebra: Indeed, if the residue class $\check{x}_0 \in \check{E}$ of the vector $x_0 \in E$ is distinct from zero, then x_0 does not lie in M, consequently $\{xx_0 + y : x \in E,\ y \in M\}$ is an ideal which contains M properly, hence must coincide with E. In particular e can be represented in the form $e = xx_0 + y$, $y \in M$. If we pass to residue classes, we obtain from this equation that $\check{x}_0$ has the inverse $\check{x}$. Proposition 45.2 implies now that $\check{E}$ is one-dimensional, from where it follows that codim $M = 1$ (Exercise 3 in §20). ■

The results obtained so far suggest to let the set Φ, by means of which we associated with every $x \in E$ a function $F_x : \Phi \to \mathbf{K}$ (§107), not consist of all multiplicative linear forms (as we have planned at the beginning), but only of all non-trivial ones, i.e., to choose $\Phi = E^\times$. The homomorphism H of algebras which associates with every $x \in E$ the function

$$F_x : \begin{cases} E^\times \to \mathbf{C} \\ f \mapsto f(x) \end{cases}$$

will be called the *Gelfand homomorphism* of the Banach algebra E. We know that it is injective if and only if $E^\times$ is total. We now turn to the question whether—or under what conditions—$E^\times$ is total. From now on we assume that the commutative Banach algebra E is complex.

It follows from Propositions 108.1 and 108.3 that $x \in E$ is non-invertible if and only if there exists an $f \in E^\times$ such that $f(x) = 0$. From here we obtain the assertion

(108.1) $\qquad x \in E$ is invertible $\Leftrightarrow f(x) \neq 0$ for all $f \in E^\times$.

If now $f(x) = 0$ for all $f \in E^\times$, then we have $f(\lambda e - x) = \lambda \neq 0$ for all $\lambda \neq 0$ and all $f \in E^\times$; because of (108.1) it follows that $\lambda e - x$ is invertible for all $\lambda \neq 0$. Consequently the spectral radius $r(x) = \lim \|x^n\|^{1/n}$ of x vanishes: x is quasi-nilpotent. Let conversely x be quasi-nilpotent. Then for $f \in E^\times$ and $n \in \mathbf{N}$ we have $|f(x)| = |f(x^n)|^{1/n} \leqq (\|f\| \, \|x^n\|)^{1/n} = \|x^n\|^{1/n}$ and therefore $f(x) = 0$. Thus:

(108.2) $\qquad f(x) = 0$ for all $f \in E^\times \Leftrightarrow x$ is quasi-nilpotent.

We leave to the reader the simple proof that the vector x in a commutative, complex Banach algebra E is quasi-nilpotent if and only if $e - yx$ is invertible for every $y \in E$, i.e., if x lies in the radical of E (the concept of a radical has been introduced already in §52). If we denote by $Q(E)$ the set of all quasi-nilpotent elements of E and by $\mathrm{rad}(E)$ the radical of E, then because of Proposition 108.3 and (108.2) we can say:

(108.3) $\quad \mathrm{rad}(E) = Q(E) =$ the intersection of all maximal ideals in E.

E is called a *semi-simple algebra* if $\mathrm{rad}(E) = \{0\}$. With this terminology we can formulate the result of our considerations as follows:

Proposition 108.4 *Let E be a commutative, complex Banach algebra. $E^\times$ is total, and the Gelfand homomorphism H of E is injective if and only if E is semi-simple.*

The weak topology $\sigma(E', E)$ induces on $E^\times$ a topology which is called the *Gelfand topology* of $E^\times$. We have noted in §107 that it coincides with the initial topology for the family of maps $F_x: E^\times \to \mathbf{C}$, $x \in E$. We now have:

Proposition 108.5. *Let E be a commutative, complex Banach algebra. $E^\times$ equipped with the Gelfand topology is a* compact *Hausdorff space.*

Proof. $\sigma(E', E)$ is Hausdorff (Theorem 82.1), the same is then also true for the Gelfand topology (Exercise 5 in §81). The compactness of $E^\times$ will be ensured if we can show that $E^\times$ is a $\sigma(E', E)$-closed subset of the unit ball $K' := \{f \in E' : \|f\| \leqq 1\}$ of E', which is $\sigma(E', E)$-compact according to Proposition 102.1. The function $L_x: K' \to \mathbf{C}$, defined by $L_x(f) = f(x)$, is $\sigma(E', E)$-continuous, consequently for all x, y in E the sets $\{f \in K' : (L_{xy} - L_x L_y)(f) = 0\}$ and thus also their intersection

$$\bigcap_{x, y \in E} \{f \in K' : (L_{xy} - L_x L_y)(f) = 0\} = \bigcap_{x, y \in E} \{f \in K' : f(xy) - f(x)f(y) = 0\}$$

are weakly closed in K'; but this intersection is clearly the set M of all multiplicative linear forms on E. Since the restriction Λ_e of L_e to M is weakly continuous and

$$\Lambda_e(f) = f(e) = \begin{cases} 1 & \text{for } f \in E^{\times} \\ 0 & \text{for } f = 0, \end{cases}$$

hence $E^{\times} = \Lambda_e^{-1}(\{1\})$, it follows that $E^{\times}$ is a weakly closed subset of M and thus also a weakly closed subset of K'. ■

§109. The Gelfand representation theorem

Let E be again a commutative, complex Banach algebra. The image Hx of $x \in E$ under the Gelfand homomorphism H has been denoted so far by F_x. The conventional notation is $\hat{x}$, which we shall use from now on. We equip $E^{\times}$ with the Gelfand topology and denote by $C(E^{\times})$ the algebra of all complex-valued, continuous functions on the compact Hausdorff space $E^{\times}$ (Proposition 108.5). $C(E^{\times})$ will be equipped with the supremum-norm and be made thus into a Banach algebra. We list once more the two maps with which we have to do:

$$H: \begin{cases} E \to C(E^{\times}) \\ x \mapsto \hat{x} \end{cases}, \qquad \hat{x}: \begin{cases} E^{\times} \to \mathbf{C} \\ f \mapsto f(x) \end{cases}.$$

The following *Gelfand representation theorem* is for the greater part only a summary and reformulation of our results obtained so far:

Theorem 109.1. *Let E be a commutative, complex Banach algebra. Then the Gelfand homomorphism $H: E \to C(E^{\times})$ has the following properties:*

(a) *H is a homomorphism of algebras from E into $C(E^{\times})$.*
(b) *$\|\hat{x}\| \leqq \|x\|$ for all $x \in E$; in particular H is continuous.*
(c) *$\hat{e}(f) \equiv 1$ on $E^{\times}$.*
(d) *$x \in E$ is invertible $\Leftrightarrow \hat{x}(f) \neq 0$ for all $f \in E^{\times}$.*
(e) *$\hat{x}(E^{\times}) = \sigma(x)$.*
(f) *$\|\hat{x}\| = r(x)$.*
(g) *$\operatorname{rad}(E) = \{x \in E : \hat{x} = 0\} = N(H)$.*
(h) *H is injective $\Leftrightarrow$ E is semi-simple $\Leftrightarrow$ $E^{\times}$ is total.*

Proof. (a) and (b) have already been discussed in §107.
(c) is trivial.
(d) is nothing but (108.1).
(e) follows from (c) and (d) because $\lambda \notin \hat{x}(E^{\times}) \Leftrightarrow (\lambda\hat{e} - \hat{x})(f) \neq 0$ for all $f \in E^{\times}$.
(f) is obtained immediately from (e).
(g) is only a reformulation of (108.2) combined with (108.3).
(h) repeats Proposition 108.4. ■

Because of Proposition 108.3 one uses in the Gelfand theory instead of the set $E^\times$ frequently the set $\mathfrak{M}$ of all maximal ideals of E. The bijection $f \mapsto N(f)$ transfers the Gelfand topology of $E^\times$ onto $\mathfrak{M}$ and $\mathfrak{M}$ becomes so a compact Hausdorff space. The value of the function $\hat{x}: \mathfrak{M} \to \mathbf{C}$ associated with the vector $x \in E$ is computed at the point $M \in \mathfrak{M}$ as follows: one writes x in the form $x = \lambda e + y$, $y \in M$; then $\hat{x}(M) = \lambda$.

We now present one of the first and best known application of the Gelfand theory.

Let W be the set of all complex-valued functions x defined on the boundary of the unit disk, which can be expanded into an absolutely convergent trigonometric series:

$$(109.1) \qquad x(e^{i\varphi}) = \sum_{-\infty}^{\infty} \alpha_n e^{in\varphi} \quad \text{with} \quad \alpha_n \in \mathbf{C}, \ \sum_{-\infty}^{\infty} |\alpha_n| < \infty, \quad 0 \leqq \varphi < 2\pi$$

(e is the basis of the natural logarithms). If we define the arithmetic operations pointwise and the norm by

$$\|x\| := \sum_{-\infty}^{\infty} |\alpha_n|,$$

then W becomes a commutative, complex Banach algebra with the unit element $e^{i\varphi} \mapsto 1$, the so-called *Wiener algebra*. Now let $f \in W^\times$. Then we have—with a harmless laxity in the notation—

$$(109.2) \qquad f(x) = \sum_{-\infty}^{\infty} \alpha_n f(e^{in\varphi}) = \sum_{-\infty}^{\infty} \alpha_n \lambda^n, \qquad \text{where} \quad \lambda := f(e^{i\varphi}).$$

Obviously $\lambda \neq 0$, $|\lambda| \leqq 1$ and $|1/\lambda| = |f(e^{-i\varphi})| \leqq 1$, hence $|\lambda| = 1$ and so $\lambda = e^{i\psi}$ with a certain ψ, $0 \leqq \psi < 2\pi$. Because of (109.1) and (109.2) we can thus say: For every $f \in W^\times$ there exists a real number ψ, $0 \leqq \psi < 2\pi$, such that

$$\hat{x}(f) = f(x) = \sum_{-\infty}^{\infty} \alpha_n e^{in\psi} = x(e^{i\psi}).$$

With the aid of Theorem 109.1d we obtain from here immediately *Wiener's theorem*:

If the function $x \in W$ does not assume the value 0 *then it is invertible in W, i.e., $1/x(e^{i\varphi})$ can be expanded into an absolutely convergent trigonometric series*:

$$\frac{1}{x(e^{i\varphi})} = \sum_{-\infty}^{\infty} \beta_n e^{in\varphi} \quad \text{with} \quad \sum_{-\infty}^{\infty} |\beta_n| < \infty.$$

Under the hypotheses of Theorem 109.1 the Gelfand homomorphism H is obviously an isomorphism of normed spaces from E into $C(E^\times)$, i.e., we have $\|\hat{x}\| = \|x\|$ for all $x \in E$, if and only if $\|x\| = r(x)$ for all $x \in E$ (Theorem 109.1f). With reference to the terminology of §54, we call an element x of E *normaloid* if $r(x) = \|x\|$. For a normaloid x we have because of $\|x\| = r(x) \leqq \|x^n\|^{1/n} \leqq \|x\|$

always $\|x^n\| = \|x\|^n$, and in particular $\|x^2\| = \|x\|^2$. Conversely, if this last equation holds for all $x \in E$, then $\|x^4\| = \|(x^2)^2\| = \|x\|^4$, and in general $\|x^{2^n}\| = \|x\|^{2^n}$, from where it follows immediately that $r(x) = \lim\|x^{2^n}\|^{1/2^n} = \|x\|$.

We state the result of this consideration:

Proposition 109.1. *The Gelfand homomorphism* $x \mapsto \hat{x}$ *of a commutative, complex Banach algebra* preserves norms *if and only if one of the following equivalent conditions is satisfied*:

(a) *Every* $x \in E$ *is normaloid.*
(b) *For every* $x \in E$ *we have* $\|x^2\| = \|x\|^2$.

Exercise

On the complex Banach algebra $E := C[a, b]$ we define for every $t \in [a, b]$ a multiplicative linear form f_t by $f_t(x) := x(t)$. The map $t \mapsto f_t$ is a homeomorphism from $[a, b]$ into $E^\times$, so that $E^\times$ can be identified (as a topological space) with $[a, b]$, and consequently $C(E^\times)$ with $C[a, b]$. If we make this identification, then the Gelfand homomorphism H of $C[a, b]$ is simply the identity.

§110. The representation of commutative B^*-algebras

We now raise the question, under what conditions is the Gelfand homomorphism $H: E \to C(E^\times)$ of the commutative, complex Banach algebra E *bijective* and *norm-preserving*, in other words: We ask, when is the Banach algebra E—except for the notation of its elements—nothing but the Banach algebra of all continuous, complex-valued functions on the compact Hausdorff space $E^\times$?

In the examination of this question a new structural element of the field **C** introduces itself, which we have so far hardly considered at all: the *conjugation* $\alpha \mapsto \bar{\alpha}$. That this will play a role in connection with the question, when does the subalgebra $H(E)$ coincide with the whole algebra $C(E^\times)$, can be conjectured on the basis of the *Stone–Weierstrass theorem*, which we now state for later use:

Let T be a compact Hausdorff space and R a subalgebra of the Banach algebra $C(T)$ with the following properties: R is closed, separates the points of T, contains a constant function $\neq 0$ and contains with every function $h: t \mapsto h(t)$ also its conjugate $\bar{h}: t \mapsto \overline{h(t)}$. Then $R = C(T)$.

The reader will find a proof of this fundamental theorem in [14].

We will now first assume that H is a norm-preserving isomorphism from E onto $C(E^\times)$. We denote the conjugation in $C(E^\times)$ by K, thus $Kh = \bar{h}$ for $h \in C(E^\times)$. For K the following rules are valid:

$$K(g + h) = Kg + Kh, \quad K(\alpha h) = \bar{\alpha}K(h), \quad K(gh) = K(g)K(h), \quad K^2h = h,$$

$$\|Kh\| = \|h\|.$$

The conjugation in $C(E^\times)$ generates by means of the bijection H a 'conjugation' in E if we associate with every element $x \in E$ the element $x^* := H^{-1}KHx \in E$. This correspondence has obviously the following properties:

(I1) $(x + y)^* = x^* + y^*$,
(I2) $(\alpha x)^* = \bar{\alpha} x^*$,
(I3) $(xy)^* = y^* x^*$,
(I4) $x^{**} = x$,
(I5) $\|x^* x\| = \|x\|^2$.

A self-map $x \mapsto x^*$ of a complex, not necessarily commutative Banach algebra E, which satisfies (I1) through (I4), is called an *involution*. If also (I5) is valid, then E is called a *B^*-algebra*. A homomorphism of algebras $A: E_1 \to E_2$ between two Banach algebras with involution is called a **-homomorphism* if it preserves the involution, i.e., if $A(x^*) = (Ax)^*$ for all $x \in E_1$. A **-isomorphism* is an injective *-homomorphism.

In a Banach algebra with involution we have $0^* = 0$ and $e^* = e$.

If one introduces in the complex algebra $C(T)$, T a compact Hausdorff space, or in $\mathscr{L}(F)$, F a complex Hilbert space, the involution $h \mapsto \bar{h}$ or $A \mapsto A^*$, respectively (where A^* is the adjoint operator), then $C(T)$ and $\mathscr{L}(F)$ become obviously B^*-algebras (for $\mathscr{L}(F)$ see Propositions 67.1 and 67.4).

Our considerations so far show that a commutative, complex Banach algebra E, whose Gelfand homomorphism $H: E \to C(E^\times)$ is a norm-preserving bijection, is in a natural way a B^*-algebra. We focus therefore our attention on such algebras.

An element x of a B^*-algebra is said to be *normal* if $xx^* = x^*x$, *self-adjoint* if $x^* = x$. In a commutative B^*-algebra every element is normal. Every $x \in E$ can be written in the form

(110.1) $$x = x_1 + ix_2 \quad \text{with self-adjoint} \quad x_1, x_2;$$

we only have to set

$$x_1 := \frac{x + x^*}{2}, \qquad x_2 := \frac{x - x^*}{2i}.$$

Basic for the further investigation is:

Proposition 110.1. *In a B^*-algebra E the following assertions are valid:*

(a) *$\|x^*\| = \|x\|$ for all $x \in E$.*
(b) *Every normal element is normaloid, i.e., $r(x) = \|x\|$ holds.*
(c) *Every self-adjoint element has a real spectrum.*

Proof. (a) From (I5) it follows that $\|x\|^2 = \|x^*x\| \leqq \|x^*\|\,\|x\|$ and thus $\|x\| \leqq \|x^*\|$. But then we have also conversely $\|x^*\| \leqq \|x^{**}\| = \|x\|$.

(b) For a normal x we obtain because of (I5) the equality $\|x^2\|^2 = \|(x^*)^2x^2\| = \|(xx^*)^*(xx^*)\| = \|xx^*\|^2 = \|x\|^4$, hence $\|x^2\| = \|x\|^2$. Since also x^2 is normal, it follows that $\|x^4\| = \|(x^2)^2\| = \|x^2\|^2 = \|x\|^4$, in general $\|x^{2^n}\| = \|x\|^{2^n}$ and thus $r(x) = \|x\|$.

(c) Let $x = x^*$ and $\alpha + i\beta \in \sigma(x)$, α, β real. Then

$$(\alpha + i\beta)e - x = (\alpha + i(\beta + \lambda))e - (x + i\lambda e)$$

is not invertible for any $\lambda \in \mathbf{R}$, hence $\alpha + i(\beta + \lambda) \in \sigma(x + i\lambda e)$ and thus because of (44.6) and (I5)

$$\begin{aligned}\alpha^2 + (\beta + \lambda)^2 &= |\alpha + i(\beta + \lambda)|^2 \leqq \|x + i\lambda e\|^2 = \|(x + i\lambda e)^*(x + i\lambda e)\| \\ &= \|(x - i\lambda e)(x + i\lambda e)\| = \|x^2 + \lambda^2 e\| \leqq \|x\|^2 + \lambda^2.\end{aligned}$$

Thus we have $\alpha^2 + \beta^2 + 2\beta\lambda \leqq \|x\|^2$ for all real λ, from where it follows immediately that $\beta = 0$. ■

We can now prove without effort the *Gelfand–Neumark theorem*:

Theorem 110.1. *The Gelfand homomorphism H of a commutative B^*-algebra E is a norm-preserving $*$-isomorphism onto the B^*-algebra $C(E^\times)$.*

Proof. Every $x \in E$ is normal, hence also normaloid (Proposition 110.1). Consequently H is an isomorphism of normed spaces (Proposition 109.1). The Gelfand image $\hat{y} = Hy$ of a self-adjoint $y \in E$ is because of $\hat{y}(E^\times) = \sigma(y) \subset \mathbf{R}$ (Theorem 109.1e and Proposition 110.1c) a real-valued function. If we decompose $x \in E$ according to (110.1), then we obtain $H(x^*) = H(x_1 - ix_2) = Hx_1 - iHx_2 = \overline{Hx_1} - i\overline{Hx_2} = \overline{Hx_1 + iHx_2} = \overline{Hx} = (Hx)^*$, thus H is a $*$-isomorphism. From the two assertions proved so far, it follows that the subalgebra $H(E)$ of $C(E^\times)$ is closed and that it contains with every function also its conjugate function. By Theorem 109.1c there is a non-zero constant function in $H(E)$. Finally $H(E)$ separates the points of $E^\times$: From $\hat{x}(f) = \hat{x}(g)$ for all $x \in H(E)$ it follows that $f(x) = g(x)$ for all $x \in E$ and so $f = g$. The Stone–Weierstrass theorem stated at the beginning of this section ensures now the surjectivity of H. ■

The Gelfand–Neumark theorem characterizes the commutative B^*-algebras as the B^*-algebras of all continous functions on appropriate compact Hausdorff spaces. We can consider this important theorem in another light: Though we speak always of a *map* when the elements of a set E are associated with certain elements of another set, we shall get 'true'—even if more or less distorted—images of a *structured* set E (e.g., an algebra) only via *structure-preserving* maps (homomorphisms). The simplest homomorphisms of an algebra are its multiplicative functionals, and the Gelfand–Neumark theorem now states in a loose formulation: *From the totality of the simplest 'true' images (or of the simplest 'true' mappings) of a commutative B^*-algebra E one can already obtain an exact, completely undistorted image of E*; for this one has only to construct the space of all continuous functions on the set $E^\times$ equipped with the Gelfand topology.

Let A be a normal endomorphism of a complex Hilbert space. The Banach algebra generated by A, A^* and I (the closure—for the operator-norm—of all polynomials in A, A^*) is a commutative B^*-algebra. From Theorem 110.1, applied to this algebra, one can obtain a spectral theorem for A. In particular one obtains in this way a new approach to the spectral theorem for symmetric operators (Theorem 79.1). The reader will find more details e.g., in [4].

Bibliography

A. Texts and Handbooks

[1] Banach, S.: Théorie des Opérations Linéaires. Monografje Matematyczne, I. Warszawa 1932.

[2] Bourbaki, N.: Eléments de Mathématiques. Livre V. Espaces Vectoriels Topologiques. Actualités Scientifiques et Industrielles 1189, 1229. Hermann, Paris 1966, 1955.

[3] Dieudonné, J.: Treatise on Analysis I–VI. Academic Press, New York—San Francisco—London 1960–1978.

[4] Dunford, N.; Schwartz, J. T.: Linear Operators I–III. Interscience, New York–London 1958, 1963, 1971.

[5] Edwards, R. E.: Functional Analysis: Theory and Applications. Holt, Rinehart and Winston, New York 1965.

[6] Hille, E.; Phillips, R.: Functional Analysis and Semigroups. American Mathematical Society Colloquium Publications 31. Providence, R.I. 1957.

[7] Horváth, J.: Topological Vector Spaces and Distributions I. Addison-Wesley, Reading, Mass. 1966.

[8] Köthe, G.: Topological Vector Spaces I, II. Grundlehren der math. Wissenschaften 159, 237. Springer, Berlin—Heidelberg—New York 1969, 1979.

[9] Kolmogorov, A. N.; Fomin, S. J.: Elements of the Theory of Functions and Functional Analysis I, II. Graylock, Rochester, N.Y. 1957, 1961.

[10] Riesz, F.; Sz.-Nagy, B.: Functional Analysis. Frederick Ungar, New York 1955.

[11] Robertson, A. P.; Robertson, W. J.: Topological Vector Spaces. 2nd ed. Cambridge Univ. Press 1973.

[12] Rudin, W.: Functional Analysis. McGraw-Hill, New York 1973.

[13] Schaefer, H. H.: Topological Vector Spaces. 3rd ed. Graduate Texts in Mathematics 3. Springer, Berlin—Heidelberg—New York 1971.

[14] Simmons, G. F.: Introduction to Topology and Modern Analysis. McGraw-Hill, New York 1963.

[15] Taylor, A. E.; Lay, D. C.: Introduction to Functional Analysis. 2nd ed. Wiley, New York—Chichester—Brisbane—Toronto 1980.

[16] Wloka, J.: Funktionalanalysis und Anwendungen. De Gruyter, Berlin—New York 1971.

[17] Yosida, K.: Functional Analysis. Grundlehren der math. Wissenschaften 123. 5th ed. Springer, Berlin—Heidelberg—New York 1978.

[18] Zaanen, A. C.: Linear Analysis. 4th printing. North-Holland, Amsterdam-Groningen 1964.

B. Expository articles

[19] Devinatz, A.: The deficiency index problem for ordinary selfadjoint differential operators. Bull. Amer. Math. Soc. **79** (1973) 1109–1127.

[20] Dieudonné, J.: Recent developments in the theory of locally convex vector spaces. Bull. Amer. Math. Soc. **59** (1953) 495–512.

[21] Dolph, C. L.: Recent developments in some non-self-adjoint problems of mathematical physics. Bull. Amer. Math. Soc. **67** (1961) 1–69.

[22] Dunford, N.: Spectral theory. Bull. Amer. Math. Soc. **49** (1943) 637–651.

[23] Dunford, N.: A survey of the theory of spectral operators. Bull. Amer. Math. Soc. **64** (1958) 217–274.

[24] Gamelin, T. W.: The algebra of bounded analytic functions. Bull. Amer. Math. Soc. **79** (1973) 1095–1108.

[25] Garkavi, A. L.: The theory of best approximation in normed linear spaces. Progress Math. **8** (1970) 83–150.

[26] Graves, L. M.: Topics in the functional calculus. Bull. Amer. Math. Soc. **41** (1935) 641–662. Errata, ibid. **42** (1936) 381–382.

[27] Halmos, P. R.: Ten problems in Hilbert space. Bull. Amer. Math. Soc. **76** (1970) 887–933 (Correction of misleading misprints in Math. Rev. **42** (1971) #5066).

[28] Hewitt, E.: The Rôle of Compactness in Analysis. Amer. Math. Monthly **67** (1960) 499–516.

[29] Hilbert, D.: Wesen und Ziele einer Analysis der unendlich vielen unabhängigen Variablen. Rend. Circ. Mat. Palermo **27** (1909) 59–74; Gesammelte Abhandlungen III, 56–72.

[30] Horváth, J.: An Introduction to Distributions. Amer. Math. Monthly **77** (1970) 227–240.

[31] Hyers, D. H.: Linear topological spaces. Bull. Amer. Math. Soc. **51** (1945) 1–21, 1001.

[32] Kaplansky, I.: Topological rings. Bull. Amer. Math. Soc. **54** (1948) 809–826.

[33] Khavin, V. P.: Spaces of analytic functions. Progr. Math. **1** (1968) 75–167.

[34] Krachkovskii, S. N.; Dikanskii, A. S.: Fredholm operators and their generalizations. Progr. Math. **10** (1971) 37–72.

[35] McArthur, C. W.: Developments in Schauder basis theory. Bull. Amer. Math. Soc. **78** (1972) 877–908.

[36] Murray, F. J.: The analysis of linear transformations. Bull. Amer. Math. Soc. **48** (1942) 76–93.

[37] Myers, S. B.: Normed linear spaces of continuous functions. Bull. Amer. Math. Soc. **56** (1950) 233–241.

[38] Nachbin, L.: Recent developments in infinite dimensional holomorphy. Bull. Amer. Math. Soc. **79** (1973) 625–640.

[39] Phillips, R. S.: Semi-groups of operators. Bull. Amer. Math. Soc. **61** (1955) 16–73.

[40] Taylor, A. E.: Analysis in complex Banach spaces. Bull. Amer. Math. Soc. **49** (1943) 652–669.

[41] Taylor, A. E.: Notes on the history of the uses of analyticity in operator theory. Amer. Math. Monthly **78** (1971) 331–342, 1104.

[42] Weyl, H.: Ramifications, old and new, of the eigenvalue problem. Bull. Amer. Math. Soc. **56** (1950) 115–139.

C. Monographs and specialized literature

I. Topological, in particular normed linear spaces and their operators

[43] Adasch, N.: Der Graphensatz in topologischen Vektorräumen. Math. Zeitschr. **119** (1971) 131–142.

[44] Bonsall, F. F.; Duncan, J.: Numerical Ranges of Operators on Normed Spaces and

of Elements of Normed Algebras. London Mathematical Society Lecture Note Series 2. Cambridge University Press 1971.

[45] Bonsall, F. F.; Duncan, J.: Numerical Ranges II. London Mathematical Society Lecture Note Series 10. Cambridge University Press 1973.

[46] Brace, J. W.; Richetta, P. J.: The approximation of linear operators. Trans. Amer. Math. Soc. **157** (1971) 1–21.

[47] Conway, J. B.: The inadequacy of sequences. Amer. Math. Monthly **76** (1969) 68–69.

[48] Davie, A. M.: The approximation problem for Banach spaces. Bull. London Math. Soc. **5** (1973) 261–266.

[49] Day, M. M.: Normed Linear Spaces. 2nd ed. Ergebnisse der Mathematik 21. Springer, Berlin—Göttingen—Heidelberg 1962.

[50] De Wilde, M.: Closed Graph Theorems and Webbed Spaces. Research Notes in Mathematics 19. Pitman, London—San Francisco—Melbourne 1978.

[51] Douglas, R. G.: Banach Algebra Techniques in Operator Theory. Pure and Applied Mathematics 49. Academic Press, New York—London 1972.

[52] Dvoretzky, A.; Rogers, C. A.: Absolute and unconditional convergence in normed linear spaces. Proc. Nat. Acad. Sci. U.S.A. (3) **36** (1950) 192–197.

[53] Enflo, P.: A counterexample to the approximation problem in Banach spaces. Acta math. **130** (1973) 309–317.

[54] Goldberg, S.: Unbounded Linear Operators, Theory and Applications. McGraw-Hill, New York 1966.

[55] Hoffman, K.: Banach Spaces of Analytic Functions. Prentice-Hall, Englewood Cliffs, N.J. 1962.

[56] Husain, T.: The Open Mapping and Closed Graph Theorems in Topological Vector Spaces. Oxford Mathematical Monographs. Oxford University Press 1965.

[57] James, R. C.: A non-reflexive Banach space isometric with its second conjugate space. Proc. Nat. Acad. Sci. U.S.A. **37** (1951) 174–177.

[58] Kato, T.: Perturbation theory for nullity, deficiency and other quantities of linear operators. J. Analyse Math. **6** (1958) 261–322.

[59] Kato, T.: Perturbation Theory for Linear Operators. Grundlehren der math. Wissenschaften 132. Springer, Berlin—Heidelberg—New York 1966.

[60] Lax, P. D.: Symmetrizable linear transformations. Comm. Pure and Appl. Math. **7** (1954) 633–647.

[61] Lindenstrauss, J.; Tzafriri, L.: Classical Banach Spaces I, II. Ergebnisse der Mathematik 92, 97. Springer, Berlin—Heidelberg—New York, 1977, 1979.

[62] Murray, F. J.: On complementary manifolds and projections in spaces L_p and l_p. Trans. Amer. Math. Soc. **41** (1937) 138–152.

[63] Pietsch, A.: Operator Ideals. North-Holland, Amsterdam—New York 1980.

[64] Radjavi, H.; Rosenthal, P.: Invariant Subspaces. Ergebnisse der Mathematik 77. Springer, Berlin—Heidelberg—New York 1973.

[65] Singer, I.: Bases in Banach Spaces I. Grundlehren der math. Wissenschaften 154. Springer, Berlin—Heidelberg—New York 1970.

[66] Stummel, F.: Diskrete Konvergenz linearer Operatoren I, II. Math. Ann. **190** (1970/71) 45–92; Math. Zeitschr. **120** (1971) 231–264.

[67] Wilansky, A.: Modern Methods in Topological Vector Spaces. McGraw-Hill, New York 1978.

II. Inner product spaces and their operators

[68] Akhiezer, N. I.; Glazman, I. M.: Theory of Linear Operators in Hilbert Space I, II. Frederick Ungar, New York 1961, 1963.

[69] Beals, R.: Topics in Operator Theory. Chicago Lectures in Mathematics. The University of Chicago Press, Chicago, Ill.—London 1971.

[70] Berberian, S. K.: Lectures in Functional Analysis and Operator Theory. Graduate Texts in Mathematics 15. Springer, New York—Heidelberg—Berlin 1974.
[71] Dixmier, J.: Sur les bases orthonormales dans les espaces préhilbertiens. Acta Sci. Math. Szeged **15** (1953) 29–30.
[72] Dixmier, J.: Les Algèbres d'Opérateurs dans l'Espace Hilbertien (Algèbres de von Neumann). Gauthier-Villars, Paris 1957.
[73] Dixmier, J.: C* Algebras. North-Holland, Amsterdam—New York 1977.
[74] Fillmore, P. A.: Notes on Operator Theory. Mathematical Studies 30. Van Nostrand Reinhold, New York—London—Melbourne 1970.
[75] Fillmore, P. A.; Williams, J. P.: On operator ranges. Advances in Math. **7** (1971) 254–281.
[76] Friedrichs, K. O.: Spectral Theory of Operators in Hilbert Space. Applied Mathematical Sciences 9. Springer, New York—Heidelberg—Berlin 1973.
[77] Halmos, P. R.: Introduction to Hilbert Space and the Theory of Spectral Multiplicity. Chelsea, New York 1951.
[78] Halmos, P. R.: A Hilbert Space Problem Book. Graduate Texts in Mathematics 19. Springer, New York—Heidelberg—Berlin 1974.
[79] Helmberg, G.: Introduction to Spectral Theory in Hilbert Space. North-Holland Series in Applied Mathematics 6. North-Holland, Amsterdam—New York 1975.
[80] Jordan, P.; von Neumann, J.: On inner products in linear, metric spaces. Ann. of Math. (2) **36** (1935) 719–723.
[81] Maurin, K.: Methods of Hilbert Spaces. 2nd ed. Monografje Matematiczne 45. Polska Akademia Nauk, Warszawa 1972.
[82] Stone, M. H.: Linear Transformations in Hilbert Space and their Applications to Analysis. American Mathematical Society Colloquium Publication 15. Providence, R.I. 1970.
[83] Sz.-Nagy, B.: Spektraldarstellung Linearer Transformationen des Hilbertschen Raumes. Ergebnisse der Mathematik 39. Springer, Berlin—Heidelberg—New York 1967.
[84] Sz.-Nagy, B.; Foiaş, C.: Harmonic Analysis of Operators on Hilbert Space. North-Holland, Amsterdam—London 1970.
[85] Wielandt, H.: Über die Eigenwertaufgaben mit reellen diskreten Eigenwerten. Math. Nachr. **4** (1950/51) 308–314.

III. Banach algebras

[86] Bonsall, F. F.; Duncan, J.: Complete Normed Algebras. Ergebnisse der Mathematik 80. Springer, Berlin—Heidelberg—New York 1973.
[87] Gelfand, I. M.; Raikov, D. A.; Silov, G. E.: Commutative Normed Rings. Chelsea, New York 1964.
[88] Naimark, M. A.: Normed Rings. P. Noordhoff, Groningen 1960.
[89] Rickart, Ch. E.: General Theory of Banach Algebras. Van Nostrand, Princeton—Toronto—London 1960.
[90] Sakai, S.: C*-Algebras and W*-Algebras. Ergebnisse der Mathematik 60. Springer, Berlin—Heidelberg—New York 1971.

IV. Operators with finite deficiency and with finite chains. Normal solvability

[91] Atkinson, F. V.: The normal solubility of linear equations in normed spaces. Mat. Sbornik N.S. **28** (70) (1951) 3–14 (in Russian; Math. Rev. **13** (1952) 46).
[92] Atkinson, F. V.: On relatively regular operators. Acta Sci. Math. Szeged **15** (1953) 38–56.

[93] Breuer, M.: Banachalgebren mit Anwendungen auf Fredholmoperatoren und singuläre Integralgleichungen. Bonn 1965.
[94] Breuer, M.: Fredholm theories in von Neumann algebras I, II. Math. Ann. **178** (1968) 243–258; ibid. **180** (1969) 313–325.
[95] Caradus, S. R.; Pfaffenberger, W. E.; Yood, B.: Calkin Algebras and Algebras of Operators on Banach Spaces. Lecture Notes in Pure and Appl. Math. 9. Dekker, New York 1974.
[96] Cordes, H. O.: On a generalized Fredholm theory. J. Reine Angew. Math. **227** (1967) 121–149.
[97] Deprit, A.: Contributions à l'étude de l'algèbre des applications linéaires continues d'un espace localement convexe séparé; Théorie de Riesz—théorie spectrale Acad. Roy. Belg. Classe Sci. Mém. Coll. 8° (31) (1959) no. 2.
[98] Dieudonné, J.: La dualité dans les espaces vectoriels topologiques. Ann. Sci. Ecole Normale Sup. (3) **59** (1942) 107–139.
[99] Dieudonné, J.: Sur les homomorphismes d'espaces normés. Bull. Sci. Math. (2) **67** (1943) 72–84.
[100] Feldman, I.; Gohberg, I. C.; Markus, A. S.: Normally solvable operators and ideals associated with them. Amer. Math. Soc. Transl. (2) **61** (1967) 63–84.
[101] Gohberg, I. C.; Krein, M. G.: The basic propositions on defect numbers, root numbers, and indices of linear operators. Amer. Math. Soc. Transl. (2) **13** (1960) 185–264.
[102] Gramsch, B.: Ein Schema zur Theorie Fredholmscher Endomorphismen und eine Anwendung auf die Idealkette der Hilberträume. Math. Ann. **171** (1967) 263–272.
[103] Gramsch, B.: Über analytische Operatorfunktionen und Indexberechnung. Studia Math. **38** (1970) 313–317.
[104] Gramsch, B.: Meromorphie in der Theorie der Fredholmoperatoren mit Anwendungen auf elliptische Differentialoperatoren. Math. Ann. **188** (1970) 97–112.
[105] Heuser, H.: Über Operatoren mit endlichen Defekten. Inaug.-Diss. Tübingen 1956.
[106] Heuser, H.: Zur Eigenwerttheorie einer Klasse symmetrischer Operatoren. Math. Zeitschr. **74** (1960) 167–185.
[107] Heuser, H.: Algebraic theory of Atkinson operators. Revista Colombiana Mat. **2** (1968) 137–143. (The erroneous review in Zbl. **193** (1970) 96 was corrected in Zbl. **207** (1971) 452).
[108] Kroh, H.: Fredholmoperatoren in dualisierbaren Algebren. Inaug.-Diss. Frankfurt/M. 1970.
[109] Kroh, H.: Saturierte Algebren. Math. Ann. **211** (1974) 175–182.
[110] Lay, D. C.: Spectral analysis using ascent, descent, nullity and defect. Math. Ann. **184** (1970) 197–214.
[111] Neubauer, G.: Über den Indcx abgeschlossener Operatoren in Banachräumen I, II. Math. Ann. **160** (1965) 93–130; ibid. **162** (1965/66) 92–119.
[112] Nikolskij, S.: Linear equations in normed linear space. Izvestija Akad. Nauk SSSR **7** (3) (1943) 147–166 (in Russian; Math. Rev. 5 (1944) 187).
[113] Pietsch, A.: Zur Theorie der σ-Transformationen in lokalkonvexen Vektorräumen. Math. Nachr. **21** (1960) 347–369.
[114] Przeworska-Rolewicz, D.; Rolewicz, S.: Equations in Linear Spaces. Monografje Matematyczne 47. Polska Akademia Nauk, Warszawa 1968.
[115] Schaefer, H. H.: Über singuläre Integralgleichungen und eine Klasse von Homomorphismen in lokalkonvexen Räumen. Math. Zeitschr. **66** (1956) 147–163.
[116] Schechter, M.: Basic theory of Fredholm operators. Ann. Sci. Norm. Sup. Pisa, Sci. fis, mat., III Ser. **21** (1967) 261–280.
[117] Taylor, A. E.: Theorems on ascent, descent, nullity and defect of linear operators. Math. Ann. **163** (1966) 18–49.
[118] Wendland, W.: Die Fredholmsche Alternative für Operatoren, die bezüglich eines bilinearen Funktionals adjungiert sind. Math. Zeitschr. **101** (1967) 61–64.

[119] Yood, B.: Difference algebras of linear transformations on a Banach space. Pacific J. Math. **4** (1954) 615–636.

V. Compact and Riesz operators

[120] Altman, M. S.: On linear functional equations in locally convex spaces. Studia Math. **13** (1953) 194–207.
[121] Aronszajn, N.; Smith, K.: Invariant subspaces of completely continuous operators. Ann. Math. **60** (1954) 345–350.
[122] Atkinson, F. V.: A spectral problem for completely continuous operators. Acta Math. Acad. Sci. Hung. **3** (1952) 53–59.
[123] de Bruyn, G. F. C.: Asymptotic properties of linear operators. Proc. London Math. Soc. (3) **18** (1968) 405–427.
[124] de Bruyn, G. F. C.: Addendum to 'Asymptotic properties of linear operators'. Proc. London Math. Soc. (3) **19** (1969) 191–192.
[125] de Bruyn, G. F. C.: Riesz Properties of Linear Operators. J. London Math. Soc. **44** (1969) 460–466.
[126] Calkin, J. W.: Two-sided ideals and congruences in the ring of bounded operators in Hilbert space. Ann. of Math. (2) **42** (1941) 839–873.
[127] Gohberg, I. C.; Krein, M. G.: Introduction to the Theory of Linear Nonselfadjoint Operators. American Mathematical Society, Translations of Mathematical Monographs 18. Providence, R.I. 1969.
[128] Gohberg, I. C.; Krein, M. G.: Theory and Applications of Volterra Operators in Hilbert Space. American Mathematical Society, Translations of Mathematical Monographs 24. Providence, R.I. 1970.
[129] Köthe, G.: Zur Theorie der kompakten Operatoren in lokalkonvexen Räumen. Port. Math. **13** (1954) 97–104.
[130] Lindenstrauss, J.: Extension of Compact Operators. Mem. Amer. Math. Soc. **48** (1964).
[131] Lomonosov, V. J.: Invariant subspaces for the family of operators which commute with a completely continuous operator. Functional Anal. and Appl. **7** (1973) 213–214.
[132] Reid, W. T.: Symmetrizable completely continuous linear transformations in Hilbert space. Duke Math. J. **18** (1951) 41–56.
[133] Riesz, F.: Über lineare Funktionalgleichungen. Acta Math. **41** (1918) 71–98.
[134] Ringrose, J.: Compact Non-Self-Adjoint Operators. London 1971.
[135] Ruston, A. E.: Operators with a Fredholm theory. J. London Math. Soc. **29** (1954) 318–326.
[136] Schatten, R.: Norm Ideals of Completely Continuous Operators. 2nd ed. Ergebnisse der Mathematik 27. Springer, Berlin—Heidelberg—New York 1970.
[137] Schauder, J.: Über lineare, vollstetige Funktionaloperationen. Studia Math. **2** (1930) 183–196.
[138] Schechter, M.: Riesz operators and Fredholm perturbations. Bull. Amer. Math. Soc. **74** (1968) 1139–1144.
[139] Williamson, J. H.: Compact linear operators in linear topological spaces. J. London Math. Soc. **29** (1954) 149–156.

VI. Integral operators

[140] Calderón, A. P.: Singular integrals. Proc. Sympos. Pure Math. **10**, Chicago (Ill.) 1966, 18–55, Amer. Math. Soc., Providence, R.I. 1967.
[141] Fredholm, I.: Sur une classe d'équations fonctionnelles. Acta Math. **27** (1903) 365–390.

[142] Grothendieck, A.: La théorie de Fredholm. Bull. Soc. Math. France **84** (1956) 319–384.
[143] Hellinger, E.; Toeplitz, O.: Integralgleichungen und Gleichungen mit unendlich vielen Unbekannten. Enzyklopädie d. Math. Wiss. II. C. 13. Leipzig 1928.
[144] Hilbert, D.: Grundzüge einer Allgemeinen Theorie der Linearen Integralgleichungen. Teubner, Leipzig 1912, 1924; Chelsea 1953.
[145] Jörgens, K.: Lineare Integraloperatoren. Mathematische Leitfäden. Teubner, Stuttgart 1970.
[146] Luxemburg, W. A. J.; Zaanen, A. C.: Compactness of integral operators in Banach function spaces. Math. Ann. **149** (1962/63) 150–180.
[147] Noether, F.: Über eine Klasse singulärer Integralgleichungen. Math. Ann. **82** (1921) 42–63.
[148] Prössdorf, S.: Einige Klassen Singulärer Gleichungen. Birkhäuser, Basel–Stuttgart 1974.
[149] Schmeidler, W.: Integralgleichungen mit Anwendungen in Physik und Technik. Leipzig, 1950.
[150] Smithies, F.: Integral Equations. Cambridge Tracts in Mathematics and Mathematical Physics 49. Cambridge University Press 1958.

VII. Differential operators

[151] Browder, F. E.: Functional analysis and partial differential equations I, II. Math. Ann. **138** (1959) 55–79. ibid **145** (1962) 81–226.
[152] Collatz, L.: Eigenwertaufgaben mit Technischen Anwendungen. Akademische Verlagsgesellschaft, Leipzig 1963.
[153] Garabedian, P. R.; Schiffman, M.: On solution of partial differential equations by the Hahn–Banach Theorem. Trans. Amer. Math. Soc. **76** (1954) 288–299.
[154] Hellwig, G.: Differential Operators of Mathematical Physics. Addison-Wesley, Reading, Mass. 1967.
[155] Hörmander, L.: Linear Partial Differential Operators. 3rd ed. Grundlehren der math. Wissenschaften 116. Springer, Berlin—Göttingen—Heidelberg 1969.
[156] Neumark, M. A.: Linear Differential Operators. Frederick Ungar, New York 1967.
[157] Schäfke, F. W.; Schneider, A.: S-hermitesche Rand- und Eigenwert probleme I, II. Math. Ann. **162** (1965) 9–26. ibid **177** (1968) 67–94.
[158] Schechter, M.: Spectra of Partial Differential Equations. North-Holland Series in Applied Mathematics 14. North-Holland, Amsterdam—London 1971.
[159] Stummel, F.: Rand- und Eigenwertaufgaben in Sobolewschen Räumen. Lecture Notes in Mathematics 102. Springer, Berlin—Heidelberg—New York 1969.
[160] Voigt, A.; Wloka, J.: Hilberträume und Elliptische Differentialoperatoren. Bibliographisches Institut, Mannheim—Wien—Zürich 1975.
[161] Weidman, J.: Zur Spektraltheorie von Sturm-Liouville-Operatoren. Math. Zeitschr. **98** (1967) 268–302.

VIII. Fixed point theorems and iteration procedures

[162] Bonsall, F. F.: Lectures on Some Fixed Point Theorems of Functional Analysis. Tata Institute, Bombay 1962.
[163] Deimling, K.: Nichtlineare Gleichungen und Abbildungsgrade. Universitext. Springer, Berlin—Heidelberg—New York 1974.
[164] Patterson, W. M.: Iterative Methods for the Solution of a Linear Operator Equation

in Hilbert Space. A Survey. Lecture Notes in Mathematics 394. Springer, Berlin—Heidelberg—New York 1974.

[165] Smart, D. R.: Fixed Point Theorems. Cambridge University Press 1974.

IX. Theory of approximation

[166] Akhieser, N. I.: Theory of Approximation. Frederick Ungar, New York 1956.

[167] Lorentz, G. G.: Approximation of Functions. Holt, Rinehart and Winston, New York—Chicago—San Francisco—Toronto—London 1966.

[168] Schönhage, A.: Approximationstheorie. De Gruyter, Berlin—New York 1971.

[169] Singer, I.: Best Approximations in Normed Linear Spaces by Elements of Linear Subspaces. Grundlehren der math. Wissenschaften 171. Springer, Berlin—Heidelberg—New York 1970.

X. Some further applications of functional analysis

[170] Collatz, L.: Funktionalanalysis und Numerische Mathematik. Grundlehren der math. Wissenschaften 120. Springer, Berlin—Heidelberg—New York 1968.

[171] Kantorovich, L. V.: Functional Analysis and Applied Mathematics. NBS Rep. 1509. Los Angeles, Calif. 1952.

[172] Neumann, J. von: Mathematische Grundlagen der Quantenmechanik. Grundlehren der math. Wissenschaften 38. Springer, Berlin 1932, 1968; Dover, New York 1943.

[173] Porter, W. A.: Modern Foundations of Systems Engineering. Macmillan, New York—London 1966.

[174] Sobolev, S. L.: Applications of Functional Analysis in Mathematical Physics. American Mathematical Society, Translations of Mathematical Monographs 7. Providence, R.I. 1963.

[175] Zeller, K.: FK-Räume in der Funktionentheorie I, II. Math. Zeitschr. **58** (1953) 288–305, 414–435.

[176] Zeller, K.; Beckmann, W.: Theorie der Limitierungsverfahren. Ergebnisse der Mathematik 15. 2nd ed. Springer—Berlin—Heidelberg—New York 1970.

D. Auxiliary tools

[177] Ahlfors, L.: Complex Analysis. 2nd ed. McGraw-Hill, New York 1966.

[178] Courant, R.; Hilbert, D.: Methods of Mathematical Physics I. Wiley-Interscience, New York 1953.

[179] Dugundji, J.: Topology. Allyn and Bacon, Boston, Mass. 1966.

[180] Hardy, G. H.; Littlewood, J. E.; Pólya, G.: Inequalities. 3rd ed. Cambridge University Press 1959.

[181] Hewitt, E.; Stromberg, K.: Real and Abstract Analysis. Graduate Texts in Mathematics 25. Springer, New York—Heidelberg—Berlin 1975.

[182] Loomis, L. H.: The converse of the Fatou Theorem for positive harmonic functions. Trans. Amer. Math. Soc. **53** (1943) 239–250.

[183] Pólya, G.; Szegö, G.: Problems and Theorems in Analysis I. Grundlehren der math. Wissenschaften 193. Springer, Berlin—Heidelberg—New York 1972.

[184] Royden, H. L.: Real Analysis. 2nd ed. MacMillan, New York 1971.

[185] Rudin, W.: Principles of Mathematical Analysis. 2nd ed. McGraw-Hill, New York 1964.

Index

absolute convergence 50, 188
absolute value 33
  discrete 34
absolutely convex 341
  hull 357
absorbing 327
adjoint 263
admissible region 200
admissible topology 355, 364
$\mathscr{A}(E)$-complement 115
$\mathscr{A}(E)$-direct sum 115
$\mathscr{A}(E)$-projectable 115
algebra 30
  commutative 30
  normed 51
  semi-simple 386
algebraic basis 26
algebraic complement 27
algebraic dual 73
approximation, in $C[a, b]$ 154
  in inner product spaces 245, 247
  in normed spaces 232
  in reflexive spaces 253
  in strictly convex spaces 234
  in the square mean 231
  in uniformly convex spaces 249
Atkinson operator 121, 159, 221, 255
Atkinson region 217

Baire category theorem 138
balanced 327
  core 327
  hull 327
ball 12
Banach algebra 51, 106, 182, 381
Banach space 35
Banach's fixed-point theorem 16, 380
barrel 367
barrelled space 368
basic set 260
basis, of a filter 316
  of neighborhoods 313
    of zero 327, 340
B-bounded set 347
Bessel identity 245, 288
Bessel inequality 245, 256, 288
best approximation 232
bidual 169, 369
bijective 2, 31
bilinear 76
  form 76
    bounded 78
    canonical 77
    continuous 78, 141
  system 76
bipolar 357
  theorem 357
boundary, condition 20
  eigenvalue problem 21, 283, 296
  value problem 21
bounded dilation 41
bounded sequence of symmetric operators 267
bounded set 12, 346, 367
Brouwer fixed-point theorem 374
B^*-algebra 389

canonical basis of neighborhoods of zero 340
canonical homomorphism 102, 104, 333
canonical imbedding 79
canonical injection 102, 104, 334
canonical surjection 316
Cantor intersection theorem 137
cartesian product 3, 13, 28, 40, 140, 316 333, 346

category theorem 138
Cauchy convergence criterion 11, 258
Cauchy–Hadamard formula 193
Cauchy integral formulas 191
Cauchy multiplication 199
Cauchy–Schwarz inequality 4
Cauchy sequence 11, 312, 345, 347
Cauchy series 46
Cauchy's integral theorem 190
Čebišov approximation problem 232, 234
Čebišov polynomials 246
chain-finite 162
chain-length 162
circularly isolated 206
closed graph 140
  theorem 141
closed hull 314
closed set 14, 54, 314
closure 14, 314
  point 314
codimension 28, 103
compact 58, 360
  operator 63, 69, 166, 169, 195, 211, 279, 355
comparable 4
complement 1
  algebraic 27
  topological 112, 335, 336, 338
complementary 112, 335, 336, 338
complete hull 67, 68, 262
complete metric space 11
completely continuous 63, 69
completion 67, 68, 262
complexification 28
component 3, 27
  of an open set 212
composition 2
conjugable operator 85, 89, 320, 365, 371
conjugate (conjugate operator) 85, 89, 336
conjugate number 145, 151
conjugate operator (*see also* Dual operator) 85, 89, 336, 353
conjugation of a Hilbert space 261
continuous bilinear form 141
continuous image 41
continuous inverse 44, 48, 63, 105, 124, 140, 155, 311, 335
  of the conjugate operator 157, 353
continuous linear form 73, 105, 335, 345
  on finite-dimensional spaces, l^p, $C[a, b]$, c, (c_0), (s) 151, 312
continuous map 41, 314
converge 9
convergence, absolute 50, 188
  componentwise 9
  in quadratic mean 240
  in the sense of the norm 42
  metric 9
  pointwise 9, 42, 326
  unconditional 50, 256, 258
  uniform 42, 326
convergent sequence 9, 314, 345
convergent series 46
convex 233, 340
  combination 341
  hull 341
convexity modulus 251
convolution 55
Courant maximum-minimum principle 294
Courant minimum-supremum principle 294

deficiency 107
  value 223
dense 65
derivative of a vector-valued function 186
diameter 137
difference set 1
dimension 26, 27
direct sum 27, 29
Dirichlet's boundary value problem 255
discrete absolute value 34
discrete metric 9
disk of convergence 192
distance 7, 155
  functional 342
division algebra 185
dual 73, 334
  algebraic 73
  operator (*see also* Conjugate operator) 86, 89, 135, 156, 157, 171, 174, 353
  system 79, 319
  topological 73, 334

E-bounded 324, 325, 347
eigensolution, of a boundary value problem 21
  of an eigenvalue problem 296
  of an operator 188
eigenspace 177
eigenvalue, approximate 274
  of a boundary value problem 21
  of an eigenvalue problem 296
  of an operator 188

eigenvector of an operator 177
endomorphism 29
E^+-conjugable 85
E^+-conjugate 85
E^+-normally solvable 133
E^+-saturated algebra of operators 114
E^+-weak Cauchy sequence 187
E^+-weak convergence 187
ε-neighborhood 188
equation of a hyperplane 351
equicontinuous 359
essential singularity 192
essential spectrum 222
exponential function 198
extension theorem 82, 129, 131, 133
extremal point 373
extremal subset 373

family 2
F-bounded 324, 325, 347
filter 316
  basis 316
  generated by a basis 317
  of neighborhoods 14, 312
final topology 315
finest locally convex topology 345
finite deficiency 107
finite-dimensional operator 54
finite support 144
fixed-point 16
  theorems 16, 374, 376, 377, 378, 380
foot of a perpendicular 246
form, bilinear 76
  linear 73
Fourier coefficient 231, 256
Fourier series 232, 257
Fréchet–Riesz representation theorem 260
Fréchet space 368
Fredholm alternative 99
Fredholm integral equation 19, 53, 72, 98, 244
Fredholm operator in an operator algebra 114, 122, 215, 223
  on a Fréchet space 372
  on a normed space 112, 113, 114, 118, 126, 158, 169, 179, 211
  on a topological vector space 338, 350
Fredholm point 212
Fredholm radius 215
Fredholm region 211
function 2
  linear 25
  space 9

functional 2
  linear 73
  sublinear 133

gauge 342
Gauss approximation problem 245
Gelfand homomorphism 385
Gelfand representation theorem 387
Gelfand topology 386
generate 25
Gram determinant 239
Gram–Schmidt orthogonalization process 246
graph 140
  closed 140
greatest lower bound of topologies 313
Green operator 298
Green's function 21, 283

Hahn–Banach extension theorem 129, 131, 133, 319
half-space 356
Hamel basis 26
Hausdorff criterion 314
Hausdorff space 313
heat equation 240
Hermite polynomials 246
hermitian form 267
hermitian matrix 224
Hilbert dimension 260
Hilbert space 238
Hölder inequality 4
holomorphic 190
homeomorphic topological spaces 315
homeomorphism 315
homomorphism 29
  canonical 102
  of algebras 106
  theorem 102
*-homomorphism 390
H-orthogonal 291
hull, absolutely convex 357
  balanced 327
  closed 14
  complete 67, 262
  convex 341
  linear 25
hyperplane 28, 351

ideal 70
  maximal 393
  proper 393

idempotent 32, 113, 204
identity 30
image 2, 31, 155
  chain 160
  continuous 41
  deficiency 107
inclusion theorem 295
index 108
  theorem 108
induced 65
infimum of topologies 313
infinite series 46
initial topology 315
initial value problem 379
  in Banach spaces 200
injection, canonical 102
injective 2
inner product 236
  space 236
integral transformation 20
integration path 189
integro-differential equation 301
intersection of topologies 314
interval of convergence 192
invariant subspace 32
inverse in an algebra 51, 53
inverse map 2, 31, 122, 124, 335
inverse, relative 113
invertible 51
involution 390
isolated points of the spectrum 207
isometric 65
  isomorphism 67
isometry 65
isomorphic normed spaces 68
isomorphism 31
  of normed spaces 67
*-isomorphism 390
iterate 2, 15

Jacobi polynomials 246
$\mathscr{J}$-commutes 218

kernel 19
Kolmogorov's normability theorem 346
Köthe space 326
Kronecker's symbol 1

Laguerre polynomials 246
Laurent expansion 192
least upper bound of topologies 314
left dual system 79, 319
left-inverse 51, 55, 122, 124, 335
left-invertible 51, 55, 122
Legendre polynomials 246, 247
lemma of F. Riesz 57
length, of image chain 160, 214
  of nullchain 160, 212
limit, of a sequence 9, 314
  point 363
linear differential operator 20, 282, 296
linear form 73
  bounded 73, 105
  continuous 73, 105, 335, 345
  multiplicative 83, 384
  on a topological vector space 335
linear functional 73
linear hull 25
linear image 29
linear manifold 351
linear map (*see also* Operator) 29
linear space 24
linear subspace 25
linear transformation (*see also* Operator) 29
linearly dependent 26
linearly independent 26
Lipschitz-condition 15, 16, 380
locally convex vector space 340
locally holomorphic 190
Lomonosov 378
lower bound of a symmetric operator 276

Mackey topology 364
map (*see also* Operator) 2
  bounded 41
  closed 140
  continuous 41, 314, 317
  contracting 16
  homeomorphic 315, 317
  identity 2, 30
  inverse 2, 31
  isometric 65
  linear 29
  nuclear 74
  open 104, 316
  sequentially continuous 314
  uniformly continuous 69
map-norm 73, 196
matrix-norm 196
matrix transformation 151, 195
maximal element 4
maximum-metric 7, 8
maximum principle 193
method of orthogonal projection 255

metric 7
canonical 8, 34, 35
continuous 41
discrete 9
equivalent 13
euclidean 7
induced 65
space 7
stronger 13
translation-invariant 33
vector space 311
weaker 13
ninimal modulus 155
ninimal polynomial 179
ninimizing sequence 55, 232, 247, 248
249, 252, 253
Minkowski functional 342
Minkowski inequality 4
noment problem 154
nonotone sequence of symmetric
operators 267
Montel space 372
nultiplicative linear form 83, 382
nultiplicity 177

neighborhood 14, 312
axioms 312
of zero in a normed space 59
open 14, 313
Neumann series 19, 47, 52, 60, 195
nilpotent 185
orm 34
canonical 38, 237
equivalent 44, 142
induced 38
of a linear map 73, 196
of algebras 106
of an operator 42
of the dual operator 135
stronger 44
orm-preserving 131
map 382
ormable 346
ormal algebra of operators 121
ormal element 390
ormal equations 247
ormal operator 272
ormalized function 155
ormalized vector 243
ormally solvable 133, 135, 264, 273, 277,
350, 355
ormaloid element 388
ormaloid operator 223
ormed algebra 51
normed vector space 34
nowhere differentiable continuous function
149
nuclear map 74
nullchain 160
nulldeficiency 107
nullspace 31

open cover 360
ε-neighborhood 188
mapping principle 138, 311, 369
set 14, 53, 313
operator (*see also* Map) 29
adjoint 263
algebraic dual 84, 89, 126
Atkinson 121
bidual 171, 173
bounded 41, 374
chain-finite 162
compact 63, 69, 166, 169, 195, 211,
279, 355
completely continuous 63
conjugable 85, 89, 320, 365, 371
conjugate 85, 89, 336, 353
continuous 41, 334, 344
dual 86, 89, 135, 156, 157, 171, 174,
353
finite-dimensional 54, 72
Fredholm 112, 113, 114
fully symmetric 269
fully symmetrizable 269, 291
idempotent 32
meromorphic 225
normal 272
normaloid 223
of finite rank 54, 72
paranormal 229
positive 265
precompact 176, 296
quasicompact 216
relatively regular 111, 113, 114, 119,
335, 338, 355
self-adjoint 267
spectrally normaloid 227
symmetric 254, 265, 269, 280, 291, 301,
392
symmetrizable 269, 290
unitary 277
Wielandt 288, 300
with finite deficiency 107
order 4
for symmetric operators 267
of a pole 192
ordered set 4

orthogonal complement 254
decomposition 254
projector 254
sequence 243
series 256
space 134, 243
system 243
orthogonality, in inner product spaces 239
in normed spaces 244
orthogonally closed 134, 172, 350
orthonormal basis 239, 258
orthonormal sequence 240, 243
orthonormal system 243, 288
maximal 259

parallelogram law 236, 237
Parseval identity 258, 260
participation of an eigenvalue 291
partition 316
path integral 189
Peano's existence theorem 379
perpendicular 239
projector 235
perturbation theorem 158
Φ-ideal 121, 215, 223
Φ_0-ideal 223
Picard–Lindelöf existence theorem 380
point 7, 24
of closure 314
spectrum 181
approximate 274, 277
pure 275
polar 356
pole 192, 209, 211
power of a map 2
power series, everywhere convergent 192
in Banach algebras 194
in Banach spaces 187
nowhere convergent 192
precompact map 176, 296
precompact set 362
prehilbert space 236
preimage 2
principle of uniform boundedness 142, 368
product, of locally convex spaces 346
of maps 2, 29
of metric spaces 13
of normed spaces 140, 333
of topological vector spaces 333
of valued spaces 40
topology 316, 326
projection 3
projector 3, 31
proper value 177
proper vector 177
Pythagoras 243

quadrangle-inequality 12
quadratic form 266
quadrature formulas 148
quasi-nilpotent 195
quotient algebra 105
quotient norm 103, 333
quotient semi-norm 346
quotient space 102, 316, 333, 346
quotient topology 316

radical 219, 386
radius of convergence 192
Rayleigh quotient 294
reducing pair of subspaces 33
reducing subspace of a Hilbert space 270
reflexive normed space 170, 172, 253, 371
reflexive topological vector space 370
region 190
regular 51
value 177
Reid's inequality 268
relative inverse 113
relative neighborhood 315
relatively closed 315
relatively compact 58, 360
relatively open 315
relatively regular element 113
relatively regular operator 111, 113, 114, 119, 335, 338, 355
relatively sequentially compact 58
removable singularity 192
representation theorem for normed spaces 363
residue class 101
resolvent 181, 182
equation 183
operator 181
set 181, 182
resolving kernel 23
resolving transformation 23
Riesz, F. 260
ideal 219
lemma of 57
operator 217, 255
point 217
projector 205
radius 217
region 217
representation theorem 154
theory of compact operators 166

right dual system 79, 319
right-inverse 51, 55, 122, 124, 335
right-invertible 51, 55, 122
root test 48

saturated algebra of operators 114
saturated family of semi-norms 344
scalar 1
Schauder fixed-point theorems 374, 377, 378
Schwarz inequality 237
  generalized 265
Schwarz quotient 292
segment joining two points 233, 372, 373
segment spectrum 275
  pure 275
self-adjoint element 390
self-adjoint operator 267
self-map 2
semi-Fredholm operator 159, 221, 255
semi-inner product 265
  continuous 269
semi-norm 37, 128
  continuous 326, 342, 344
semi-reflexive 370
separable 174
separate 382
separated 313
sequence 2
  Cauchy 11
  linear 25
  space 9
sequentially compact 58
sequentially complete 345
series, Cauchy 46
  convergent 46
  infinite 46
sesquilinear form 262
set of definition 2
set of indices 2
sign 153
simple pole 192
singular integral equations 108
singular value of an operator 177
span 25
spectral disk 184
spectral family 302, 305
spectral mapping theorem 202
spectral point 184
spectral projector 205
spectral radius 184
spectral set 203
spectral theorem 307, 392
spectrum 181, 182

continuous 183
essential 222, 274
peripheral 225
residual 183
square root of a positive operator 305
Stieltjes integral 306
S-topology 325
strictly convex space 234
strong topology 325, 365, 367
Sturm–Liouville eigenvalue problem 283
sublinear functional 133, 351
subspace 25, 38, 65, 315, 332, 345
  invariant 32
  linear 25
sum of subspaces 27, 29
summable family 258
summation method 147
supremum-metric 8
supremum of topologies 314
surjective 2
symmetric operator 254, 265, 269, 280, 291, 301, 392
symmetric set 327
Szegö's convergence theorem 148

target set 2
theorem, of Alaoglu–Bourbaki 363
  of Arzelà–Ascoli 73
  of Baire 138, 149
  of Banach–Schauder 138
  of Banach–Steinhaus 143
  of Bolzano–Weierstrass 57
  of Fréchet–Riesz 260
  of Gelfand 387
  of Gelfand–Neumark 391
  of Hahn–Banach 129, 131, 133, 349
  of Hellinger–Toeplitz 267
  of Kolmogorov 346
  of Krein–Milman 374
  of Liouville 191
  of Mackey 366
  of Mackey–Arens 365
  of Montel 372
  of Peano 379
  of Picard–Lindelöf 380
  of Pythagoras 243
  of F. Riesz 154
  of Schauder 174
  of Stone–Weierstrass 389
  of Szegö 148
  of Tikhonov 361, 378
  of Toeplitz 147
  of Weyl 295
  of Wiener 388

Tikhonov's fixed-point theorem 378
Toeplitz permanence theorem 147
topological complement 112, 335, 336, 338
topological product 316
topological space 312
topological vector space 321
topology 312
  admissible 355, 364
  chaotic 313
  coarser 313
  discrete 313
  final 315
  finer 313
  generated by a family of maps 315
  generated by a family of semi-norms 323
  Hausdorff 313
  induced 315
  initial 315
  metric 312
  metrizable 324
  of pointwise convergence 324
  of uniform convergence 324
  product 316, 346
  quotient 316, 346
  relative 315, 345
  separated 313
total family of semi-norms 323
total order 4
total sequence of semi-norms 311, 323
total set of linear forms 382
total variation 37
totally bounded 173, 176, 362
totally ordered set 4
transfer of topologies 315
transformation (*see also* Operator *and* Map) 29
translation-invariant 33
triangle inequality 7, 33, 34
  strict 234

unconditional convergence 50, 256, 258
uniformly convex space 249
unit element 30
unit vector 26
upper bound 4
  of a symmetric operator 276

valued space 33
vector 24
vector space 24
  complex 24
  locally convex 340
  metric 311
  normed 34
  real 24
  topological 321
  topology 321
    generated by semi-norms 323, 340
    generated by $\mathfrak{S}$ 324, 340
Volterra integral equation 49
Volterra integral transformation 49
Volterra operator 185

weak Cauchy sequence 187
weak Cauchy series 187
weak convergence 187, 193, 252, 261, 320
  of operators 193, 326
weak topology 318, 320, 325, 332, 340
weakly continuous 320
weakly differentiable 187
weakly locally holomorphic 190
weakly sequentially compact 371
weakly sequentially complete 253
Weyl comparison theorem 295
Weyl spectrum 222
Wielandt operator 288, 300
Wiener algebra 388

Zero map 29
Zorn's lemma 4